Soil and Drought

Drought, a serious global issue, is being aggravated by climate change. Both pedological and agronomic droughts are major risk factors with adverse effects on agronomic productivity, food and nutritional security, and human wellbeing. This volume in the *Advances in Soil Sciences* series provides research information regarding case studies from diverse agro-ecoregions around the world and lists examples of effective management of drought at farm, state, national, regional, and global scales.

Features:

- Considers processes, factors, and causes of pedological/agronomic droughts.
- Discusses effects of global warming on soil drought and describes management options to enhance drought resilience of agricultural soils.
- Focuses on specific case studies along with review of a variety of tools and techniques designed to mitigate drought and reduce its impact on agronomic productivity.
- Includes information on soil health and its effects on drought.

In addition to highlighting the scientific accomplishments of Dr. Bobby A. Stewart, this book is a major contribution to the global issue of drought management and its dynamics in relation to soil properties under changing climate. It is reference material for researchers, students, practitioners, and policymakers in soil science, agronomy, ecology and management of natural resources with specific focus on adaptation and mitigation of climate change, restoration of soil health, strengthening of biodiversity and promoting the strategy for advancing the "Sustainable Development Goals" or Agenda 2030 of the United Nations.

Advances in Soil Science

Series Editors:
Rattan Lal
B.A. Stewart

Soil Erosion and Carbon Dynamics
E.J. Roose, R. Lal, C. Feller, B. Barthès, and B. A. Stewart

Soil Quality and Biofuel Production
R. Lal and B. A. Stewart

World Soil Resources and Food Security
R. Lal and B. A. Stewart

Soil Water and Agronomic Productivity
R. Lal and B. A. Stewart

Principles of Sustainable Soil Management in Agroecosystems
R. Lal and B. A. Stewart

Soil Management of Smallholder Agriculture
R. Lal and B. A. Stewart

Soil-Specific Farming: Precision Agriculture
R. Lal and B. A. Stewart

Soil Phosphorus
R. Lal and B. A. Stewart

Urban Soils
R. Lal and B. A. Stewart

Soil Nitrogen Uses and Environmental Impacts
R. Lal and B. A. Stewart

Soil and Climate
Rattan Lal and B.A. Stewart

Food Security and Soil Quality
R. Lal and B. A. Stewart

Soil Degradation and Restoration in Africa
Rattan Lal and B.A. Stewart

Soil and Fertilizers: Managing the Environmental Footprint
Rattan Lal

The Soil-Human Health-Nexus
Rattan Lal

Soil Organic Matter and Feeding the Future: Environmental and Agronomic Impacts
Rattan Lal

Soil Organic Carbon and Feeding the Future: Basic Soil Processes
Rattan Lal

Soil and Drought: Basic Processes
Rattan Lal

For more information about this series, please visit: www.crcpress.com/Advances-in-Soil-Science/
book-series/CRCADVSOILSCI

Soil and Drought

Basic Processes

Edited by
Rattan Lal

Distinguished University Professor of Soil Science, School of Environment and
Natural Resources (SENR), College of Food, Agricultural and Environmental
Sciences (CFAES), The Ohio State University
Director, CFAES Rattan Lal Center for Carbon Management and Sequestration
(Lal Carbon Center)

CRC Press is an imprint of the
Taylor & Francis Group, an **informa** business

First edition published 2024
by CRC Press
2385 NW Executive Center Drive, Suite 320, Boca Raton FL 33431

and by CRC Press
4 Park Square, Milton Park, Abingdon, Oxon, OX14 4RN

CRC Press is an imprint of Taylor & Francis Group, LLC

© 2024 selection and editorial matter, Rattan Lal; individual chapters, the contributors

Reasonable efforts have been made to publish reliable data and information, but the author and publisher cannot assume responsibility for the validity of all materials or the consequences of their use. The authors and publishers have attempted to trace the copyright holders of all material reproduced in this publication and apologize to copyright holders if permission to publish in this form has not been obtained. If any copyright material has not been acknowledged please write and let us know so we may rectify in any future reprint.

With the exception of Chapter 11, no part of this book may be reprinted or reproduced or utilised in any form or by any electronic, mechanical, or other means, now known or hereafter invented, including photocopying and recording, or in any information storage or retrieval system, without permission in writing from the publishers.

Chapter 11 of this book is available for free in PDF format as Open Access from the individual product page at www.routledge.com. It has been made available under a Creative Commons Attribution-Non Commercial-No Derivatives 4.0 license.

For permission to photocopy or use material electronically from this work, access www.copyright.com or contact the Copyright Clearance Center, Inc. (CCC), 222 Rosewood Drive, Danvers, MA 01923, 978-750-8400. For works that are not available on CCC please contact mpkbookspermissions@tandf.co.uk

Trademark notice: Product or corporate names may be trademarks or registered trademarks and are used only for identification and explanation without intent to infringe.

Library of Congress Cataloging-in-Publication Data
Names: Lal, R., editor. Title: Soil and drought: basic processes / Rattan Lal.
Other titles: Advances in soil science (Boca Raton, Fla.) ; v. 20.
Description: First edition. | Boca Raton, FL : CRC Press, 2024. |
Series: Advances in soil sciences ; v. 20 | Includes bibliographical references and index. |
Summary: "Drought, a serious global issue, is being aggravated by climate change.
Both pedological and agronomic droughts are a major risk factor with adverse effects on agronomic productivity, food and nutritional security, and human wellbeing. This volume in the Advances in Soil Sciences series: Considers processes, factors, and causes of pedological/agronomic droughts. Discusses effects of global warming on soil drought Describes management options to enhance drought resilience of agricultural soils Presents soil moisture management options to mitigate drought Includes information on soil health and its effects on drought Explains innovative options in measurement of soil moisture content "– Provided by publisher.
Identifiers: LCCN 2023014367 | ISBN 9781032286747 (hardback) |
ISBN 9781032286754 (paperback) | ISBN 9781003297932 (ebook)
Subjects: LCSH: Dry farming. | Soil moisture. | Soil moisture conservation. |
Droughts. Classification: LCC SB110 .S673 2024 |
DDC 631.5/86–dc23/eng/20230825
LC record available at https://lccn.loc.gov/2023014367

ISBN: 9781032286747 (hbk)
ISBN: 9781032286754 (pbk)
ISBN: 9781003297932 (ebk)

DOI: 10.1201/b22954

Typeset in Times
by Newgen Publishing UK

Dedication

This volume is dedicated to Dr. Bobby A. Stewart who was the Founding Editor-In-Chief of Advances in Soil Sciences. During his distinguished career, Dr. Stewart made iconic contributions to dryland farming.

Contents

Foreword by Jean L. Steiner ix
Foreword by B.A. Stewart xi
Preface xiii
About the Editor xv

Chapter 1   Drought Hazard to Dryland Farming in Arid Region 1

Rattan Lal

Chapter 2   Soil Water Conservation for Dryland Farming 11

Paul W. Unger, Robert C. Schwartz, R. Louis Baumhardt, and Qingwu Xue

Chapter 3   Distinctive Dryland Soil Carbon Transformations: Insights from Arid
Rangelands of SW United States 39

*Jean L. Steiner, Carolina B. Brandani, Adrian Chappell, Jose Castaño-Sanchez,
Mikaela Hoellrich, Matthew M. McIntosh, Shelemia Nyamuryekung'e,
Nicole Pietrasiak, Alan Rotz, and Nicholas P. Webb*

Chapter 4   Crop Nutrition Management for Semiarid Areas of Sub-Saharan Africa with
Increasingly Variable Climate 83

*Charles Wortmann, Aliou Faye, Maman Garba, Idriss Serme, and
Zachary P. Stewart*

Chapter 5   Soil–Plant–Water–Environment Interaction in Dryland Agriculture 108

Sushil Thapa, Qingwu Xue, and Rajan Ghimire

Chapter 6   Managing Soil and Water Resources by Tillage, Crop Rotation, and
Cover Cropping 129

Paul Bradley DeLaune, Katie Lynn Lewis, and Joseph Alan Burke

Chapter 7   Managing Drought Stress in Agro-Ecosystems of Latin America and the
Caribbean Region 157

*Stoécio Malta Ferreira Maia, Carlos de Oliveira Galvão,
Thalita Fernanda Abbruzzini, Julio Campo, José Antonio Marengo Orsini,
Carlos Eduardo Pellegrino Cerri, and Teogenes Senna de Oliveira*

Chapter 8   Conservation Agriculture: Water Use Efficiency in Dryland Agriculture ... 181

D.C. Reicosky

Chapter 9 Physiological Mechanisms for Improving Crop Water Use Efficiency in the US Southern Great Plains 236

Qingwu Xue, Sushil Thapa, Shuyu Liu, Jourdan Bell, Thomas H. Marek, and Jackie Rudd

Chapter 10 Improving Water Storage through Effective Soil Organic Matter Management Strategies under Dryland Farming in India 256

Ch. Srinivasarao, S. Rakesh, G. Ranjith Kumar, M. Jagadesh, K.C. Nataraj, R. Manasa, S. Kundu, S. Malleswari, K.V. Rao, J.V.N.S. Prasad, R.S. Meena, G. Venkatesh, P.C. Abhilash, J. Somasundaram, and R. Lal

Chapter 11 Ancient Infrastructure Offers Sustainable Agricultural Solutions to Dryland Farming *(Online only)* 285

Matthew C. Pailes, Laura M. Norman, Christopher H. Baisan, David M. Meko, Nicolas Gauthier, Jose Villanueva-Diaz, Jeff Dean, Jupiter Martínez, Nicholas V. Kessler, and Ron Towner

Index 287

Foreword by Jean L. Steiner

As I prepare this foreword, the world's population has just passed 8 billion people and is projected to grow throughout this century to over 10 billion people. At the same time, the United Nations Climate Change Conference COP27 reaffirmed their commitment to limit global temperature rise to 1.5°C above pre-industrial levels and importantly established a mechanism for "loss and damage" funding for vulnerable countries hit by climate disasters. Further, international drought maps indicate severe drought on all continents of the globe. With the pressures of increasing human population facing increasing challenges of climate change, the threats to global food security and the environment have never been greater, making this volume focused on dryland agriculture and drought management timely.

This volume is dedicated to Dr. B.A. Stewart, and I can think of no individual who has contributed more than Dr. Stewart to advancing our understanding of the critical challenges that people in the world's drylands face. For over 40 years I have benefited from his vision and leadership and been inspired by his tireless work to build scientific knowledge and capacity while communicating the urgent needs of dryland agriculture.

Dr. Stewart worked for more than 40 years with the US Department of Agriculture's Agricultural Research Service, with more than 25 of those years as the Director of the Conservation and Production Research Laboratory at Bushland, Texas. Under his leadership, the Bushland laboratory grew in national and international stature along with his ever-expanding network. His international reach began in 1983 when he led a People-to-People tour to China, establishing relationships that lasted throughout his career. In 1984, he was selected by USAID to serve as a Scientific Liaison Officer to the International Center for Agricultural Research in Dryland Areas. His international work culminated in the 1988 International Conference on Dryland Farming which was organized to commemorate the 50th anniversary of the Bushland research station and which attracted 460 attendees from 52 countries and 36 US states.

In 1975, Dr. Stewart became the Editor-in-Chief of the *Soil Science Society of America Journal* and then served as President of the Soil Science Society of America in 1980–1981. Starting in the mid-1980s, Dr. Stewart took on the role of Editor of the Advance in Soil Science Series, developing an astonishing 20 volumes in 7 years, published by Springer-Verlag. Subsequently, 25 volumes were published with Routledge, Taylor & Francis Group from 1992 to present, with Dr. Rattan Lal joining Dr. Stewart as Co-Editor from 2011 to 2019, and Dr. Lal continuing as Editor of this invaluable series.

When Dr. Stewart left USDA-ARS in 1993, rather than rest on his remarkable laurels, he accepted the position as Distinguished Professor of Agriculture and the first Director of the Dryland Agriculture Institute at West Texas A&M University at Canyon. During his 25 years with the Dryland Institute, he was sought out as a teacher and trained a remarkable 77 graduate students who continue his legacy, including 32 from the USA and others from 15 other countries: China, India, Cote d'Ivoire, Zimbabwe, Argentina, Kenya, Nepal, Uzbekistan, Japan, Pakistan, Korea, Brazil, Ethiopia, Ghana, and Rwanda.

Throughout a long and varied career, some things remained constant – a deep curiosity, a passion for research, and compassion for others. Those of us from all over the world who were fortunate to work with him will remain forever grateful for his words of wisdom, humor, and dedication to making life better for people in the world's drylands. It is an honor to introduce readers to this book prepared in honor of Dr. B.A. Stewart and to recognize, with gratitude, the tremendous impact that Dr. Stewart's guidance, support, and insights have had on the trajectory of my career.

Jean L. Steiner
Senior Science Coordinator
Sustainable SW Beef Project, NMSU
& Adjunct Professor, Kansas State University
December 2022

Foreword by B.A. Stewart

Drought simply means that it is significantly drier than normal and occurs worldwide in favorable rainfall areas as well as low rainfall areas. Rainfall effectiveness depends largely on the quality of the soil on which it falls. The quality of soils is generally lower in low rainfall regions because of lower soil organic matter content making dryland farming difficult even in normal rainfall years, and particularly challenging in drought years. In recent years, climate change has made matters even worse, making improved soil and water conservation practices vital.

Dr. Lal has assembled scholarly reviews from the United States, Latin America, Africa, and China authored by renowned scientists to address how improved soil and water conservation practices can improve crop production in drought-prone areas. Because of Dr. Lal's worldwide experience in both arid and humid regions, he was able to select and inspire these authors to write concise and well-documented papers that can be particularly helpful to researchers and change agents worldwide.

Without question, Dr. Lal is one of the world's most qualified and best-known soil scientists. I first met Dr. Lal on April 10, 1984. I remember this well because it was during the most trying international trip of my career. During much of the 1980s when I was Director of the USDA Conservation and Production Laboratory at Bushland, TX, I was asked by USAID to visit the International Center for Agricultural Research in Dryland Areas (ICARDA) in Syria once a year and report back to USAID officers in Washington, D.C. In 1984, USAID asked me to also visit the International Institute for Tropical Agriculture (IITA) on my way back from Syria to observe some no-tillage experiments. Unfortunately, I was delayed in Rome for several hours because of cancelled flights and did not arrive in Lagos, Nigeria until about 3:00 a.m. My bags were nowhere to be found. A person from IITA did meet me inside customs but said it was not safe to take me to the guest house during the night because of a recent incident with the Deputy Director of IITA being robbed going from airport to the guest house. He told me to stay inside customs until he came back in the morning. Since I had no bags to claim and could not leave customs, I simply remained as the only person in customs. When the person came back to get me, we left for the guest house for breakfast, and then drove to the Research Center about 2 hours away. It never occurred to me that I left the room without ever officially passing through customs, so I had no papers showing that I was in Nigeria. That is not all. Nigeria was also having a currency crisis and I could not change dollars for local currency. I arrived at the Research Center with no money, no bags, and no papers showing I was in Nigeria. I did not know a single person but was soon introduced to Dr. Lal who I had never met or even heard of at the time. While the Research Center provided me lodging and food, I had no money to buy clothes, toiletries, and other necessities. Dr. Lal took me under his care, and we were soon observing his tillage experiments. He also introduced me to agroforestry which was quite a departure from my background in dryland farming in the almost treeless U.S. Great Plains. We discussed about my missing bags and more importantly how to obtain required signed forms for getting back through customs. I finally told Dr. Lal: "If you will just get me out of this country, you will be my friend forever." I will not state how I finally got signed forms other than to say it was anything but an approved practice. Dr. Lal demonstrated originality, tenacity, dedication, and nerve beyond reason. Also, as I was leaving, we got permission to enter a huge building at the airport containing more luggage than I can describe, and I was able to spot my bags because of large colored straps around them. That was my introduction to Dr. Lal, and we kept in close contact, and after he moved to Ohio State University in 1987, we cooperated on many projects and traveled together on several international trips. The wonderful part of our relationship is that we could, and often did, have different opinions on an issue, but always with respect, and generally felt that we both gained from the discussion. Words cannot express how much I learned from Dr. Lal. In retrospect, my worst international trip turned out to be my best.

The first volume of Advances in Soil Science was initiated in 1982 and published in 1984. I invited Dr. Lal to write a paper soon after I met him at ITTA and then he co-edited a volume on Soil Degradation followed by a volume on Soil Restoration. I later invited him to be co-editor of the series and when I retired in 2018, Dr. Lal assumed the editorship of the series.

For this volume of Advances in Soil Science, Dr. Lal invited experts from some of the largest dryland farming regions of the world. This is especially timely because while sustainable production in these areas has always been difficult, climate change is making it even more so. The scholarly reviews by these authors from different regions will be a valuable resource for planning future studies. Although the regions are vastly different, they have similar challenges, and some solutions may be applicable to other regions.

In summary, I have had a long and fruitful relationship with Dr. Lal and was privileged to observe his steady growth from a young scientist at ITTA to a Distinguished Professor at Ohio State University. As a scientist, he has authored more than 2,000 publications, and as a leader his achievements include President of Soil Science Society of America, President of International Union of Soil Sciences, and recipient of the 2020 World Food Prize.

B.A. Stewart

Preface

Drought, a process of combination of low soil moisture reserve and heat wave leading to high evapotranspiration demand with adverse effects on crop growth and agronomic yield, is a widespread problem in global drylands. Furthermore, the problem is exacerbated by present and projected anthropogenic global warming through increase in intensity and frequency of drought. Anthropogenic global warming affects all types of drought: meteorological, hydrological, pedological, agronomic, ecological, and social. While crop growth is strongly and adversely affected by pedological (decline of plant-available water capacity in soil) and agronomic drought (low plant-available soil moisture reserves at critical stages of growth), both of these are manageable through implementation of soil and crop management options for the site-specific conditions.

Pertinent among soil management technologies is conservation agriculture based on no-till farming in conjunction with the retention of crop residue mulch and complex crop solutions. Another option is farming whereby four to six plants are grown together with wide spacing between clumps. Drip sub-fertigation is another pertinent option. Dr. Bobby A. Stewart, along with many of his colleagues who have contributed to chapters to this volume (Dr. Paul Unger, Dr. Jean Steiner, Dr. Thapa, Dr. Wortmann, Dr. Debaune, and many others), was involved in developing and promoting these technologies for adaptation to and mitigation of both pedological and agronomic drought.

Because of his scientifically innovative and practically applicable contributions to adaptation and mitigation of pedological and agronomic drought in global drylands, this volume of Advances in Soil Science is dedicated to Dr. Bobby A. Stewart. This timely and pertinent book provides research information about case studies from diverse agro-ecoregions around the world and lists examples of effective management of drought at farm, state, national, regional, and global scales. The volume is a reflection of a very distinguished and long career (spanning over six decades) of research contributions. Dr. Stewart and other authors with whom he has had close professional relationships have focused on specific case studies along with review of a variety of modern tools and techniques designed to mitigate drought and reduce its impact on agronomic productivity.

Authors, who have conducted research in diverse agro-ecoregions, support the conclusion that risks of drought can be reduced through the adoption of regional and site-specific best management practices. Hence, this book makes a major contribution to the important topic of drought and its management. In addition to highlighting the scientific accomplishments of Dr. Bobby A. Stewart, this book is a major contribution to the global issue of drought management and its dynamics in relation to soil properties under changing climate. This book provides reference material for researchers, students, practitioners, and policymakers in soil science, agronomy, ecology, and management of natural resources, with specific focus on adaptation and mitigation of climate change, restoration of soil health, strengthening of biodiversity, and promoting the strategy for advancing the "Sustainable Development Goals" or Agenda 2030 of the United Nations.

I thank the authors for sharing their knowledge and wisdom, and for timely submission and revisions. I thank the help received from the staff of the CFAES Rattan Lal Center for Carbon Management and Sequestration for managing the flow of the manuscript and for checking format, references, style, and submitting it to the publisher. I also thank Ms. Randy Brehm and Mr. Tom Connelly of Taylor & Francis for their help and support in the timely publication of this book.

Rattan Lal
1st February 2023
Columbus, OH

About the Editor

Rattan Lal, PhD, is Distinguished University Professor and Director of the CFAES Rattan Lal Center for Carbon Management and Sequestration, The Ohio State University, and Adjunct Professor at the University of Iceland and IARI, New Delhi. He was President of the WASWAC (1987–1990), ISTRO (1988–1991), SSSA (2006–2008), and the IUSS (2017–2018). He researches soil C sequestration, conservation agriculture, soil health, soil erosion and C dynamics, soil structure, eco-intensification, soil restoration, and soils of the tropics. He has authored 2,885 journal articles, authored/edited more than 100 books, mentored 390 researchers, has h-index of 172, and total citations of 129,861. In a Stanford study (Ioannidis et al. 2019, 2020), he is ranked #111 globally among the world's top 2% of scientists and #1 among scientists in Agronomy and Agriculture. In 2023 Ranking of Best Scientists in Plant Science and Agronomy, Research.com ranked him no. 1 in the world and in the United States. He holds IICA's Chair in Soil Science and Goodwill Ambassador in Sustainable Development. He is a member of the 2021 U.N. Food System Summit Science Committee, Action Track 3 and Coalition 4 Soil Health. He received the 2018 GCHERA World Agriculture Prize, 2018 Glinka World Soil Prize, 2019 Japan Prize, 2019 IFFCO Award, 2020 Arrell Global Food Innovation Award, the 2020 World Food Prize, and the 2021 Padma Shri Award, India.

1 Drought Hazard to Dryland Farming in Arid Region

Rattan Lal
CFAES Rattan Lal Center for Carbon Management and Sequestration,
The Ohio State University, Columbus, OH 43210 USA

CONTENTS

1.1 Global Drylands 1
1.2 Dryland under Changing Climate 2
1.3 Types of Drought 3
1.4 Inter-Connectivity among Types of Water 4
1.5 Drought Management in Drylands 4
1.6 Carbon Sequestration Potential in Global Drylands 5
1.7 It Is All about Water 8
1.8 Research and Development Priorities 8
References 9

1.1 GLOBAL DRYLANDS

Drylands are regions with aridity index (AI= mean annual precipitation/mean annual evapotranspiration) values of <0.65. These regions cover 41% of the terrestrial surface (Feng and Fu, 2013; Maestre et al., 2021). Plaza et al. (2018) reported that drylands cover 66.7 million (M) km^2 or 45% of Earth's land surface and as much as 70% of dryland areas are located in developing countries. Drylands include hyper-arid regions with AI of <0.05(6.6%), arid regions with AI of 0.05–0.20 (10.6%), semi-arid regions with AI of 0.20–0.50 (15.2%), and dry sub-humid regions with AI of 0.5–0.65 (8.7). Drylands, Earth's largest biome, contain 44% of global cropland areas, 50% of global livestock (Maestre et al., 2021), and are a source of major ecosystem services for its inhabitants (Maestre et al., 2022). In comparison, Plaza et al. (2018) reported that in global drylands 11% or 7.6 M km^2 is used for cropland and 30% or 20.0 M km^2 for pastures. However, drylands are characterized by abrupt changes in plant productivity in relation to dynamical patterns with regard to climatic, topographical, edaphic and anthropogenic factors (Bertugo et al., 2022). Rainfed agriculture, the dominant form of farming globally, is widely practiced in drylands. Globally, as much as 80% of world's agriculture land area is rainfed and it produces 65–70% of world's food staples. However, drought and water scarcity have affected and will adversely affect food production in dryland farming. In addition, all regions of the world have experienced a warming process since the 1970s, but the fastest warming region is the mid-latitudes of the northern hemisphere (IPCC, 2013, Chen et al., 2015). Among biomes which are vulnerable to global warming, arid regions are most sensitive to anthropogenic climate change (Lioubimtseva, 2021). Despite the harsh environment and ecologically fragile conditions, a large proportion of 2.5 billion poor people live in global drylands. Sustainable Development Goal (SDG) number 1 (End Poverty) and number 2 (Zero Hunger) of the agenda 2030 of the United Nations may not be on track for people living in global dryland regions.

Drylands are home to 38–39% of the global human population (Plaza et al., 2018; Maestre et al., 2021). High population is causing tremendous stress on these fragile ecosystems, leading to desertification and the attendant land degradation (Bertugo et al., 2022). Drylands are also the regions

DOI: 10.1201/b22954-1

experiencing rapid growth in population and the attendant increase in urbanization. Of the 1692 cities, 35% (586) are located in dry regions (U.N., 2014; European Commission, 2023). Of these, 198 cities (11%) are located in dry sub-humid, 251 (15%) in semiarid, 108 (6%) in arid, and 29 (2%) in hyper-arid regions (U.N., 2014). By 2100, population (million) of some cities in arid regions may increase to 74 for Dar Es Salaam, 57 for both Delhi and Khartoum, 56 for Niamey, 50 for Kabul, 49 for Karachi, and 41 for both Lilongwe and Cairo. Thus, planning for food and nutritional security through promotion of sustainable farming in urban centers such as home gardens (Lal, 2020b; Lal et al., 2020) is an important strategy.

Thus, global drylands are important to achieving sustainability of current and future human population. In this context, review and synthesis of the research and development of dryland farming innovations advanced by Dr. Bobby A. Stewart are of critical importance. Thus, this special volume of Advances in Soil Science is dedicated to Dr. Bobby A. Stewart and to recognize his contributions to advancing science and practice of sustainable management of dryland agriculture. Dr. Stewart devoted his illustrious career spanning over six decades to sustainable management of dryland agriculture. His research has been devoted to identification of soil and crop management practices to strengthen ecosystem services for human well-being and nature conservancy.

1.2 DRYLAND UNDER CHANGING CLIMATE

Global drylands are centers of biodiversity (Maestre et al., 2021). Because these ecosystems are limited by water, dryland are prone to climate change. The magnitude of impact, however, will depend on future trends in rainfall at regional scale (Hanan et al., 2021). These regions, covering an area of 60.9 M km^2 between 1961 and 1990, may expand by as much as 10% because of the anthropogenic climate change. Feng and Fu (2013) reported that global drylands have expanded since the 1950s and will expand by another 5.8×10^6 km^2 by the end of the 21st century compared to the area under 1961–1990. Thus, climate change and sustainability are important concerns of these regions (Liu et al., 2021). The projected dryland expansion will increase the population affected by water scarcity and land degradation (Feng and Fu, 2013). Drylands in Africa and Asia, each containing 23 M km^2 of drylands (15% of the global dryland area), are prone to major environmental perturbations, such as drought, water stress, extreme rainfall events, dust storms, and heat wave (Prăvălie, 2016). Desertification is an escalating concern in global drylands (Bestelmeyer et al., 2015). It is widely believed that dry edges of the tropics may be shifting toward poles at about 50 km per decade. Additionally, the anthropogenic climate change is increasing potential evapotranspiration. Thus, an increase in aridity of drylands may also aggravate risks of desertification and related soil degradation processes. Chen et al. (2015) also reported that the northwest arid regions of China have experienced a sharp increase in both temperature and precipitation since the 1960s. In northwestern China, Chen and colleagues observed a dramatic rise in winter temperatures in the northwest arid region of China. Chen et al. (2020) observed that temperatures rose at a rate of 0.31°C per decade during 1961–2017.

Modeling studies show that, for the dry season (April to September) and by the 2050s, North Africa and some parts of Egypt, Saudi Arabia, Iran, Syria, Jordan and Israel may receive reduced rainfall by 20–25% less than the present amount (Ragab and Prudhomme, 2002). A similar increase in annual temperature of 1.75–2.5°C may occur in Southern Africa (Angola, Namibia, Mozambique, Zimbabwe, Zambia, Botswana, and South Africa). Similar trends of increase in temperatures are observed in the Thar Desert (India, Pakistan, and Afghanistan) but along with a decrease in an average annual precipitation by 5–25% and also in the Aral Sea Basin (Kazakhstan, Turkmenistan, and Uzbekistan) but with an increase in annual rainfall by 5–20%. Ragab and Prudhomme (2002) observed an increase in average temperatures by 1–1.5° C in the northern parts of Australia, with slightly higher trends during the summer than in the winter. However, the average rainfall in Australia is projected to decrease by 20–25% in the South and 5–10% in the north.

Climate change is also affecting the semi-arid regions of the world. Semi-arid regions are characterized by low annual precipitation ranging from 20% to 50% of the potential plant water demand (Scholes, 2020). These regions occupy 15.2% of the global land area and are home to more than 1.1 billion people and have a high variability in annual precipitation. Predominant agriculture in these regions is rainfed, and even a small shift in temperature or rainfall pattern can have disastrous impacts on agriculture through changes in soil water regime as is observed in Mexico (Herrera-Pantoja and Hiscock, 2015). These regions are also undergoing rapid urbanization and land use changes. Huang et al. (2015) observed that global semi-arid regions have been expanding over 60 years, from 1948 to 2008, and this expansion accounts for more than half of total dryland expansion. Guan et al. (2021) observed that more rainfall will occur in semi-arid regions of East Asia, which could drastically change ecological environment over the Loess Plateau of China.

Climate change in semi-arid regions may also aggravate the related extreme events including drought and heat wave. Furthermore, these regions, constrained by water availability and dry climate (Guan et al., 2021), are experiencing warming at rates higher than the global mean rate overland. High rates of warming are leading to high evaporative demand, and, consequently, reduced and even more variable rainfall. These regions are at high risks of soil degradation by erosion (water and wind), salinization, depletion of soil organic carbon (SOC), and other degradation processes (Hermann and Hutchinson, 2016).

Change in climate may increase the intensity and severity of soil erosion by wind and water, salinization, compaction and crusting, and decline in overall soil functionality. Increase in aridity may reduce soil moisture storage in deep soil layers and increase susceptibility to drought during the growing season. Reduction in soil moisture storage may also lead to a major shift in vegetation zones (biomes or ecoregions) and decline in ecosystem services necessary for human well-being and nature conservancy. Decline in soil moisture storage may also increase soil temperature (or soil warming) with cascading effects on increase in evaporation, acceleration of soil drying, increase in pedological and agronomic droughts, and decrease in agronomic productivity, especially of cereals (i.e., wheat, barley, and maize) and other food staples.

1.3 TYPES OF DROUGHT

Understanding the type of drought (Lal, 2014) is essential for the identification and implementation of land use and soil/crop/water management practices to mitigate the effects of drought on dryland farming. Changes in the hydrological cycle led to meteorological and hydrological drought (Figure 1.1). In the American Southwest, hydrological drought is caused by declining flows in Colorado River and Rio Grande due to increasing temperature. Overpeck and Udall (2020) observed that flow reduction will also happen in Columbia River as well as in rivers along the Sierra Nevada in California, and in the Missouri River basin. These changes are likely to increase risks of drought and more arid environments across an expanding swath of Americas Southwest. Other parts of America may also witness more frequent and severe arid events that are characterized by extreme dry spells, intense and inter-annual drought (Overpeck and Udall, 2020).

Soil degradation, caused by expansion of global drylands, may also aggravate intensity and severity of drought which affect crop growth and yield. Important among these droughts (Figure 1.1) are pedological (caused by reduction in soil water storage capacity) and agronomic (low availability of soil water in the root zone during the critical stages of crop growth) leading to severe reductions in agronomic productivity, decline in use efficiency of inputs, and reduction in nutritional quality of food. Adverse effects of pedological and agronomic droughts are often aggravated by ecological drought (low water availability because of land use conversion) and sociological drought (demand of a community for water exceeding supply under specific conditions), especially under the prevalent conditions of rapid population growth and urbanization.

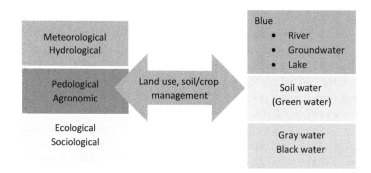

FIGURE 1.1 Linking type of drought with the source of water to manage the drought hazard.

1.4 INTER-CONNECTIVITY AMONG TYPES OF WATER

Similar to understanding different types of drought and their patterns of occurrence for site-specific conditions, it is also important to know the availability of different types of water (Figure 1.1) and how to manage their inter-connectivity (Lal, 2014). The overall strategy is to enhance the use efficiency of finite water resources and minimize the losses by evaporation, runoff, and other processes. Principal types of water are blue (river, ground water, precipitation, lakes, and ocean), brown (soil water), grey (urban water), and black (water contaminated by human waste). The brown water or soil water storage can be partitioned into total water and green water. The green water is the plant available water capacity of the root zone, and which is estimated as the difference between field moisture capacity and the permanent wilting point expressed on volumetric basis for each soil layer. The green water storage capacity of soil depends on soil texture, structure, clay minerals, and soil organic matter (SOM) content. An increase in SOM content can increase the green water storage capacity (Lal, 2020a). Changing blue water into green water through supplemental irrigation (drip sub irrigation) and conserving precipitation in the root zone (reducing losses by evaporation and runoff) are important strategies. Some of the blue water resources in the arid region (ground water or lake water) are brackish and must be used prudently to minimize the risks of soil salinization. With rapid urbanization, it is critical to reuse grey and black water (following sanitary procedures and purification) for promoting urban agriculture so that some fresh produce (vegetables) can be grown within the city limits. In this regard, home gardening can play an important role in advancing food security and minimizing the risks of disruption in food supply chains such as those cause by the COVID-19 pandemic (Lal et al., 2020).

1.5 DROUGHT MANAGEMENT IN DRYLANDS

Changing climate, degrading soil, increasing population, and growing urbanization are impacting water resources, especially soil water resources of agro-ecosystems, in drylands. Increase in population is decreasing the per capita availability of finite water resources. Western Asia and Northern Africa (WANA) region is characterized by low per capita water resources, and the finite water resources continue to be misused (El-Beltagy and MadKour, 2012).

Water resources are most critical determinants of agronomic productivity of these regions. Furthermore, intensification of the hydrological cycle and increase in frequency of extreme events related to global warming are exacerbating threats of pedological and agronomic droughts in these regions. Resource-poor and marginalized small land holder farmers are most vulnerable to water scarcity in arid regions (El-Beltagy and MadKour, 2012). Thus, there is a strong need for improving the water use efficiency, and for exploring nonconventional sources of water (i.e., desalinization of brackish or seawater, reuse of grey and black water, and water harvesting and recycling from

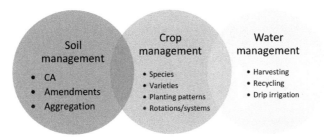

FIGURE 1.2 Managing the soil–crop–water nexus to maximize the water use efficiency and sustain productivity (CA refers to conservation agriculture).

urban structures). Such innovations are essential to enable vulnerable communities to accomplish sustainable use of fragile and finite resources in arid lands and save resources for future generations (El-Beltagy and MadKour 2012). Adoption of effective soil-water management practices is critical to building resilience against drought and floods (Cornelis et al. 2019). Identification of site-specific practices, based on the soil-water-crop-nexus (Figure 1.2), is the critical component to fine-tune management practices under site-specific conditions.

- Species
- Varieties
- Planting Patterns
- Rotations/Systems
- Harvesting
- Recycling
- Drip Irrigation
- CA
- Amendments
- Aggregation

1.6 CARBON SEQUESTRATION POTENTIAL IN GLOBAL DRYLANDS

Soils of drylands contain two forms of carbon: SOC and soil inorganic carbon (SIC). In general, stocks of SIC in global drylands are more than those of SOC. To 1-m depth, world soils contain 1500 petagram (Pg) of SOC and 940 Pg of SIC (Jobbagy and Jackson, 2000; Eswaran et al., 2000). To 2-m depth, world soils contain 1993 Pg of SOC and 2300 Pg of SIC (Song et al., 2022). Despite relatively low SOC stock in topsoil and vegetation, world's drylands contain 30% of global C stock in above and below ground biomass and soil (30 cm) (Hanan et al., 2021). The magnitude of stock in above and below ground plant biomass is often less than 100 Mg C /ha, and those of SOC to 30 cm depth is less than 50 Mg C/ha, with the exception of those in high-altitude drylands. In some cases, however, the terrestrial C stock in drylands (vegetation plus soil) can be as much as 200 Mg C/ ha(Hanan et al., 2021).Estimates of above-ground and below-ground C stock at global scale are 790 Pg and 614 Pg, representing 22% and 38% of the global totals in these pools, respectively (Hanan et al., 2021).The data in Tables 1.1 and 1.2 show the magnitude of SOC and SIC for different depths and in diverse ecoregions of global drylands.

The data in Table 1.1 show that the ratio of SOC stock in dryland versus humid regions (expressed as percent) is 46–49% and that of dryland versus global stock is 32 to 33 % for 0–2 m depth. In comparison, the SIC stock in dryland is 3.9 to 5.4 times more than SOC. However, the ratio of SIC stock in dryland versus global (expressed as percent) ranges from 79% to 84%. The ratio of total C stock (sum of SIC and SOC) for dryland versus humid regions (expressed as percent) ranges from

TABLE 1.1

Total Biomass and Soil Carbon Stocks in Global Drylands

Biome	AI	SOC Stock (Pg C)			SIC Stock (Pg C)			Total Soil C Stock (Pg C)			Ratio of Total SOC : SIC Stock (0–2m)
		0–0.3m	0–1m	0–2m	0–0.3m	0–1m	0–2m	0–0.3m	0–1m	0–2m	
Hyper arid	<0.05	11	22	31	20	65	127	31	87	158	0.24
Arid	0.05–0.20	45	91	127	63	241	487	108	332	674	0.28
Semi-arid	0.20–0.50	100	190	259	48	207	456	148	397	715	0.57
Dry sub-humid	0.50–0.65	91	167	228	15	66	168	106	233	396	1.35
Dryland	<0.65	248	470	646	145	578	1237	393	1048	1683	0.52
Humid	>0.65	502	955	1401	28	107	321	530	1062	1722	4.36
Global		750	1425	2047	173	685	1558	923	2110	3505	1.54
Dryland:Humid (%)		49.4	49.2	46.1	518	540	385	74.1	98.7	97.7	–
Dryland:Global (%)		33.0	33.0	31.6	83.8	84.3	79.4	42.5	49.7	48.0	–

Source: Adapted and synthesized from Hanan et al. (2021) and Scientific Reports.

TABLE 1.2

Amount of Soil Organic and Soil Inorganic Carbon Stocks in Different World Biomes

Biome	0–30 cm Depth			0–100 cm Depth			SIC as % of Total	
	SOC	SIC	Total	SOC	SIC	Total	0–30 cm	0–100 cm
Tundra	54.0	9.5	63.5	158.8	16.4	175.2	15.0	9.4
Cold Parklands	16.1	5.1	21.2	35.7	15.3	51.0	24.0	30.0
Forest Tundra	74.2	8.5	82.7	180.2	14.4	194.6	10.3	7.3
Boreal Forest	152.1	12.6	164.7	345.9	34.9	380.8	7.7	9.2
Cool Desert	9.6	14.4	24.0	19.5	49.6	69.1	60.0	71.8
Steppe	35.3	19.0	54.3	69.1	73.1	142.2	35.0	51.4
Temperate Forest	63.9	11.3	75.2	120.3	31.9	152.2	15.0	21.0
Hot Desert	36.6	71.3	107.9	66.1	231.6	297.7	66.0	77.8
Chaparral	21.8	15.0	38.8	42.5	53.7	96.2	40.7	55.8
Warm Temperate Forest	15.9	2.8	18.7	30.4	6.3	36.7	15.0	17.2
Tropical Semi-arid	27.0	24.3	51.3	53.1	52.2	135.3	47.4	60.8
Tropical Dry Forest	61.1	18.7	79.8	118.3	55.9	174.2	23.4	32.1
Tropical Seasonal Forest	70.8	6.8	77.6	133.0	21.1	154.1	8.8	13.7
Tropical Rain Forest	45.7	2.7	48.4	89.0	8.6	97.6	5.6	8.8
All Ecosystems	684.1	222.0	906.1	1462.0	695.0	2157.0	24.5	32.2

Source: Recalculated from Batjes (1996).

74% in the surface 0–0.3 m layer to 98 % in the top 1 and 2 m layers. In contrast, the ratio of total C stock in dryland versus global ranges from 43% in the surface to ~50% in the subsurface layers up to 2-m depth (Table 1.1).

Similar to data in Table 1.1, Plaza et al. (2018) also reported that dryland soils store 646 Pg of SOC up to 2-m depth.

Data on relative magnitude of SIC and SOC stocks for 0–0.3 m and 0–1 m depth are reported by Batjes (1996) and shown in Table 1.2. The SIC stocks in 0–0.3 m depth and 0–1 m depth are more than SOC stocks in cool desert and hot desert biomes. SIC stock expressed as percent of the total stock for 1 m depth (last column in Table 1.2) is more than 50% in cool desert, hot desert, steppe, chaparral, and tropical semi-arid climate. Expressed to 0.3 m depth, SIC is more than 50% of the total C stock for cool desert and hot desert and, more than40% for chaparral and semi-arid tropics, and more than 30% for steppe (Table 1.2). Thus, the potential for formation of secondary carbonates and leaching of bicarbonates by using good-quality irrigation water is high for soils of agroecosystems of these regions.

The data in Tables 1.1 and 1.2 indicate that in addition to SOC, soils of drylands have potential to sequester SIC as secondary carbonates (Naorem et al., 2022). Conversion to a restorative land use and adoption of best management practices can lead to sequestration of carbon in soil (SOC and SIC) and vegetation (above and below ground) with a potential impact on draw down of atmospheric CO_2 and offsetting part of anthropogenic emissions. Sequestration of SOC in drylands may also be limited by relative scarcity of essential nutrients drylands. Plaza et al. (2018) reported that for 0–0.5 m soil depth the ratio for essential nutrients decreases from humid to hyper-arid for 14 to 8 for C: N, 38 to 5 for C:P, and 2.8 to 0.8 for N:P. Thus, judicious management of water and that of N and P are critical to enhancing SOC contents in soils of these ecosystems. Irrigation with a good-quality water can also lead to leaching of bicarbonates into the subsoil. Similar to its effects on SOC, land use change can also have a significant impact on SIC and formation of secondary carbonates (An et al., 2019).

1.7 IT IS ALL ABOUT WATER

Green water supply in the root zone is a strong determinant of the agronomic productivity of crops (and livestock) in drylands. Drought stress (pedological and agronomic droughts) is a widespread problem, and its intensity and frequency/duration may be aggravated by current and projected climate change. Dryland ecoregions are also prone to drought/floods syndrome, such as those experienced in Pakistan in 2022. Thus, an effective strategy is to conserve the precipitation where it falls by enhancing water infiltration capacity of the soil and reducing evaporation and building resilience against drought and flood (Cornelis et al., 2019). Grain yield of cereals in dryland regions is given by Eq. (1.1) (Stewart and Lal, 2018):

$$GY = ET \times T/ET \times 1/TR \times HI \qquad \text{(Eq. 1.1)}$$

where GY is the dry grain yield in kg/ha, ET is the evapotranspiration per kg/ha, T/ET is the fraction of ET transpired by crop, TR is the transpiration ratio (kg of water transpired to produce 1 kg of above-ground biomass), and HI is the harvest index (kg of dry grain/kg of above-ground biomass). Parameters in Eq. (1.1) show that GY is directly proportional to ET, T, and HI. Thus, agronomic productivity in dryland farming is all about water (Stewart and Lal, 2018). Thus, increasing the supply of green water in the root zone is the most pertinent strategy to enhancing and sustaining agronomic productivity.

1.8 RESEARCH AND DEVELOPMENT PRIORITIES

This and the companion volume of advances in soil science are aimed at collation and synthesis of available information on science and practice of managing drylands for sustainable management of soil, water and crops, livestock, and ecosystems. Each of the chapters in this book is focused on themes pertinent to these issues and identifies priorities for the specific challenge in research, education, and training. Some generic recommendations that are pertinent to sustainable management of drylands are listed below:

a. **Evaluate status of soil degradation** by physical, chemical, biological, and ecological parameters in relation to processes, factors, and causes to advance frontiers of knowledge and identify site-specific technologies to protect, restore, and sustainably manage fragile soil resources in harsh environments.

b. **Identify key parameters or indicators of soil health** and establish critical limits of these parameters for specific land use and farming/cropping systems.

c. **Determine stock and flux of carbon** between soil and atmosphere and measure the rate of carbon sequestration (SOC and SIC) for diverse land uses, soils, and farming/cropping systems.

d. **Estimate the societal value of carbon stored in the terrestrial biosphere** by rapid and cost-effective methods so that farmers can be rewarded through payments for ecosystem services.

e. **Determine the relationship between soil health and agronomic productivity** and assess the impact of gain in SOC on crop yield and saving of inputs (irrigation water, fertilizer use, and energy consumption) so that more can be produced from less.

f. **Establish protocol to restore degraded soils and desertified ecosystems** so that the concept of zero net soil degradation (SDG #15) can be implemented at different scales.

g. **Promote awareness about the importance of soil health** among policymakers, general public, and land managers in relation to human well-being and nature conservancy following the "one health concept" which states that the health of soil, plants, animals, people, ecosystems, and planetary processes is one and indivisible.

h. **Strengthen human resources to protect, restore, and sustainably manage** soil, water, and biodiversity and revise curricula at different levels (kindergarten to graduate schools) to promote education in this discipline.

i. **Involve private sector** to facilitate translation of science into action (Lal, 2020c) and address issues of global significance (e.g., global warming, expansion of dryland, depletion of water resources, and pollution of water).

j. **Enhance adoption of best management practices** which upscale nutrition-sensitive agriculture to address malnutrition and hidden hunger in global drylands.

REFERENCES

An, H., Wu, X., Zhang, Y., & Tang, Z. (2019). Effects of land-use change on soil inorganic carbon: A meta-analysis. *Geoderma, 353*, 273–282.

Batjes, N. H. (1996). Total carbon and nitrogen in the soils of the world. *European Journal of Soil Science, 47*(2), 151–163.

Berdugo, M., Gaitán, J. J., Delgado-Baquerizo, M., Crowther, T. W., & Dakos, V. (2022). Prevalence and drivers of abrupt vegetation shifts in global drylands. *Proceedings of the National Academy of Sciences, 119*(43), e2123393119.

Bestelmeyer, B. T., Okin, G. S., Duniway, M. C., Archer, S. R., Sayre, N. F., Williamson, J. C., & Herrick, J. E. (2015). Desertification, land use, and the transformation of global drylands. *Frontiers in Ecology and the Environment, 13*(1), 28–36.

Chen, Y., Fam, Y., Li, Z., & Wang, H. (2015). Progress and prospects of climate change impacts on hydrology in the arid region of northwest China. *Environmental Research.* doi: 10.1016/j.envres.2014.12.029

Chen, Y., Zhang, X., Fang, G., Li, Z., Wang, F., Qin, J., & Sun, F. (2020). Potential risks and challenges of climate change in the arid region of northwestern China. *Regional Sustainability 1*(1), 20–30.

Cornelis, W., Waweru, G., & Araya, T. (2019). Building resilience against drought and floods: The soil water management perspective. *Sustainable Agriculture Reviews 29*, 125–142.

El-Beltagy, A., & Madkour, M. (2012). Impact of climate change on arid lands agriculture. *Agriculture & Food Security 1*, Article #3. https://doi.org/10.1186/2048-7010-1-3

Eswaran, H., Van den Berg, E., Reich, P., & Kimble, J. (1995). Global soil carbon resources. In R. Lal et al. (Eds) Soils and Global Change, CRC Publishers, Boca Raton, FL: 27–43.

European Commission. (2023). *Aridity and Urban Population.* European Commission Joint Research Centre. https://wad.jrc.ec.europa.eu/aridityurban

Feng, S., & Fu, Q. (2013). Expansion of global drylands under a warming climate. *Atmospheric Chemistry and Physics, 13*(19), 10081–10094.

Guan, X., Zhu, K., Huang, X., Zhang, X., & He, Y. (2021). Precipitation changes in semi-arid regions in East Asia under global warming. *Frontiers in Earth Science, 11*. https://doi.org/10.3384/feat2021762348

Hanan, N. P., Milne, E., Aynekulu, E., Yu, Q., & Anchang, J. (2021). A role for drylands in a carbon neutral world?. *Frontiers in Environmental Science,* doi:10.3389/fenvs.2021.786087

Hermann, S. M., & Hutchinson, C. F. (2016). The scientific basis: Links between land degradation drought and desertification. In P. M. Johnson, K. Mayrand, & M. Paquin (Eds) Governing Global Desertification, Routledge, London: 31–46.

Herrera-Pantoja, M., & Hiscock, K. M. (2015). Projected impacts of climate change on water availability indicators in semi-arid regions of central Mexico. *Environmental Science & Policy,* doi:10.10/6/J.envsci201506020

Huang, J., Ji, M., Xie, Y., Wang, S., He, Y., & Ran, J. (2015). Global semi-arid climate change over last 60 years. *Climate Dynamics.* doi: 10-1007/8-382-015-2638-8

IPCC. (2013). Working group contribution to the IPCC Fifth Assessment Report. Climate Change: The Physical Science Basis. Summary for Policy Makers, Oxford Univ. Press.

Jobbágy, E. G., & Jackson, R. B. (2000). The vertical distribution of soil organic carbon and its relation to climate and vegetation. *Ecological Applications, 10*(2), 423–436.

Lal, R. (2014). The Nexus of Soil, Water and Waste. Lecture Series – No. 1 Dresden: United National University Institute for Integrated Management of Material Fluxes and of Resources. UNU-FLORES. 1–17.

Lal, R. (2020a). Soil organic matter and water retention. *Agronomy Journal, 112*(5),3265–3277.

Lal, R. (2020b). Home gardening and urban agriculture for advancing food and nutritional security in response to the COVID-19 pandemic. *Food Security.* https://doi.org/10.1007/s12571-020-01058-3

Lal, R. (2020c). The role of industry and the private sector in promoting the "4 per 1000" initiative and other negative emission technologies. *Geoderma, 378,* 114613.

Lal, R., Brevik, E. C., Dawson, L., Field, D., Glaser, B., Hartemink, A. E., ... & Sánchez, L. B. R. (2020). Managing soils for recovering from the COVID-19 pandemic. *Soil Systems, 4,* 46. doi: 10.3390/soilsystems4030046

Lioubimtseva, E. (2004). Climate change in arid environments: Revisiting the past to understand the future. *Progress in Physical Geography, 28*(4), 502–530.

Liu, Z., Chen, Z., Yu, G., Zhang, T., & Yang, M. (2021). A bibliometric analysis of carbon exchange in global drylands. *Journal of Arid Land, 13,* 1089–1102.

Maestre, F. T., Benito, B. M., Berdugo, M., Concostrina-Zubiri, L., Delgado-Baquerizo, M., Eldridge, D. J., ... & Soliveres, S. (2021). Biogeography of global drylands. *New Phytologist, 231*(2), 540–558. https://doi.org/10.1111/nph.17395

Maestre, F. T., Le Bagousse-Pinguet, Y., Delgado-Baquerizo, M., Eldridge, D. J., Saiz, H., Berdugo, M., & Gross, N. (2022). Grazing and ecosystem service delivery in global drylands. *Science, 378*(6622), 915–920. doi: 10.1126/science.abq4062

Naorem, A., Jayaraman, S., Dalal, R. C., Patra, A., Rao, C. S., & Lal, R. (2022). Soil inorganic carbon as a potential sink in carbon storage in dryland soils—A review. *Agriculture, 12*(8), 1256.

Overpeck, J. T., & Udall, B. (2020). Climate change and the aridification of North America. *PNAS, 117*(22), 11856–11858. https://doi.org/10.1073/pnas.2006323117

Plaza, C., Zaccone, C., Sawicka, K., Méndez, A. M., Tarquis, A., Gascó, G., ... & Maestre, F. T. (2018). Soil resources and element stocks in drylands to face global issues. *Scientific Reports, 8*(1), 13788.

Prăvălie, R. (2016). Drylands extent and environmental issues. A global approach. *Earth-Science Reviews, 161,* 259–278.

Ragab, R., & Prudhomme, Christal (2002). SW-soil and water: Climate change and water resources management in arid and semi-arid regions: Prospective and challenges for the 21st century. *Biosystems Engineering, 81*(1), 3–34.

Scholes, R. J. (2020). The future of semi-arid regions: A weak fabric unravels. *Climate, 8*(3), 43. https://doi.org/10.3390/cli8030043

Song, X. D., Yang, F., Wu, H. Y., Zhang, J., Li, D. C., Liu, F., ... & Zhang, G. L. (2022). Significant loss of soil inorganic carbon at the continental scale. *National Science Review, 9*(2), nwab120.

Stewart, B. A., & Lal, R. (2018). Increasing world average yields of cereal crops: It's all about water. *Advances in Agronomy, 151,* 1–44.

United Nations. (2014). Global location of big cities. Department of Economic and Social Affairs, Population Division: World Urbanization Prospects. New York, USA.

2 Soil Water Conservation for Dryland Farming

Paul W. Unger[1], Robert C. Schwartz[2], R. Louis Baumhardt[2], and Qingwu Xue[3]

[1]Soil Scientist, USDA Agricultural Research Service, Conservation and Production Research Laboratory, Bushland, TX 79012 (now, 3603 Thurman St., Amarillo, TX 79109)
[2]Research Soil Scientist, USDA Agricultural Research Service, Conservation and Production Research Laboratory, Bushland, TX 79012
[3]Texas A&M AgriLife Research & Extension Center at Amarillo, Amarillo, TX 79106

CONTENTS

2.1 Introduction 11
2.2 Precipitation Variability 13
2.3 Water Capture and Infiltration 13
    2.3.1 Runoff 15
    2.3.2 Deep Soil Loosening and Soil Surface Alterations 15
2.4 Soil Water Evaporation 17
    2.4.1 Tillage and Residue Effects on Evaporation Processes 18
2.5 Redistribution and Deep Drainage of Soil Water 19
2.6 Weed Control 20
2.7 Cover Crops 21
2.8 Fallow and Dryland Crop Intensification 22
2.9 Growing Season Interventions 24
    2.9.1 Plant Density, Spacing, and Clumping 25
    2.9.2 Climate-Informed Management 26
2.10 Climate Change Effects 28
    2.10.1 Climate Change Effect on Wheat Yield 28
    2.10.2 Strategies for Adapting to Climate Change 30
References 31

2.1 INTRODUCTION

Dryland farming generally is defined as agriculture where the lack of available water limits crop and/or pasture production in a part of the year (Stewart et al., 2006). In this report, we deal with crop production without irrigation in semiarid regions where available water often is the most limiting factor for good crop yields. Precipitation in semiarid regions is highly variable. Where dryland agriculture occurs, it usually ranges from about 250–500 millimeters (mm) per year (Brengle, 1982; Stewart et al., 1985), but grain production is possible, although quite risky, in some regions with precipitation amounts of 150–200 mm (Dregne, 2006).

Dryland farming is widely used in many countries. In North America, dryland farming is highly important in the Canadian Prairies and in the United States (US) Great Plains, the US Pacific

DOI: 10.1201/b22954-2

Northwest, the US Southwest, and parts of the US Intermountain Areas (Cannell and Dregne, 1983). Because the amount and seasonal distribution of precipitation in these regions is highly variable from year to year, soil water conservation is critical for maximizing the available stored soil water at planting so that crops can produce at their potential (Willis, 1983). As a result, strong efforts usually are made to improve dryland crop yields in semiarid regions by increasing soil water storage from any precipitation that may occur.

Because primary tillage for weed control and seedbed preparation significantly modifies soil surface characteristics and consequently water fluxes at this boundary, the intensity and method of tillage practice greatly influences water conservation in dryland cropping systems. When early settlers came to the semiarid parts of the US, they used clean tillage practices that were used in higher precipitation areas of the country. The tillage implements used included moldboard plows or one-way disk plows that buried most plant residues (Johnson et al., 1983). Clean tillage usually relied on frequent disc plowing at a shallow depth. Such tillage created a "dust mulch" (Campbell, 1907) for which the goal was to reduce water evaporation from exposed soil areas (Patil et al., 2013). It was believed that the fine materials entering soil capillaries and wide cracks where water evaporation occurs would reduce that evaporation. According to James (1945), however, water conservation was not achieved by using a dust mulch. Soil water contents were higher at 15-, 45-, and 90-cm depths under straw mulch than under the dust mulch. At the 15-cm depth, the dust mulch was no more effective than bare soil for conserving soil water. The primary benefit of the repeated tillage was weed control, not evaporation at the soil surface (Rowe-Dutton, 1957). Other problems related to dust mulching were greater water runoff, greater soil erosion, greater air pollution, and lower crop yields (Chalker-Scott, 2008).

Use of clean tillage combined with a major drought contributed to the Dust Bowl that occurred in the US in the 1930s and clearly showed that clean tillage was not suitable for use in the semiarid region of the US. As a result of that drought, farmers struggled to find ways to control the wind erosion. An Oklahoma farmer, Fred Hoeme, developed a heavy duty chisel plow that became recognized as "The Plow that Saved the Plains." That plow resulted in large erosion-resistant soil aggregates and retained crop residues on the surface. Those conditions protected the soil from erosion and increased soil water storage (Stewart et al., 2010). It was, in fact, the increased wheat (*Triticum aestivum* L.) yields resulting from greater infiltration and water conservation when Fred Hoeme moved a road-scarifier overland to his equipment lot that led to the development of this improved chisel plow (Peterson, 1948).

In response to the Dust Bowl, a concerted research effort by the United States Department of Agriculture (USDA) and State Experiment Stations of the western US was initiated to develop and adapt stubble mulch tillage to combat wind erosion and improve precipitation capture as soil water (Johnson, 1950). Based on a blade plow developed by Charles Noble, several stubble mulch tillage implements were adapted and tested throughout the late 1940s and 1950s. Stubble mulch tillage undercuts the surface at depths of about 8–12 cm (Johnson, 1950; Jones and Popham, 1997), thereby retaining most residues on the surface and also loosening the soil to improve water infiltration. The surface residues reduce soil water evaporation (Jones et al., 2003; Unger, 1988), help control erosion, and usually result in improved soil water storage (Johnson et al., 1983; Stewart et al., 2010; Unger and McCalla, 1980; Woodruff et al., 1966).

With the use of herbicides, it became feasible to reduce the intensity and frequency of tillage or eliminate it entirely in modern crop production. Water conservation under dryland cropping is, in general, improved during fallow periods under no tillage and often under reduced tillage practices compared with conventional tillage (Lindstrom et al., 1974; Unger and Wiese, 1979; Unger, 1984; Norwood et al., 1990; Jones et al., 1994; Schlegel et al., 2018a). Adoption of no tillage in the US Great Plains has increased considerably during the past three decades, principally in the Northern and Central Plains regions (Hansen et al., 2012).

Successful water management under dryland farming conditions invariably entails conserving soil water during fallow periods for subsequent use by dryland crops and employing management

Soil Water Conservation for Dryland Farming

interventions during the growing season that have the potential to improve crop productivity. In this chapter, we address current and past research developments highlighting the effectiveness of management interventions that influence components of the soil water balance and crop water use and yield during the growing season of dryland crops.

2.2 PRECIPITATION VARIABILITY

Year to year precipitation variability is common in semiarid regions and usually ranges from 250 to 500 mm annually (Brengle, 1982; Stewart et al., 1985). At Bushland, TX, for example, annual precipitation ranged from 171 mm in 2011 to 827 mm in 1960. Annual average precipitation at Bushland is about 472 mm. Annual precipitation totals, however, do not necessarily indicate what the available soil water content may be for use by a dryland crop. Rather, stored soil water at planting largely depends on the amount and distribution of precipitation that occurs during fallow periods (Unger, 1978a).

In a study evaluating precipitation storage during fallow (Unger, 1978a), precipitation from harvest of winter wheat until grain sorghum (*Sorghum bicolor* (L) Moench) planting (11 months later) was 303 mm in 1973–1974, 436 mm in 1974–1975, and 216 mm in 1975–1976. Stored soil water at sorghum planting as a fraction of fallow precipitation, however, was 0.076, 0.20, and 0.50 for the respective fallow years. The greater fraction of stored soil water at planting for 1975–1976 resulted from 170 mm precipitation during the final two months of the fallow period, indicating that total precipitation for a certain period does not necessarily influence soil water content at crop planting time.

At Bushland, precipitation during fallow after wheat from 1984 to 1989 ranged from 390 to 676 mm. During fallow after sorghum from 1985 to 1989, precipitation ranged from 291 to 578 mm (Unger, 1994). These large ranges indicate the effects of precipitation variability that potentially affects stored water at planting and crop yields. At a research site near Bushland, monthly precipitation ranged from 0 to 150 mm from late 1994 until December 1997. Low amounts usually occurred in the same months, but the high amount occurred in a different month each year, indicating precipitation variability that affected soil water content at crop planting time and crop yields (Unger, 1999).

2.3 WATER CAPTURE AND INFILTRATION

Effective dryland crop management invariably requires the use of practices that improve infiltration (capture) of precipitation to increase stored soil water. Because soil surface characteristics largely influence infiltration rates, tillage and residue management practices that modify this zone have a major influence on precipitation capture. Tillage alters the continuity, size, and extent of soil pores and consequently influences soil hydraulic properties. However, increased infiltration rates that usually result from tillage are short lived due to reconsolidation and crust formation (Baumhardt and Schwartz, 2004; Moret and Arrue, 2007). Residues reduce raindrop impact on the soil, thereby reducing dispersion and slowing the formation of a surface crust (seal). Consequently, infiltration is typically increased with increasing residue cover (Duley and Kelly, 1939; Baumhardt and Lascano, 1996; Baumhardt et al., 2012). However, the effectiveness of residue for increasing infiltration rates can interact with tillage and residue characteristics (Unger, 1992). Baumhardt and Lascano (1996) demonstrated that under a disk tilled surface with wheat straw, cumulative infiltration increased with increasing residue cover up to 2.5 Mg ha^{-1} with diminishing effects at greater residue rates. In addition, for wheat straw consisting of both standing and flat components, they found that the amount of standing residue had an insignificant effect on increasing infiltration. In a subsequent study using simulated rainfall, Baumhardt et al. (2012) showed that while cumulative infiltration did not vary with respect to no tillage and stubble mulch tillage with about 95% residue cover,

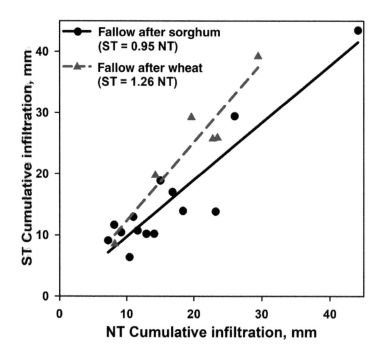

FIGURE 2.1 Relationship between event-based cumulative infiltration under no-tillage (NT) and stubble-mulch tillage (ST) during fallow periods in Bushland, TX, from 2007 to 2010. Cumulative depth of storm events ranged from 10 to 55 mm and average storm intensities ranged from 1 to 28 mm hr^{-1} (Schwartz et al., 2019).

cumulative infiltration at 1 hour was lower for no tillage than with stubble mulch tillage. In eastern Washington where most of the precipitation is received during the winter months, no tillage resulted in greater runoff and reduced stored soil water compared with tilled surfaces under conditions of soil freezing followed by spring rainfall (Lindstrom et al., 1974).

Infiltration throughout fallow periods under natural precipitation events would be expected to differ considerably from simulated rainfall. Components of the soil water budget can be estimated indirectly through observed changes in stored soil water at high temporal resolution using electromagnetic sensors such as time-domain reflectometry (Schwartz et al., 2008). This approach was used by Schwartz et al. (2019) to calculate event-based estimates of infiltration throughout the fallow periods of a wheat–sorghum–fallow rotation under no tillage and stubble mulch tillage in Bushland, TX. For precipitation events with cumulative infiltration greater than 8 mm, cumulative infiltration under stubble mulch tillage was similar to no tillage during fallow after sorghum (Figure 2.1).

However, during fallow after wheat, cumulative infiltration with stubble mulch tillage was significantly greater (26%) than with no tillage. Cumulative infiltration as a fraction of storm precipitation was greater under wheat fallow (92%) compared with sorghum fallow (76%) due in part to greater residue cover (41% and 33% during wheat and sorghum fallow, respectively) and also greater protection afforded by wheat stubble. However, similar to the results of Baumhardt et al. (2012), cumulative infiltration under stubble mulch tillage responded more to increased residue cover compared with no tillage. In contrast to these results in Texas, cumulative infiltration under no tillage in Tribune, KS, averaged 59% greater compared with stubble mulch tillage, regardless of the fallow phase in this wheat–sorghum–fallow rotation (Schwartz et al., 2019). These differences in tillage and residue effects on infiltration are likely attributable to the greater propensity of the Pullman clay loam in Texas to form thick structural crusts compared with the Richfield silt loam in Kansas.

2.3.1 Runoff

Runoff observations from gauged large plots or watersheds comparing tillage effects in semiarid regions are limited, although useful in examining aggregate effects across the landscape. Extensive investigations of runoff in Texas for a fine-textured clay loam being used for a wheat–sorghum–fallow rotation were made by Jones et al. (1994) and Baumhardt et al. (2012; 2017; 2020) in graded terraces (2–4 ha) equipped with flumes. Their result showed that no tillage management resulted in similar or increased runoff compared with stubble mulch tillage. As a fraction of the total mean runoff during the 3-year rotation, 40% occurred during fallow after sorghum followed by the next largest contribution of 25% under fallow after wheat. Runoff under no-tillage management exceeded stubble mulch tillage in all phases of the rotation, except for the sorghum growing season (Baumhardt et al., 2017). Using 90-m^2 runoff plots, Tullberg et al. (2001) evaluated controlled traffic and tillage effects on runoff in a wheat–sorghum–fallow–maize (*Zea mays* L) dryland rotation on a Vertisol in southeastern Queensland. They demonstrated that controlled traffic reduced runoff by an average of 63 mm yr^{-1} across all tillage systems compared to wheel-tracked areas. No tillage management generated less runoff than stubble mulch tillage independent of wheel traffic in years with significant precipitation. The increase in stored soil water under controlled traffic treatments correspondingly increased grain yields by an average of 14%.

Because the observed runoff as a fraction of rainfall decreases with increasing scale, observed runoff from large plots or watersheds is not necessarily a meaningful criterion in assessing infiltration and stored soil water at smaller scales. The mechanisms affecting this scale dependence can be attributed to temporal properties associated with rainfall events, run-on, and the spatial variability of infiltration (van De Giesen et al., 2000; Moreno-de las Heras et al., 2010; Chen et al., 2016). Schwartz et al. (2019) observed greater heterogeneity in cumulative infiltration under no tillage compared with stubble mulch tillage that likely resulted from surface crusting and areas of run-on associated with residue patches. Lal (1997) observed differing scale effects on runoff for no tillage and conventionally tilled plots with varying slope lengths.

2.3.2 Deep Soil Loosening and Soil Surface Alterations

Fragipans, hardpans, plowpans, clay pans, and dense clay horizons are present in some soils. Such pans and horizons usually limit water penetration into soils and also plant root exploration. By disrupting such pans or layers, water penetration and crop yields usually are increased (Busscher et al., 2000, 2001; Unger, 1993; Unger and Stewart, 1983). The use of some tillage implements causes formation of plow soles (pans) that restrict water flow and root penetration, thereby limiting crop growth and yields under some conditions (Eck and Unger, 1985). The problem due to plow soles varies across the country and can usually be eliminated by using chiseling or subsoiling implements. In northern areas, freezing may eliminate plow pans. The paraplow is a heavy straight chisel with a shank that bends 45° to lift soil at working depths of 0.30–0.43 m to fracture compacted layers without disturbing surface residue (Figure 2.2).

Paraplow effects on rain infiltration, water storage, penetration resistance, density, and crop yield were quantified under conventional stubble mulch tillage and no tillage conditions on a clay loam soil (Baumhardt and Jones, 2002a, 2002b). Although paratillage (and single sweep tillage) soil loosening resulted in significantly lower penetration resistance and bulk density only in no tillage plots, paratillage did not affect infiltration of simulated rain, soil water storage during fallow, or yield of the subsequent crop.

More complete soil disturbance approaching profile modification that also disrupts dense subsoil layers for improved water movement, root exploration, and crop performance was evaluated with soil profile modifying to about a 0.7-m depth with a 1.0-m single blade large moldboard plow in 1971 (Baumhardt et al., 2008). After 4 and 31 years, deep plowed soils had lower soil density

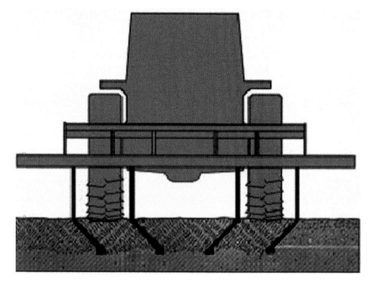

FIGURE 2.2 Diagram of a paraplow configured with four shank-shatter plate assemblies faced toward the center and adjusted for total loosening.

FIGURE 2.3 The Pullman clay loam is slowly permeable resulting in standing water several days following large rain events (left), but a sustained (>30 years) benefit of deep plowing was increased permeability and drainage (right).

and penetration resistance at lower depths, but not near the surface compared with an unmodified profile. Steady ponded infiltration measured for practically saturated profiles after a 56 mm rain event during the fourth year, 1975, averaged 1.2 mm hr^{-1} for the unplowed conditions compared with 8.1 mm hr^{-1} with deep-plowing or nearly a sevenfold increase. The difference in ponded water infiltration into drier soil profiles measured in 2002 (31 years after tillage) revealed the unplowed control still required 1.5 longer time for amounts of water exceeding 100 mm to infiltrate than that for deep plowed soil. Deep plowing resulted in long-term sorghum yields averaging 2.86 and 2.61 Mg ha^{-1} for plowed and unplowed soil, respectively, that were significantly different (Baumhardt et al., 2008). They concluded that about 90% of that yield difference was during years benefitting from improved aeration because of drainage of excess precipitation at the beginning of the growing season (Figure 2.3), and not due to increased rooting to explore larger soil volumes for water or nutrients.

Soil surface alterations achieved by a variety of practices can be utilized to create digressional storage or increase the saturated conductivity to reduce runoff and increase stored soil water (Unger et al., 2010). Contouring involves creating ridges across the slope of the land. Contouring is best suited for gently sloping land and potentially may provide water conservation benefits that improve crop yield. Because of aggregate dispersion and resultant surface sealing, water ponded in furrows may evaporate before it enters some soils (Unger et al., 2010). Another problem with contour tillage is the current use of large farming equipment. Cultural operations when using large equipment often are done without regard for slope of the land. For example, those operations sometimes are done across terraces made earlier to help control erosion on sloping fields (personal observation, P.W. Unger).

Increased surface detention storage to prevent loss of storm water as runoff can be done using basins formed by placing small dams in furrows of a ridged field to form furrow dikes after crop emergence (Jones and Baumhardt, 2003). With furrow diking, precipitation or irrigation retained in the furrows gains a longer opportunity time for the water to enter the soil, thus increasing the potential for greater crop yields (Gerard et al., 1984; Nuti et al., 2009; Sui et al., 2016) with a few exceptions due to seasonal rainfall patterns (Baumhardt et al., 1993; Bryant et al., 2019). Furrow diking was used in the US starting in 1930, but its use was abandoned by 1950 because of various problems – diking difficulty, weed control, performing cultural operations, and limited yield benefits (Jones and Clark, 1987).

2.4 SOIL WATER EVAPORATION

With significant wetting of the soil, soil water evaporation is initially controlled by available atmospheric energy (Stage 1) and thereafter by a falling rate stage (Stage 2) whereby water transmission to the surface is limited by soil hydraulic properties and vapor transport through a drying surface. A third stage, which may not necessarily be distinct from Stage 2 (van Bavel and Hillel, 1976), is usually associated with a low, nearly constant evaporation rate from very dry soil (Idso et al., 1974). Non-steady state atmospheric conditions, finite wetting depths of shallow precipitation events, wetting history, and nonisothermal water transport limits these evaporation stage conceptualizations under field conditions (Jackson et al., 1973; Brutsaert and Chen, 1996; Brutsaert, 2014, Schwartz et al., 2019). Upward nighttime water flow toward the surface can result in Stage 2 transitioning to Stage 1 (Tolk et al., 2015) and likewise Stage 3 transitioning to Stage 2 (Idso et al., 1974). Tolk et al. (2015) also demonstrated that Stage 1 and Stage 2 may occur simultaneously because of surface heterogeneity likely arising from micro-depressions.

Since the actual rate of evaporation is controlled by both climatic conditions and soil properties, accurate prediction of soil water evaporation requires coupled soil and atmospheric models to describe the exchanges of energy and water at the soil interface (van Bavel and Hillel, 1976; Vanderborght et al., 2017). For field applications, simplifications to these process-based models are necessary because of uncertainties in estimated surface resistances arising from nonuniform arrangements of surface residues, an assortment of residue and soil surface characteristics, as well as changing soil surface characteristics resulting from tillage and subsequent reconsolidation (Schwartz et al., 2010). Gardner (1959) demonstrated that the rate of evaporation during the second stage could be described as a desorption process that scaled with the square root of time with the rate constant (β) as a function of diffusivity. Boesten and Stroosnijder (1986) showed that cumulative evaporation was likewise proportional to the square root of cumulative potential evapotranspiration (ET_o). Remarkably, this relationship describes field evaporation rates reasonably well for bare soils (Black et al., 1969; Jackson et al., 1976; van Bavel and Hillel, 1976) and soils with surface residue (Steiner, 1989; Schwartz et al., 2019) provided that cumulative evaporation is evaluated over daily time increments (Figure 2.4). Besides utilizing the square root of time relationship to approximate evaporation (e.g., Ritchie, 1972), it has also been useful in comparing residue and tillage treatment effects on soil water evaporation (Steiner, 1989; Schwartz et al., 2019).

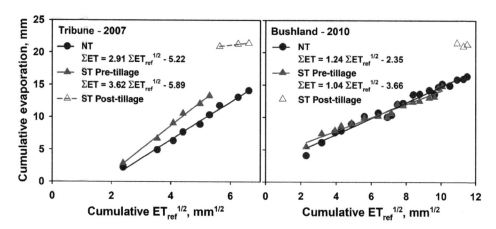

FIGURE 2.4 Cumulative daily evaporation as a function of the square root of reference evapotranspiration for periods without precipitation under no tillage (NT) and stubble-mulch tillage (ST) in Bushland, Texas (23 April–13 May) and Tribune, Kansas (28 April–6 May). The effect of sweep tillage on evaporation is also shown for stubble-mulch fields (Adapted from Schwartz et al., 2019).

2.4.1 TILLAGE AND RESIDUE EFFECTS ON EVAPORATION PROCESSES

Because they alter the physical environment at the soil surface, tillage and surface residues have profound effects on the soil water energy balance and consequently soil water evaporation. Exposing moist soil to the atmosphere via a sweep tillage operation results in an evaporative loss of 1–8 mm of stored soil water, the magnitude depending on soil water content near the surface and the potential ET (Good and Smika, 1976; Schwartz et al., 2010). However, immediately after tillage, the rate of soil water loss through evaporation is considerably reduced (Figure 2.4) until the next precipitation event (Schwartz et al., 2019). In and of itself, tillage increases the random roughness of the surface reducing albedo, and thereby increasing the net radiant energy received at the surface (Allmaras et al., 1977; Schwartz et al., 2010). Schwartz et al. (2010) demonstrated that in the absence of residues, differences in albedo between a no-tillage and a recently tilled stubble mulch tillage surface resulted in 3.2 MJ m^{-2} d^{-1} of additional net radiation under stubble mulch tillage for a day with 29 MJ m^{-2} d^{-1} solar irradiance. Greater net radiant energy under stubble mulch tillage explained most of the greater daily evaporation (0.1–1.5 mm d^{-1}) compared with no tillage in this study. However, differences in net radiation and soil water evaporation between no tillage and stubble mulch tillage surfaces were indistinguishable 50 days after tillage, likely due to rainfall-mediated aggregate destruction and reconsolidation of the surface.

Surface residues influence soil water evaporation through their effects on albedo and net radiation balance, heat exchanges with the atmosphere, and modification of diffusive transport of water vapor to the atmosphere (Bristow et al., 1986; Bristow, 1988; Steiner, 1994; Horton et al., 1996). Because of the complex way in which the residue, soil, and atmosphere interact under field conditions, it is difficult to separate the effects of tillage from that of residue (Steiner, 1994). Daytime net radiation on residue-covered surfaces is typically smaller compared with moist bare soils because of a lower albedo associated with residues (approximately 0.2 for wheat straw). However, as the soil surface dries, albedo of the bare soil surface increases and may exceed that for a residue-covered surface and result in less daytime net radiation under bare soil (Bristow, 1988). Under a residue cover compared with bare soil, near surface soil temperature fluctuations are dampened, thereby resulting in lower daytime temperatures under the residue cover (McCalla and Duley, 1946; Unger, 1978b; Smika, 1983; Bristow, 1988). However, some of the temperature dampening may result from greater soil heat capacity under wetter, residue-covered soils. The magnitude of evaporation from

Soil Water Conservation for Dryland Farming

residue-covered soils depends on the vapor density at the soil surface as well as diffusive and advective resistance to vapor transport through the residue layer (Bristow et al., 1986; Horton et al., 1996). Noting that vapor density at the soil surface depends on the competing effects of soil temperature and soil water potential (Horton et al., 1996), cooler and wetter conditions at a soil–residue interface compared to a bare soil may not necessarily have lower vapor densities at the surface.

Thickness of a residue layer, in general, has the greatest impact on evaporation reduction (Unger and Parker, 1976; Steiner, 1989) because it is most closely associated with the diffusive resistance of the mulch layer (Flury et al., 2008). The orientation or architecture of the residue will also influence the net radiation, temperature, and evaporation under a residue covered soil (Bristow, 1988; Smika, 1983). Smika (1983) found standing stubble was more effective than flat straw in reducing soil water evaporation, likely as a result of reduced turbulent mixing near the soil surface (Aase and Siddoway, 1980). Moreover, standing stubble that is isolated from the soil surface will persist for longer periods because of reduced rates of decomposition (Steiner et al., 1999).

Under field conditions and extended fallow periods, the effectiveness of crop residues in reducing evaporation during fallow periods under prolonged drying has, in general, been shown to be limited (Fischer, 1987; Unger et al., 1991). The presence of residues will shorten Stage 1 evaporation, but prolong Stage 2 evaporation because of net energy at the soil surface combined with greater near-surface soil water contents that maintain capillary conductivity (Steiner, 1994). Schwartz et al. (2019) demonstrated that evaporation rates under no tillage and stubble mulch tillage at two locations were principally dependent on near-surface soil water contents that tended to be greater under no tillage. Notwithstanding, greater soil water contents near a residue-covered surface in conjunction with timely precipitation prior to planting afford better soil water conditions and a longer window of opportunity for dryland crop establishment.

2.5 REDISTRIBUTION AND DEEP DRAINAGE OF SOIL WATER

The rate and extent of redistribution of soil water away from the soil layers that are subject to evaporation can influence long-term storage and availability of water to crops during the growing season. The redistribution process is governed largely by soil texture. However, other factors and management interventions can conserve water for use by crops. In a study evaluating soil water storage through several phases of a wheat–sorghum–fallow rotation at two locations, Schwartz et al. (2019) showed that lasting increases in stored soil water during the fallow periods occurred for precipitation amounts greater than 25 mm or for a series of closely spaced events. Soil water redistribution to deeper in the profile (e.g., > 1 m) may reduce available soil water to crops with sparse root systems at these depths (Moroke et al., 2005; Tolk and Evett, 2012). Although water potential gradients that develop around crop roots can facilitate water extraction from soil zones lacking roots, the time scale of this process can exceed the reproductive phase of the crop.

Deep drainage past the rooting zone of crops (1.5–5 m) during the growing season and fallow periods has long been a concern when quantifying the soil water balance (McGowan and Williams, 1980; Stone et al., 2011; Evett et al., 2012). Water loss due to growing season drainage is typically considered negligible under dryland crop production (Nielsen et al., 2002; Williams et al., 2015). However, this assumption can potentially bias estimated seasonal crop water use, especially for longer growing seasons. Moroke (2002) calculated Darcy-flux estimates of drainage by using measured water contents at 2.1 and 2.3 m and estimated hydraulic properties. During the growing season, 19- and 6-mm cumulative drainage were calculated for cowpea [*Vigna unguiculata* (L.) Walp] and sorghum, respectively. However, for sunflower (*Helianthus annuus* L.), a deep rooted annual, there was likely upward water flow that supplied the crop with an additional 24 mm water during the growing season.

Schwartz et al. (2019) also used the Darcy-flux method to determine drainage throughout a wheat–sorghum fallow rotation. Drainage during the 11-month sorghum and wheat fallow periods,

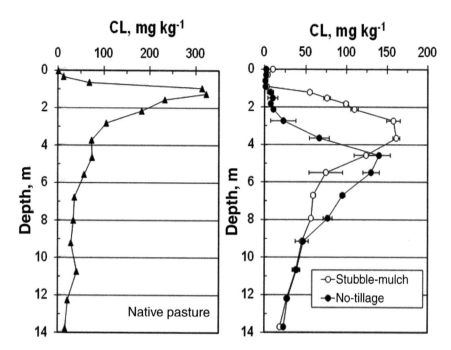

FIGURE 2.5 Chloride concentration, mg kg^{-1}, plotted with depth for nearby native pasture and for no tillage and stubble mulch tillage residue management sites.

both averaged 28 mm, whereas drainage during the growing season of sorghum and wheat averaged 12 and 9 mm, respectively. Assuming a unit hydraulic gradient (gravity-driven flow) and using an estimated value of hydraulic conductivity, Baumhardt and Jones (2002a) calculated annualized drainage rate of 17 mm. In a sandy loam soil of southwest NSW, Australia, O'Connell et al. (2003) reported deep drainage rates of up to 31 mm yr^{-1} under a wheat–pea (*Pisum sativum* L.)–fallow rotation. The above calculated drainage rates are based on short-term studies and do not consider that drainage usually occurs during episodic periods of above average rainfall. Scanlon et al. (2008) and Baumhardt et al. (2017) evaluated drainage rates utilizing displacement of chloride in boreholes under dryland cropping as compared with natural grassland ecosystems. This approach provides a time-integrated assessment of drainage, including infrequent periods of above average precipitation (Figure 2.5). Using this approach for a wheat–sorghum fallow rotation, Baumhardt et al. (2017) calculated annualized deep drainage of 2 mm and 14 mm for stubble mulch tillage and no tillage, respectively. For deep (0.7 m) moldboard plowed cropland, drainage ranged from 58 to 81 mm yr^{-1}.

2.6 WEED CONTROL

Successful dryland crop production requires that decisions should be focused on implementing practices that can increase the fraction of precipitation used for transpiration by the economic crop (Stewart, 2016). All plants require water for their growth and reproduction activities. As a result, water use by weeds and volunteer crop plants must be controlled for successful crop production. Weeds often remove water from greater soil depths than crops (Abouziena et al., 2014–2015), thus often causing water stress for crops.

An early reason for using any type of tillage method was to control weeds that used soil water. When herbicides became available to control (or eliminate) weeds, various limited tillage practices, also known as chemical fallow, conservation tillage, or ecofallow often were used to control weeds (Wiese, 1983). Other advantages of these systems were reduced erosion, labor, fuel use, and crop production cost. The ultimate system of chemical fallow is no tillage, provided that complete control of unwanted vegetation (e.g., weeds or plants from previous crops) is achieved and sufficient residues are present on the surface to enhance soil water storage (Unger and Wiese, 1972). With no tillage, unwanted plants are controlled with herbicides and the next crop is planted through the existing surface residues remaining from the previous crop. An increase in herbicide-resistant weeds, rising costs of herbicide applications, and a decline in available herbicides for control of certain weed classes may reduce herbicide use in the future, especially for dryland production with low rates of return (Lyon et al., 1996). Weed control strategies that employ rotations and greater use of competitive crops or cover crops, and precision application technology may offer alternative means to control weeds with reduced herbicide use.

A problem with volunteer crop plants is that they usually germinate and grow at the same time that planted crops germinate and grow (Unger and Box, 1972). As a result, volunteer plants compete with crops for soil water. The problem is especially severe when continuous cropping is done and under no tillage conditions. Any applied herbicide would also adversely affect the planted crop. Under conditions other than no tillage, some volunteer plants can be culled with various types of tillage (Fowler, 1972).

2.7 COVER CROPS

Cover crops can have certain advantages when used within dryland cropping systems such as protection against wind and water erosion, increasing soil organic carbon, and recovery of nitrate being leached during periods of significant rainfall. However, cover crops also extract greater quantities of soil water and at deeper soil depths compared with soil water evaporation under fallow and thereby can reduce available water at planting and subsequent crop yields (Lyon et al., 2007; Nielsen et al., 2016; Holman et al., 2018). Herein, we consider cover crops used to increase residue cover and not harvested for forage or grain, which is considered as fallow reduction and cropping intensification later in this chapter. In semiarid regions where the principal objective is to increase cover and erosion protection, Unger and Vigil (1998) recommended that the cover crop should be terminated when sufficient plant cover is attained, yet early enough to permit increases in stored soil water prior to planting the main crop. Oftentimes, the use of conservation tillage with crop residue retention can provide many of the benefits of cover crops without increasing risks associated with a decrease in soil water at planting (Unger and Vigil, 1998).

Research documenting water use by cover crops demonstrated that stored soil water available at planting time depends to a large extent on the growing season of the cover crops as well as their time of termination. In a six-year study of a legume cover crop-winter wheat rotation in Colorado (Nielsen and Vigil, 2005), soil water at wheat planting was reduced by 55–104 mm compared with a fallow–wheat rotation for legume termination dates ranging from early June to August. In the study, wheat yield was strongly correlated with water content at planting, increasing by 15 kg ha^{-1} for each additional mm of soil water. In a similar five-year study comparing wheat–fallow to wheat–cover crop rotations in southwestern Kansas (Holman et al., 2018), available soil water at planting was reduced by 43–92 mm by cover crops, depending on the cover crop species (spring and fall) and if the cover crop was harvested for forage. Cover crops significantly reduced soil water at all depths compared to fallow, except at the surface (0–0.31 m). Lastly, biomass generated by cover crops was correlated with water use, with the lowest amount of soil water at planting associated with the cover crops that produced the greatest biomass. Holman et al. (2018) concluded that retaining fallow or intensifying the rotation with a forage crop, such as winter or spring triticale (*Triticale hexaploide* Lart.), resulted in the greatest net returns.

Fall planted cover crops terminated in the spring prior to cotton (*Gossypium hirsutum* L.) establishment may offset transpirational soil water losses incurred during the spring and winter by reducing soil water evaporation during the early part of the growing season, which can be substantial under cotton because of the sparse canopy cover (Lascano and Baumhardt, 1996). In a study of cover crop effects on soil water and cotton yield, DeLaune (2015) and DeLaune et al. (2020) demonstrated that soil water at planting was greater under no tillage and conventional tillage compared with fall planted cover crops under no tillage during the first study year. However, in the following year, soil water at planting was similar among all treatments because of significant precipitation during May. Across all years, cover crops did not significantly reduce lint yields, although precipitation in May coinciding with cotton planting was significantly above average for the 2014–2016 study years that likely recharged soil water lost to transpiration during the winter and spring (DeLaune et al., 2020).

In a summary of research documenting the influence of cover crops on yield, the studies tabulated by Nielsen et al. (2016) suggest that in the vast majority of cases, cover crops depress the yield of subsequent crops. Dryland studies evaluated by Nielsen et al. (2016) at locations within the Southern and Central US Great Plains (Colorado, Kansas, and Texas) show that winter cover crops planted in the fall depress yields of subsequent summer crops by an average of 27%. Likewise, winter wheat yields are also depressed by 25% following summer cover crops terminated from mid- to late-summer. For the Northern US Great Plains, yield of spring wheat or durum wheat (*Triticum durum* Desf.) within a cover crop – wheat rotation averaged 80% of fallow–wheat rotations. In contrast, fall-planted cover crops had little or no effect on subsequent corn yield (average 2% yield reduction) in South Dakota and Nebraska (Reese et al., 2014; Thompson et al., 2014). However, in both studies, successful establishment and survival of the cover crops during the winter was problematic and consequently the cover crops transpired little and likely had little or no influence on soil water contents at planting.

2.8 FALLOW AND DRYLAND CROP INTENSIFICATION

Summer fallow periods are often necessary because of depletion of soil water by crops during a rotational phase and the corresponding seasonal precipitation patterns result in conditions of inadequate stored soil water at the beginning of the next growing season. The principal objective of the fallow period is to maximize stored soil water available to the following crop and reduce risk associated with dryland crop production (Stewart et al., 2016). The total soil water that can be stored and subsequently available to plants depends to a large extent on soil texture and depth as well as the rooting depth of the subsequent crop. Plant available water ranges from about 80 mm per m depth of a sandy soil to about 160 mm for the same depth of a clay loam or silt loam soil (Unger, 1971). Therefore, crops on sandy soils undergo stress due to a water shortage quicker than those on clayey or silty soils. Rooting depth of the crop influences how much of the stored soil water is accessible. Sunflowers use soil water to depths of 2.1–3.3 m (Stone et al., 2002; Jones and Unger, 1977; Moroke et al., 2005) and sorghum uses water to depths from about 1.65 m to 2.50 m (Rachidi et al., 1993; Stone et al., 2002; Moroke et al., 2005). In contrast, water use by cowpea (*Vigna unguiculata* L.) was confined to less than 1.5 m soil depths with little or no evidence of roots below 1 m. Water extraction to greater depths by sunflower than by grain sorghum occurred at Bushland, TX (Unger, 1984; Moroke et al., 2005).

There are few studies examining the competing effects of soil water evaporation and infiltration on stored soil water throughout the fallow period. Those studies are important because short-term observations examining hydrological processes in isolation may be inconsistent with the net effect of soil water evaporation and infiltration on stored soil water observed throughout the entire fallow period. In a study evaluating soil water storage within a winter wheat–sorghum–fallow rotation in western Kansas and Texas (Schwartz et al., 2019), 37 mm greater infiltration with no tillage than with stubble mulch tillage during a 90-day summer fallow period at the Kansas location was offset by greater evaporation rates because of greater near-surface soil water contents with no tillage. Consequently, the change in stored soil water was similar with both tillage systems. Nonetheless, stored soil water at 0–2.0 m with no tillage averaged 61 mm greater than with stubble mulch tillage throughout this period. Greater infiltration with no tillage compared with stubble mulch tillage combined with lower evaporation rates during the winter fallow period at the Kansas location was more effective at increasing stored soil water prior to sorghum planting. At the Bushland, TX, location, cumulative infiltration, evaporation, and stored soil water (0–2.0 m depth) for no tillage and stubble mulch tillage were similar throughout all summer fallow periods. For the fallow after wheat in the spring of 2010 prior to sorghum planting, cumulative infiltration with stubble mulch tillage exceeded that with no tillage; however, this gain in stored soil water was offset by lower evaporation with no tillage. Wuest (2018) examined differences in evaporation and stored soil water throughout summer fallow periods in eastern Washington for varying wheat straw residue levels and differing surface soil mulches using microlysimeters weighed at periodic intervals. Stored soil water was considerably greater immediately after rainfall events for microlysimeters covered with surface residues. However, during extended periods with limited precipitation, residue had little effect on cumulative evaporation and stored soil water. Stored soil water increases of 4–10 mm depended largely on the recency of the final rainfall event.

The effect of fallow on stored soil water is typically characterized in terms of a "fallow efficiency" that is defined as the increase in stored soil water as a percentage of precipitation received during the fallow period. Greater fallow efficiencies result in greater available soil water contents at planting that can contribute significantly to subsequent crop yields (Unger and Baumhardt, 1999; Stone and Schlegel, 2006; Nielsen et al, 2009; Schlegel et al., 2018b). Under winter wheat–fallow or spring wheat–fallow rotations with conventional inversion tillage, fallow efficiencies ranged from 0% to 20% (Johnson, 1950; Greb, 1979; Farahani et al., 1998; Nielsen and Vigil, 2010; Stewart, 2016), depending on the fallow period and location. As stubble mulch tillage was adopted and tillage intensity reduced or eliminated, improved fallow efficiencies of 10–40% could be attained principally because of the reduced evaporation associated with greater residue cover (Greb, 1979; Peterson et al., 1996).

Compared to conventional tillage, reduced tillage and no tillage practices normally increase fallow efficiencies and available soil water contents for the following crop. For a six-year study at Bushland, TX, examining tillage effects on stored soil water at planting in a wheat–sorghum–fallow rotation, Baumhardt and Jones (2002a) showed that fallow efficiencies ranged from 4% to 33% and – 9% to 37% for fallow periods extending from sorghum harvest to wheat planting and wheat harvest to sorghum planting, respectively. Mean fallow efficiencies were 7% greater under no tillage compared to stubble mulch tillage; however, paratillage operations once every 3 years had no effect on fallow efficiency. Baumhardt et al. (2017) examined the 30-year record (1983–2013) of graded terrace watersheds also in a wheat–sorghum–fallow rotation and likewise found that fallow efficiencies under no-tillage exceeded those of stubble–mulch tillage by 5%, resulting an additional 33- and 23-mm stored soil water under fallow after wheat and sorghum, respectively. In a study examining 12 tillage systems in a wheat–sorghum–fallow rotation with varying tillage intensities and herbicide applications (sweep tillage as required to no tillage), Unger (1994) found that mean stored soil water during fallow and water use of wheat and sorghum were not influenced by tillage. Mean fallow efficiencies (6-year) were 27% and 17% for fallow after wheat and sorghum, respectively. Grain yield of sorghum was greater for some tillage levels; however, wheat grain yield was not influenced by tillage.

Tanaka and Aase (1987) demonstrated that over-winter fallow efficiencies for winter wheat–fallow and spring wheat–fallow can be large (36–71%); however, fallow efficiencies during the summer period were considerably smaller (13–28%). In this study, chemical fallow stored more water compared with sweep tillage on winter wheat–fallow, but not for spring wheat–fallow rotations, likely due to insufficient residue production by spring wheat. For an 11-month fallow period extending from wheat harvest to sorghum planting, Unger (1978a) demonstrated that as straw mulch levels increased from 0 to 12 Mg ha^{-1}, average precipitation stored as soil water at planting doubled. However, for a straw mulch level of 2 Mg ha^{-1} normally achievable in this region, precipitation stored as soil water was 50% greater than that measured for bare soils. Inadequate residue production and persistence is problematic under dryland production (Stewart, 2016). In some areas within the US Southern Great Plains (e.g., Bushland, TX), fallow efficiencies rarely exceed 30% even under no tillage management (Jones and Popham, 1997; Baumhardt and Jones, 2002a; Baumhardt et al., 2017).

Because a large proportion of precipitation is lost primarily due to evaporation during fallow periods, the practice of fallow, even with conservation tillage or no tillage, often results in inefficient use of precipitation for grain production. Conservation tillage and no tillage practices, however, improve both fallow efficiency and in-season water use efficiency, thereby making possible the intensification of dryland cropping rotations by reducing the time in fallow. Most of the annual precipitation in the US Great Plains occurs during the months of May to September, a period that coincides with periods of low fallow efficiencies (Farahani et al., 1998; Baumhardt and Salinas-García, 2006). Consequently, introduction of summer crops to produce a wheat–summer crop–fallow rotation system has been successful in increasing the proportion of annual precipitation for use by crops and annualized grain yield compared to wheat–fallow rotations (Norwood et al., 1990; Norwood, 1994; Jones and Popham, 1997; Peterson et al., 1996; Anderson et al., 1999). Norwood et al. (1990) compared wheat–fallow with wheat–sorghum–fallow rotations under no tillage, conservation tillage, and reduced tillage practices at two locations in western Kansas and found that wheat yields did not differ under wheat–fallow and wheat–fallow–sorghum rotations. In addition, the wheat–sorghum–fallow rotation exhibited greater yields when combined with reduced tillage intensity. Jones and Popham (1997) examined similar rotations at Bushland, TX, under no tillage and stubble mulch tillage and found that grain yields were not affected significantly by tillage in any cropping system, likely because stored soil water at planting was small in comparison to seasonal evapotranspiration. In contrast, wheat–sorghum–fallow increased annualized grain yields by 70% without reducing wheat yields compared with the wheat–fallow rotation. In a study by Anderson et al. (1999) evaluating eight rotations consisting of winter wheat, maize, proso millet (*Panicum miliaceum*), and sunflower at Akron, CO, annualized grain yield of continuously cropped wheat–maize–millet and wheat–millet was nearly double that of wheat–fallow. Observed annualized grain yield for other rotations such as wheat–maize–fallow and wheat–maize–millet–fallow was at least 60% greater than the wheat–fallow rotation. More importantly, intensification of the rotation did not necessarily increase yield variability. For instance, yield variability for continuous wheat–millet was less than for the wheat–fallow system. However, rotations with maize and sunflower tended to increase yield variability of rotations.

2.9 GROWING SEASON INTERVENTIONS

Tillage and residue practices implemented during fallow periods can influence crop yields principally because of their effect on plant available water at planting and the proportion of in-season precipitation redirected to transpiration. However, during the growing season, management interventions that influence water use and yield of crops and consequently water use efficiency are largely based on crop management strategies. Two strategies related to water management include (i) increasing the proportion of evapotranspiration utilized for crop transpiration and (ii) limiting water use during the vegetative phase so that there is sufficient plant available water during the reproductive and grain filling periods (Stewart and Wei-li, 2015; Stewart and Peterson, 2015). Choosing crops and hybrids that can avoid or survive dehydration while minimizing yield losses associated with periods of water stress is also important for optimizing water use efficiency. Crops can avoid dehydration stress by developing an extensive root system, reducing transpiration through mechanisms such as leaf rolling in sorghum and maize, or by adjustment of the timing of crop development stages in response to water stress. Dehydration tolerance such as that expressed by the post-inflorescence stay green trait in maize and sorghum delays leaf senescence, thereby maintaining transpiration and resulting in greater seed weight (Rauf et al., 2015).

2.9.1 Plant Density, Spacing, and Clumping

A widely used strategy in dryland cropping is the use of low plant population densities to reduce water use during the vegetative stage, thereby increasing plant available water during the reproductive and grain filling periods. This strategy may increase the harvest index, but conversely can reduce leaf area and canopy cover, thereby increasing evaporation at the expense of transpiration. This may explain why many studies have shown limited success in using low plant densities (Stewart and Wei-li, 2015). Uniform planting practices spatially distribute seedlings to minimize any competition for nutrients, water, and light. The density for a uniformly planted crop could be governed by canopy dimensions when avoiding potential overlap for maximum light interception within the target row spacing, provided the soil supplies adequate nutrients and water. Under dryland conditions, the optimum plant density depends on a balance between available soil water and the crop yield potential. In a study evaluating planting geometries, Bandaru et al. (2006) compared grain yield of uniformly planted dryland grain sorghum at a density of 3.6 and 5.4 plants m^{-2}. Sorghum planted in no tillage residues at the lower density had sufficient available water for numerically larger grain yields and associated harvest indexes that were significant for all but the highest water level and limited water stress.

One dryland cropping intervention that is distinctive from uniformly planted rows (Figure 2.6) produces a clump geometry of seedlings by planting multiple (three to four) seeds together in a single hill (Bandaru et al., 2006).

FIGURE 2.6 Grain sorghum planted at a fixed density in rows 0.76 m apart using clumps (left) or uniform spacing (right) after favorable June rains, >100 mm, followed by about 7 mm rain by mid-August. Uniformly spaced sorghum plants exhibit severe leaf rolling and poor seed head development due to water-deficit stress in contrast to less stressed clumped sorghum with flowering seed heads.

Those clumps resemble maize stands of the historical methods of planting in use by native Americans of the desert US Southwest (Stewart and Lal, 2012) due to water scarcity of semiarid climates. As observed by native Americans, dryland and minimally irrigated (50- and 100-mm) maize grain yields were significantly greater with clumps than uniformly planted rows in five of six comparisons (Kapanigowda et al., 2010). Their corresponding water use increased with increasing irrigation but did not differ for either planting geometry. Kapanigowda et al. (2010) concluded that growing maize with clump geometry reduced tillering, from 1.43 for rows to 0.28 for clumps when averaged across irrigation levels, and conserved water for producing grain.

This alternate, clump planting geometry, reduces tillering in grain sorghum possibly because of reduced light and consequently assimilate production (Lafarge et al., 2002). Greater competition between clumped plants for water will also decrease production of assimilates used for tiller development compared with uniform row planting as shown in Table 2.1.

Removal or clump reduction of tillers reduced biomass at all water levels compared with uniform planting at 42 DAP, but grain yield and harvest index increased without tillers as a benefit of delayed water use, except with added runoff (Bandaru et al., 2006). As shown in Table 2.1 and concluded by Bandaru et al. (2006), clump benefits decreased as yield increased with improved water availability. Clump impact on tillering and the resulting improved harvest index and yield may be due to competition for light and water resources or more efficient use due to improved microclimate effects that lowers the vapor pressure deficit within the crop canopy (Thapa et al., 2017a).

2.9.2 CLIMATE-INFORMED MANAGEMENT

Early forecasting of temperature and precipitation anomalies from the historical mean caused by climate-driven events such as the El Niño-Southern Oscillation (ENSO) or Artic Oscillation (AO) could provide actionable information to producers with regard to planting, timing of planting, crop

TABLE 2.1
Grain Sorghum Response to Planting Geometries in 0.76 m Row Width on Tillering 28 Days after Planting (DAP), above Ground Biomass 42 DAP, Grain Yield, and Harvest Index of Grain Sorghum Grown on No-Till Terrace Watershed Areas Receiving Rain and Rain Minus or Plus Runoff

Planting Geometry	Plant Tillers 28 DAP	Biomass, 42 DAP	Grain Yield	Harvest Index
S		—— kg ha^{-1} ——		
Rain minus runoff				
Uniform	2.2a	2716a	2270b	0.29b
Uniform TR	removed	1550c	2645ab	0.40a
Clump-3 75	0.5b	1831b	2891a	0.38a
Clump-4 100	0.5b	1622c	3011a	0.40a
Rain				
Uniform	2.3a	2924a	2690c	0.32a
Uniform TR	removed	1634d	2904b	0.39a
Clump-3 75	0.7b	1897b	3338ab	0.40a
Clump-4 100	0.6b	1754c	3479a	0.42a
Rain plus runoff				
Uniform	2.3a	3047a	4812a	0.41b
Uniform TR	removed	1786c	4222b	0.46a
Clump-3 75	0.9b	2088b	4968a	0.46a
Clump-4 100	0.8b	1880c	4807a	0.46a

Source: Data are after Bandaru et al. (2006).

Soil Water Conservation for Dryland Farming

and hybrid selection, and other interventions that depend on crop available soil water. The success of forecasted precipitation anomalies tends to be both regionally and temporally specific and is not easily generalized (Rosenberg et al., 1997). The La Niña climatic conditions exhibit greater winter precipitation in eastern Australia and northern US and Canada, but grades to below average precipitation in the US Southern Great Plains and US southeast (Stone et al., 1996). The ENSO-generated weather pattern influences winter precipitation over the US Southern Great Plains (Ropelewski and Halpert, 1996), thereby strongly influencing wheat yields in the region (Mjelde and Keplinger, 1998; Mauget and Upchurch, 1999; Baumhardt et al., 2016). Using computer crop simulation, Mauget et al. (2009) compared the productivity of various dual-purpose wheat and grazing scenarios on the US Southern High Plains based on the ENSO phase. They concluded that crop inputs optimized for ENSO forecast, such as best optimizing grain production over grazing during wet El Niño conditions, may increase profitability compared to when no forecast information is available.

In an analysis of a long-term (1954–2011) field experiment in the Texas High Plains, Baumhardt et al. (2016) determined that during the La Niña phase, growing season precipitation was lower than during the El Niño phase for wheat (P = 0.04) and grain sorghum (P = 0.08), but the corresponding lower grain yields for both crops during the La Niña phase did not differ significantly from the El Niño phase (Table 2.2).

Marek et al. (2018) likewise found that continuous simulation mean wheat yields of 2,830 kg ha^{-1} for El Niño and 2,300 kg ha^{-1} for La Niña were not significant in response to the ENSO phase effects on grain yield at the same location. Mauget and Upchurch (1999) and Mjelde and Keplinger (1998) also concluded that ENSO anomalies are more correlated to winter wheat yields compared with summer crop yields in the Texas High Plains. Using long-term precipitation records at Bushland, TX, Baumhardt et al. (2016) showed that the Oceanic Niño Index (ONI) classification for the 3-month period from June through July was able to correctly identify 78% and 68% of the mature phase La Niña and El Niño, respectively, occurring from September through November, thereby permitting timely decisions for wheat planting and management. In contrast, using the 3-month period from April through June correctly classified only 61% and 53% of the La Niña and El Niño mature phases. Consequently, the use of ONI to forecast growing season conditions during the

TABLE 2.2
Wheat and Grain Sorghum Growing-Season Precipitation Plus Grain Yield and Percentage Departure from Trend (PDT) for Those Years (n) Classified as El Niño, Neutral, or La Niña Phase Using Oceanic Niño Index Periods of September–November, SON

	WHEAT		
	N	**Precipitation, mm**	**Grain Yield, kg ha^{-1}**
	19	261 a	1710
	21	250 a	1330
	17	200 b	1130
		——— P > F ———	
	0.04	>0.10	

	GRAIN SORGHUM		
	N	**Precipitation, mm**	**Grain Yield, kg ha^{-1}**
	19	262	2820
	21	236	2520
	17	193	2010
		——— P > F ———	
	0.08	>0.10	

Source: CPC, 2015.

summer was not useful for sorghum planting decisions in this region. Pu et al. (2016) and Fernando et al. (2019) suggest that the diminished reliability of ENSO in predicting precipitation anomalies during the summer months in the US Southern Great Plains may be a result of summertime convection and land surface feedbacks.

Simulation models often are used to evaluate ENSO effects on crop growth in lieu of sufficiently long (e.g., >30 years) field experiments. The ENSO impact is largely diminished during June–August, except for La Niña that is more subject to ocean currents (Okumura and Deser, 2010) rather than to seasonal conditions (Galanti and Tziperman, 2000). Nevertheless, ENSO effects on summer crops like cotton and improved water management are often the topic for crop simulations. ENSO impact on climate consistently increases precipitation for El Niño and neutral phases over the drier La Niña phase, resulting in greater simulated yields of cotton lint (Baumhardt et al., 2014) and sorghum grain (Baumhardt et al., 2019). They concluded that dryland management strategies to reduce risk during drier El Niño ENSO phases may vary the planting geometry and plant population while elsewhere nutrient management is considered. Cotton planted in narrow rows and at higher populations may benefit more from increased precipitation; however, when all climatic conditions are combined, modeled yields did not favor narrow row spacing (Baumhardt et al., 2018).

2.10 CLIMATE CHANGE EFFECTS

2.10.1 CLIMATE CHANGE EFFECT ON WHEAT YIELD

Climate change, particularly rising temperatures, has a significant effect on crop production globally. From 1980 to 2016, the five-year moving average global land and ocean temperature increased from about 14°C to 14.7°C. For the US, the increase was from about 11°C to 12°C (Stewart et al., 2018). In addition, extreme weather events are increasing as well due to climate change (Tack et al., 2015). Winter wheat represents 70–80% of total US wheat production and ranks as the third largest crop in terms of value and acreage (Vocke and Ali, 2013). In the US Southern High Plains, winter wheat is widely grown under dryland and irrigated conditions primarily for grain and secondarily for forage production (Howell et al., 1995). Winter wheat requires an optimum temperature of about 20.3°C for shoot growth and 21°C for anthesis and grain-filling period (Porter and Gawith, 1999). However, water stress coupled with high temperature, especially at the later growth stages, is the major environmental factor limiting wheat yield in the US Southern High Plains (Thapa et al., 2017b). A combination of these stresses has a significant detrimental effect on the growth and productivity of crops compared to each stress alone (Prasad et al., 2011).

In the US Central and Southern Great Plains, the effect of climate change on winter wheat production has been studied using field trials and climatic data (Tack et al., 2015; Stewart et al., 2018; Shrestha et al., 2020) and simulation models (Kothari et al., 2019; Obembe et al., 2021). Stewart et al. (2018) investigated the effect of climate change on winter wheat yield in the US Great Plains based on the county yield records and National Oceanic and Atmospheric Administration (NOAA) weather data from 1939 to 2016. The study area consisted of North Central Texas, Oklahoma, Kansas, Nebraska, South Dakota, and North Dakota as subareas, roughly 1,500 km north-south and 400–600 km east-west directions. They found that the average annual temperatures increased in all subareas, but more in the northern than in the southern subareas. Since 1990, average grain yields for North Central Texas decreased and Oklahoma remained somewhat steady. In contrast, Kansas average yields showed an upward trend, and Nebraska, South Dakota, and North Dakota showed significant increases. The increased yields in northern states were likely due to the warmer temperatures reducing winter killing and providing more growing degree days during critical growth stages in the northern areas where historically the climate was too cold for optimum winter wheat growth. Shrestha et al. (2020) examined the climate and wheat variety trial data at Bushland, TX, from 1940 to 2016 under both dryland and irrigated conditions. Wheat genotypes were divided into three groups: Kharkof as a check cultivar, the highest yielding cultivar for each year, and all other cultivars. Mean temperature during wheat growing season (September 15 to June 15) increased at a rate of 0.35°C per decade since 1980. Consequently, seasonal growing degree days (GDD) and extreme degree days (EDD) increased at a rate of 69°C d and 19°C d per decade. From 1940 to 2016, wheat heading date has shifted earlier by 16 days for dryland wheat. Early heading may have benefits for heat escape but also has the disadvantage of reducing precipitation prior to heading – a loss of 25 mm when heading shifted earlier by 16 days for dryland wheat (Figure 2.7). The average yields increased steadily from about 1960 to 1980, plateaued until the late 1980s, and then declined significantly from the 1990s to 2016. Increase in maximum temperature, frequencies of maximum temperature greater than or equal to 31°C, and minimum temperature during the grain filling period had a significant negative effect on grain filling days and yield. Precipitation during the vegetative growth (November, January, and March) and grain filling (May) had a significant positive effect on dryland wheat yields. Tack et al. (2015) analyzed a unique data set combining Kansas wheat variety trial from 1985 to 2013 with location-specific weather data. The results showed that the effect of temperature on wheat yield varied across the whole September–May growing season. The largest drivers of yield loss were freezing temperatures in the fall and extreme heat events in the spring. The overall effect of warming on yields is negative, even after accounting for the benefits of reduced exposure to freezing temperatures. This strongly suggests that the changing climate is making the area less suitable for wheat grain production because the reproduction and grain filling growth stages are becoming less synchronized with the precipitation patterns.

Obembe et al. (2021) examined the impact of climate change on winter wheat production in Kansas from 1981 to 2007, using county-level production data on winter wheat coupled with fine-scale weather data. Their results showed that climate change affected both wheat harvested yield and acreage. However, the effect of climate change on wheat production was mainly from yield loss, but the effect on acreage was much smaller.

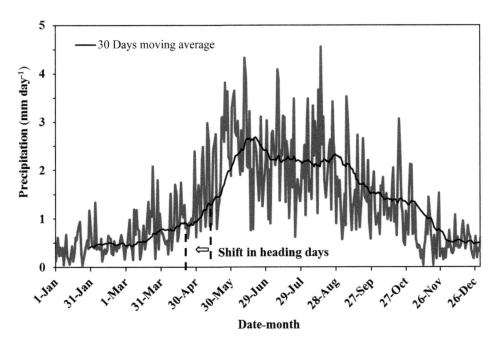

FIGURE 2.7 Average daily precipitation distribution during 1940–2016 and average heading date shift from May 14 to April 29 attributing to 25 mm of precipitation loss during vegetative period for all cultivars under dryland conditions at Southern Regional Performance Nursery, Bushland, TX (Shrestha et al., 2020).

2.10.2 Strategies for Adapting to Climate Change

Dryland wheat production is largely determined on soil water availability at planting as well as during growing season (Musick et al., 1994; Unger, 2001; Xue et al., 2012). However, wheat production under water-limited conditions also faces more challenges due to increasing temperatures and extreme weather events. Therefore, developing adaptation strategies to climate change is of particular importance. Because the wheat heading date has shifted to becoming earlier under rising temperatures (Shrestha et al., 2020), one legitimate practice could be to delay the sowing date. However, sowing date of winter wheat did not affect the heading or flowering date in central Great Plains (Streck et al., 2003; Xue et al., 2004). In addition, late sowing after the middle of October decreased wheat yield (Xue et al., 2004). In the US Southern Great Plains, there is a wide range of sowing dates for winter wheat (August–November) because of dryland and irrigated cropping systems either with or without grazing, and highly variable weather conditions. For grain production, early October is normally considered as the optimum sowing date. However, the sowing date may be delayed to November due to dry weather conditions (Winter and Musick, 1993). Sowing either early in August or late in November reduced wheat yield and water use as compared to sowing at the optimum date (Winter and Musick, 1993). Since most of the sowing date studies were conducted in the 1980s and the 1990s in the US Southern Great Plains, more studies are needed for dryland wheat responses to sowing date. Shrestha et al. (2020) showed that the advance of heading dates in newly developed wheat cultivars was faster than for the Kharkof cultivar that has not been changed genetically (Winter et al., 1988). Breeding cultivars with early heading may avoid the higher temperatures that increase rapidly during May. However, new cultivars with earlier heading may not lengthen the grain filling if the temperature continues to increase (Stewart et al., 2018; Shrestha et al., 2020).

Winter wheat is commonly grown as a dual-use crop (forage and grain) in the US Southern Great Plains (Lollato et al., 2017). The dual-use wheat crop is used for grazing by cattle during the winter and early spring, and then the cattle are removed so that the plants can produce grain. In years when prospects for grain production are not favorable or perhaps when wheat prices are low and cattle prices are high, grazing will continue, and grain production will be forfeited. Growing winter wheat for both forage and grain can increase overall profitability compared to forage- or grain-only systems (Lollato et al., 2017). Vocke and Ali (2013) reported that the wheat harvest-to-planted ratios for Texas and Oklahoma have decreased from about 0.8 in 1980 to less than about 0.6 from 2000 to 2012. As temperatures have increased in recent years, however, the percentage of planted area that is not harvested for grain has increased dramatically in some southern areas. The changes in acreage for dual-use wheat might also be related to grain and cattle prices. As grain production is more challenging due to increasing temperatures, harvesting wheat for forage may have less risk in the climate change scenarios in the future. As such, dual-use wheat production can be another strategy for adapting to climate change (Stewart et al., 2018).

Genetic improvements have increased wheat yield significantly. However, with increasing temperature, drought becomes an important factor limiting wheat yields in the US Southern Great Plains. Therefore, improving drought tolerance can be a strategy for adapting climate change. Xue et al. (2014) found a strong positive correlation between single-stem dry weight at anthesis stage and the amount of carbon (C) reserve remobilization to increase yield at Bushland, TX. Pradhan et al. (2014) and Thapa et al. (2018) reported that cooler canopy temperature during the daytime helped some newer cultivars producing higher yield under drought conditions. Thapa et al. (2017b) found that more recent cultivars were able to extract more water from greater soil profile depths and had higher evapotranspiration, biomass, and yield than the older cultivars during the historic drought years of 2011 and 2012 in the Texas High Plains. They demonstrated that effective use of water from deeper soil profile was important for greater yield under dryland conditions in a semiarid environment. These results indicated some of the potential areas for future plant breeding, especially in the southern regions where wheat production has been becoming more challenging during recent years due to the rising temperature coupled with severe drought.

REFERENCES

Aase, J.K., and F.H. Siddoway. 1980. Stubble height effects on seasonal microclimate, water balance, and plant development of no-till winter wheat. *Agric. Meteorol.* 21:1–20.

Abouziena, H.F., H.M. El-Saeid, and A.A. El-Said Amin. 2014–2015. Water loss by weeds: A review. *Int. J. ChemTech Res.* 7:323–336.

Allmaras, R.R., E.A. Hallauer, W.W. Nelson, and S.D. Evans. 1977. Surface energy and soil thermal property modifications by tillage-induced soil structure. *Minn. Agric. Exp. Stn. Tech. Bull.* 306.

Anderson, R.L, B.A. Bowman, D.C. Nielsen, M.F. Vigil, R.M. Aiken, and J.G. Benjamin. 1999. Alternative crop rotations for the Central Great Plains. *J. Prod. Agric.* 12:95–99.

Bandaru, V., B.A. Stewart, R.L. Baumhardt, S. Ambati, C.A. Robinson, and A. Schlegel. 2006. Growing dryland grain sorghum in clumps to reduce vegetative growth and increase yield. *Agron. J.* 98:1109–1120.

Baumhardt, R.L., G.L. Johnson, J.R. Dockal, D.K. Brauer, R.C. Schwartz, and O.R. Jones. 2020. Precipitation, runoff, and yields from terraces drylands with stubble-mulch and no tillage. *Agron. J.* 112:3295–3305.

Baumhardt, R.L., G.L. Johnson, and R.C. Schwartz. 2012. Residue and long-term tillage and crop rotation effects on simulated rain infiltration and sediment transport. *Soil Sci. Soc. Am. J.* 76:1370–1378.

Baumhardt, R.L., and O.R. Jones. 2002a. Residue management and tillage effects on soil-water storage and grain yield of dryland wheat and sorghum for a clay loam in Texas. *Soil Tillage Res.* 68:71–82.

Baumhardt, R.L., and O.R. Jones. 2002b. Residue management and paratillage effects on some soil properties and rain infiltration. *Soil Tillage Res.* 65:19–27.

Baumhardt, R.L., O.R. Jones, and R.C. Schwartz. 2008. Long-term effects of profile modifying deep plowing on soil properties and crop yield. *Soil Sci. Soc. Am. J.* 72:677–682.

Baumhardt, R.L., and R.J. Lascano. 1996. Rain infiltration as affected by wheat residue amount and distribution in ridged tillage. *Soil Sci. Soc. Am. J.* 60:1908–1913.

Baumhardt, R.L., S.A. Mauget, P.H. Gowda, and D.K. Brauer. 2014. Modeling cotton lint yield response to irrigation management as influenced by El Niño–Southern Oscillation. *Agron. J.* 106:1559–1568.

Baumhardt, R.L., S.A. Mauget, R.C. Schwartz, and O.R. Jones. 2016. El Niño southern oscillation effects on dryland crop production in the Texas High Plains. *Agron. J.* 108:736–744.

Baumhardt, R.L., and J. Salinas-Garcia. 2006. Dryland agriculture in Mexico and the U.S. Southern Great Plains. In *Dryland Agriculture*, Gary A. Peterson, Paul W. Unger, and William A. Payne, eds. American Society of Agronomy, Madison, WI, pp. 341–364.

Baumhardt, R.L., and R.C. Schwartz. 2004. Crusts/Structural. In *Encyclopedia of Soils in the Environment*, Daniel Hillel, ed. Elsevier Ltd., Oxford. vol. 1, pp. 347–356.

Baumhardt, R.L., R.C. Schwartz, O.R. Jones, B.R. Scanlon, R.C. Reedy, and G.W. Marek. 2017. Long-term conventional and no-tillage effects on field hydrology and yields of a dryland crop rotation. *Soil Sci. Soc. Am. J.* 81:200–209.

Baumhardt, R.L., R.C. Schwartz, G.W. Marek, and J.M. Bell. 2018. Planting geometry effects on the growth and yield of dryland cotton. *Agric. Sci.* 9:99–116.

Baumhardt, R.L., R.C. Schwartz, G.W. Marek, and J.E. Moorhead. 2019. Modeled El Niño southern oscillation effects on grain sorghum under varying irrigation strategies and cultural practices. *Agron. J.* 111:1913–1922.

Baumhardt, R.L., C.W. Wendt, and J.W. Keeling. 1993. Tillage and furrow diking effects on water balance and yield of sorghum and cotton. *Soil Sci. Soc. Am. J.* 57:1077–1083.

Black, T.A., W.R. Gardner, and G.W. Thurtell. 1969. The prediction of evaporation, drainage, and soil water storage for a bare soil. *Soil Sci. Soc. Am. Proc.* 33:655–660.

Boesten, J.J.T.I., and L. Stroosnijder. 1986. Simple model for daily evaporation from fallow tilled soil under spring conditions in a temperate climate. *Netherlands J. Agric. Sci.* 34:75–90.

Brengle, K.G. 1982. *Principles and Practices of Dryland Farming*. Dept. Agron., Colorado State Univ., Colorado Assoc. Univ. Press, Boulder, CO.

Bristow, K.L. 1988. The role of mulch and its architecture in modifying soil temperature. *Aust. J. Soil Res.* 26:269–280.

Bristow, K.L., G.S. Campbell, R.I. Papendick, and L.F. Elliott. 1986. Simulation of heat and moisture transfer through a surface residue-soil system. *Agric. For. Meteorol.* 36:193–214.

Brutsaert, W. 2014. Daily evaporation from drying soil: Universal parameterization with similarity. *Water Resources Res.* 50:3206–3215.

Brutsaert, W., and D. Chen. 1996. Diurnal variation of surface fluxes during thorough drying (or severe drought) of natural prairie. *Water Resources Res.* 32:2013–2019.

Bryant, C.J., L.J. Krutz, R.C. Nuti, C.C. Truman, M.A. Locke, L. Falconer, R.L. Atwill, C.W. Wood, and G.D. Spencer. 2019. Furrow diking as a mid-Southern USA irrigation strategy: Soybean grain yield, irrigation water use efficiency, and net returns above furrow diking costs. *Crop, Forage & Turfgrass Manage.* 5, 180076. doi:10.2134/cftm2018.09.0076

Busscher, W.J., J.R. Frederick, and P.J. Bauer. 2000. Timing effects of deep tillage on penetration resistance and wheat and soybean yield. *Soil Sci. Soc. Am. J.* 64:999–1003.

Busscher, W.J., J.R. Frederick, and P.J. Bauer. 2001. Effect of penetration resistance and timing of rain on grain yield of narrow-row corn in a Coastal Plain loamy sand. *Soil Tillage Res.* 63:15–24.

Campbell, H.H. 1907. *A Complete Guide to Scientific Agriculture as Adapted to the Semi-Arid Regions.* The Campbell Soil Culture Co., Lincoln, NE.

Cannell, G.H., and H.E. Dregne. 1983. Regional setting. In *Dryland Agriculture*, H.E. Dregne and W.O. Willis, eds. American Society of Agronomy, Madison, WI, pp. 3–17.

Chalker-Scott, L. 2008. Dust mulches – myth, miracle or marketing? *Master Gardener*, Summer. pp. 1–5.

Chen, L., S. Sela, T. Svoray, and S. Assouline. 2016. Scale dependence of Hortonian rainfall-runoff processes in a semiarid environment. *Water Resour. Res.* 52:5149–5166.

CPC: Climate Prediction Center. 2015. U.S. seasonal drought outlook archive, SON (September, October, November).

DeLaune, P.B. 2015. Impact of cover crops on Texas rolling plains cotton production. 2015. In *2015 Beltwide Cotton Conferences*, San Antonio, TX, Jan. 5–7, 2015. pp. 387–391.

DeLaune, P.B., P. Mubvumba, Y. Fan, and S. Bevers. 2020. Agronomic and economic impacts of cover crops in Texas rolling plains cotton. *Agrosyst. Geosci. Environ.* doi.org/10.1002/agg2.20027

Dregne, H.E. 2006. Historical perspective. In *Dryland Agriculture*, second edition, G.A. Peterson, P.W. Unger, and W.A. Payne, eds., *Agron. Monogr.* 23:27–38.

Duley, F.L., and L.L. Kelly. 1939. Effect of soil type, slope, and surface conditions on intake of water. *Univ. Nebraska Agric. Exp. Stn. Res. Bull.* 112. Lincoln, NE.

Eck, H.V., and P.W. Unger. 1985. Soil profile modification for increasing crop production. *Adv. Soil Sci.* 1:65–100.

Evett, S.R., R.C. Schwartz, T.A. Howell, R.L. Baumhardt, and K.S. Copeland. 2012. Can weighing lysimeter ET represent surrounding field ET well enough to test flux station measurements of daily and sub-daily ET? *Adv. Water Resour.* 50:79–90.

Farahani, H.J., G.A. Peterson, and D.G. Westfall. 1998. Dryland cropping intensification: A fundamental solution to efficient use of precipitation. *Adv. Agron.* 64:197–223.

Fernando, D.N., S. Chakraborty, R. Fu, and R.E. Mace. 2019. A process-based statistical seasonal prediction of May–July rainfall anomalies over Texas and the Southern Great Plains of the United States. *Clim. Serv.* 16:100133. https://doi.org/10.1016/j.cliser.2019.100133

Fischer, R.A. 1987. Responses of soil and crop water relations to tillage. In *Tillage, New Directions in Australian Agriculture,* Cornish, P.S. and J.E. Pratley, eds. *Australian Soc. Agron.*, Inkata Press, Melbourne, pp. 194–221.

Flury, M., J.B. Mathison, J.Q. Wu, W.F. Schillinger, and C.O. Stöckle. 2008. Water vapor diffusion through wheat straw residue. *Soil Sci. Soc. Am. J.* 73:37–45.

Fowler, Lehman. 1972. Experience with no-tillage – Winrock Farms. *Proc. No-Tillage Systems Symposium*, Columbus, OH, pp. 108–112.

Galanti, E., and E. Tziperman. 2000. ENSO's phase locking to the seasonal cycle in the fast-SST, fast-wave, and mixed-mode regimes. *J. Atmos. Sci.* 57:2936–2950.

Gardner, W.R. 1959. Solutions to the flow equation for the drying of soils and other porous media. *Soil Sci. Soc. Am. Proc.* 20:317–320.

Gerard, C.J., P.D. Sexton, and D.M. Conover. 1984. Effects of furrow diking, subsoiling, and slope position on crop yields. *Agron. J.* 76:945–950.

Good, L.G., and D.E. Smika. 1976. Chemical fallow for soil and water conservation in the Great Plains. *J. Soil Water Conserv.* 33:89–90.

Greb, B.W. 1979. Reducing drought effect on croplands in the west-central Great Plains. *USDA Info. Bull. No. 420*, U.S. Gov. Print. Office, Washington, DC.

Hansen, N.C., B.L. Allen, R.L. Baumhardt, and D.J. Lyon. 2012. Research achievements and adoption of no-till, dryland cropping in the semi-arid U.S. Great Plains. *Field Crops Res.* 132:196–203.

Holman, J.D., K. Arnet, J. Dille, S. Maxwell, A. Obour, T. Roberts, K. Roozeboom, and A. Schlegel. 2018. Can cover or forage crops replace fallow in the semiarid Central Great Plains? *Crop Sci.* 58:932–944.

Horton, R., K.L. Bristow, G.J. Kluitenberg, and T.J. Sauer. 1996. Crop residue effects on surface radiation and energy balance – Review. *Theor. Appl. Climatol.* 54:27–37.

Howell, T.A., J.L. Steiner, A.D. Schneider, and S.R. Evett. 1995. Evapotranspiration of irrigated winter wheat – southern High Plains. *Trans. Am. Soc. Agric. Eng.* 38:745–756.

Idso, S.B., R.J. Reginato, R.D. Jackson, B.A. Kimball, and F.S. Nakayama. 1974. The three stages of drying of a field soil. *Soil Sci. Soc. Am. Proc.* 38:831–837.

Jackson, R.D., S.B. Idso, and R.J. Reginato. 1976. Calculation of evaporation rates during the transition from energy-limiting to soil limiting phases using albedo data. *Water Resour. Res.* 12:23–26.

Jackson, R.D., B.A. Kimball, R.J. Reginato, and F.S. Nakayama. 1973. Diurnal soil-water evaporation: Time-depth-flux patterns. *Soil Sci. Soc. Am. Proc.* 37:505–509.

James, E. 1945. Effect of certain cultural practices on moisture conservation on a piedmont soil. *J. Am. Soc. Agron.* 37:945–952.

Johnson, W.C. 1950. Stubble-mulch farming on wheat-lands of the Southern High Plains. *USDA Circ. 860.* US Govt. Print. Off., Washington, DC. 18 pp.

Johnson, W.C., E.L. Skidmore, B.B. Tucker, and P.W. Unger. 1983. Soil conservation: Central Great Plains winter wheat and range region. In *Dryland Agriculture.* H.E. Dregne and W.O. Willis, eds., *Agron. Monogr.* 23:197–217. *Am. Soc. Agron.*, Madison, WI.

Jones, O.R., and R.L. Baumhardt. 2003. Furrow dikes. In *Encyclopedia of Water Science.* B.A. Stewart and T.A. Howell, eds. Marcel-Dekker Inc., New York, NY, pp. 317–320.

Jones, O.R., and R.N. Clark. 1987. Effects of furrow dikes on water conservation and dryland crop yields. *Soil Sci. Soc. Am. J.* 51:1307–1314.

Jones, O.R., V.L. Hauser, and T.W. Popham. 1994. No-tillage effects on infiltration, runoff, and water conservation on dryland. *Trans. Am. Soc. Agric. Eng.* 37:473–479.

Jones, O.R., and T.W. Popham. 1997. Cropping and tillage systems for dryland grain production in the Southern High Plains. *Agron. J.* 89:222–232.

Jones, O.R., and P.W. Unger. 1977. Soil water effects on sunflowers in the Southern High Plains. *Sunflower Forum Proc.*, vol. 2, Fargo, ND. pp. 12–16.

Jones, O.R., P.W. Unger, R.L. Baumhardt, and Brent Bean. 2003. Sorghum tillage in the Texas High Plains. L-5439. Texas Coop. Extension, The Texas A&M University System, College Station, TX.

Kapanigowda, M., B.A. Stewart, T.A. Howell, H. Kadasrivenkata, and R.L. Baumhardt. 2010. Growing maize in clumps as a strategy for marginal climatic conditions. *Field Crop Res.* 118:115–125.

Kothari, K., S. Ale, A. Attia, N. Rajan, Q. Xue, and C.L. Munster. 2019. Potential climate change adaptation strategies for winter wheat production in the Texas High Plains. *Agric. Water Manage.* 225:105764. doi.org/10.1016/j.agwat.2019.105764

Lafarge, T.A., I.J. Broad, and G.L. Hammer. 2002. Tillering in grain sorghum over a wide range of population densities: Identification of a common hierarchy for tiller emergence, leaf area development and fertility. *Ann. Bot. (London)* 90:87–98.

Lal, R. 1997. Soil degradative effects of slope length and tillage methods on alfisols in western Nigeria. I. Runoff, erosion and crop response. *Land Degrad. Dev.* 8:201–219.

Lascano, R.J., and R.L. Baumhardt. 1996. Effects of crop residue on soil and plant water evaporation in a dryland cotton system. *Theor. Appl. Climatol.* 54:69–84.

Lindstrom, M.J., F.E. Koehler, and R.I. Papendick. 1974. Tillage effects on fallow water storage in Eastern Washington dryland region. *Agron. J.* 66:312–316.

Lollato, R.P., D. Marburger, J.D. Holman, P. Tomlinson, D. Presley, and J.T. Edwards. 2017. *Dual Purpose Wheat Management for Forage and Grain Production.* Kansas State University, Manhattan. www.bookstore.ksre.ksu.edu/pubs/MF3375.pdf

Lyon, D.J., S.D. Miller, and G.A. Wicks. 1996. The future of herbicides in weed control systems of the Great Plains. *J. Prod. Agnc.* 9:209–215.

Lyon, D.J., D.C. Nielsen, D.G. Felter, and P.A. Burgener. 2007. Choice of summer fallow replacement crops impacts subsequent winter wheat. *Agron. J.* 99:578–584.

Marek, G.W., R.L. Baumhardt, D.K. Brauer, P.H. Gowda, S.A. Mauget, and J.E. Moorhead. 2018. Evaluation of the Oceanic Niño Index as a decision support tool for winter wheat cropping systems in the Texas High Plains using SWAT. *Comput. Electron. Agric.* 151:331–337.

Mauget, S.A., and D.R. Upchurch. 1999. El Niño and La Niña related climate and agricultural impacts over the Great Plains and Midwest. *J. Prod. Agric.* 12:203–215.

Mauget, S.A., J. Zhang, and J. Ko. 2009. The value of ENSO forecast information to dual-purpose winter wheat production in the U.S. southern High Plains. *J. Appl. Meteorol. Climatol.* 48:2100–2117.

McCalla, T.M., and F.L. Duley. 1946. Effect of crop residues on soil temperature. *J. Am. Soc. Agron.* 38:75–89.

McGowan, M., and J.B. Williams. 1980. The water balance of an agricultural catchment. 1. Estimation of evaporation from soil water records. *J. Soil Sci.* 31:217–230.

Mjelde, J.W., and K. Keplinger. 1998. Using the Southern Oscillation to forecast Texas winter wheat and sorghum crop yields. *J. Climate.* 11:54–60.

Moreno-de las Heras, M., J.M. Nicolau, L. Merino-Martín, and B.P. Wilcox. 2010. Plot-scale effects on runoff and erosion along a slope degradation gradient. *Water Resour. Res.* 46, W04503, doi:10.1029/2009WR007875

Moret, D., and J.L. Arrúe. 2007. Dynamics of soil hydraulic properties during fallow as affected by tillage. *Soil Tillage Res.* 96:103–113.

Moroke, T.S. 2002. Root distribution, water extraction patterns, and crop water use efficiency of selected dryland crops under differing tillage systems. PhD. diss., Texas A&M Univ., College Station, TX.

Moroke, T.S., R.C. Schwartz, K.W. Brown, and A.S.R. Juo. 2005. Soil water depletion and root distribution of three dryland crops. *Soil Sci. Soc. Am. J.* 69:197–205.

Musick, J.T., O.R. Jones, B.A. Stewart, and D.A. Dusek. 1994. Water-yield relationships for irrigated and dryland wheat in the U.S. Southern Plains. *Agron. J.* 86:980–986.

Nielsen, D.C., D.J. Lyon, R.K. Higgins, G.W. Hergert, J.D. Homan, and M.F. Vigil. 2016. Cover crop effect on subsequent wheat yield in the Central Great Plains. *Agron. J.* 108:243–256.

Nielsen, D.C., and M.F. Vigil. 2005. Legume green fallow effect on soil water content at wheat planting and wheat yield. *Agron. J.* 97:684–689.

Nielsen D.C., and M.F. Vigil. 2010. Precipitation storage efficiency during fallow in wheat-fallow systems. *Agron. J.* 102:537–543.

Nielsen, D.C., M.F. Vigil, R.L. Anderson, R.A. Bowman, J.G. Benjamin, and A.D. Halvorson. 2002. Cropping system influence on planting water content and yield of winter wheat. *Agron. J.* 94:962–967.

Nielsen, D.C., M.F. Vigil, and J.G. Benjamin. 2009. The variable response of dryland corn yield to soil water content at planting. *Agric. Water Manage.* 96:330–336.

Norwood, C.A. 1994. Profile water distribution and grain yield as affected by cropping system and tillage. *Agron. J.* 86:558–563.

Norwood, C.A., A.J. Schlegel, D.W. Morishita, and R.E. Gwin. 1990. Cropping system and tillage effects on available soil water and yield of grain sorghum and winter wheat. *J. Prod. Agric.* 3:356–362.

Nuti, R.C., M.C. Lamb, R.B. Sorenson, and C.C. Truma. 2009. Agronomic and economic response to furrow diking tillage in irrigated and non-irrigated cotton (*Gossypium hirsutum* L.). *Agric. Water Manage.* 96:1078–1084.

Obembe, O.S., N.P. Hendricks, and J. Tack. 2021. Decreased wheat production in the USA from climate change driven by yield losses rather than crop abandonment. *PLOS ONE* | https://doi.org/10.1371/journal.pone.0252067

O'Connell, M.G., G.J. O'Leary, and D.J. Connor. 2003. Drainage and change in soil water storage below the root-zone under long fallow and continuous cropping sequences in the Victorian Mallee. *Aust. J. Agric. Res.* 54:663–675.

Okumura, Y.M., and C. Deser. 2010. Asymmetry in the duration of El Niño and La Niña. *J. Climate.* 23:5826–5843. https://doi.org/10.1175/2010JCLI3592.1

Patil, S.S., T.S. Kelkar, and S.A. Bhalerao. 2013. Mulching: A soil and water conservation practice. *Res. J. Agric. Forestry* 1:26–29.

Peterson, E. 1948. This plow saves rain: The Graham-Hoeme machine looks like it's here to stay even though its grandpappy was a road scarifier. *Farm J.* 72:89–90.

Peterson, G.A., A.J. Schlegel, D.L. Tanaka, and O.R. Jones. 1996. Precipitation use efficiency as affected by cropping and tillage systems. *J. Prod. Agric.* 9:180–186.

Porter, J. and M. Gawith. 1999. Temperatures and the growth and development of wheat: A review. *European J. Agron.* 10:23–36.

Pradhan, G.P., Q. Xue, K.E. Jessup, J.C. Rudd, S. Liu, R.N. Devkota, and J.R. Mahan. 2014. Cooler canopy contributes to higher yield and drought tolerance in new wheat cultivars. *Crop Sci.* 54:2275–2284.

Prasad, P., S. Pisipati, I. Momcilovic, and Z. Ristic. 2011. Independent and combined effects of high temperature and drought stress during grain filling on plant yield and chloroplast EF-Tu expression in spring wheat. *J. Agron. Crop Sci.* 197:430–441.

Pu, B., R. Fu, R.E. Dickinson, and D.N. Fernando. 2016. Why do summer droughts in the Southern Great Plains occur in some La Niña years but not others? *J. Geophys. Res. Atmos.* 121:1120–1137.

Rachidi, F., M.B. Kirkham, L.R. Stone, and E.T. Kanemasu. 1993. Soil water depletion by sunflower and sorghum under rainfed conditions. *Agric. Water Manage.* 24:49–62.

Rauf, S., J.M. Al-Khayri, M. Zaharieva, P. Monneveux, and F. Khalil. 2016. Breeding strategies to enhance drought tolerance in crops, Chapter 11. In *Advances in Plant Breeding Strategies: Agronomic, Abiotic and Biotic Stress Traits*, J.M. Al-Khayri et al. eds. Springer, Cham, 707 pp.

Reese, C.L., D.E. Clay, S.A. Clay, A.D. Bich, A.C. Kennedy, S.A. Hansen, and J. Moriles. 2014. Winter cover crops impact on corn production in semiarid regions. *Agron. J.* 106:1479–1488.

Ritchie, J.T. 1972. Model for predicting evaporation from a row crop with incomplete cover. *Water Resour. Res.* 8:1204–1213.

Ropelewski, C.F., and M.S. Halpert. 1996. Quantifying Southern Oscillation–precipitation relationships. *J. Climate* 9:1043–1059.

Rosenberg, N.J., R.C. Izaurralde, R.A. Brown, R.D. Sands, M. Tiscareño-Lopéz, D. Legler, and R. Srinivasan. 1997. Sensitivity of North American agriculture to ENSO-based climate scenarios and their socioeconomic consequences: Modeling in an integrated assessment. PNNL, US DOE. 148 pp.

Rowe-Dutton, P. 1957. *The Mulching of Vegetables*. Common Wealth Bureau of Horticulture and Plantation Crops. Bucks.

Scanlon, B.R., R.C. Reedy, R.L. Baumhardt, and G. Strassberg. 2008. Impact of deep plowing on groundwater recharge in a semiarid region: Case study, High Plains, Texas. *Water Resour. Res.* 44:W00A10, doi:10.1029/2008WR006991

Schlegel, A.J., Y. Assefa, L.A. Haag, C.R. Thompson, and L.R. Stone. 2018a. Long-term tillage on yield and water use of grain sorghum and winter wheat. *Agron. J.* 110:269–280.

Schlegel, A.J., F.R. Lamm, Y. Assefa, and L.R. Stone. 2018b. Dryland corn and grain sorghum yield response to available soil water at planting. *Agron. J.* 110:236–245.

Schwartz, R.C., R.L. Baumhardt, and S.R. Evett. 2010. Tillage effects on soil water redistribution and bare soil evaporation throughout a season. *Soil Tillage Res.* 110:221–229.

Schwartz, R.C., R.L. Baumhardt, and T.A. Howell. 2008. Estimation of soil water balance components using an iterative procedure. *Vadose Zone J.* 7:115–123.

Schwartz, R.C., A.J. Schlegel, J.M. Bell, R.L. Baumhardt, and S.R. Evett. 2019. Contrasting tillage effects on stored soil water, infiltration and evapotranspiration fluxes in a dryland rotation at two locations. *Soil Tillage Res.* 190:157–174.

Shrestha, R., S. Thapa, Q. Xue, B.A. Stewart, B.C. Blaser, E.K. Ashiadey, J.C. Rudd, and R.N. Devkota. 2020. Winter wheat response to climate change under irrigated and dryland conditions in the US southern High Plains. *J. Soil Water Conserv.* 75:112–122, doi:10.2489/jswc.75.1.112

Smika, D.E. 1983. Soil water change as related to position of wheat straw mulch on the soil surface. *Soil Sci. Soc. Am. J.* 47:988:991.

Steiner, J.L. 1989. Tillage and surface residue effects on evaporation from soils. *Soil Sci. Soc. Am. J.* 53:911–916.

Steiner, J.L. 1994. Crop residue effects on water conservation. In *Managing Agricultural Residues*, P.W. Unger, ed. CRC Press, New York, NY. pp. 41–76.

Steiner, J.L, H.H. Schomberg, P.W. Unger, and J. Cresap. 1999. Crop residue decomposition in no-tillage small-grain fields. *Soil Sci. Soc. Am. J.* 63:1817–1824.

Stewart, B.A. 2016. *Dryland Farming. Reference Module in Food Science*. Elsevier. https://doi.org/10.1016/B978-0-08-100596-5.02937-1

Stewart, B.A., R.L. Baumhardt, and S.R. Evett. 2010. Major advances of soil and water conservation in the Southern Great Plains. In *Soil and Water Conservation Advances in the United States*, T.M. Zobeck and W.F. Schillinger, eds. *Soil Sci. Soc. Am. Spec. Publ.* 60:103–129.

Stewart, B.A., P. Koohafkan, and K. Ramamoorthy. 2006. Dryland agriculture defined and its importance to the world. In *Dryland Agriculture*, second edition, G.A. Peterson, P.W. Unger, and W.A. Payne, eds. *Agron. Monogr.* 23:1–26.

Stewart, B.A., and R. Lal. 2012. Manipulating crop geometries to increase yields in dryland areas. In *Soil Water and Agronomic Productivity*, R. Lal and B.A. Stewart, eds. Taylor & Francis, Boca Raton, FL. pp. 409–428.

Stewart, B.A., and G.A. Peterson. 2015. Managing green water in dryland agriculture. *Agron. J.* 107:1544–1553.

Stewart, B.A., S. Thapa, Q. Xue, and R. Shrestha. 2018. Climate change effect on winter wheat (*Triticum aestivum* L.) yield in the US Great Plains. *J. Soil Water Conserv.* 73:601–609, doi:10.2489/ jswc.73.6.601

Stewart, B.A., P.W. Unger, and O.R. Jones. 1985. Soil and water conservation in semiarid regions. In *Soil Erosion and Conservation*. S.A. El-Swaify, W.C. Moldenhauer, and A. Lo, eds. *Soil Conserv. Soc. Am.*, Ankeny, IA. pp. 328–337.

Stewart, B.A., and L. Wei-li. 2015. Strategies for increasing the capture, storage, and utilization of precipitation in semiarid regions. *J. Integrative Agric.* 14:1500–1510.

Stone, L.R., D.E. Goodrum, A.J. Schlegel, M.N. Jaafar, and A.H. Khan. 2002. Water depletion depth of grain sorghum and sunflower in the Central High Plains. *Agron. J.* 94:936–943.

Stone, L.R., N.L. Klocke, A.J. Schlegel, F.R. Lamm, and D.J. Tomsicek. 2011. Equations for drainage component of the field water balance. *Appl. Eng. Agric.* 27:345–350.

Stone, L.R., and A.J. Schlegel. 2006. Yield-water supply relationships of grain sorghum and winter wheat. *Agron. J.* 98:1359–1366.

Stone, R.C., G.L. Hammer, and T. Marcussen. 1996. Prediction of global rainfall probabilities using phases of the Southern Oscillation Index. *Nature* 384:252–255.

Streck, N.A., A. Weiss, Q. Xue, and P.S. Baenziger. 2003. Improving predictions of developmental stages in winter wheat: A modified Wand and Engel model. *Agric. Forest Meteorol.* 115:139–150.

Sui, Y., Y. Ou, B. Yan, X. Xu, A.N. Rousseau, and Y. Zhang. 2016. Assessment of micro-basin tillage as a soil and water conservation practice in the black soil region of Northeast China. *PLoS ONE* 11 (3): e0152313. doi:10.1371/journal.pone.0152313

Tack, J., A. Barkley, and L.L. Nalley. 2015. Effect of warming temperatures on US wheat yields. *Proc. Nat. Acad. Sci. US A.* 112(22): 6931–6936.

Tanaka, D.I., and J.K. Aase. 1987. Fallow method influences on soil water and precipitation storage efficiency. *Soil Tillage Res.* 9:307–316.

Thapa, S., B.A. Stewart, Q. Xue, and Y. Chen. 2017a. Manipulating plant geometry to improve microclimate, grain yield, and harvest index in grain sorghum. *PLoS ONE* 12(3): e0173511. https://doi.org/10.1371/journal.pone.0173511

Thapa, S., Q. Xue, K.E. Jessup, J.C. Rudd, S. Liu, R.N. Devkota, and J. Baker. 2018. Canopy temperature depression at grain filling correlates to winter wheat yield in the U.S. Southern High Plains. *Field Crops Res.* 217:11–19.

Thapa, S., Q. Xue, K.E. Jessup, J.C. Rudd, S. Liu, G.P. Pradhan, R.N. Devkota, and J. Baker. 2017b. More recent wheat cultivars extract more water from greater soil profile depths to increase yield in the Texas High Plains. *Agron. J.* doi:10.2134/ agronj2017.02.0064

Thompson, L., C. Burr, K. Glewen, G. Lesoing, J. Rees, and G. Zoubek. 2014. *Cover Crop Studies from 2014 Show Varied Results*. University of Nebraska, Lincoln, NE. https://cropwatch.unl.edu/5-cover-crop-studies-2014-show-varied-results

Tolk, J.A., and S.R. Evett. 2012. Lower limits of crop water use in three soil textural classes. *Soil Sci. Soc. Am. J.* 76:607–616.

Tolk, J.A., S.R. Evett, and R.C. Schwartz. 2015. Field-measured, hourly soil water evaporation stages in relation to reference evapotranspiration rate and soil to air temperature ratio. *Vadose Zone J.* 14. doi:10.2136/vzj2014.07.0079

Tullberg, J.N., P.J. Ziebarth, and Y. Li. 2001. Tillage and traffic effects on runoff. *Aust. J. Soil Res.* 39:249–257.

Unger, P.W. 1971. Knowing your soil and its water needs. *Am. Hortic. Mag.* 50:114–117.

Unger, P.W. 1978a. Straw-mulch rate effect on soil water storage and sorghum yield. *Soil Sci. Soc. Am. J.* 42:486–491.

Unger, P.W. 1978b. Straw mulch effects on soil temperatures and sorghum germination and growth. *Agron. J.* 70:858–864.

Unger, P.W. 1984. Tillage and residue effects on wheat, sorghum, and sunflower grown in rotation. *Soil Sci. Soc. Am. J.* 48:885–891.

Unger, P.W. 1988. Residue management for dryland farming. In *Challenges in Dryland Agriculture--A Global Perspective,* P.W. Unger, W.R. Jordan, T.V. Sneed, and R.W. Jensen, eds. *Proc. Int. Conf. on Dryland Farming,* Amarillo/Bushland, Texas, August 1988. Texas Agric. Exp. Stn., College Station, TX. pp. 483–489.

Unger, P.W. 1992. Infiltration of simulated rainfall: Tillage system and crop residue effect. *Soil Sci. Soc. Am. J.* 56:283–289.

Unger, P.W. 1993. Paratill effects on loosening of a Torrertic Paleustoll. *Soil Tillage Res.* 26:1–9.

Unger, P.W. 1994. Tillage effects on dryland wheat and sorghum production in the Southern Great Plains. *Agron. J.* 86:310–314.

Unger, P.W. 1999. Conversion of Conservation Reserve Program (CRP) grassland for dryland crops in a semiarid region. *Agron. J.* 91:753–760.

Unger, P.W. 2001. Alternative and opportunity dryland crops and related soil conditions in the southern Great Plains. *Agron. J.* 93:216–226.

Unger, P.W., and R.L. Baumhardt. 1999. Factors related to dryland grain sorghum yield increases, 1939 through 1997. *Agron. J.* 91:870–875.

Unger, P.W., and J. Box. 1972. Technical problems associated with minimum tillage. *Proc. Coop. Conserv. Workshop,* College Station, TX. 21 pp.

Unger, P.W., M.B. Kirkham, and D.C. Nielsen. 2010. Soil and water conservation advances in the United States. In *Water Conservation for Agriculture*, T.M. Zobeck and W.F. Schillinger, eds. Soil Sci. Soc. Am. Spec. Publ. 60:1–45.

Unger, P.W., and T.M. McCalla. 1980. Conservation tillage systems. *Adv. Agron.* 33:1–58.

Unger, P.W., and J.J. Parker. 1976. Evaporation reduction from soil with wheat, sorghum, and cotton residues. *Soil Sci. Soc. Am. J.* 40:298–300.

Unger, P.W., and B.A. Stewart. 1983. Soil management for efficient water use: An overview. In *Limitations to Efficient Water Use in Crop Production*, H.M. Taylor, W.R. Jordan, and T.R. Sinclair, eds. Am. Soc. Agron., Crop Sci. Soc. Am., Soil Sci. Soc. Am., Madison, WI. pp. 419–460.

Unger, P.W., B.A. Stewart, J.F. Parr, and R.P. Singh. 1991. Crop management and tillage methods for conserving soil and water in semi-arid regions. *Soil Tillage Res.* 20: 219–240.

Unger, P.W., and M.F. Vigil. 1998. Cover crop effects on soil water relationships. *J. Soil Water Conserv.* 53:200–207.

Unger, P.W., and A.F. Wiese. 1972. Experiences with minimum tillage. *Proc. 5th Annual Texas Conf. on Insect, Plant Disease, Weed and Brush Control,* College Station, TX. pp. 115–123.

Unger, P.W., and A.F. Wiese. 1979. Managing irrigated winter wheat residues for water storage and subsequent dryland grain sorghum production. *Soil Sci. Soc. Am. J.* 43:582–588.

van Bavel, C.H.M., and D.I. Hillel. 1976. Calculating potential and actual evaporation from a bare soil surface by simulation of concurrent flow of water and heat. *Agric. Meteorol.* 17:453–476.

van De Giesen, N.C., T.J. Stomph, and N. de Ridder. 2000. Scale effects of Hortonian overland flow and rainfall-runoff dynamics in a West African catena landscape. *Hydrol. Processes* 14:165–175.

Vanderborght, J., T. Fetzer, K. Mosthaf, K.M. Smits, and R. Helmig. 2017. Heat and water transport in soils and across the sol-atmosphere interface: 1. Theory and different model concepts. *Water Resour. Res.* 53:1057–1079.

Vocke, G., and M. Ali. 2013. *U.S. wheat production practices, costs, and yields: Variations across regions.* Washington, DC: USDA Economic Research Service. www.ers.usda.gov/webdocs/publications/43783/39923_eib116.pdf?v=41516

Wiese, A.F. 1983. Weed control. In *Dryland Agriluture*, H.E. Dregne and W.O. Willis, eds. *Am. Soc. Agron., Crop Sci. Soc. Am., Soil Sci. Soc. Am.*, Madison, WI. pp. 463–488.

Williams, J.D., S.B. Wuest, and D.S. Robertson. 2015. Soil water and water-use efficiency in no-tillage and sweep tillage winter wheat production in Northeastern Oregon. *Soil Sci. Soc. Am. J.* 79:1206–1212.

Willis, W.O. 1983. Water conservation: Introduction. In *Dryland Agriculture*, H.E. Dregne and W.O. Willis, eds. *Agron. Monogr.* 23:21–24. Am. Soc. Agron., Madison, WI.

Winter, S.R., and J.T. Musick. 1993. Wheat planting date effects on soil water extraction and grain yield. *Agron. J.* 85:912–916.

Winter, S.R., J.T. Musick, and K.B. Porter. 1988. Evaluation of screening techniques for breeding drought-resistant winter wheat. *Crop Sci*. 28:512–516.

Woodruff, N.P., C.R. Fenster, W.W. Harris, and M. Lundquist. 1966. Stubble-Mulch tillage and planting in crop residue in the Great Plains. *Trans. Am. Soc. Agric. Eng*. pp. 849–852. doi:10.13031/2013.40115

Wuest, S. 2018. Surface effects on water storage under dryland summer fallow: A lysimeter study. *Vadose Zone J*. 17:160078. doi:10.2136/vzj2016.09.0078

Xue, Q., W. Liu, and B.A. Stewart. 2012. Improving wheat yield and water use efficiency under semi-arid environment – The US Southern Great Plains and China's Loess Plateau. In *Advances in Soil Science: Soil Water and Agronomic Productivity*, R. Lal and B.A. Stewart, eds. Taylor & Francis, , Boca Raton, FL.

Xue, Q., J.C. Rudd, S. Liu, K.E. Jessup, R.N. Devkota, and J.R. Mahan. 2014. Yield determination and water use efficiency of wheat under water-limited conditions in the U.S. Southern High Plains. *Crop Sci*. 54:34–47.

Xue, Q., A. Weiss, and P.S. Baenziger. 2004. Predicting phenological development in winter wheat. *Climate Res*. 25:243–252.

3 Distinctive Dryland Soil Carbon Transformations

Insights from Arid Rangelands of SW United States

Jean L. Steiner[1], Carolina B. Brandani[2], Adrian Chappell[3], Jose Castaño-Sanchez[4], Mikaela Hoellrich[5], Matthew M. McIntosh[4], Shelemia Nyamuryekung'e[1], Nicole Pietrasiak[5], Alan Rotz[6], and Nicholas P. Webb[4]

[1]Animal and Rangeland Sciences Department, New Mexico State University, Las Cruces, NM 88003, USA

[2]Texas A&M AgriLife Research and Extension Center, 6500 W. Amarillo Blvd, Amarillo, TX, 79106, USA

[3]School of Earth & Environmental Science, Cardiff University, Cardiff, Wales, UK

[4]USDA-ARS Jornada Experimental Range, 2995 Knox St, Las Cruces, NM 88003, USA

[5]Plant and Environmental Sciences Department, New Mexico State University, Las Cruces, NM, USA

[6]USDA-ARS, Pasture Systems and Watershed Management Research Unit, University Park, PA, USA

CONTENTS

3.1 Introduction 40
3.2 Carbon and Nitrogen Stocks and Dynamics in Rangelands....................................44
    3.2.1 Soil Organic and Inorganic Carbon in Arid and Semiarid Rangelands 44
    3.2.2 Woody-Shrub Encroachment into Grassland...45
        3.2.2.1 Aboveground...45
        3.2.2.2 Belowground 46
    3.2.3 Fire 47
    3.2.4 Increasing Atmospheric CO_2 Concentration (CO_2 Fertilization) 47
    3.2.5 Distribution and Role of Biocrusts in Stocks, Flows, and Ecosystem Processes 48
    3.2.6 Soil Inorganic Carbon Sequestration 50
    3.2.7 Soil Aggregation 50
3.3 Modeling the Carbon Cycle for Arid Rangelands 51
    3.3.1 Plant Processes...52
    3.3.2 Animal Processes 53
    3.3.3 Soil Components and Processes..54
    3.3.4 Other Aspects of the C Cycle of Semiarid and Arid Ecosystems 55

DOI: 10.1201/b22954-3

3.4 Role and Impact of Soil Redistribution in Carbon Cycling 56
    3.4.1 Lateral Soil Organic Carbon Erosion by Wind 57
    3.4.2 Vertical Emission of Soil Organic Carbon 58
    3.4.3 Implications of Omitted Soil Organic Carbon Erosion for Carbon Cycling 59
3.5 Adaptation and Mitigation Options 62
    3.5.1 Plant Community Composition – Abiotic and Biotic Drivers 62
    3.5.2 Rangeland Forage Quality Interaction with Grazer-Derived CH_4 63
    3.5.3 Biocrust Community Changes in Rangelands 63
    3.5.4 Adaptation Strategies 64
        3.5.4.1 Stocking Rates/Animal Numbers 64
        3.5.4.2 Animal Size/Species 65
        3.5.4.3 Animal Breed 65
        3.5.4.4 Novel Tools and External Inputs 66
    3.5.5 Novel Ecological Restoration Strategies (Mitigation) 67
3.6 Arising Insights into Soil C Dynamics in Arid Rangelands 68
3.7 Acknowledgments 69
References 69

3.1 INTRODUCTION

Drylands occupy over 40% of the global land surface, supporting about 35% of the world's population (Figure 3.1). Ruminant livestock grazing on dry rangelands and grasslands support food and livelihoods systems in these regions. Almost ten years ago, the Millennium Ecosystem Assessment (SCBD, 2013) highlighted the important levels of biodiversity and abundance of direct and indirect ecosystem services of dryland regions but emphasized that these values are underrecognized. Sustaining ecosystem services requires maintaining the resource base, but the attention to dryland systems has focused on direct production benefits and has undervalued the indirect benefits

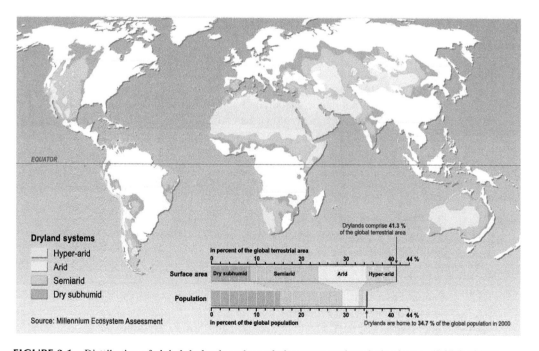

FIGURE 3.1 Distribution of global drylands and population supported on dryland areas (SCBD (2013).

Distinctive Dryland Soil Carbon Transformations

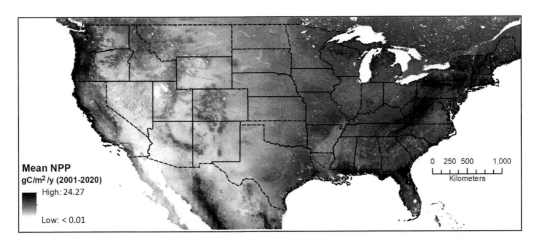

FIGURE 3.2 Map of long-term (2001–2020) mean MODIS net primary productivity (NPP; g C/m2/y1).

of climate regulation, water supply and filtering, nutrient cycling, biodiversity, and others (Havstad et al., 2007; Yahdijan et al., 2015).

Ecosystems store large amounts of carbon (C) which can contribute to global greenhouse gas concentrations if degraded (Lal et al., 2012). Soil C stocks represent an important portion of these ecosystem storage stocks (Lal, 1997; Jackson et al., 2017). Because dryland ecosystems are in arid and largely hot climates, the concentration of soil C is low, but because the land area is vast, dryland soils contain about 36% of the global C stock (Trumper et al., 2008). The C stocks in vegetation and soils have been degraded in much of the global drylands, indicating a potential to rebuild C stocks in these regions (Trumper et al., 2008). However, with high populations on fragile landscapes, combined with increasing temperatures and projected precipitation declines in these already dry regions, it will be increasingly difficult to sustain or rebuild C stocks.

In the continental USA (CONUS), the net primary productivity (NPP) exhibits a pronounced pattern of low NPP in the western part of the country, particularly in the southwest quadrant (Figure 3.2). Gedefaw et al. (2021) found that New Mexico grassland and shrubland productivity decreased from 1984 to 2015 by 2.2% and 4.5%, respectively, and that the eastern parts of the state experienced a significant decrease. Godde et al. (2020) projected that mean herbaceous biomass would decrease from 2000 to 2050 across global rangelands by 4.7% (under Representative Concentration Pathway (RCP) 8.5), while inter- and intra-annual variability would increase. Western USA rangelands are characterized by heterogeneity across multiple scales (Photo Plate 3.1A–1E). Management options to sustain or rebuild rangeland C stocks have common drivers across regions of the globe, but successful outcomes depend on the soils, topography, climate, vegetation, and agricultural systems in a particular place (Herrick et al., 2012; Peters et al., 2015).

In drylands, soil respiration (SR) is large relative to NPP and can be approximated as a function of NPP following Raich and Schlesinger (1992). SR can be subtracted from NPP to provide a straightforward long-term steady-state approximation of the net ecosystem exchange (NEE; Chapin et al., 2005). Using MODIS NPP, we predicted SR and estimated NEE (Figure 3.3) across the CONUS which shows a clear separation of SR and NEE between the humid east and the arid west.

The Southwest USA is a major rangeland region, where beef cattle production is a significant agricultural enterprise occupying a large portion of the landscape. A history of soil degradation combined with extended drought over the past 20 years (Figure 3.4) presents great challenges in the region. Williams et al. (2022) reported that the ongoing megadrought in the Western USA is the worst in 1200 years, indicating that climate change has led to drying of the regional climate.

PHOTO PLATE 3.1 Rangeland landscapes, ecosystem components and processes. A) Blue grama and Sand drop seed dominated grassland in the Rio Puerco Watershed, central New Mexico; B) Sagebrush expansion into grasslands in central New Mexico; C) Creosote dominated rangeland in the Chihuahuan Desert; D) Blue grama, a C4 perennial grass; E) Sage brush, a C3 shrub; F) Caliche layer in a black grama grassland (photo credit: Curtis Monger); G) A piece of the petrocalcic laminar cap from the Picacho alluvial surface near Las Cruces, NM; H) Biological soil crusts with dark pigmented lichens in front of Tobosa bunch grasses in a Jornada Basin rangeland I) Cryptic appearance of light cyanobacterial soil crust from the Chihuahuan Desert J) Physical soil crusts with distinct vesicular pores formed by soil physical processes; K) Wind erosion of rangeland in northern New Mexico.

In the face of increasing climate pressures, the challenge is ever greater. However, there is also a rich ranching heritage which residents and communities in the region are motivated to sustain. Additionally, there are extensive research resources to identify critical processes that drive change in C stocks as well as management strategies to direct the change in the positive direction (Havstad et al., 2007; Peters et al., 2006; Peters et al., 2014a, b).

While the focus of this chapter is on soil C dynamics, the soil C is inextricably linked with the ecosystem C dynamics which differ in significant ways in arid rangelands compared to more humid regions (SCBD, 2013). The objectives of this chapter are to provide insight into C dynamics in semiarid to arid rangelands, highlight contrasts between key processes in dryland compared to more humid regions, and identify management practices to adapt to and sustain the natural resource base in rangeland systems. While the chapter will focus on the context of the southwestern USA, the principles are relevant throughout the global arid rangelands.

Distinctive Dryland Soil Carbon Transformations

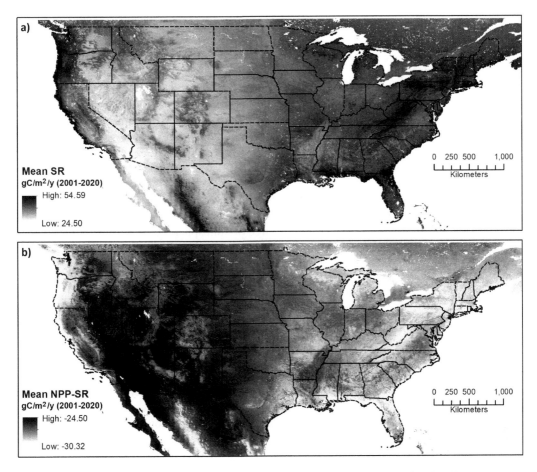

FIGURE 3.3 Maps of long-term (2001–2020) mean soil respiration (SR; g C/m2/y1) and net ecosystem exchange (NEE; g C/m2/y1) where SR = 1.24(NPP)+24.5 (Raich and Schlesinger, 1992) and NEE = NPP-SR (g C/m2/y1).

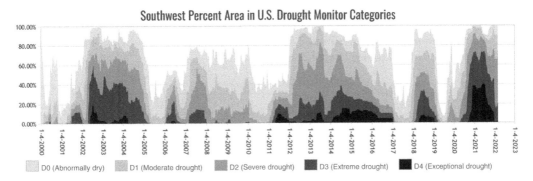

FIGURE 3.4 Fraction of Southwest USA in different stages of drought (2000–2021). Source: National Drought Monitor (https://droughtmonitor.unl.edu/).

3.2 CARBON AND NITROGEN STOCKS AND DYNAMICS IN RANGELANDS

3.2.1 SOIL ORGANIC AND INORGANIC CARBON IN ARID AND SEMIARID RANGELANDS

Soils play an important role in the global C cycle, through the capacity to act as a source and sink for carbon dioxide (CO_2) emissions. Relative to soils of humid ecosystems that contain predominantly organic C (SOC), arid and semiarid rangelands are reservoirs of the dual pool of soil C – SOC and inorganic C (SIC) (Table 3.1; Lal et al., 2021; Monger et al., 2014; Plaza et al., 2018).

SIC is mainly restricted to soils of arid and semiarid regions, which globally hold 78% and 14%, respectively (Eswaran et al., 2000). Estimates of inorganic C as soil carbonate (~940 Pg C; Pg = petagram = 1 billion metric tons) (Eswaran et al., 2000) and as bicarbonate in groundwater (~1404 Pg C) exceed organic C which include 594 Pg C contained in land plants and SOC (~1530 Pg C) (Monger et al., 2015). Although global SIC is large, this pool of C has been considered inactive (controlled by glacial-interglacial cycles; millennial-scale (Monger et al., 2015) and has not been expected to become a major sink or source of atmospheric CO_2 over the next centuries (Schlesinger, 2002), unless we experience major soil loss that could expose the SIC.

Although SOC in the CONUS has very low concentrations in the southwestern quadrant relative to more humid regions (Figure 3.5), the storage of organic C amounts to 54% of the terrestrial C pool, and twice the atmospheric C pool of 760 Pg (Monger et al., 2015). For the world's dryland soils it is also significant, with 646 Pg of organic C storage up to 2 m, representing 32–36 % of the global SOC pool (Plaza et al., 2018; Trumper et al., 2008), with approximately 385 Pg of SOC in the top 1 m that correspond one-quarter of the world's SOC (Jobbágy and Jackson, 2000).

The potential of C sequestration in USA dryland soils ranges from 38×10^{-3} to 56×10^{-3} Pg C yr^{-1} (Lal, 2004). However, the SOC in arid regions and the potential sequestration are driven by three major factors: extreme aridity, parent material, and soil age. Aridity (annual precipitation <0.2 evaporative demand) limits plant production (World Resources Institute, 2005), resulting in landscapes with low biomass input and lack of soil physical protection. Water deficit also limits soil development, leading to the dominance of characteristic young soils with coarse texture, low fertility, low SOM, and low C:N and C:P ratios (Augusto et al., 2017; Cotrufo et al., 2021; Plaza et al., 2018). Thus, drought is the major constraint on C and nitrogen (N) biogeochemical processes in arid rangeland ecosystems. In a meta-analysis evaluating drivers of plant nutrient limitations in terrestrial ecosystems, Augusto et al. (2017) found that N limitation was best explained by climate, with ecosystems under harsh conditions, such as arid rangelands, being

---

TABLE 3.1

Soil Organic C, Inorganic C, and Total N Stocks (Mean±Standard Deviation) in Hyperarid, Arid, Semiarid, Dry Subhumid, Dry, and Humid Areas

	Organic C			Inorganic C			Total N		
					Pg				
Land	0-0.3 m	0-1 m	0-2 m	0-0.3	0-1 m	0-2 m	0-0.3 m	0-1 m	0-2 m
Hyperarid	11±1	22±1	31±1	20±2	65±3	127±5	1.3±0.1	2.9±0.1	4.5±0.2
Arid	45±3	91±3	127±3	63±2	241±5	487±9	4.9±0.2	10.9±0.2	17.3±0.3
Semiarid	100±2	190±3	259±3	48±2	207±4	456±7	9.3±0.1	19.6±0.2	30±0.2
Dry subhumid	91±3	167±4	228±6	15±1	66±2	168±4	7.1±0.2	14.3±0.2	21.4±0.3
Dry subhumid	248±6	470±7	646±9	145±4	578±8	1237±15	22.6±0.4	47.7±0.5	73.2±0.6
Humid	502±12	955±19	1401±36	28±3	107±5	321±9	36.1±0.7	72.6±1.1	111.1±1.6
Global	750±15	1425±21	2047±39	173±5	686±10	1558±19	58.6±0.8	120.4±1.3	184.2±1.9

Source: Plaza et al. (2018).

Distinctive Dryland Soil Carbon Transformations 45

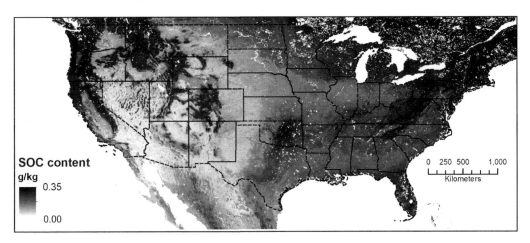

FIGURE 3.5 Map of soil organic carbon (SOC; g C/kg) content 0–0.3 m depth from SoilGrids (Hengl et al., 2017).

more N-limited than ecosystems under more humid climates. Cotrufo et al. (2021) hypothesized that C input-limited systems (such as arid and semiarid rangelands) have N in excess of demand, which results in a decoupling of the N and C cycles which opens the N cycle to losses from the soil. This framework corroborates Augusto et al.'s (2017) findings in that climate would be the first overarching control under extreme climates. These characteristics have driven the misconception that drylands have little impact on global biogeochemical cycles or energy balance (Hoover et al., 2020). However, drylands are highly sensitive to sporadic and small rainfall events that lead to a rapid increase in the biological activity demonstrated through SR, CO_2 efflux, and plant growth (Petrie et al., 2015).

3.2.2 Woody-Shrub Encroachment into Grassland

Warming temperatures, shifting precipitation regimes, and extended droughts in the USA arid and semiarid ecosystems have led to transitions in rangeland plant communities, with woody-shrub plant communities encroaching into grassland settings (e.g., black grama replaced by mesquite in New Mexico) with subsequent changes in soil C dynamics (Barger et al., 2011; Connin et al., 1997; Petrie et al., 2015). Woody-shrub and grasslands above- and below-ground biomass differ in productivity, standing biomass, rooting patterns, and chemical quality of litter inputs. These characteristics along with edaphic and climatic conditions determine the dynamic decomposition of plant biomass into C storage in the soil profile (SOC), influence turnover (mean residence time) as well as affect C quality and forms in different fractions of SOM.

3.2.2.1 Aboveground

Although the responses of arid and semiarid ecosystems are highly variable, enhanced ecosystem C sequestration is expected in woody vegetation (Barger et al., 2011; Neff et al., 2009; Petrie et al., 2015). Overall, woody encroachment into grasslands tends to increase SOC by 1% per year (Neff et al., 2009). In the Northern Chihuahuan Desert, shrublands exhibited 49 g C m^{-2} yr^{-1} (2007 to 2011) sequestration of C compared to 31 g C m^{-2} yr^{-1} in grassland (Petrie et al., 2015). A notable difference between the plant communities was the sensitivity to soil moisture. The grasslands were up to two times more productive and a large C sink during wetter periods than shrubland. However, during dry periods, greater respiration resulted in loss of C while shrublands sequestered C even in extremely dry conditions. Nevertheless, Barger et al. (2011) found woody encroachment in North

American arid rangelands impacted ecosystem C that differed by ecoregion, specifically regarding mean annual precipitation (MAP). In the absence of disturbance, woody encroachment in semiarid to subhumid ecosystems (MAP >336 mm) resulted in net C gains with an average increase of 0.7 g m^2 yr^{-1} per mm precipitation across MAP zones >336 mm. In contrast, the arid ecosystems of the Chihuahuan Desert in New Mexico (MAP <336 mm) exhibited negative to neutral change in ecosystem C in response to woody encroachment (Barger et al., 2011; Knapp et al., 2008) which contrasts with Petrie et al.'s (2015) findings.

3.2.2.2 Belowground

The most important biogeochemical consequence of shrub expansion into grasslands may be the modifications to rooting depth and distribution, and altered hydrology, with production of woody biomass below the rooting zone of perennial grasses (Connin et al., 1997; Jobbágy and Jackson, 2000). The maximum rooting depth observed for desert species was 9.5±2.4 m, with deep root habits common in woody and herbaceous species across most of the terrestrial biomes (Canadell et al., 1996). Although deep soil C (>20 cm) is a paramount component of the soil C pool, the dynamics of deep soil C (Mathieu et al., 2015) and the role of changing root litter quality and turnover time with woody expansion (Boutton et al., 2009; Connin et al., 1996) are poorly understood. Increasing soil C sequestration is expected due to the greater input of low-quality biomass (high C/N and chemical complexity) to deeper soil layers where overall residence times may be up to several thousand years (Rumpel and Kogel-Knabner, 2011). Subsoil microbes depend more on SOM as a C source than fresh plant inputs, and therefore deep SOM is thought to be primarily from microbial-derived C compounds (Rumpel and Kogel-Knabner, 2011). However, the chemical composition of SOM in subsoils is specific by soil class and more strongly influenced by pedological processes/traits (e.g., texture and mineralogy) than climate (Cotrufo et al., 2021; Mathieu et al., 2015; Rumpel and Kogel-Knabner, 2011).

Jobbagy and Jackson (2000) observed that the percentage of SOC was greatest in shrubland compared to grasslands and forests (P< 0.05) and the vegetation type was more important than the direct effect of precipitation. Woody plant expansion of grassland has resulted in significant C sequestration over the past century in the Rio Grande Plains of southern Texas, with the oldest wooded areas ranging from 3000 to 4500 g C m^{-2}, while grassland SOC was 1000 g C m^{-2}, which reflected in an increase of 200% to 350% in the upper 15 cm of the soil profile. Observed SOC accumulation rates ranged from 12 to 43 g C m^{-2} yr^{-1} and accounted for 55% to 70% of C accumulation in the belowground system. Specifically, litter accounted for less than 10%, and roots accounted for 25% to 45% of the observed C accumulation (Boutton et al., 2009). In the Chihuahuan Desert where C4 grassland was encroached on by C3 creosote bush, Throop et al. (2012) found that the soil C concentration varied with the microsite. Shrub subcanopies have about twice the C as in intercanopy spaces or remnant grasslands. Estimated SOC accumulation rates from creosote encroachment (4.79 g C m^{-2} year^{-1} under canopies and 1.75 g C m^{-2} year^{-1} when intercanopy losses were considered) were lower than reported for higher productivity *Prosopis* systems, but still represent a potentially large regional C sink.

Woody plant roots generally have slower turnover times compared to grass and forb roots because of high concentrations of metabolites such as lignin, suberin, cutin, tannins, and others. Lignin has a complex chemical form and needs a specialized microbial community to start the decomposition process. Boutton et al. (2009) found an unexpected higher root lignin concentration in the remnant grassland compared to shrub-invaded grassland in the Rio Grande Plains of southern Texas. However, litter in wooded areas contained up to seven times greater concentrations of lignin and cutin- and suberin-derived substituted fatty acids than grassland litter, indicating that litter in wooded areas is biochemically more resistant to decomposition than grassland litter. Mun and Whitford (1997) observed similar root lignin concentrations and decomposition rates in mixed shrub and grass vegetation in the Chihuahuan Desert.

3.2.3 Fire

Fire on rangeland, natural or as a management practice, may also influence long-term soil C storage through deposition of partially combusted biomass and black carbon (Ansley et al., 2006). Rangelands are often managed with fire to increase forage production and nutritive value while preventing shrub expansion. Black carbon is highly resistant to decomposition and thus contributes to the recalcitrant fraction of soil C, representing a long-term soil C storage pool. In grasslands and savannas, approximately 0.6–4% of biomass C is converted to black carbon by fire (Skjemstad et al., 1996).

Increased plant growth that usually results from fire can compensate for the loss of aboveground C with little or no change in SOC (Follett et al., 2001). However, small plant cover following fire can accelerate runoff and erosion (Lal, 2004) depending on the edaphoclimatic condition, specifically in the context of drought that delays the recovery of grasses post-fire, thereby increasing susceptibility to soil erosion (Bestelmeyer et al., 2021).

In arid grasslands, prescribed fire frequency may be small because of detrimental side effects (Fuhlendorf et al., 2011) such as reduced vegetative cover and the subsequent soil degradation, contrasting with humid ecoregions, where fire has been required to maintain the ecosystem integrity (Bestelmeyer et al., 2021). While shrubs resprouted quickly following fire, perennial grasses in southern New Mexico did not recover from prescribed fire after 5 years, even with above-average rainfall, resulting in increased susceptibility to soil erosion (Bestelmeyer et al., 2021). These results indicate caution in considering prescribed fire in mesic and more arid grasslands (Lohmann et al., 2014), especially in the Chihuahuan Desert (Bestelmeyer et al., 2021). In the northern Chihuahuan Desert, Wang et al. (2019) hypothesized that fire facilitates remobilization of nutrient-enriched soil from shrub microsites to grass and bare microsites but did not detect significant differences in total soil C and total soil N among the grassland–shrubland transition zone one year after the fire. Overall, soil responses recovered in burnt rangelands within two years (Fuhlendorf et al., 2011).

In a mixed-grass savanna in north-central Texas, Ansley et al. (2006) found that the SOC at 0–20 cm depth was 2044 g C m^{-2} in unburned compared to 2393–2534 g C m^{-2} in seasonal and frequency fire treatments, implying rates of C accumulation of 58–82 g C m^{-2} yr^{-1} over 6 years. Although SOC increased, black carbon did not change, even though black carbon comprised about 13–17% of the SOC pool in both burned and unburned treatments. The increased soil C in summer fire treatments was attributed to shifts in community composition (lower Δ ^{13}C of SOC) toward greater relative productivity by C3 species.

3.2.4 Increasing Atmospheric CO_2 Concentration (CO_2 Fertilization)

In addition to changes in SOC due to vegetation shifts and fire, increases in atmospheric CO_2 concentration ($[CO_2]$) are expected to alter plant productivity, affecting the quantity of C inputs to soil, and contributions of roots and shoots to SOC (Berryman et al., 2020). Under elevated $[CO_2]$, arid and semiarid ecosystems may play more important roles in global C cycles (Evans et al., 2014; Koyama et al., 2019; Tfaily et al., 2018). Ecosystems have shown more response to elevated $[CO_2]$ when NPP is mostly limited by water availability (Koyama et al., 2019; Tfaily et al., 2018). Production of aboveground litter, root litter, and rhizodeposition was stimulated under elevated $[CO_2]$ in the Free Air CO_2 Enrichment (FACE) Facility in the Nevada Desert, which was reflected in greater SOC and total N in the 0.4 to 0.8 m soil layer (Koyama et al., 2019). The authors concluded that in arid ecosystems under elevated $[CO_2]$, shrubs should play a major role in determining C sink capacity in desert soils. Additionally, they anticipate that N limitation for vascular plants is unlikely due to the sustained net N supply observed under shrubs and in-plant interspace that had extensive biological soil crusts where reduced $\delta^{15}N$ by 0.4‰ suggested elevated $[CO_2]$ stimulated N_2-fixation. Evans et al. (2014) reported that over a decade at the same facility, ecosystem organic C and total N were significantly higher under elevated than ambient $[CO_2]$ due to organic C and total N in soils. Koyama et al. (2018) suggested that stimulated aboveground litter production in wet years might be a major mechanism driving greater SOC under elevated [CO_2] and that increased soil microbial necromass production might be another mechanism behind greater SOC.

3.2.5 Distribution and Role of Biocrusts in Stocks, Flows, and Ecosystem Processes

Arid rangelands are characterized by patchy vascular plant cover and up to 98% of the ground can comprise exposed soil and rocky terrain (Pietrasiak et al., 2013, 2014). However, these open spaces are not devoid of life: much of this area can be colonized by photosynthetically active microorganisms in association with heterotrophic microbes, largely as part of biological soil crusts (biocrust hereafter) which represent major players in the dryland carbon cycle (Belnap and Lange, 2003; Evans and Johansen, 1999; Pietrasiak et al., 2011a; Stovall et al., 2022; Weber et al., 2016). Biocrusts are formed when microbial communities and their byproducts combine with inorganic soil particles into a living aggregate found at the soil surface. They represent mini ecosystems in which cyanobacteria, eukaryotic algae, lichens, and bryophytes function as the primary producers supporting a biodiversity-rich microbial community of bacteria, archaea, free-living fungi, and microscopic animals (Belnap et al., 2016; Büdel et al., 2016; Omari et al., 2022; Pombupa et al., 2020).

Biocrust cover, biomass, and community composition in Southwestern USA landscapes vary considerably resulting in heterogeneous distributions and functioning across scales (Bowker et al., 2016). At the broadest scales biocrust abundance and composition vary by climate (Bowker et al., 2016; Rivera-Aguilar et al., 2006). Moisture availability and mode, seasonality, and air temperatures are major drivers. In general, cyanobacterial crusts dominate the plant interspaces in arid to hyperarid ecoregions while bryophytes and lichen crust abundances increase in semiarid ecoregions (Belnap et al., 2001, 2003; Bowker et al., 2016; Weber et al., 2016). Regionally, soil parent material is an important biocrust predictor. For example, in the Mojave and northern Sonoran Deserts granitic parent material had the greatest biocrust cover and richest community composition while dolomite and quartzite-derived parent material had low or no cover (Pietrasiak et al., 2011a, 2014). In the Colorado Plateau ecoregion, Bowker et al. (2005, 2006; Bowker and Belnap, 2008) found that crust variability exists between sedimentary parent materials with gypsiferousd-erived soils having highest biocrust species richness and cover followed by limestone and sandstone, while shale had the lowest richness and cover. Within a landscape dominated by a particular parent material, crust cover and composition vary with landform and its associated geomorphic dynamics and edaphic heterogeneity of physical (soil texture, gravel content, and surface cover of rocks) and chemical variables (pH, salt, and nutrient content) (Belnap et al., 2014; Bowker et al., 2005, 2006; Pietrasiak et al., 2011a, b, 2014; Williams et al., 2013). At the patch scale, microclimatic conditions, bioturbation frequency as well as aeolian processes are important for biocrust presence and composition (Bowker et al., 2005, 2006; Pietrasiak et al., 2011a, b, 2014; Williams et al., 2013). All these spatial and temporal factors in turn affect biocrust-mediated ecosystem functionality including C and N dynamics.

Biocrusts carry out a variety of ecosystem functions in arid rangelands, such as building soil organic matter, stabilizing the soil, redistributing rainfall runoff, facilitating water infiltration and retention, modulating seed germination, and providing food sources for associated microflora and fauna (Belnap et al., 2001, 2003, 2016; Evans and Johansen, 1999; Weber et al., 2016). Biocrusts are also responsible for 25% of all terrestrial N-fixation (Elbert et al., 2012; Weber et al., 2016) and an estimated 9% of the total dryland NPP (Sancho et al., 2016; Zhao et al., 2005).

They directly and indirectly contribute to the dryland C cycling via organic and inorganic processes. Despite mounting evidence that biocrusts are critical for maintaining environmental stability of arid rangelands, local and regional estimates of their contribution to rangeland C stocks and fluxes are sparse and have large uncertainties.

Biocrusts represent dynamic pools of SOC. Biocrusts exchange C most directly via photosynthesis and respiration (Figure 3.6). Biocrust lichens, bryophytes, eukaryotic algae, and cyanobacteria are all autotrophic microorganisms capable of performing oxygenic photosynthesis (Figure 3.6). As a result, carbon dioxide is converted to organic C which can build up over time as living biomass via organismal growth and can be stored in recalcitrant SOM if left undisturbed (Zhang et al., 2018). Further, photosynthetically produced organic C can be sequestered in exopolysaccharide material, excreted sugar polymers, that aid in crust aggregate formation, stabilization, and water retention (Figure 3.6; Mager and Thomas, 2011; Rossi et al., 2022).

Biocrust C fixation rates vary largely by geography, biocrust types, and time of desiccation recovery (Grote et al., 2010; Hoellrich et al., 2023; Pietrasiak et al., 2013; Tamm et al., 2018). Globally, C fixation rates of light cyanobacterial, dark cyanobacterial, lichen crusts, and moss crusts vary substantially from each other and reach values as high as 3.03, 5.00, 10.89, and 9.04 µmol CO_2 m^{-2} s^{-1}, respectively (Grote et al., 2010; Pietrasiak, et al., 2013; San José et al., 1991). Within a single site, rates may also vary between biocrust types (Hoellrich et al., 2023; Tamm et al., 2018). In Hoellrich et al. (2023), average net fixation rates in light cyanobacteria, lichen, and moss crusts after 24 h incubation were 1.51±1.11, 4.45±1.58, and -0.91±0.63 µmol CO_2 m^{-2} s^{-1}, respectively, where the underlying average C fixation rate for each crust type was 3.95±0.96, 8.01±0.84, and 6.92±0.34 µmol CO_2 m^{-2} s^{-1}. Additionally, between four sites within 100 km of one another, the maximum net fixation rates of *Peltula* lichen crusts varied greatly (1.35±0.6, –0.13±1.58, 2.05±1.24, and 4.45±1.58 µmol CO_2 m^{-2} s^{-1}, Hoellrich et al., 2023).

FIGURE 3.6 Key organic and inorganic processes influencing dryland carbon cycling, where EPS is exopolysaccharide material.

Heterotrophic microbial SR, on the other hand, represents a major contributor to soil C losses (Figure 3.6). In arid rangelands SR can occur at much higher rates compared to C uptake via photosynthesis, especially at elevated temperatures. For example, Cable et al. (2008) showed that at temperatures above 30°C photosynthetic rates declined while respiration rates continued to increase up to temperatures of 60–70°C (Cable et al., 2008). For positive net CO_2 uptake in biocrusts to occur, moisture availability, the ratio of photoautotrophic to heterotrophic microbial populations, and the period of recovery time these communities have after rehydration are important (Hoellrich et al., 2023). Biocrust microbes are poikilohydric, therefore they are only active when water is available and must undergo a recovery period once hydrated, before photosynthesis can reach peak capacity (Graham et al., 2006; Satoh et al., 2002; Wu et al., 2017). In Hoellrich et al. (2023), biocrust C fixation rates generally increased the longer samples were incubated before gas exchange measurements were performed. Furthermore, most samples did not achieve net positive C uptake until at least 6 h after the point of rehydration.

3.2.6 SOIL INORGANIC CARBON SEQUESTRATION

In arid rangelands inorganic calcium carbonate ($CaCO_3$) precipitates can form thick layers known as caliche below the soil surface at shallow depth (Figure 3.6, Schlesinger, 2017) (Photo Plate 3.1F–1G.). In general, CO_2 stemming from root and microbial respiration as well as entering the soil from the atmosphere can combine with water to form bicarbonate (Figure 3.6). Bicarbonate can then react with free calcium ions released by weathering to precipitate as $CaCO_3$ when the soil dries out and equilibrium is reached. These biogeochemical processes represent complex chemical interactions between C uptake during calcium carbonate dissolution, calcium ions and bicarbonate leaching to shallow subsurface depth, and $CaCO_3$ re-precipitation and CO_2 re-evolution with loss to the atmosphere as the soil dries out where the waterfront stops (Schlesinger, 2017).

There is potential for promoting SIC sequestration in the subsoil/deep soil of semiarid and arid rangelands through manipulation of Ca. Manipulations of soil fungi and bacteria can also be used to promote the mineralization of calcite and foster SIC sequestration (Monger et al., 2015). Unlike SOC that achieves equilibrium ("steady-state") with organic residue input, SIC can theoretically accumulate up to the soil profile depth as a petrocalcic horizon, which continues to accumulate until the soil is degraded by erosion processes or until a climatic change occurs (Monger, 2014). Monger and Martinez-Rios (2001) reported that USA grazing lands contain about 6–7% of the total carbonate-C in world soils relative to the global value of 930 Pg carbonate-C.

Biocrusts can also be involved in inorganic C pathways (Photo Plate 3.1H–1J). Yet, the role of biocrust in this process is less well understood. However, they represent hot zones of microbial activity where respiration rates are high when wet and are known to trap dust that could contribute calcium and carbonate-rich minerals to a landscape. This process of pedogenic/geological carbon sequestration operates at very low rates yet can lead to a large buildup of inorganic carbon in the long term, acting as a carbon sink in non-calcareous soils with rates estimated at 0.12–0.42 g C m^{-2} yr^{-1} (Schlesinger, 1985). Another often overlooked pathway of C sequestration is the formation of calcium oxalate. Biocrust microbes such as mycorrhizal and lichen fungi, cyanobacteria, and other bacteria can produce organic acids including oxalic acid, to acquire phosphorus and other nutrient ions including calcium via bioweathering (Allen et al., 1996; Briones et al., 1997; Gadd et al., 2014). Oxalic acid can react with free calcium ions to form calcium oxalate which then can get decomposed via bacterial oxalotrophy, finally resulting in caliche formation (Figure 3.6).

3.2.7 SOIL AGGREGATION

Carbonates contribute indirectly to the SOC accumulation and occlusion of SOC, as carbonate can increase the soil Ca^{2+} concentration, which increases soil aggregation and structural stability (Baldock and Skjemstad, 2000; Rowley et al., 2018). Carbonate ions are also capable of reprecipitation with Ca^{2+} forming secondary $CaCO_3$ crystals that cement aggregates (Fernandez-Ugalde et al., 2014), decreasing aggregate porosity and the accessibility of intra-microaggregate SOC to decomposers. Zuberer et al. (1996) reported that much of the C in New Mexico desert soils was in relatively unavailable forms. Fernandez-Ugalde et al. (2014) showed that carbonates had a positive effect on occluded SOC stocks. However, there is a lack of information about the amount and stability of occluded SOC, in soils that differ in texture, mineralogy, and organic inputs (Rowley et al., 2018).

Soil aggregation is also driven by biological mechanisms and dryland plants and microbes can indirectly contribute to carbon cycling due to their essential role in soil aggregate stability. In biological soil aggregation, fine roots, fungal hyphae, filamentous bacteria, or the excretion of labile extracellular polysaccharides/polymeric substances by microorganisms and roots bind soil particles together and occlude SOC residue within the aggregate (Six et al., 2004). Glomalin which is a fungal-derived material has also been found in interspaces, although in lower concentrations than under plants (Bird et al., 2002). Herrick et al. (2001) observed a sequential increase in the glomalin concentration from interspace to black grama to mesquite sites and a positive correlation of glomalin with aggregate stability, soil C:N ratios, and carbonate C. They estimated the proportion of total organic C in the top 10 cm represented by glomalin, ranged from 7% in the interspaces, 10% under black grama, and 12% under the mesquite.

Biocrust organisms exhibit multiple traits that promote aggregate formation and soil stability. Cyanobacteria, eukaryotic algae, and free-living fungi excrete sticky exopolysaccharides that bind inorganic particles together (Cania et al., 2020; Mahapatra and Banerjee, 2013; Rossi and De Philippis, 2015,). Aggregation is enhanced by the filamentous growth forms of many biocrust microbes weaving through the soil forming a web of sticky threads (Fick et al., 2019; Rossi and De Philippis, 2015). Cyanobacteria are visible as hanging threads of soil-on-soil surface fragments (Belnap and Gillette, 1998). Bryophytes and lichens further strengthen soil stability through their root-like structures (Lan et al., 2012). Keeping the dryland topsoil in place sustains and maintains SOC in the plant interspace (Belnap and Gillette, 1998). The aggregated topsoil further protects the underlying subsurface soil from erosion as well as shielding the inorganic carbonates from weathering.

Estimation of aggregate stability using the "soil stability kit" (Herrick et al., 2001) is used for rangeland assessment and monitoring (Karl et al., 2003), Rangeland National Resources Inventory (Spaeth et al., 2003), and state-and-transition models (Herrick et al., 2001). This low-cost method meets the criteria of repeatability and is positively related to soil properties that are associated with grassland persistence, including resistance to erosion, water infiltration rates, and microbial activity (Bestelmeyer et al., 2006; Herrick et al., 2001). Overall, results have shown higher soil stability under vegetation compared to within plant interspaces at the soil surface and subsurface in sandy, semiarid rangelands (Bestelmeyer et al., 2006; Bird et al., 2002, 2007; Herrick et al., 2001). As continuous grass cover is fragmented, exposure of the soil surface to raindrop impact, wind, and overland flow across interconnected bare areas results in erosion and decreased soil aggregate stability (Bestelmeyer et al., 2006).

3.3 MODELING THE CARBON CYCLE FOR ARID RANGELANDS

Mathematical modeling provides a tool for improving our understanding of the many processes and their interactions making up an ecosystem such as arid rangelands. Arid rangelands are characterized by high spatial heterogeneity and fragmentation, often varying across scales ranging from sub-meter to thousands of hectares, which affect soil C distribution (Bestelmeyer et al., 2006; Bird et al., 2002; Brown et al., 2010). Models are developed for a wide range of scales from small components such as biocrusts to full plant animal systems to regional and global carbon dynamics.

One approach to address the issue of scale in arid rangelands is the use of modeling units of the ecotypes integrated framework proposed by Herrick et al. (2006). Each ecological site is defined based on physiographic features, climate, plant communities, animal use, and hydrology. Moreover, based on the relationship between NPP and soil C in rangelands (Parton et al., 2000), the integration of remote sensed biophysical estimations (NPP, NDVI, and LAI) with process-based models could significantly improve temporal and spatial representation. Although this concept has been applied to corn and soybeans (Bandaru et al., 2022), application to arid rangeland environments remains challenging. In arid rangelands the use of vegetation indices is challenging because the vegetation is not sufficiently "green."

Although models exist across all scales, few have been developed and calibrated specifically for arid rangeland systems. For exploring the C dynamics of the Chihuahuan Desert ecosystem, cattle and their interaction with the soil and plant communities are important components and processes that must be considered. Cattle on rangeland form a natural cycle where C is taken from the atmosphere, cycled through the soil/plant/animal system, and returned to the atmosphere as CO_2 (Figure 3.7). Photosynthesis uses sunlight, CO_2, and water to form carbohydrates in the plant, which are in various forms including cellulose, hemicellulose, starch, pectin, sugars, and polysaccharides. The concentrations of these plant components vary by species and as the plant matures, affecting digestibility of the plant material or quality of the feed for the animal.

3.3.1 PLANT PROCESSES

Many models have been developed to represent the growth and development of forage crops. Some are specific to a plant species while most are more general for modeling a forage type such as warm or cool season grasses, legumes, or forbs. These types may also be combined to represent a mixed sward (Corson et al., 2007). Photosynthesis is modeled as a function of solar radiation, shading within the plant canopy, [CO_2], ambient temperature, and temperature, moisture, or N stresses on the plant.

In a review of agro-ecosystem models, Brilli et al. (2017) pointed out limitations for grassland in representing grazing management, plant use, nutrient return from grazing animals and animal performances (weight growth, milk production, and calving). A major limitation in simulating rangelands is that most models do not consider monthly and year-to-year change in plant species composition such as the shift from grass to shrub and C3 to C4 grasses. These shifts affect nutrient dynamics, biomass structure, water utilization, seasonality, and biomass allocation (Parton et al., 1994, 1993). This is an important point for arid environments where year-to-year changes resulting from drought are common and long-term shifts resulting from grazing management and climate change may cause shrub expansion (Petrie et al., 2015). State-and-transition models (Bestelmeyer et al., 2017) focus on describing the change in species composition and structure of ecosystems as a function of climate and land use, but current agro-ecosystem models do not include these relationships (Parton et al., 2000).

Few forage-livestock production models have been developed specifically for U.S. rangelands (Ma et al., 2019). An older model called Simulation of Production and Utilization of Rangelands (SPUR) was found to effectively represent rangeland production systems (Carlson and Thurow, 1996). Theory and functions from this model have been used in other models such as the Great Plains Framework for Agricultural Resource Management (GPFARM; McMaster et al., 2002) and the Integrated Farm System Model (IFSM; Corson et al., 2006), but the SPUR model is no longer active.

Forage-livestock models have been satisfactorily applied to rangeland conditions, but extensive calibration, evaluation, and application have not been conducted in the USA rangelands (Descheemaeker et al., 2018; Moore et al., 2014; Rotz et al., 2019; Zilverberg et al., 2017). A potential weakness in applying this type of model to rangelands is the difficulty to represent moisture,

N availability, and stresses on plants. With proper calibration, rangeland productivity and forage availability for animal feed can be represented, but there is opportunity for developing models that better represent plant productivity and feed quality for the semiarid rangelands (Ma et al., 2019). Such models, like the Australian Grass Production (GRASP) model and its spatial implementation, AussieGRASS, have demonstrated the utility of forage-livestock models for understanding climate-management interactions on semi-arid rangelands and short-term forecasting (e.g., Carter et al., 2000; McKeon et al., 2009; Webb et al., 2012a).

3.3.2 Animal Processes

Within the C cycle, cattle consume C stored as plant carbohydrates where the amount consumed and the digestibility influence animal growth and performance. During digestion, a substantial portion of the C is transformed to CO_2 with some transformed to CH_4 and other volatile organic compounds (VOCs). These gases are returned to the atmosphere, primarily through respiration and eructation (Figure 3.7). Some C is maintained or used by the animal in growth, gestation, and milk production. The remainder is excreted in feces and urine.

Cattle models are well developed to represent feed intake and performance for a wide range of diets (NAS, 2016). Ruminant models normally represent cattle fed with a balanced diet to meet their energy, protein, and mineral needs. Cattle on rangeland are not receiving this type of diet, so many current models do not represent these conditions well. Current models can be calibrated to represent rangeland conditions, but ruminant models more specific to the forage quality and availability and influence of cattle genetics are needed.

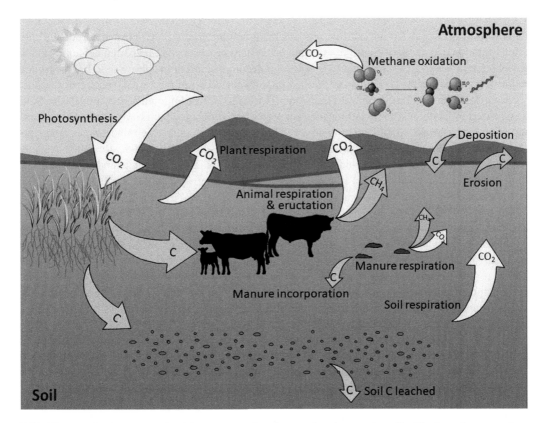

FIGURE 3.7 Cattle on rangeland form a natural carbon cycle where carbon dioxide from the atmosphere cycles through the soil/plant/animal system back to carbon dioxide in the atmosphere.

Cattle models may predict CO_2 and CH_4 emissions, normally as functions of total feed intake, energy intake, fiber concentration in the diet and fiber digestibility (Ellis et al., 2010). Most of these models have been developed from cattle-fed balanced diets to maximize production. To predict CH_4 emission from cattle on the range, it is important to use a model that is influenced by fiber content and digestibility of the diet (Rotz, 2018). Gathering accurate CH_4 emission data for model development and evaluation from cattle on rangeland is challenging which has inhibited the development of emission models specific to rangeland cattle. Little work has been done to measure or model non-methane VOCs from cattle in any environment. Carbon emitted in this form is considered small with minimal effect on the C cycle.

After feces and urine are excreted, organisms in the manure continue to breakdown C molecules and respire CO_2 along with minor amounts of CH_4. For grazing cattle, CH_4 emitted from manure is normally ignored (Chianese et al., 2009). For cattle maintained in a barn or other housing facility, manure CH_4 emissions can be substantial, particularly if that manure is stored for long periods under anaerobic conditions (Rotz, 2018). Over time the manure C excreted by grazing cattle that is not volatilized infiltrates or becomes incorporated in the soil.

The carbon in the form of CO_2 and $-CH_4$ emitted from cattle and their manure returns to the atmosphere. Since the CO_2 is returning to its original source, it has no net effect on the atmosphere. Compared to CO_2, CH_4 is a short-lived gas in the atmosphere. Most CH_4 oxidizes in the atmosphere forming CO_2 and water (Wuebbles and Hayhoe, 2002). The half-life of emitted CH_4 is about 8 years; although this CH_4 does not have a long-term effect on the atmosphere, it is more efficient in trapping radiation than CO_2 and thus has a very large short-term warming effect.

3.3.3 Soil Components and Processes

Carbon from decomposing animal manure and plant residue forms SOC when incorporated into soil. Formation and decay of SOC are essential ecosystem processes that help regulate atmospheric greenhouse gasses of CO_2, N_2O, and CH_4. Through decomposition of the above- and below-ground residues, about half of the C originating through photosynthesis is returned as CO_2 back to the atmosphere through the respiration of soil organisms (Horwath, 2015). Following decomposition, SOC is stabilized in labile, resistant, and recalcitrant forms based on age and turnover rates of nutrients and microbial biomass. Stabilization is driven, among other factors, by the molecular components of the inputs (lignin to cellulosic ratio and C to N ratio), and soil texture which affects the capacity to preserve SOC in stable organomineral complexes (Horwath, 2015). In rangelands, most SOC is from decomposition of dead roots in a recalcitrant form with a slow turnover rate (Follett, 2000; Parton et al., 2000).

The most widely used ecosystem models share a soil organic C cycle model where SOC is compartmentalized in 3–7 conceptual pools. The division among pools is based on stabilization mechanisms, bioavailability, and biochemical and kinetic parameters (Brilli et al., 2017; Parton et al., 2015). Most models have at least three main soil C pools: active (live soil microbes), slow (resistant plant materials), and passive (soil-stabilized plant and microbial material very resistant to decomposition). Some models include plant litter pools that are subdivided between surface and below ground pools. In the most detailed models, these pools are subdivided into structural (lignin, cellulose, and hemi-cellulose) and metabolic (protein, lipids, starches, and nucleic acid) pools (Izaurralde et al., 2006; Li et al., 1994; Parton et al., 2015). Transformations among pools, SOC formation and decomposition, together with N transformations (denitrification, nitrification, and immobilization) are modeled using kinetic reactions influenced by litter/humus molecular composition, inorganic and organic N availability, soil properties (e.g., texture, bulk density), temperature, and water content (Izaurralde et al., 2006; Parton et al., 2015).

Most soil models were developed primarily for croplands and secondly for improved pasturelands with little application to rangelands. The Century and DayCent models have been applied to arid and semiarid conditions including grasslands worldwide (Parton et al., 1993), in western north America (West et al., 1994), semiarid Sudan (Ardö and Olsson, 2003), Central Chile (Stolpe et al., 2008), southwestern USA (Brown et al., 2010), and semiarid Chaco, Argentina (Baldassini and Paruelo, 2020). The EPIC model has been applied in Arizona and New Mexico (De Steiguer, 2008) and the semiarid region of Central East Kazakhstan (Causarano et al., 2011). The authors identified model limitations for these conditions in representing spatial and temporal variation of SOC with more data needed (SOC and flux tower measurements) to improve model parametrization, calibration, and evaluation at point and spatial scales. Other identified concerns were for information on the interactions of temperature, moisture, and substrate quality on C decomposing organisms, N cycling, and salinization (West et al., 1994); land use history due to long residence of some SOC pools (Ardö and Olsson, 2003); and non-equivalency between modeled and measured values that can occur when theoretical pools of models are not easily quantifiable in the laboratory. Parton et al. (2015) noted that the lack of direct linking of conceptual pools of SOM with measurable fractions is a limitation across current soil C models. With a new modeling approach, the MEMS model simulates two commonly measured SOC pools (particulate and mineral-associated organic matter; Zhang et al., 2021).

3.3.4 OTHER ASPECTS OF THE C CYCLE OF SEMIARID AND ARID ECOSYSTEMS

Soil C dynamics play out across a range of temporal and spatial scales in arid rangelands, and many of these processes are not addressed in current ecosystem models due to lack of process-level understanding or lack of data to develop and parameterize models across scales of time and space. In arid conditions, a substantial portion of organic matter is decomposed by abiotic processes (Horwath, 2015). These processes include drying/wetting cycles with a short-term increase in soil CO_2 effluxes after wetting and plant litter physical degradation (photodegradation) by ultraviolet radiation especially in areas exposed because of animal grazing (Adair et al., 2008). The abiotic factors and their relationship with litter inputs must be considered as upscaling factors in C decomposition models to improve modeling of arid grasslands (Hosseiniaghdam, 2021). Barnes et al. (2015) has proposed a conceptual model of dryland decomposition by ultraviolet radiation.

A missing component in modeling soil-plant-livestock systems in hot deserts is the representation of the biocrust. Organisms making up the biocrust contribute to maintaining soil fertility by fixing C and N which is mostly leaked to surrounding soils (Belnap, 2003). These processes have not been modeled, but West et al. (1994) presented a conceptual model that included the surface crust in the soil C cycle.

Dryland soils contain as much or more SIC, in the form of soil carbonates, as SOC (Lal, 2004). Although SIC is considered to develop over centuries, remaining unchanged by rangeland management practices, Wang et al. (2016) in the Chihuahuan Desert of New Mexico found an increase in SIC over the last 150 years as native C4 grasses have been replaced with C3 desert shrubs. The Soil-Landscape Inorganic Carbon (SLIC) model was developed with field data from the Mojave Desert to represent those processes (Hirmas et al., 2010), but this component has not been integrated into production or larger scale models.

The soil fauna (ants and termites) are important contributors to soil turnover in arid and semiarid environments. In addition to physical disturbance, they frequently change the chemical composition of the soil (Whitford, 2010). In northern Chihuahuan Desert grasslands, Filser et al. (2016) quantified the annual soil moved by ants to range from 21.3 to 85.8 kg ha^{-1}. Grandy et al. (2016) and Fry et al. (2019) proposed the inclusion of soil fauna in biogeochemical models to improve the representation of the C cycle.

New approaches based on data science could provide other soil C estimations. Wang et al. (2018) presented a promising model to predict SOC stocks using machine learning algorithms

(MLA) with topographic, climate, and remotely sensed data in semiarid rangelands of eastern Australia. Artificial intelligence (AI)-based tools and MLA have the potential to produce models based on "big" spatiotemporal data sets to fill temporal and spatial agricultural information gaps (Bestelmeyer et al., 2020).

Wind erosion is currently represented in some models using empirical and process-based approaches. Within soil C models, only EPIC and APEX simulate wind erosion. Modeling these processes is limited in that erodibility and climate factors remain constant during the year, so the relationship between field erosion and individual windstorms is not considered (Jarrah et al., 2020). Edwards et al. (2022) developed a more process-based Aeolian EROsion (AERO) model for rangeland conditions at the plot scale (100 m × 100 m). They found the model to effectively represent temporal variability in aeolian transport rates for rangeland and provide robust assessments of land health, changes in air quality and the impacts of land management activities.

3.4 ROLE AND IMPACT OF SOIL REDISTRIBUTION IN CARBON CYCLING

During the 1930s, the USA Great Plains region witnessed many tens of (and perhaps up to a hundred in one year) "black blizzards" in the Dust Bowl era (Figure 3.8). Massive dust storms in the region carried vast amounts of topsoil containing C and nutrients thousands of kilometers to the East Coast and out into the Atlantic Ocean (Lee and Gill, 2015). The devastating vegetation removal and accelerated soil erosion ushered in a new era of soil conservation (Bennett and Chapline, 1928). Only 6 years after the USA Dust Bowl, Jenny's (Jenny, 1941) influential work on soil formation was published but did not include soil erosion as a soil forming factor. The geography of soil and its distribution of organic C in the landscape was understood through Dokuchaev's factors influencing soil formation $S = \text{f}(cl, o, r, p, t, …)$, where cl is the climate, o represents organisms, r is the relief, p is the parent material, and t is the time, leaving open the inclusion of other factors (state variables). The work enabled an important simplification of soil which influenced C cycling and the omission of soil erosion in the early C cycling models (e.g., RothC; Chappell et al., 2016). The omission of soil erosion is explained implicitly throughout Jenny's book, where erosion is repeatedly associated with the removal and destruction of soil. Paradoxically, soil erosion and its complementary deposition, and therefore the spatiotemporal redistribution of soil, play a central role in the catena concept, soil geomorphology (Gerrard, 1993) and the formation of soil and amount and type of soil organic C.

Soil erosion influences the C cycle by affecting (a) lateral fluxes of C between different soils by sediment transport in wind and water, and to fluvial systems and the Neritic zone of continental shelves (Raymond and Bauer, 2001); (b) vertical fluxes of C dust emission to the atmosphere (Chappell et al., 2014, 2019; Webb et al., 2012b) and C transport from terrestrial to aquatic and marine environments (Jickells et al., 2005); and (c) the balance between release of greenhouse gases (e.g., CO_2) that influence global warming and C stored in soils (Van Oost et al., 2007). These interactions vary across landscapes with soil and site potential (i.e., productivity) and the attenuating effects of vegetation on soil erosion (Brown et al., 2010). The interactions are then further moderated by land management, disturbances, and the ecological state of rangeland systems, with plant and soil microbial community responses influencing spatial and temporal patterns of above- and below-ground net primary production (e.g., Throop and Lajtha, 2018), rates of C storage and release (Brown and McLeod, 2011; Svejcar et al., 2008), and the magnitude and frequency of soil erosion and deposition (Webb et al., 2014; Williams et al., 2016).

Both wind and water erosion are important agents of soil, nutrient, and C redistribution in dryland ecosystems (Berhe et al., 2018). Globally, most research into the interactions between soil erosion and the C cycle has focused on the role of water erosion (Lal, 2003). That research has included extensive debate about whether water erosion is a net C source or sink – described as the "soil C erosion paradox" – and it is apparent that whether erosion results in a net loss or accumulation of soil C depends on the spatial and temporal scales at which the interactions are studied

FIGURE 3.8 A "black blizzard" approaching Rolla, KS on 6 May 1935 (Image from the FDR Digital Archives).

(Sanderman and Berhe, 2017; Van Oost and Six, 2022). Remarkably little is known about wind erosion of soil organic and inorganic C and how the resultant "C dust emission" influences C cycling in arid rangelands and downwind ecosystems (Webb et al., 2012b). Aeolian (wind-driven) soil erosion and dust emission occur extensively across the southwest USA (Hennen et al., 2022) and are important positive feedback promoting ecosystem changes (e.g., grassland to shrubland transitions) that have been regarded as desertification (Bestelmeyer et al., 2018; Ravi et al., 2010). In a study of erosion processes across different southwest USA ecosystems, Breshears et al. (2003) projected that wind erosion rates at desert shrubland sites could be c. 33 times larger than water erosion rates. Given the potential importance of wind erosion for C cycling, and paucity of information, we focus here on what is known about the interactions and identify key knowledge gaps.

3.4.1 Lateral Soil Organic Carbon Erosion by Wind

Lateral redistribution of soil organic C by wind occurs naturally where protection of the soil surface by vegetation is sparse (Gillette, 1999). However, when vegetation is abruptly removed, for example, due to wildfire and clearing for cultivation, or changes with invasive plants and/or loss of endemic species, C erosion by wind is typically accelerated. For example, Li et al. (2007) measured 3–5 times increase in aeolian sediment transport and up to 25% loss of TOC and TN from soils over 3 years (with 60% of the losses occurring in the first year) in grass removal treatments in a Chihuahuan Desert ecosystem. The majority of sediment is transported laterally in a process known as saltation, in which typically aggregated particles and sand grains hop repeatedly across the surface (Figure 3.9). This process occurs intermittently in localized occurrences with large frequency and small magnitude events, or more regionally with small frequency and large magnitude sand or dust storms (Lee and Tchakerian, 1995). The SOC entrained and transported in these lateral sediment fluxes has been found to be much smaller than in vertical sediment fluxes (i.e., dust emission) because the large saltating particles have a smaller SOC content than finer silt and clay-sized dust (Boon et al., 1998; Li et al., 2007, 2009; Sterk et al., 1996; Webb et al., 2013).

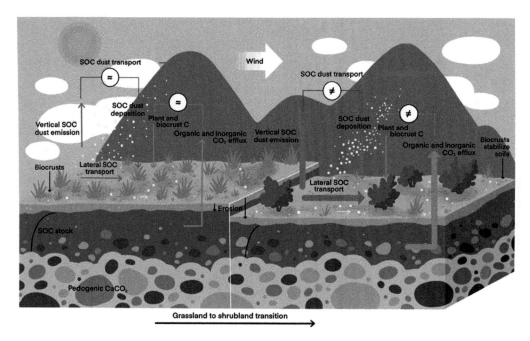

FIGURE 3.9 Illustration of the effects of wind erosion on carbon cycling across a grassland to shrubland transition.

Sediment, nutrients, and C transported in saltation may move a few meters or tens of meters during strong wind events. Plant shape influences wind speeds at the soil surface and the ability of plants to trap sediment and C moving in saltation (e.g., Liu et al., 2021). Plants with canopies close to the soil surface are typically most effective in trapping material and increasing the spatial heterogeneity of C by forming localized "islands of fertility" within depleted soils (Okin et al., 2006). In addition, wind may redistribute leaf litter before it decomposes, further promoting heterogeneity of soil C at the plant-interspace scale (Throop and Belnap, 2019). Conversely, fire has been found to promote redistribution of soil C and N from shrub microsites to nutrient-depleted sites, reducing their spatial heterogeneity (Wang et al., 2019). Due to the generally short sediment transport distances of saltating grains, lateral C erosion by wind tends to mostly result in local C redistribution, rather than net ecosystem losses, on time scales of seasons to *c.* 100 years (Petrie et al., 2015). However, it has not been established how lateral C erosion processes vary among different soils and vegetation communities, with most research having been conducted in small scale (*c.* 1 ha) experiments on sandy soils with honey mesquite (*Prosopis glandulosa* Torr.) dominated plant communities (e.g., Li et al., 2007, 2009; Wang et al., 2019). It has also not been established whether SOC stocks in dryland systems stabilize following erosion and ecological state transitions, or if soil erosion results in a net long-term decline in soil C in some ecosystem states (e.g., shrub-dominated plant communities with large unvegetated gaps that promote wind and water erosion).

3.4.2 Vertical Emission of Soil Organic Carbon

Wind-eroded, saltating sediments bombard the soil surface causing abrasion that selectively removes fine silt and clay-sized particles – called dust – and lofts them vertically into the airstream (Shao, 2004). Dust may be suspended in the atmosphere and, depending on its size and the wind speed, can travel many tens to hundreds of kilometers before being deposited (Shao et al., 2011). Soil carbon associated with dust may, therefore, be lost from sites experiencing significant dust emission (Photo Plate 3.1K). In fact, because soil organic C is typically associated with the silt and clay soil particle size fractions (Du et al., 2021), C dust emission is the main mechanism through which net wind-driven losses of soil C occur in arid rangelands (Chappell et al., 2014). At the continental scale, C dust emission can directly reduce the terrestrial SOC stock (Chappell et al., 2014), with wet and dry deposition of eroded material potentially increasing the marine SOC stock (Jickells et al., 2005) reaching beyond the Neritic zone of continental shelves, which typically restrict the dispersion of water-borne SOC (Raymond and Bauer, 2001).

Research in the southwest USA, and elsewhere around the world, has demonstrated significant variability in the eroded SOC content of dust (Webb et al., 2012b). It has been established that this variability results from the spatial and temporal variability of SOC content in different soils and in the efficiency of dust emission by abrasion during saltation, which is also a function of soil texture (Webb et al., 2013). Wind-blown sediments often show a pattern of increasing SOC content with increasing height, which is expressed as an enrichment ratio of dust C content to soil C content in the labile pool (Chappell et al., 2014). Measured SOC contents of dust have had enrichment factors ranging from 1 to >7, with greater enrichment associated with emissions from sandy soils that have smaller SOC contents than loamy and clayey soils but greater abrasion and dust emission efficiency (Webb et al., 2013). Chappell et al. (2014) demonstrated that C enrichment could be estimated from the ratio of fine mineral particles in dust to parent soils, enabling gross C dust emissions to be modeled at the continental scale. Recently, Du et al. (2021) showed that C dust emission modeling could be simplified with knowledge of how soil C is distributed across soil particle size classes. However, little information is available on C distribution among dry soil aggregate sizes that would enable local to regional quantification of C dust emission from southwest USA ecosystems. Available estimates of wind-driven C erosion in the US produced by Chappell et al. (2019) have excluded dust emission to quantify the contribution of saltation to gross soil C erosion (Figure 3.10). Experimental results of Li et al. (2007) suggest that C enrichment of dust is likely to vary following vegetation change and subsequent changes in surface soil texture and labile C within plant interspaces due to wind erosion. It is also likely that temporal variability in C dust emission arises from variability in the abrasion efficiency of saltation due to differences in wind erosivity among erosion events that affects dust particle size distributions (Webb et al., 2021).

As wind erosion acts at the soil surface, the immediate effect of C dust emission is to reduce the labile soil organic C pool, thus reducing the input to stable soil C pools. Severe and/or long-term soil redistribution by saltation and removal by dust emission may expose deeper soil horizons, eroding the more stable soil organic and inorganic C pools and accelerating CO_2 efflux through mineralization (Webb et al., 2012b). However, to our knowledge no research has directly addressed these interactions with wind erosion. Furthermore, while it is well established that mineral dust may be transported and deposited long distances from source soils, the fate of eroded C remains largely undescribed. Webb et al. (2012b) suggested a portion could be mineralized and released to the atmosphere through photo-degradation during transport (i.e., C source), while C dust deposited in more humid terrestrial environments and aquatic and marine systems is more likely to be a C sink. Addressing these knowledge gaps will be critical for establishing the net effects of wind erosion on dryland ecosystems. What is currently clear is that mitigating soil erosion will be necessary to avoid net soil organic and inorganic C losses.

3.4.3 Implications of Omitted Soil Organic Carbon Erosion for Carbon Cycling

The omission of SOC erosion in C cycling models is expected to be up to 0.3–1.0 Pg C y^{-1} indicating an uncertainty of –18% to –27% globally and +35% to –82% regionally relative to the long-term (2000–2010) terrestrial C flux of several land surface models (LSMs; Chappell et al., 2016). Depending on how the SOC is eroded initially, the SOC may be transported (a) only small distances (field scale) and become buried reducing decomposition and CO_2 emission; and (b) over large distances (continents), it may be trapped in vegetation and deposited in marine environments where

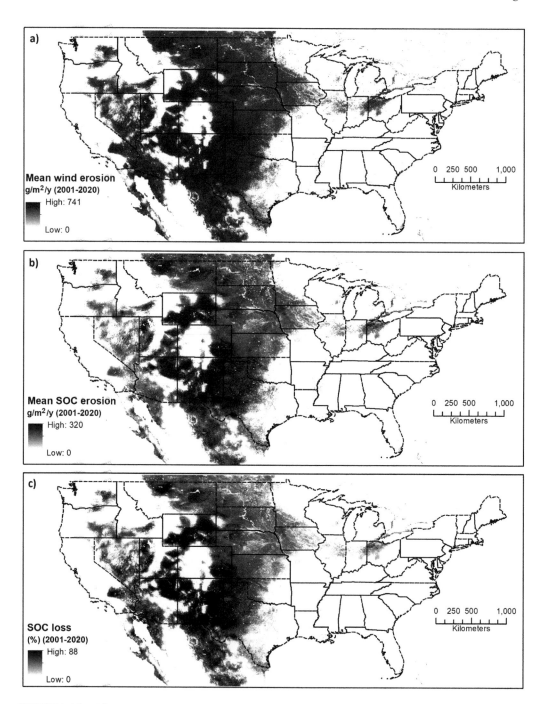

FIGURE 3.10 Wind erosion, mean SOC erosion, and gross SOC loss (2001–2020) (after Chappell et al., 2019).

it may become a net sink. The debate about whether soil erosion is a net source or sink of C (van Oost et al., 2007) has focused on humid, temperate environments where water erosion dominates and arguably is a small contributor where decomposition dominates the reduction in SOC stock. In arid rangelands, where SOC decomposition is restricted by the absence of moisture, wind erosion and dust emission could be the dominant control on reduced SOC stock.

Distinctive Dryland Soil Carbon Transformations

Here we introduce a ratio (NEEd) of the proportion of the long-term net ecosystem exchange (NEE=NPP–SR) that is diverted (d) away from the atmosphere by erosion (Ce) to provide NEEd = Ce/(NPP–SR)s. Using the information presented in the previous sections, the NPP (Figure 3.2) and the SR produce the NEE (Figure 3.3) which for drylands produces NEE=NPP-SR = 50–86.5 = –36.5 g C m^{-2} y^{-1}. With SOC erosion (Ce) by wind at around 10 g C m^{-2} y^{-1} the NEEd = Ce / NEE = 10/–36.5 = –0.36 which means that around 36% of the NEE is redistributed in the landscape and diverted away from its direct path to atmospheric CO$_2$. For comparison, more humid agricultural systems have much larger NEE = 475–618 = –143.25 g C m^{-2} y^{-1} and the SOC erosion by wind might be smaller than that by water up to 30 g C m^{-2} y^{-1} (van Oost et al., 2007) but which results in NEEd = 30 / –143.25 = –0.21 g C m^{-2} y^{-1}. The proportion of SOC lost from the exchange is around 20%. This NEEd of SOC erosion by water in more humid systems is smaller than the NEEd of SOC erosion by wind in arid rangelands, indicating that wind erosion is a more efficient diversion of SOC away from direct contribution to atmospheric CO$_2$. We extended this NEEd ratio across North America using the available information provided in the previous section to produce the NEEd of SOC erosion by wind (Figure 3.11). There are three regional "hotspots" with the largest SOC erosion by wind in Wyoming, much of New Mexico, and southern Colorado. Significantly, a large proportion of the SOC-eroded area diverts more than 5% of the NEE away from the atmosphere.

The mainly agricultural western Great Plains and the eastern North American Deserts have much smaller NEEd than those central regions where most SOC occurs, and wind erosion occurs. These results indicate that erosion by wind is an important determinant in C cycling and without its inclusion, models are very likely to overestimate the amount of soil organic matter, which is decomposing, and the amount of SOC directed to the atmosphere (Chappell et al., 2016). These results raise the issue that the debate about C erosion being a source or sink is unlikely to be reconciled without the consideration of C erosion by wind and C dust emission.

In developing approaches to mitigate and adapt to changing controlling factors, we must account for changes in SOC erosion. Sanderman and Chappell (2013) showed that if modest levels of soil erosion by water (1 t ha^{-1} y^{-1}) are excluded from C accounting, the combined uncertainty in C sequestration rates (0.3–1.0 t CO$_2$ ha^{-1} yr^{-1}) is similar to expected C sequestration rates for many management options. Therefore, including SOC redistribution by wind and water is critical to the success of C monitoring or trading schemes. Including soil erosion by water and wind in large-scale

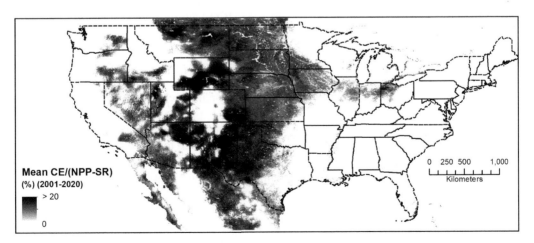

FIGURE 3.11 Long-term (2001–2020) mean soil organic carbon (SOC) erosion (Ce) ratio (%) with the net ecosystem exchange (NEE): Ce/NEE where NEE=NPP-SR is the net primary productivity and soil respiration, respectively.

modeling should reduce uncertainty in SOC flux estimates with implications for CO_2 emissions, mitigation and adaptation strategies, and interpretations of trends and variability in global SOC cycling.

3.5 ADAPTATION AND MITIGATION OPTIONS

3.5.1 PLANT COMMUNITY COMPOSITION – ABIOTIC AND BIOTIC DRIVERS

Changes in species abundance can have a detrimental effect on the stability of ecosystem functions and processes especially if it involves "keystone" species (Power et al., 1996). Biodiversity is an important component in ecosystem stability due to the multiple pathways that individual species can utilize resources or offering of alternative energy pathways through trophic interaction (Willig and Presley, 2018). The buffering of the ecosystem to disturbance with high biodiversity is due to having similar functional species within the community composition across different gradients of disturbance (Chapin et al., 1997). However, if different functional species populate on different gradients of the disturbance, the ecosystem is likely vulnerable to changes (Chapin et al., 1997). Hence, the importance of management practices that would maintain biodiversity is recommended for the system's resilience. However, the current and future climatic trends are hypothesized to create a new disturbance regime especially in arid and semiarid regions pushing the ecosystem away from its coping range creating a future of high uncertainty (Briske et al., 2005; Willig and Presley, 2018).

Climate change trends caused by an increase in atmospheric $[CO_2]$, with subsequent increase in temperature and variability in precipitation leaning toward a reduction, could have profound effects on the arid southwest US (Briske et al., 2005; Polley et al., 2013). The notable influence of the increase in $[CO_2]$ is the overall increase in NPP that would potentially affect SOC formation through litter or BNPP. However, the change $[CO_2]$ could disrupt species composition at a site. For instance, difference in photosynthetic pathways and the carbon dioxide saturation concentration between C3 (mostly woody species and broad-leaf species (forbs)) and C4 (mostly include grass species) plants, would ultimately lead to dominance of C3 plants from the increased $[CO_2]$ (Hatfield et al., 2011; Izaurralde et al., 2011; Morgan et al., 2007; Polley et al., 2013). Another projected change due to elevated $[CO_2]$ is the increase in plant water use efficiency due to the decrease in stomatal conductance (Polley et al., 2013; Wand et al., 1999). The improved moisture regime due to the increase in $[CO_2]$ might in turn affect the soil biota community that are responsible for nutrient cycling. However, in arid rangelands, the subsequent shift in warming and altered precipitation patterns might negate the overall effect of the increase in $[CO_2]$ due to the increase in vapor pressure deficit that might lead to soil drying through increased evapotranspiration rate (McKeon et al., 2009; Polley et al., 2013). Additionally, the indirect effect of shrub encroachment is an overall reduction in vegetation cover, hence increasing patchiness in distribution of the vegetation that might increase the heterogeneity of soil nutrients. The disruption of connectivity on the landscape increases vulnerability of erosion on interspace areas, hence creating a self-reinforcement for the shrub dominance within the system (Briske et al., 2005; Okin et al., 2015).

The changes that have pushed ecosystems into new stability or instability with their subsequent indirect effects altering community composition are referred to as ecological transformation (Crausbay et al., 2022). These ecological transformations set new precedence for land managers with ecosystems so profoundly altered in community compositions that management actions alone are not adequate to reverse the ecosystems back to historical conditions. In turn, natural resource managers have developed the resist-accept-direct (RAD) framework, a paradigm to address this nonstationary assumption (Millar et al., 2007; Schuurman et al., 2022). Under this framework, managers can resist the trajectory of the ecological transformation by actively intervening in internal and external drivers in hopes of restoring ecosystem community, structure, or functions to acceptable conditions (Millar et al., 2007; Schuurman et al., 2022). However, such intervention might carry a large financial burden. Under the acceptance response, managers allow changes to proceed autonomously. However, due to the high uncertainty of the alternative state, managers face substantial risk of such response (Crausbay et al., 2022; Schuurman et al., 2022). The third option entails intervention in the trajectory of the ecosystem in hopes of steering toward a more stable desired outcome that has maintained ecosystem functionality (Schuurman et al., 2022). Forecasting future ecosystem changes in arid and semiarid rangelands is challenging due to complex abiotic and biotic factors that must simultaneously be accounted for. However, management approaches that adopt a holistic framework including goals of maintaining or improving biodiversity and moderating land use intensity with science-based approaches might improve ecosystem services derived from maintaining C sequestration. To fully address these factors, management must address multiple objectives, under an adaptive management model that puts emphasis on monitoring.

3.5.2 Rangeland Forage Quality Interaction with Grazer-Derived CH_4

Reductions in perennial grass and forb production have strong implications for limiting carrying capacity of grazing ungulates (either wildlife or livestock; McIntosh et al., 2019). Emerging evidence suggests that increased atmospheric $[CO_2]$ could reduce nutrient quality of perennial rangeland grasses (Augustine et al., 2018; Craine et al., 2017). In a seven-year study, Augustine et al. (2018) elevated ambient temperature and atmospheric CO_2 levels to 600 ppm; they measured *in vitro* dry matter digestibility (IVDMD) of three dominant shortgrass prairie species and found that warming and CO_2 together yielded variable results, but that CO_2 independently increased forage production by 38% but simultaneously decreased digestibility and N. Craine et al. (2017) evaluated diet quality of cattle for 10 eco-climatic regions and found a marked reduction in crude protein contents in the second versus first half of the study period across all locations (1994–2015). These authors also detected a strong relationship between dietary crude protein and the Palmer drought severity index, suggesting that increased drought frequency caused by climate change has already led to reduced forage quality and protein stress in western cattle populations over the past two decades. Reduced forage production (McIntosh et al., 2019) and lower quality forage (Augustine et al., 2018; Craine et al., 2017) could imply that protein-stressed grazers will need to eat more lower quality feeds, which could exacerbate already stressed grasslands, leading to less stable soil–plant ecosystems, in addition to causing ruminant animals to emit more methane (Figure 3.12). As daily intake increases or quality of feed is decreased or ambient temperature increases, CH_4 production is increased in the rumen (Shibata and Terada, 2010).

3.5.3 Biocrust Community Changes in Rangelands

Global climate change and rangeland management practices are two major stressors that can alter biocrust abundance and distribution, thereby altering carbon cycling. Climate change forces conditions out of the ranges of parameters habitable for extant species, while intensive grazing rangeland practices facilitate continuous mechanical impacts via physical trampling of biocrusts, resulting in accelerated surface erosion and soil loss. Both physical disturbance and manipulations of climate – like altered warming and watering regimes – have been shown to shift biocrust communities toward early successional states (Ferrenberg et al., 2015). Early biocrust successional stages are characterized by lower organic C, total N, and exopolysaccharide content (Cantón et al., 2020), as well as reduced soil aggregate stability, lower carbon cycling potential, and water holding capacity, compared to later successional crust types (Hoellrich et al., 2023; Kuske et al., 2012; Pietrasiak et al., 2013). It is therefore important to consider biocrusts and their ecosystem services as an integral part of rangeland health assessments, needed to guide conservation and sustainability-based rangeland management practices (Stovall et al., 2022).

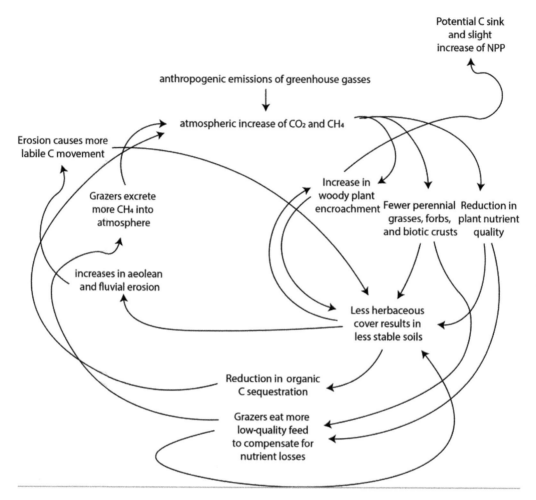

FIGURE 3.12 Theoretical negative feedback loop of C cycle under external influence of climate change on edaphic and biotic phenomenon in southwestern rangelands. Note that most scenarios result in organic and labile losses of C, with the exception of woody shrub encroachment. Woody shrub encroachment may have a net neutral effect on C because, although woody plants like mesquite are likely to slightly increase NPP as atmospheric CO_2 levels rise and can sequester C. However, the overall loss of a herbaceous layer and fewer stabilizing biotic crusts are likely to cause wind erosion corridors and less stable soils, resulting in less opportunities for below-ground C capture.

3.5.4 Adaptation Strategies

3.5.4.1 Stocking Rates/Animal Numbers

Extensive research has revealed that light-conservative stocking rates (20–35% vegetation use rates) are most sustainable on southwestern rangelands (Holechek, 1988 and sources therein). Several studies have compared these stocking rates with heavier grazing regimes (e.g., > 36%), and results have consistently shown decreased forage production, increased woody plant production, and decreased cow productivity (Holechek et al., 1998; Khumalo and Holechek, 2005; Khumalo et al., 2007, Navarro et al., 2002; Thomas et al., 2015). For rangelands with multiple species grazing (e.g., two or more livestock species and/or various wildlife), stocking rates may need more critical scrutiny, because they represent maximum vegetation use by all grazing animals. Fortunately, these stocking rates are currently *de facto* among US land management agencies including the Bureau of Land Management and the US Forest Service which monitor and set stocking rates on over half of all southwestern rangelands. Research from a long-term grazing study in the northern Great Plains suggests that light (10% in this case) or no grazing (0% vegetation use rates) had more positive effects on SOC, total N, microbial biomass, respiration, and N mineralization rates compared to heavily (50% vegetation use) stocked pastures (Ingram et al., 2008).

Emerging evidence suggests that abiotic factors like edaphic conditions, as they relate to climate change, may have an overriding effect on plant productivity compared to stocking rates – meaning that although these vegetation use regimes (light or conservative) may be sustainable, local environmental conditions will affect evapotranspiration rates, soil water holding capacities, and nutrient cycling, such that loss of herbaceous vegetation and increases in woody plant encroachment will be unavoidable (Ingram et al., 2008; McIntosh, 2021). Researchers have proposed several novel strategies for adapting to and mitigating such ecological changes, which could be paired with appropriate stocking rate protocols to preserve or improve soil stability despite worsening aridity.

3.5.4.2   Animal Size/Species

Holechek et al. (2020) predict that as climate change worsens, western livestock producers may need to convert their enterprises from cattle to sheep and goat or wildlife production depending on local aridity severity. While beef cattle are the primary species raised in the region, hotter and drier conditions could lead to an unsuitable environment (Holechek et al., 2020). Sheep and goats, which are smaller, more heat adapted, and better able to consume browse plants, may be better suited to such conditions with the added benefits of needing fewer outside C sources (e.g., supplemental feed), and potentially exert a lighter environmental footprint resulting in less soil degradation (e.g., smaller size = less compaction; Estell et al., 2012; Holechek et al., 2020). Considering that there is increased public concern for sustainably raised meat, wildlife could fill a useful niche, wherein they are best suited to local conditions because of increased diet breadth, large home ranges, and morphological advantages like increased thermotolerance compared to livestock counterparts (Holechek and Valdez, 2018; Holechek et al., 2020). Matching the correct animal to landscapes will become critically important for sustainably managing provisioning services while maintaining or improving plant and soil conditions (Rook et al., 2004; Scasta et al., 2016).

3.5.4.3   Animal Breed

Differences within species merit closer attention, as heritage or adapted breeds/biotypes could serve as critical adaptation resources to a hotter and increasingly grassless world (Estell et al., 2012; Scasta et al., 2016). Several studies have revealed the merits of adopting *Bos indicus* or *indicus*-based cross cattle like Brahman, Braford, or Brangus, which are hypothesized to yield market-sized calves with a lighter environmental footprint compared to conventional breeds (e.g., Angus or Hereford; Herbel and Nelson, 1966; McIntosh et al., 2023; Russell et al., 2012; Winder et al., 1996). Multiple lines of evidence suggest that heritage Raramuri Criollo (RC) cattle, a biotype with over 400 years of naturalization in the harsh landscapes of the Mexican Copper Canyon (Armstrong et al., 2022; Anderson et al., 2015), could offer producers a means of meeting sustainability and production goals on southwestern rangelands in the face of a rapidly dwindling forage supply and a hotter, drier climate (Nyamuryekung'e et al., 2022; Peinetti et al., 2011; Spiegal et al., 2019). Compared to conventional breeds (e.g., Angus), RC cows and steers have been documented traveling farther per day and exploring larger areas over multiple seasons (McIntosh et al., 2021; Nyamuryekung'e et al., 2022; Peinetti et al., 2011; Spiegal et al., 2019), ecosystems (Chihuahuan Desert, California chaparral, and Colorado Plateau; Duni et al., 2023), and physiological states (Nyamuryekung'e et al., 2021, 2022). Likewise, RC cows have a lighter body weight (McIntosh et al., 2020) and tend toward more even grazing distribution and fewer re-visitation rates during critical periods of the year when rangelands are most vulnerable to overgrazing – suggesting a potentially lighter footprint on soils and vegetation (Spiegal et al., 2019). Raramuri Criollo cows, compared to conventional counterparts, express a unique "follower" mothering style, which allows cows more freedom for daily exploration (Nyamuryekung'e et al., 2020, 2021). Raramuri Criollo also tend to exhibit better heat tolerance by maintaining/regulating internal temperatures, possibly because of body size, hair length, hide color, and behavioral tendencies and they are less likely to adjust behaviors to accommodate physiological demands like resting or shade seeking – both of which could increase patch-scale vegetation and soil overuse and degradation (McIntosh et al., 2020; Nyamuryekung'e et al., 2021). A diet comparison study determined that Raramuri Criollo consumed less black grama, a critical and threatened forage resource (McIntosh et al., 2019), and more mesquite, an invasive shrub, compared to Angus counterparts (Estell et al., 2023). Maintaining a more intact herbaceous layer will contribute to C and other nutrient cycling and reduce bare ground areas which are prone to erosion. The use of RC genetics may present opportunities related to reduced supplemental feed inputs (reducing C and N inputs from outside regions) and potentially reduce overhead costs compared to conventional counterparts (McIntosh et al., 2021; Torell et al., 2023). This assemblage of studies suggests that RC genetics may be well suited to emerging climate conditions on southwestern ranches because of their unique foraging behaviors and physiology, different resource requirements, and theoretical ability to produce more weaned beef per pound of cow versus conventional biotypes. Their lighter size and unique foraging behaviors could offer producers a lower impact animal, which could be used to conserve fragile vegetation.

3.5.4.4 Novel Tools and External Inputs

Emerging effects of climate change have caused policymakers to implement mitigation strategies for alleviating greenhouse gas emissions, specifically on landscapes where provisioning ecosystem services dominate. Carbon sequestration policies have been introduced to provide incentives for land managers to alter their practices in order to build C stocks on rangelands (Booker et al., 2013). A substantial amount of research has revealed that land use changes can affect soil C sequestration across different ecological contexts (Conant et al., 2016). For instance, strategies like improved grazing management, fertilization of plants, inclusion of legumes and improved grass species into pastures, pasture irrigation, and conversion from cropland to a less intensive land use base (e.g., native species) have all tended to increase soil C sequestration, albeit at variable rates (Conant et al., 2016). For C sequestration policies to be practical in application, policymakers need to ensure such strategies align closely with ecological system capacities (Booker et al., 2013). In arid and semiarid rangelands, abiotic factors play a more significant role in affecting ecosystem phenomenon than biotic factors like management interventions. Thus, policymakers seeking to implement C sequestration strategies in the southwest should consider that arid rangelands have less capacity to sequester C than their wetter counterparts (Briske et al., 2005; Vetter, 2005). One major hindrance of C sequestration is monitoring costs. Considering the agricultural workforce is already limited, such efforts could be economically unsustainable at the ranch scale, and thus policymakers should also consider these barriers to adoption if C sequestration practices are to be incorporated across regions, countries, or even continents.

Advancements in sensor technology could help land managers improve ecological and financial outcomes by providing informed, timely, adaptive, and accurate metrics to assist with managerial decision-making. A precision agriculture technology that is gaining traction on rangelands is virtual fencing – a livestock management tool which allows ranchers to virtually dictate where cattle graze (Anderson et al., 2014; Umstatter, 2011). The congregation of livestock around ecologically sensitive areas is a common detriment to sustainable rangeland management and has been a challenge to managers of arid rangelands who have attempted to alter livestock grazing distribution via conventional fencing, herding, or water placement schemes over the last century. Such practices are often futile as well as cost-prohibitive (Knight et al., 2011). Likewise, initiatives to incentivize rangeland restoration practices, in particular the protection of riparian areas, have been under significant consideration among ranchers and the scientific community for several years now (Bailey and Provenza, 2008). Virtual fencing consists in the use of animal wearable collars that employ auditory-electric

Distinctive Dryland Soil Carbon Transformations 67

pulse cues to deter animals from trespassing a virtually determined fence boundary (Anderson et al., 2014). The flexibility of the technology to allocate land and forage resources with polygons of configurable shape, size, and duration offers land managers an unprecedented opportunity to deal with the challenge of managing heterogeneous resources, both in time and across space (Anderson et al., 2014; Campbell et al., 2019; Umstatter, 2011). The advances in technology might offer land managers an opportunity for prescribing grazing for enhancing ecosystem services such as minimizing overgrazing and trampling, or for target grazing specific plant communities, which could improve several soil health metrics like water holding capacity, nutrient cycling, and aeration.

Other mitigation strategies like incorporating solar panel infrastructure, minimizing external livestock feed inputs, or reducing vehicle use (for instance, by relying on sensor technology to monitor cattle, water level, and other infrastructure), will likely reduce ranch-scale C emissions. It is unclear how these efforts will directly affect soil properties. We expect it could be quite important, however, as reductions in vehicle usage could reduce aeolian sediment movement and help stabilize labile C as demonstrated by Nauman et al. (2018).

3.5.5 NOVEL ECOLOGICAL RESTORATION STRATEGIES (MITIGATION)

Arid land restoration is complicated by several factors including access to reliable reference states, complicated heterogeneous spatiotemporal functional scales of abiotic and biotic phenomenon, and environmental or human-derived (socio-economic) limitations (Svejcar and Kildisheva, 2017). Additionally, restoration efforts can be marred by narrowly focused goals, wherein strategies applied to improve certain landscape elements or production outcomes could have unintended ecological consequences. Nevertheless, novel restoration strategies have potential to re-establish degraded soils, flora, and fauna (Svejcar and Kildisheva, 2017).

Aside from livestock management, herbicide treatment is one of the most widely used restoration tools in the southwest. Herbicides application has gained traction in recent decades because of more accessible aerial application and its potential to kill unwanted and increasingly invasive woody shrubs like mesquite. However, herbicides come with the caveat that many function as endocrine disruptors which are known to have non-target effects (Meyers et al., 2016). This presents concerns about unintended toxicological consequences that span trophic levels; thus, drawbacks need to be weighed against benefits. For instance, one Chihuahuan Desert study found significantly reduced microarthropod populations over a decade after herbicide application; this result alone could negatively impact several soil recovery processes (Kay et al., 1999).

As shrub encroachment worsens, the interstitial space between woody plants becomes increasingly bare, which could lead to aeolian erosion corridors and thus contribute to a negative feedback loop of less herbaceous ground cover, more shrub cover, and less stable soils (Rachal et al., 2015). One solution may be to utilize artificial structures to restrict sediment transport. Researchers in New Mexico evaluated connectivity modifiers (colloquially "Con-Mods"), small "+" shaped wire mesh structures which are affixed in interstitial spaces using wire stakes. Rachal et al. (2015) determined that Con-Mods could effectively capture sediment transport on desert basin sites.

Southwestern restoration efforts are primarily limited by water scarcity, hence several strategies for introducing water to rangeland plantings like irrigation, roots balls, or slow-release containers have been employed. Hydrogels, starch-based compounds that can hold several times their weight in water, have potential for restoring arid grasslands alongside reseeding efforts. In one greenhouse study, researchers reported black grama, a critical native Chihuahuan Desert forage species, increased water use and growth in relation to engorged hydrogels (Lucero et al., 2010). These authors could not detect a difference between hydrogel and control treatments in the field because of a wet growing season. However they concluded that transplanted black grama plants were successfully established in relation to adequate watering (Lucero et al., 2010).

Biocrust inoculums could be used as a novel strategy for stabilizing soils and reducing aeolian and fluvial sediment erosion. Researchers evaluated this strategy in the Great Basin and Chihuahuan

Desert and determined that inoculum type and/or habitat amelioration benefitted recovery rates of soils but found that overriding abiotic effects like location or soil texture played a greater role in recovery (Faist et al., 2020). In this case, researchers determined that finer-textured soils had the greatest potential for restoration (Faist et al., 2020).

It is likely that a complex combination of adaptation and restoration efforts would have the most profound impact on stabilizing, improving, and reducing deterioration of southwestern desert soils. Likewise, location and abiotic effects will dictate which areas are most prone to deterioration (least resilient) and which are best suited to restoration efforts (most resilient). The paucity of data surrounding these techniques merits closer attention if land managers are expected to reduce deleterious effects of climate change and drought on Southwestern soils in the immediate future.

3.6 ARISING INSIGHTS INTO SOIL C DYNAMICS IN ARID RANGELANDS

Arid rangelands occupy about 40% of the terrestrial surface and are home to about 35% of the world's population (SCBD, 2013). The Millennium Ecosystem Assessment highlighted the important levels of biodiversity and abundance of direct and indirect ecosystem services of dryland regions but emphasized that these values are under-recognized (SCBD, 2013). Sustaining the soil resource is critical to sustaining these ecosystems and the people and biodiversity they support (Lal and Stewart, 2013). Arid rangelands are characterized by low precipitation, high evaporative demand, high temperatures, and, hence, small NPP. Although soil organic C concentrations are small, because of the vast areas involved, about 36% of the global C stock is in drylands. Furthermore, unlike more humid regions, soil inorganic carbon (SIC) is an important soil C stock distinctive of drylands. While SIC has often been regarded as inactive, the potential exists to manage biogeochemical processes to sequester SIC. Because SOC concentration levels are small, arid rangeland soils have often been considered to have small biological activity. However, biocrusts are an essential characteristic of rangeland soils. Biocrusts are responsible for 25% of all terrestrial N-fixation (Elbert et al., 2012) and an estimated 9% of the total dryland NPP (Zhao et al., 2005; Sancho et al., 2016). Despite mounting evidence that biocrusts are critical for maintaining environmental stability of arid rangelands (Weber et al., 2016), local and regional estimates of their contribution to rangeland C stocks and fluxes are sparse and have large uncertainties.

Assessments of C cycling at local, regional, and global levels rely on models, but processes that are critical to understanding C dynamics in arid rangeland ecosystems and soils are lacking due to incomplete understanding, lack of comprehensive data to develop and validate new models, and lack of appreciation of the importance of the large dryland regions of the world. A major limitation in simulating rangelands is that most models do not consider monthly and year-to-year change in plant species composition such as the shift from grass to shrub and C3 to C4 grasses. These shifts affect nutrient dynamics, biomass structure, water utilization, seasonality, and biomass allocation (Parton et al., 1994, 1993) and are important for arid environments where year-to-year changes resulting from drought and long-term shifts because of management and climate change are characteristic of arid rangelands. Critical gaps in C cycling models include the role of biocrusts in productivity, emissions, soil aggregation, and biogeochemical processes involved in SOC and SIC transformations.

Additionally, C cycling models do not adequately capture erosion process effects on the heterogeneous, fragile arid landscapes. Water erosion is the process best represented in some models, but erosion by water may be a small contributor to SOC erosion in arid rangelands, compared to erosion by wind. Because wind erosion and dust emission are a major contributor to SOC erosion and removal of plant material before it decomposes, aeolian processes may dominate SOC stock reduction in arid rangelands. In humid/temperate regions NPP and SR are large and SOC erosion is likely a small proportion of loss. However, in arid rangelands water erosion is small, but wind erosion is large. With small NPP and larger SR, is SOC erosion a large proportion of NEE in arid rangelands? Modeling studies suggest that SOC cycling models without soil erosion overestimate global SOC to atmosphere by around 20% and regionally up to 85% (Chappell et al., 2016). Our simple approximations here (NEEd) suggest a similar magnitude of overestimated SOC to the atmosphere caused by the redistribution of SOC by erosion. If the above outcomes are representative of arid rangelands, then reducing SOC erosion is essential to retain, if not increase, the store of SOC in dryland soils. That is only possible by finding ways to sustain vegetation communities that minimize erosion and increase NPP.

In the face of these challenges, adaptation and mitigation strategies for rangeland management are essential. Managing multiple ecosystem services will be critical, including to sustain livelihoods of those who live on and manage the land. Sustaining ecosystem services requires maintaining the resource base, but past attention to dryland systems has focused on direct production benefits and has undervalued the indirect benefits of climate regulation, water supply and filtering, nutrient cycling, biodiversity, and others (Havstad et al., 2007; Yahdijan et al., 2015). This will require a new focus across global, national, and regional scales, as well as innovative technologies and approaches at the local level.

3.7 ACKNOWLEDGMENTS

JLS, CBB, JCS, MMM, SN, and AR were supported by USDA National Institute of Food and Agriculture Project No. 2019-69012-29853. MH's work on biocrust carbon exchange was supported by funding from the US National Science Foundation (NSF) to New Mexico State University for the Jornada Basin Long Term Ecological Research Program (DEB 12-35828). MH and NP were supported by NSF EAR-2012475 during the writing phases of this review manuscript. NPW and AC were supported by a joint grant from the NSF and UK Natural Environmental Research Council (EAR-1853853).

The first author recognizes with gratitude the tremendous impact that Dr. B. A. Stewart's guidance, support, and insights had on the trajectory of her career.

REFERENCES

Adair, E. C., Parton, W. J., Del Grosso, S. J., Silver, W. L., Harmon, M. E., Hall, S. A., Burke, I. C., Hart, S. C. 2008. Simple three-pool model accurately describes patterns of long-term litter decomposition in diverse climates. *Global Change Biology*, 14, 2636–2660. https://doi.org/10.1111/j.1365-2486.2008.01674.x

Allen, M. F., Figueroa, C., Weinbaum, B. S., Barlow, S. B., Allen, E. B. 1996. Differential production of oxalates by mycorrhizal fungi in arid ecosystems. *Biology and Fertility of Soils*, 22, 287–292.

Anderson, D. M., Estell, R. E., Gonzalez, A. L., Cibils, A. F., Torell, L. A. 2015. Criollo cattle: Heritage genetics for arid landscapes. *Rangelands*, 37, 62–67. https://doi.org/10.1016/j.rala.2015.01.006

Anderson, D. M., Estell, R. E., Holechek, J. L., Ivey, S., Smith, G. B. 2014. Virtual herding for flexible livestock management – A review. *Rangeland Journal*, 36, 205–221. https://doi.org/10.1071/RJ13092

Ansley, R. J., Boutton, T. W., Skjemstad, J. O. 2006. Soil organic carbon and black carbon storage and dynamics under different fire regimes in temperate mixed-grass savanna. *Global Biogeochemical Cycles*, 20, 1–11. https://doi.org/10.1029/2005GB002670

Ardö, J., Olsson, L. 2003. Assessment of soil organic carbon in semi-arid Sudan using GIS and the CENTURY model. *Journal of Arid Environments*, 54, 633–651. https://doi.org/10.1006/jare.2002.1105

Armstrong, E., Rodriguez Almeida, F. A., McIntosh, M. M., Poli, M., Cibils, A. F., Martínez Quintana, J. A., Félix-Portillo, M., Estell, R. E. 2022. Genetic and productive background of Criollo cattle in Argentina, Mexico, Uruguay and the United States. *Journal of Arid Environments*, 200, 104722. https://doi.org/10.1016/j.jaridenv.2022.104722

Augustine, D. J., Blumenthal, D. M., Springer, T. L., LeCain, D. R., Gunter, S. A., Derner, J. D. 2018. Elevated CO_2 induces substantial and persistent declines in forage quality irrespective of warming in mixed grass prairie. *Ecological Applications*, 28, 721–735. https://doi.org/10.1002/eap.1680

Augusto, L., Achat, D. L., Jonard, M., Vidal, D., Ringeval, B. 2017. Soil parent material-A major driver of plant nutrient limitations in terrestrial ecosystems. *Global Change Biology*, 23, 3808–3824.

Bailey, D. W., Provenza, F. D. 2008. Mechanisms determining large-herbivore distribution. In: Prins, H. H. T., van Langevelde, F. (Eds.), *Resource Ecology: Spatial and Temporal Dynamics of Foraging*, Springer, Wagenenin, NL, pp. 7–28.

Baldassini, P., Paruelo, J. M. 2020. Deforestation and current management practices reduce soil organic carbon in the semi-arid Chaco, Argentina. *Agricultural Systems*, 178, 102749. https://doi.org/10.1016/j.agsy.2019.102749

Baldock, J. A., Skjemstad, J. O. 2000. Role of the soil matrix and minerals in protecting natural organic materials against biological attack. *Organic Geochemistry*, 7–8, 697–710.

Bandaru, V., Yaramasu, R., Jones, C., Izaurralde, R. C., Reddy, A., Sedano, F., Daughtry, C. S. T., Becker-Reshef, I., Justice, C. 2022. Geo-CropSim: A Geo-spatial crop simulation modeling framework for regional scale crop yield and water use assessment. *ISPRS Journal of Photogrammetry and Remote Sensing*, 183, 34–53. https://doi.org/10.1016/j.isprsjprs.2021.10.024

Barger, N. N., Archer, S. R., Campbell, J. L., Huang, C. Y., Morton, J. A., Knapp, A. K. 2011. Woody plant proliferation in North American drylands: A synthesis of impacts on ecosystem carbon balance. *Journal of Geophysical Research: Biogeosciences*, 116, 1–17. https://doi.org/10.1029/2010JG001506

Barnes, P. W., Throop, H. L., Archer, S. R., Breshears, D. D., McCulley, R. L., Tobler, M. A. 2015. Sunlight and soil–litter mixing: Drivers of litter decomposition in drylands. In: Lüttge, U., Beyschlag, W. (Eds.), *Progress in Botany*. Springer, 76, pp. 273–302. https://doi.org/10.1007/978-3-319-08807-5_11

Belnap, J. 2003. The world at your feet: Desert biological soil crusts. *Frontiers in Ecology and the Environment*, 1, 181–189. https://doi.org/10.1890/1540-9295

Belnap, J., Büdel B., Lange O. L. 2003. Biological soil crusts: Characteristics and distribution. In: Belnap, J., Lange, O. L., editors (Eds.),. *Biological Soil Crusts: Structure, Function, and Management*. Springer, Berlin Heidelberg,; pp. 3–30. (Ecological Studies).

Belnap, J., Gillette, D. A. 1998. Vulnerability of desert biological soil crusts to wind erosion: The influences of crust development, soil texture, and disturbance. *Journal of Arid Environments*, 39, 133–142.

Belnap, J., Kaltenecker, J. H., Rosentreter, R., Williams, J., Leonard, S., Eldridge D. J. 2001. *Biological soil crusts: Ecology and management*: *TR–1730-2*, US Department of the Interior, Denver, CO.

Belnap, J., Lange, O. L. (Eds.) 2003. *Biological Soil Crusts: Structure, Function, and Management*. Springer, Berlin.

Belnap, J., Miller, D. M., Bedford, D. R., Phillips, S. L. 2014. Pedological and geological relationships with soil lichen and moss distribution in the eastern Mojave Desert, CA, USA. *Journal of Arid Environments*, 106, 45–57.

Belnap, J., Weber, B., Büdel, B. 2016. Biological soil crusts as an organizing principle in drylands. In: Weber, B., Büdel, B., Belnap, J. (Eds.), *Biological Soil Crusts: An Organizing Principle in Drylands*. Springer International Publishing, pp. 55–80 (Ecological Studies).

Bennett, H. H., Chapline, W. R. 1928. Soil erosion: A national menace. Soil Investigations, Bureau of Chemistry and Soils. Issue 33 of Circular, United States Department of Agriculture.

Berhe, A. A., Barnes, R. T., Six, J., Marin-Spiotta E. 2018. Role of soil erosion in biogeochemical cycling of essential elements: Carbon, nitrogen, and phosphorous. *Annual Review of Earth and Planetary Sciences*, 46, 521–548.

Berryman, E., Hatten, J., Page-Dumroese, D. S., Heckman, K. A., D'amore, D. V., Puttere, J., SanClements, M., Connolly, S., Perry, C. H., Domke, G. M. 2020. Soil carbon. In: Pouyat, R. V., Page-Dumroese, D. S., Patel-Weynand, T., Geiser, L. (Eds.), *Forest and Rangeland Soils of the United States Under Changing Conditions: A Comprehensive Science Synthesis*. Springer, pp. 9–32.

Bestelmeyer, B. T., Ash, A., Brown, J. R., Densambuu, B., Fernández-Giménez, M., Johanson, J., Levi, M., Lopez, D., Peinetti, R., Rumpff, L., Shaver, P. 2017. State and transition models: Theory, applications, and challenges. In: Briske, D. D. (Ed.), *Rangeland Systems: Processes, Management and Challenges*. Springer International Publishing, pp. 303–345. https://doi.org/10.1007/978-3-319-46709-2_9

Bestelmeyer, B. T., Burkett, L. M., Lister, L. 2021. Effects of managed fire on a swale grassland in the Chihuahuan Desert. *Rangelands*, 43, 181–184. https://doi.org/10.1016/j.rala.2021.05.001

Bestelmeyer, B. T., Marcillo, G., McCord, S. E., Mirsky, S., Moglen, G., Neven, L. G., Peters, D., Sohoulande, C., Wakie, T. 2020. Scaling up agricultural research with artificial intelligence. *IT Professional*, 22, 33–38. https://doi.org/10.1109/MITP.2020.2986062

Bestelmeyer, B. T., Peters, D. P. C., Archer, S. R., Browning, D. M., Okin, G. S., Schooley, R. L., Webb, N. P., 2018. The grassland-shrubland regime shift in the Southwestern United States: Misconceptions and their implications for management. *Bioscience*, 68, 678–690.

Bestelmeyer, B. T., Ward, J. P., Herrick, J. E., Tugel, A. J., 2006. Fragmentation effects on soil aggregate stability in a patchy arid grassland. *Rangeland Ecology Management*, 59, 406–415. https://doi.org/10.2111/05-180R1.1

Bird, S. B., Herrick, J. E., Wander, M. M., Murray, L. 2007. Multi-scale variability in soil aggregate stability: Implications for understanding and predicting semi-arid grassland degradation. *Geoderma*, 140, 106–118. https://doi.org/10.1016/j.geoderma.2007.03.010

Bird, S. B., Herrick, J. E., Wander, M. M., Wright, S. F. 2002. Spatial heterogeneity of aggregate stability and soil carbon in semi-arid rangeland. *Environmental Pollution*, 116, 445–455. https://doi.org/10.1016/S0269-7491(01)00222-6

Booker, K., Huntsinger, L., Bartolome, J. W., Sayre, N. F., Stewart, W. 2013. What can ecological science tell us about opportunities for carbon sequestration on arid rangelands in the United States? *Global Environmental Change*, 23, 240–251. https://doi.org/10.1016/j.gloenvcha.2012.10.001

Boon, K. F., Kiefert, L., McTainsh, G. H. 1998. Organic matter content of rural dusts in Australia. *Atmospheric Environment*, 32, 2817–2823.

Boutton, T. W., Liao, J. D., Filley, T. R., Archer, S. R. 2009. Belowground Carbon storage and dynamics accompanying woody plant encroachment in a subtropical Savanna. In: Lal, R., Follett, R. F. (Eds.), *Soil Carbon Sequestration and the Greenhouse Effect*, 2nd ed., Soil Science Society of America Monograph, Madison, WI, pp. 181–205.

Bowker, M. A., Belnap, J. 2008. A simple classification of soil types as habitats of biological soil crusts on the Colorado Plateau, USA. *Journal of Vegetation Science*, 19, 831–840.

Bowker, M. A., Belnap, J., Büdel, B., Sannier, C., Pietrasiak, N., Eldridge, D., Rivera-Aguilar, V. 2016. Controls on distribution patterns of biological soil crusts at the micro- to global scales. In: Weber, B., Büdel, B., Belnap, J. (Eds.), *Biological Soil Crusts: An Organizing Principle in Drylands*. Springer International Publishing, pp. 173–198 (Ecological Studies).

Bowker, M. A., Belnap, J., Davidson, D. W., Goldstein, H. 2006. Correlates of biological soil crust abundance across a continuum of spatial scales: Support for a hierarchical conceptual model. *Journal of Applied Ecology*, 43, 152–163.

Bowker, M. A., Belnap, J., Davidson, D. W., Phillips, S. L. 2005. Evidence for micronutrient limitation of biological soil crusts: Importance to arid-lands restoration. *Ecological Applications*, 15, 1941–1951.

Breshears, D. D., Whicker, J. J., Johansen, M. P., Pinder, J. E., 2003. Wind and water erosion and transport in semi-arid shrubland, grassland and forest ecosystems: Quantifying dominance of horizontal wind-driven transport. *Earth Surface Processes and Landforms*, 28, 1189–1209.

Brilli, L., Bechini, L., Bindi, M., Carozzi, M., Cavalli, D., Conant, R., Dorich, C. D., Doro, L., Ehrhardt, F., Farina, R., Ferrise, R., Fitton, N., Francaviglia, R., Grace, P., Iocola, I., Klumpp, K., Léonard, J., Martin, R., Massad, R. S., Recous, S., Seddaiu, G., Sharp, J., Smith, P., Smith, W. N., Soussana, J. F., Bellocchi, G., 2017. Review and analysis of strengths and weaknesses of agro-ecosystem models for simulating C and N fluxes. *Science of Total Environment*, 598, 445–470. https://doi.org/10.1016/j.scitotenv.2017.03.208

Briones, M. P. P., Hori, K., Martinez-Goss, M. R., Ishibashi, G., Okita, T. 1997. A comparison of physical properties, oxalate-oxalic acid soluble substances, protein content, and in vitro protein digestibility of the blue-green alga *Nostoc commune* Vauch from the Philippines and Japan. *Plant Foods for Human Nutrition*, 50, 287–294.

Briske, D. D., Fuhlendorf, S. D., Smeins, F. E. 2005. State-and-transition models, thresholds, and rangeland health: A synthesis of ecological concepts and perspectives. *Rangeland Ecology and Management*, 58, 1–10. https://doi.org/10.2111/1551-5028(2005)58<1:SMTARH>2.0.CO;2

Brown, J. R., Angerer, J., Salley, S. W., Blaisdell, R., Stuth, J. W. 2010. Improving estimates of rangeland carbon sequestration potential in the US Southwest. *Rangeland Ecology and Management*, 63, 147–154. https://doi.org/10.2111/08-089.1

Brown, J. R., MacLeod, N. 2011. A site-based approach to delivering rangeland ecosystem services. *The Rangeland Journal*, 33, 99–108.

Büdel, B., Dulić, T., Darienko, T., Rybalka, N., Friedl, T. 2016. Cyanobacteria and algae of biological soil crusts. In: Weber, B., Büdel, B., Belnap, J. (Eds.), *Biological Soil Crusts: An Organizing Principle in Drylands*. Springer International Publishing, pp. 55–80 (Ecological Studies).

Cable, J. M., Ogle, K., Williams, D. G., Weltzin, J. F., Huxman, T. E. 2008. Soil texture drives responses of soil respiration to precipitation pulses in the Sonoran Desert: Implications for climate change. *Ecosystems*, 11, 961–979.

Campbell, D. L. M., Lea, J. M., Keshavarzi, H., Lee, C. 2019. Virtual fencing is comparable to electric tape fencing for cattle behavior and welfare. *Frontiers in Veterinary Science*, 6. https://doi.org/10.3389/fvets.2019.00445

Canadell, J., Jackson, R. B., Ehleringer, J. B., Mooney, H. A., Sala, O. E., Schulze, E. D. 1996. Maximum rooting depth of vegetation types at the global scale. *Oecologia*, 108, 583–595.

Cania, B., Vestergaard, B., Kublik, S., Köhne, J. M., Fischer, T., Albert, A., Winkler, B., Schloter, M., Schulz, S. 2020. Biological soil crusts from different soil substrates harbor distinct bacterial groups with the potential to produce exopolysaccharides and lipopolysaccharides. *Microbial Ecology*, 79, 326–341. https://doi.org/10.1007/s00248-019-01415-6

Cantón, Y., Chamizo, S., Rodriguez-Caballero, E., Lázaro, R., Roncero-Ramos, B., Román, J. R., Solé-Benet, A. 2020. Water regulation in cyanobacterial biocrusts from drylands: Negative impacts of anthropogenic disturbance. *Water*, 12, 720. doi:10.3390/w12030720

Carlson, D. H., Thurow, T. L. 1996. Comprehensive evaluation of the improved SPUR model (SPUR-91). *Ecological Modelling*, 85, 229–240.

Carter, J. O., Hall, W. B., Brook, K. D., McKeon, G. M., Day, K. A., Paull, C. J. 2000. AussieGRASS: Australian grassland and rangeland assessment by spatial simulation. In: Hammer, G. L., Nicholls, N., Mitchell, C. (Eds.), *Applications of Seasonal Climate Forecasting in Agricultural and Natural Ecosystems*. Springer, pp. 329–349.

Causarano, H. J., Doraiswamy, P. C., Muratova, N., Pachikin, K., McCarty, G. W., Akhmedov, B., Williams, J. R. 2011. Improved modeling of soil organic carbon in a semiarid region of Central East Kazakhstan using EPIC. *Agronomy for Sustainable Development*, 31, 275–286. https://doi.org/10.1051/agro/2010028

Chapin, F. S., Walker, B. H., Hobbs, R. J., Hooper, D. U., Lawton, J. H., Sala, O. E., Tilman, D. 1997. Biotic control over the functioning of ecosystems. *Science*, 277, 500–504. https://doi.org/10.1126/science.277.5325.500

Chapin, F. S., Woodwell, G. M., Randerson, J. T., Lovett, G. M., Rastetter, E. B., Baldocchi, D. D., Clark, D. A., Harmon, M. E., Schimel, D. S., Valentini, R., Wirth, C., Aber, J. D., Cole, J. J., Goulden, M. L., Harden, J. W., Heimann, M., Howarth, R. W., Matson, P. A., McGuire, A. D., Melillo, J. M., Mooney, H. A., Neff, J. C., Houghton, R. A., Pace, M. L., Ryan, M. G., Running, S. W., Sala, O. E., Schlesinger, W. H., Schulze, E. D. 2005. Reconciling carbon-cycle concepts, terminology, and methodology. *Ecosystems*, 9, 1041–1050. doi: 10.1007/ s10021-005-0105-7

Chappell, A., Baldock, J., Sanderman, J., 2016. The global significance of omitting soil erosion from soil organic carbon cycling schemes. *Nature Climate Change*, 6, 187–191.

Chappell, A., Webb, N. P., Leys, J. F. Waters, C., Orgill, S., Eyres, M. 2019. Minimising soil organic carbon erosion is essential for land degradation neutrality. *Environmental Science and Policy*, 93, 43–52.

Chappell, A., Webb, N. P., Viscarra Rossel, R. A., Bui, E. 2014. Australian net (1950s-1990) soil organic carbon erosion: Implications for CO_2 emission and land-atmosphere modelling. *Biogeosciences*, 11, 5235–5244.

Chianese, D. S., Rotz, C. A., Richard, T. L. 2009. Simulation of methane emissions from dairy farms to assess greenhouse gas reduction strategies. *Transactions of the ASABE*, 52, 1313–1323.

Conant, R. T., Cerri, E. P., Osborne, B. B., Paustian, K. 2016. Grassland management impacts on soil carbon stocks: A new synthesis. *Ecological Applications*, 27, 662–668.

Connin, S. L., Virginia, R. A., Chamberlain, C. P. 1997. Carbon isotopes reveal soil organic matter dynamics following arid land shrub expansion. *Oecologia*, 110, 374–386.

Corson, M. S., Rotz, C. A., Skinner, R. H., Sanderson, M. A. 2007. Adaptation and evaluation of the integrated farm system model to simulate temperate multiple-species pastures. *Agricultural Systems*, 94, 502–508.

Corson, M. S., Skinner, R. H., Rotz, C. A. 2006. Modification of the SPUR rangeland model to simulate species composition and pasture productivity in humid temperate regions. *Agricultural. Systems*, 87, 169–191.

Cotrufo, F. M., Lavallee, J. M., Zhang, Y., Hansen, P. M., Paustian, K. H., Schipanski, M., Wallenstein, M. D. 2021. In-N-Out: A hierarchical framework to understand and predict soil carbon storage and nitrogen recycling. *Global Change Biology*, 27, 4465–4468. https://doi.org/10.1111/gcb.15782

Craine, J. M., Elmore, A., Angerer, J. P. 2017. Long-term declines in dietary nutritional quality for North American cattle. *Environmental Research Letters*, 12, 044019. https://doi.org/10.1088/1748-9326/aa67a4

Crausbay, S. D., Sofaer, H. R., Cravens, A. E., Chaffin, B. C., Clifford, K. R., Gross, J. E., Knapp, C. N., Lawrence, D. J., Magness, D. R., Miller-Rushing, A. J., Schuurman, G. W., Stevens-Rumann, C. S. 2022.

A science agenda to inform natural resource management decisions in an era of ecological transformation. *BioScience*, 72, 71–90. https://doi.org/10.1093/biosci/biab102

De Steiguer, J. E., 2008. Semi-arid rangelands and carbon offset markets: A look at the economic prospects. *Rangelands*, 30, 27–32. https://doi.org/10.2458/azu_rangelands_v30i2_de_steiguer

Descheemaeker, K., Zijlstra, M., Masikati, P., Crespo, O., Tui, S. H.-K. 2018. Effects of climate change and adaptation on the livestock component of mixed farming systems: A modelling study from semi-arid Zimbabwe. *Agricultural Systems*, 159, 282–295. http://dx.doi.org/10.1016/j.agsy.2017.05.004

Du, H., Li, S., Webb, N. P., Zuo, X., Liu, X. 2021. Soil organic carbon (SOC) enrichment in aeolian sediments and SOC loss by dust emission in the desert steppe, China. *Science of the Total Environment*, 798, 149189.

Duni, D. M., McIntosh, M. M., Nyamuryekung'e, S., Cibils, A. F., Estell, R. E., Gedefaw, M. G., Gonzalez, A. L., Duniway, M. C., Redd, M., Paulin, R., Steele, C. M., Perea, A., Utsumi, S. A., Spiegal, S. A. 2023. Movement, activity, and habitat use patterns of heritage vs conventional beef cattle in two southwestern ecoregions. *Journal of Arid Environments*, 213, 104975. https://doi.org/10.1016/j.jaridenv.2023.104975

Edwards, B. L., Webb, N. P., Galloza, M. S., Van Zee, J. W., Courtright, E. M., Cooper, B. F., Metz, L. J., Herrick, J. E., Okin, G. S., Duniway, M. C., Tatarko, J., Tedala, N. H., Moriasi, D. N., Newingham, B. A., Pierson, F. B., Toledo, D., Van Pelt, R. S. 2022. Parameterizing an aeolian erosion model for rangelands. *Aeolian Research*, 54, 100769. https://doi.org/10.1016/j.aeolia.2021.100769

Elbert, W., Weber, B., Burrows, S., Steinkamp, J., Budel, B, Andreae, M. O., Poschl, U. 2012. Contribution of cryptogamic covers to the global cycles of carbon and nitrogen. *Nature Geoscience*, 5, 459–462.

Ellis, J. L., Bannink, A., France, J., Kebreab, E., Dijkstra, J. 2010. Evaluation of enteric methane prediction equations for dairy cows used in whole farm models. *Global Change Biology*, 16, 3246–3256, doi: 10.1111/j.1365-2486.2010.02188.x

Estell, R. E., Havstad, K. M., Cibils, A., Anderson, D. M., Estell, R. E., Havstad, K. M., Fredrickson, E. L., Anderson, D. M., Schrader, T. S., James, D. K. 2012. Increasing shrub use by livestock in a world with less grass. *Rangeland Ecology Management*, 65, 553–562. https://doi.org/10.2307/23355244

Estell, R. E., Nyamuryekung'e, S., James, D. K., Spiegal, S. A., Cibils, A. F., Gonzalez, A. L., McIntosh, M. M., Romig, K. 2023. Diet selection of Raramuri Criollo and Angus x Hereford crossbred cattle in the Chihuahuan Desert. *Journal of Arid Environments*, 213, 104823 https://doi.org/10.1016/j.jaridenv.2022.104823

Eswaran, H., Reich, P. F., Kimble, J. M., Beinroth, F. H., Padmanabhan, E., Moncharoen, P. 2000. Global carbon stocks. In: Kimble, R. L. (Ed.), *Global Climate Change and Pedogenic Carbonates*. Lewis Publishers, pp. 15–27.

Evans, R. D., Johansen, J. R. 1999. Microbiotic crusts and ecosystem processes. *CRC Critical Reviews in Plant Sciences*, 18, 183–225.

Evans, R. D., Koyama, A., Sonderegger, D., Charlet, T. N., Newingham, B. A., Fenstermaker, L. F., Harlow, B., Jin, V. L., Ogle, K., Smith, S. D., Nowak, R. S. 2014. Greater ecosystem carbon in the Mojave Desert after ten years exposure to elevated CO_2. *Nature Climate Change*, 4, 394–397.

Faist, A. M., Antoninka, A. J., Belnap, J., Bowker, M. A., Duniway, M. C., Garcia-Pichel, F., Nelson, C., Reed, S. C., Giraldo-Silva, A., Velasco-Ayuso, S., Barger, N. N. 2020. Inoculation and habitat amelioration efforts in biological soil crust recovery vary by desert and soil texture. *Restoration Ecology*, 28, S96–S105. https://doi.org/10.1111/rec.13087

Fernández-Ugalde, O., Virto, I., Barré, P., Apesteguía, M., Enrique, A., Imaz, M. J., Bescansa, P. 2014. Mechanisms of macroaggregate stabilisation by carbonates: Implications for organic matter protection in semi-arid calcareous soils. *Soil Research*, 52, 180–192.

Ferrenberg, S., Reed, S., Belnap, J. 2015. Climate change and physical disturbance cause similar community shifts in biological soil crusts. *PNAS*, 112, 12116–12121.

Fick, S. E., Barger, N. N., Duniway, M. C. 2019. Hydrological function of rapidly induced biocrusts. *Ecohydrology*, 12, e2089. https://doi.org/10.1002/eco.2089

Filser, J., Faber, J. H., Tiunov, A. V., Brussaard, L., Frouz, J., De Deyn, G., Uvarov, A. V., Berg, M. P., Lavelle, P., Loreau, M., Wall, D. H., Querner, P., Eijsackers, H., Jiménez, J. J. 2016. Soil fauna: Key to new carbon models. *Soil*, 2, 565–582. https://doi.org/10.5194/soil-2-565-2016

Follett, R. F., 2000. Organic carbon pools in grazing land soils. In: Follett, R. F., Kimble, J. M., Lal, R. (Eds.), *The Potential of U.S. Grazing Lands to Sequester Carbon and Mitigate the Greenhouse Effect*. CRC Press, pp. 65–86. https://doi.org/10.1201/9781420032468.ch3

Follett, R. F., Kimble, J. M., Lal, R. (Eds.) 2001. *The Potential of U.S. Grazing Lands to Sequester Carbon and Mitigate the Greenhouse Effect*. Lewis Publishers.

Fry, E. L., De Long, J. R., Álvarez Garrido, L., Alvarez, N., Carrillo, Y., Castañeda-Gómez, L., Chomel, M., Dondini, M., Drake, J. E., Hasegawa, S., Hortal, S., Jackson, B. G., Jiang, M., Lavallee, J. M., Medlyn, B. E., Rhymes, J., Singh, B. K., Smith, P., Anderson, I. C., Bardgett, R. D., Baggs, E. M., Johnson, D. 2019. Using plant, microbe, and soil fauna traits to improve the predictive power of biogeochemical models. *Methods in Ecology and Evolution*, 10, 146–157. https://doi.org/10.1111/2041-210X.13092

Fuhlendorf, S. D., Limb, R. F., Engle, D. M., Miller, R. F. 2011. Assessment of prescribed fire as a conservation practice. In: Briske, D. D. (Ed.) *Conservation Benefits of Rangeland Practices: Assessment, Recommendations, and Knowledge Gaps*. USDA Natural Resources Conserv. Serv., Washington, DC, pp. 75–104.

Gadd, G. M., Bahri-Esfahanik, J., Li, Q., Rhee, Y. J., Wei, Z., Fomina, M., Liang, X. 2014. Oxalate production by fungi: Significance in geomycology, biodeterioration and bioremediation. *Fungal Biology Reviews*, 28, 36–55.

Gedefaw, M. G, Geli, H. M. E., Abera, T. 2021. Assessment of rangeland degradation in New Mexico using time series segmentation and residual trend analysis (TSS-RESTREND), *Remote Sensing*, 13(9), 1618. https://doi.org/10.3390/rs13091618

Gerrard, J. 1993. Soil geomorphology — Present dilemmas and future challenges. *Geomorphology*, 7, 61–84. Springer. ISBN 978-0-412-44180-6

Gillette, D. A., 1999. A qualitative geophysical explanation for "Hot Spot" dust emitting source regions. *Contributions to Atmospheric Physics*, 72, 67–77.

Godde, C. M., Boone, R. B., Ash, A. J., Waha, K., Sloat, L. L., Thornton, P. K., Herrero, M. 2020. Global rangeland production systems and livelihoods at threat under climate change and variability. *Environmental Research Letters*, 15 (4), art. no. 044021.

Graham, E. A., Hamilton M. P., Mishler, B. D., Rundel, P. W., Hansen, M. H. 2006. Use of a networked digital camera to estimate net CO_2 uptake of a desiccation-tolerant moss. *International Journal of Plant Sciences*, 167, 751–758.

Grandy, A. S., Wieder, W. R., Wickings, K., Kyker-Snowman, E., 2016. Beyond microbes: Are fauna the next frontier in soil biogeochemical models? *Soil Biology Biochemistry*, 102, 40–44. https://doi.org/10.1016/j.soilbio.2016.08.008

Grote, E. E., Belnap, J., Housman, D. C., Sparks, J. P. 2010. Carbon exchange in biological soil crust communities under differential temperatures and soil water contents: Implications for global change. *Global Change Biology*, 16, 2763–2774.

Hatfield, J. L., Boote, K. J., Kimball, B. A., Ziska, L. H., Izaurralde, R. C., Ort, D., Thomson, A. M., Wolfe, D. 2011. Climate impacts on agriculture: Implications for crop production. *Agronomy Journal*, 103, 351–370. https://doi.org/10.2134/agronj2010.0303

Havstad, K. M., Peters, D. P. C., Skaggs, R., Brown, J., Bestelmeyer, B., Fredrickson, E., Herrick, J., Wright, J. 2007. Ecological services to and from rangelands of the United States. *Ecological Economics*, 64, 261–268.

Hengl, T., Mendes de Jesus, J., Heuvelink, G.B., Ruiperez Gonzalez, M., Kilibarda, M., Blagotić, A., Shangguan, W., Wright, M.N., Geng, X., Bauer-Marschallinger, B. 2017. SoilGrids250m: global gridded soil information based on machine learning. *PLoS ONE*, 12, art. no. e0169748.

Hennen, M., Chappell, C., Edwards, B. L., Faist, A. M., Kandakji, T., Baddock, M. C., Wheeler, B., Tyree, G., Treminio, R., Webb, N. P. 2022. A North American dust emission climatology (2001–2020) calibrated to dust point sources from satellite observations. *Aeolian Research*, 54, 100766.

Herbel, C., Nelson, A. B. 1966. Activities of Hereford and Santa Gertrudis cattle on a Southern New Mexico range. *Journal of Range Management*, 19, 173–176.

Herrick, J. E., Bestelmeyer, B. T., Archer, S., Tugel, A. J., Brown, J. R. 2006. An integrated framework for science-based arid land management. *Journal of Arid Environments*, 65, 319–335. https://doi.org/10.1016/j.jaridenv.2005.09.003

Herrick, J. E., Brown, J. R., Bestelmeyer, B. T., Andrews, S. S., Baldi, G., Davies, J., Duniway, M., Havstad, K. M., Karl, J. W., Karlen, D. L., Peters, D. P. C., Quinton, J. N., Riginos, C., Shaver, P. L., Steinaker, D., Twomlow, S. 2012. Revolutionary land use change in the 21st century: Is (Rangeland) science relevant? *Rangeland Ecology and Management*, 65 (6), 590–598.

Herrick, J. E., Whitford, W. G., de Soyza, A. G., van Zee, J. W., Havstad, K. M., Seybold, C. A., Walton, M. 2001. Field soil aggregate stability kit for soil quality and rangeland health evaluations. *Catena*, 44, 27–35.

Hirmas, D. R., Amrhein, C., Graham, R. C. 2010. Spatial and process-based modeling of soil inorganic carbon storage in an arid piedmont. *Geoderma*, 154, 486–494. https://doi.org/10.1016/j.geoderma.2009.05.005

Hoellrich, M.R., James, D.K., Bustos, D., Darrouzet-Nardi, A., Santiago, L.S., Pietrasiak, N. 2023. Biocrust carbon exchange varies with crust type and time on Chihuahuan Desert gypsum soils. *Frontiers in Microbiology*, 14, 1128631. https://doi.org/10.3389/fmicb.2023.1128631

Holechek, J. L. 1988. An approach for setting the stocking rate. *Rangelands*, 10, 10–14.

Holechek, J. L., Geli, H. M. E., Cibils, A. F., Sawalhah, M. N. 2020. Climate change, rangelands, and sustainability of ranching in the Western United States. In: Sustainability (Switzerland). MDPI, 12,4942. https://doi.org/10.3390/su12124942

Holechek, J. L., Pleper, A. D., Herbel, C. H. 1998. *Range Management Principles and Practices*. 3rd edition. Prentice-Hall Inc.

Holechek, J. L., Valdez, R. 2018. Wildlife conservation on the rangelands of Eastern and Southern Africa: Past, Present, and Future. *Rangeland Ecology and Management*, 71, 245–258. https://doi.org/10.1016/j.rama.2017.10.005

Hoover, D. L., Bestelmeyer, B., Grimm, N. B., Huxman, T. E., Reed, S. C., Sala, O., Seastedt, T. R., Wilmer, H., Ferrenberg, S. 2020. Traversing the Wasteland: A framework for assessing ecological threats to drylands. *BioScience*, 70, 35–47. https://doi.org/10.1093/biosci/biz126

Horwath, W. 2015. Carbon cycling: The dynamics and formation of organic matter. In: Paul, E. (Ed.) *Soil Microbiology, Ecology and Biochemistry* 4th edition, Elsevier Science and Technology, pp. 339–382. https://doi.org/http://dx.doi.org/10.1016/B978-0-12-415955-6.00012-8

Hosseiniaghdam, E. 2021. Quantifying the combined effect of abiotic factors on the decomposition of organic matter in semiarid grassland soils. University of Nebraska.

Ingram, L. J., Stahl, P. D., Schuman, G. E., Buyer, J. S., Vance, G. F., Ganjegunte, G. K., Welker, J. M., Derner, J. D. 2008. Grazing impacts on soil carbon and microbial communities in a mixed-grass ecosystem. *Soil Science Society of America Journal*, 72, 939–948. https://doi.org/10.2136/sssaj2007.0038

Izaurralde, R. C., Thomson, A. M., Morgan, J. A., Fay, P. A., Polley, H. W., Hatfield, J. L. 2011. Climate impacts on agriculture: Implications for forage and rangeland production. *Agronomy Journal*, 103, 371–381. https://doi.org/10.2134/agronj2010.0304

Izaurralde, R. C., Williams, J. R., McGill, W. B., Rosenberg, N. J., Jakas, M. C. Q. 2006. Simulating soil C dynamics with EPIC: Model description and testing against long-term data. *Ecological Modelling*, 192, 362–384. https://doi.org/10.1016/j.ecolmodel.2005.07.010

Jackson, R. B., Lajtha, K., Crow, S. E., Hugelius, G., Kramer, M. G., Pineiro, G. 2017. The ecology of soil carbon: Pools, vulnerabilities, and biotic and abiotic controls. *Annual Review of Ecology, Evolution, and Systematics*, 48, 419–445.

Jarrah, M., Mayel, S., Tatarko, J., Funk, R., Kuka, K. 2020. A review of wind erosion models: Data requirements, processes, and validity. *Catena*, 187, 104388. https://doi.org/10.1016/j.catena.2019.104388

Jenny, H. 1941. *Factors of Soil Formation: A System of Quantitative Pedology*. McGraw-Hill Book Company, Inc..

Jickells, T. D., An, Z. S., Andersen, K. K., Baker, A. R., Bergametti, G., Brooks, N., Cao, J. J., Boyd, P. W., Duce, R. A., Hunter, K. A., Kawahata, H., Kubilay, N., la Roche, N., Liss, P. S., Mahowald, N., Prospero, J. M., Ridgwell, A. J., Tegen, I., Torres, R. 2005. Global iron connections between desert dust, ocean biogeochemistry, and climate. *Science*, 308, 67–71.

Jobbagy, E., Jackson, R. 2000. The vertical distribution of soil organic carbon and its relation to climate and vegetation. *Ecological Applications*, 10, 423–436.

Karl, M. G., Pyke, D. A., Tueller, P. T., Schuman, G. E., Vinson, M. R., Fogg, J. L., Shaffer, R. W., Borchard, S. J., Ypsillants, W. G., Barret Jr. R. H. 2003. Indicators for soil and water conservation on rangelands. http://sustainablerangelands.warnercnr.colostate.edu/2003Report/2003Report.htm

Kay, F. R., Hafez, H. S., Whitford, W. G. 1999. Soil microarthropods as indicators of exposure to environmental stress in Chihuahuan Desert rangelands. *Biology and Fertility of Soils*, 28, 121–128.

Khumalo, G., Holechek, J. 2005. Relationships between Chihuahuan desert perennial grass production and precipitation. *Rangeland Ecology & Management*, 58, 239–246. https://doi.org/10.2111/1551-5028(2005)58[239:RBCDPG]2.0.CO;2

Khumalo, G., Holechek, J., Thomas, M., Molinar, F. 2007. Long-term vegetation productivity and trend under two stocking levels on Chihuahuan Desert rangeland. *Rangeland Ecology and Management*, 60, 165–171. https://doi.org/10.2111/06-061R3.1

Knapp, A. K., Beier, C., Briske, D. D., Classen, A. T., Luo, Y., Reichstein, M., Smith, M. D., Smith, S. D., Bell, J. E., Fay, P. A., Heisler, J. L., Leavitt, S. W., Sherry, R., Smith, B., Weng, E. 2008. Consequences of more extreme precipitation regimes for terrestrial ecosystems. *BioScience*, 58, 811–821. https://doi.org/10.1641/B580908

Knight, K. B., Toombs, T. P., Derner, J. D. 2011. Cross-fencing on private US rangelands: Financial costs and producer risks. *Rangelands*, 33, 41–44. https://doi.org/10.2111/1551-501X-33.2.41

Koyama, A., Harlow, B., Evans, R. D. 2019. Greater soil carbon and nitrogen in a Mojave Desert ecosystem after 10 years exposure to elevated CO_2. *Geoderma*, 355, 113915. https://doi.org/10.1016/j.geoderma.2019.113915

Kuske, C. R., Yeager, C. M., Johnson, S., Ticknor, L. O., Belnap J. 2012. Response and resilience of soil biocrust bacterial communities to chronic physical disturbance in arid shrublands. *The ISME Journal*, 6, 886–897.

Lal, R. 1997. Degradation and resilience of soils. *Philosophical Transactions of the Royal Society B: Biological Sciences*, 352, 997–1010

Lal, R. 2003. Soil erosion and the global carbon budget. *Environment International*, 29, 437–450.

Lal, R. 2004. Carbon sequestration in dryland ecosystems. *Environmental Management*, 33, 528–544. https://doi.org/10.1007/s00267-003-9110-9

Lal, R., Lorenz, K., Huttl, R. F., Schneider, B. U, Von Braun, J. 2012. *Recarbonization of the Biosphere: Ecosystems and the Global Carbon*. Springer Netherlands. doi:10.1007/978-94-007-4159-1

Lal, R., Monger, C., Nave, L., Smith, P. 2021. The role of soil in regulation of climate. *Philosophical Transactions of the Royal Society B: Biological Sciences*, 376, 210084. https://doi.org/10.1098/rstb.2021.0084

Lal, R., Stewart, B. A. 2013. Soil management for sustaining ecosystem services. In: Lal, R., Stewart, B. A. (Eds.), *Principles of Sustainable Soil Management in Agroecosystems*. CRC Press, pp. 521–536. doi: 10.1201/b14972

Lan, S., Wu, L., Zhang, D., Hu, C. 2012. Successional stages of biological soil crusts and their microstructure variability in Shapotou region (China). *Environmental Earth Sciences*, 65, 77–88. https://doi.org/10.1007/s12665-011-1066-0

Lee, J. A., Gill, T. A. 2015. Multiple causes of wind erosion in the Dust Bowl. *Aeolian Research*, 19, 15–36.

Lee, J. A., Tchakerian, V. P. 1995. Magnitude and frequency of blowing dust in the Southern High Plains of the United States, 1947–1989. *Annals of the Association of American Geographers*, 85, 684–693.

Li, C., Frolking, S., Harriss, R. 1994. Modeling carbon biogeochemistry in agricultural soils. *Global Biogeochemistry Cycles*, 8, 237–254. https://doi.org/10.1029/94GB00767

Li, J., Okin, G. S., Alvarez, L., Epstein, H., 2007. Quantitative effects of vegetation cover on wind erosion and soil nutrient loss in a desert grassland of southern New Mexico, USA. *Biogeochemistry*, 85, 317–332.

Li, J., Okin, G. S., Epstein, H. E. 2009. Effects of enhanced wind erosion on surface soil texture and characteristics of windblown sediments. *Journal of Geophysical Research,* 114, G02003. doi:10,1029/2008JG000903

Liu, J., Kimura, R., Miyawaki, M., Kinugasa, T. 2021. Effects of plants with different shapes and coverage on the blown-sand flux and roughness length examined by wind tunnel experiments. *Catena*, 197, 104976.

Lohmann, D., Tietjen, B., Blaum, N., Joubert, D. F., Jeltsch, F. 2014. Prescribed fire as a tool for managing shrub encroachment in semi-arid savanna rangelands. *Journal of Arid Environments*, 107, 49–56. https://doi.org/10.1016/J.JARIDENV.2014.04.003

Lucero, M. E., Dreesen, D. R., VanLeeuwen, D. M. 2010. Using hydrogel filled, embedded tubes to sustain grass transplants for arid land restoration. *Journal of Arid Environments*, 74, 987–990. https://doi.org/10.1016/j.jaridenv.2010.01.007

Ma, L., Dernera, J. D., Harmel, R. D., Tatarkoa, J., Moore, A. D., Rotz, C. A., Augustine, D. J., Boone, R. B., Coughenoure, M. B., Beukesf, P. C., van Wijkg, M. T., Bellocchih, G., Culleni, B. R., Wilmera, H. 2019. Application of grazing land models in ecosystem management: Current status and next frontiers. *Advances in Agronomy*, 158, 173–215. https://doi.org/10.1016/bs.agron.2019.07.003

Mager, D. M., Thomas, A. D. 2011. Extracellular polysaccharides from cyanobacterial soil crusts: A review of their role in dryland soil processes. *Journal of Arid Environments*, 75, 91–97, doi: 10.1016/j.jaridenv.2010.10.001

Mahapatra, S., Banerjee, D. 2013. Fungal exopolysaccharide: Production, composition and applications. *Microbiology Insights*, 6, 1–16. doi: 10.4137/MBI.S10957

Mathieu, J. A., Hatté, C., Balesdent, J., Parent, É. 2015. Deep soil carbon dynamics are driven more by soil type than by climate: A worldwide meta-analysis of radiocarbon profiles. *Global Change Biology*, 21, 4278–4292. https://doi.org/10.1111/gcb.13012

McIntosh, M. M. 2021. Sustainable grazing management in the Chihuahuan desert: Traditional and novel approaches to adapt to a changing climate. Dissertation. NMSU.

McIntosh, M. M., Cibils, A. F., Estell, R. E., Nyamuryekung'e, S., González, A. L., Gong, Q., Cao, H., Spiegal, S. A., Soto-Navarro, S. A., Blair, A. D. 2021. Weight gain, grazing behavior and carcass quality of desert grass-fed Rarámuri Criollo vs. crossbred steers. *Livestock Science*, 249. https://doi.org/10.1016/j.livsci.2021.104511

McIntosh, M. M., Gonzalez, A. L., Cibils, A. F., Estell, R. E., Nyamuryekung'e, S., Almeida, F. A. R., Spiegal, S. A. 2020. *Archivos Latinoamericanos de Producción Animal*, 28, 3–4.

McIntosh, M. M., Holechek, J. L., Spiegal, S. A., Cibils, A. F., Estell, R. E. 2019. Long-term declining trends in chihuahuan desert forage production in relation to precipitation and ambient temperature. *Rangeland Ecology and Management*, 72, 976–987. https://doi.org/10.1016/j.rama.2019.06.002

McIntosh, M. M., Spiegal S. A., McIntosh, S. Z., Estell, R. E., Castano Sanchez, J., Steele, C. M., Elias, E. H., Bailey, D. W., Brown, J. R., Cibils, A. F., 2023. Matching beef cattle breeds to the environment for desired outcomes in a changing climate: A systematic review with meta-analysis. *Journal of Arid Environments*, 211, 104905 https://doi.org/10.1016/j.jaridenv.2022.104905

McKeon, G. M., Stone, G. S., Syktus, J. I., Carter, J. O., Flood, N. R., Ahrens, D. G., Bruget, D. N., Chilcott, C. R., Cobon, D. H., Cowley, R. A., Crimp, S. J., Fraser, G. W., Howden, S. M., Johnston, P. W., Ryan, J. G., Stokes, C. J., Day, K. A. 2009. Climate change impacts on northern Australian rangeland livestock carrying capacity: A review of issues. *Rangeland Journal*, 31, 1–29. https://doi.org/10.1071/RJ08068

McMaster, G., Ascough, J., Dunn, G., Weltz, M., Shaffer, M., Palic, D., Vandenberg, B., Bartling, P., Edmunds, D., Hoag, D., Ahuja, L. 2002. Application and testing of GPFARM: A farm and ranch decision support system for evaluating economic and environmental sustainability of agricultural enterprises. *Acta Horticulturae*, 593, 171–177. https://doi.org/10.17660/actahortic.2002.593.22

Millar, C. I., Stephenson, N. L., Stephens, S. L. 2007. Climate change and forests of the future: Managing in the face of uncertainty. *Ecological Applications*, 17, 2145–2151.

Monger, H. C. 2014. Soils as generators and sinks of inorganic carbon in geologic time. In: Hartemink, A., McSweeney, K. (Eds.), *Soil Carbon. Progress in Soil Science*. Springer, https://doi.org/10.1007/978-3-319-04084-4_3

Monger, H. C., Kraimer, R. A., Khresat, S., Cole, D. R., Wang, X., Wang, J. 2015. Sequestration of inorganic carbon in soil and groundwater. *Geology*, 43, 375–378. https://doi.org/10.1130/G36449.1

Monger, H. C., Martinez-Rios, J. J. 2001. Inorganic carbon sequestration in grazing lands. In: Follett, R. F., Kimble, J. M., Lal, R. (Eds.), *The Potential of U.S. Grazing Lands to Sequester Carbon and Mitigate the Greenhouse Effect*. Lewis Publishers, pp. 87–118.

Moore, A. D., Eckard, R. J., Thorburn, P. J., Grace, P. R., Wang, E., Chen, D. 2014. Mathematical modeling for improved greenhouse gas balances, agro-ecosystems, and policy development: Lessons from the Australian experience. *Wires Climate Change*, 5, 735–752. doi:10.1002/wcc.304

Morgan, J. A., Milchunas, D. G., Lecain, D. R., West, M., Mosier, A. R., Mooney, H. A. 2007. Carbon dioxide enrichment alters plant community structure and accelerates shrub growth in the shortgrass steppe. *PNAS*, 104, 14724–14729.

Mun, H. T., Whitford, W. 1997. Changes in mass and chemistry of plant roots during long-term decomposition on a Chihuahuan Desert watershed. *Biology and Fertility of Soils*, 26, 16–22.

Myers, J. P., Antoniou, M. N., Blumberg, B., Carroll, L., Colborn, L., Everett, L. G., Hansen, M., Landrigan, P. J., Lanphear, B. P., Mesnage, R., Vandenberg, L. N., vom Saal, F. S., Welshons, W., Benbrook, C. M. 2016. Concerns over use of glyphosate-based herbicides and risks associated with exposures: A consensus statement. *Environmental Health*, 15, 19. doi: 10.1186/s12940-016-0117-0

National Academy of Sciences (NAS). 2016. Nutrient *Requirements of Beef Cattle: 8th Revised Edition*. National Academies of Sciences, Engineering, and Medicine, Washington, DC, The National Academies Press. https://doi.org/10.17226/19014

Nauman, T. W., Duniway, M. C., Webb, N. P., Belnap, J. 2018. Elevated aeolian sediment transport on the Colorado Plateau, USA: The role of grazing, vehicle disturbance, and increasing aridity. *Earth Surface Processes and Landforms*, 43, 2897–2914. doi: 10.1002/esp.4457

Navarro, J. M., Galt, D., Holechek, J., Mccormick, J., Molinar, F.. 2002. Long-term impacts of livestock grazing on Chihuahuan Desert rangelands. *Journal of Range Management*, 55, 400–405.

Neff, J. C., Barger, N. N., Baisden, W. T., Fernandez, D. P., Asner, G. P. 2009. Soil carbon storage responses to expanding pinyon-juniper populations in southern Utah. *Ecological Applications*, 19, 1405–1416.

Nyamuryekung'e, S., Cibils, A. F., Estell, R. E., McIntosh, M., VanLeeuwen, D., Steele, C., González, A. L., Spiegal, S., Continanza, F. G. 2021. Foraging behavior of heritage versus desert-adapted commercial rangeland beef cows in relation to dam-offspring contact patterns. *Rangeland Ecology and Management*, 74, 43–49. https://doi.org/10.1016/j.rama.2020.11.001

Nyamuryekung'e, S., Cibils, A. F., Estell, R. E., Van Leeuwen, D., Spiegal, S., Steele, C., González, A. L., McIntosh, M. M., Gong, Q., Cao, H. 2022. Movement, activity, and landscape use patterns of heritage and commercial beef cows grazing Chihuahuan Desert rangeland. *Journal of Arid Environments*, 199, 104704. https://doi.org/10.1016/j.jaridenv.2021.104704

Nyamuryekung'e, S., Cibils, A. F., Estell, R. E., VanLeeuwen, D., Steele, C., Estrada, O. R., Almeida, F. A. R., González, A. L., Spiegal, S. 2020. Do young calves influence movement patterns of nursing Raramuri Criollo cows on rangeland? *Rangeland Ecology and Management*, 73, 84–92. https://doi.org/10.1016/j.rama.2019.08.015

Okin, G. S., de Las Heras, M. M., Saco, P. M., Throop, H. L., Vivoni, E. R., Parsons, A. J., Wainwright, J., Peters, D. P. C. 2015. Connectivity in dryland landscapes: Shifting concepts of spatial interactions. *Frontiers in Ecology and the Environment*, 13, 20–27. Ecological Society of America. https://doi.org/10.1890/140163

Okin, G. S., Gillette, D. A., Herrick, J. E., 2006. Multi-scale controls on and consequences of aeolian processes in landscape change in arid and semi-arid environments. *Journal of Arid Environments*. Special Issue Landscape Linkages and Cross Scale Interactions in Arid and Semiarid Ecosystems, 65, 253–275.

Omari, H., Pietrasiak, N., Ferrenberg, S., Nishiguchi, M. K. 2022. A spatiotemporal framework reveals contrasting factors shaping biocrust microfloral and microfaunal assemblages in the Chihuahuan Desert. *Geoderma*, 405, 115409.

Parton, W. J., Del Grosso, S. J., Plante, A. F., Adair, E. C., Lutz, S. M. 2015. Modeling the dynamics of soil organic matter and nutrient cycling. In: Paul, E. A. (Ed.), *Soil Microbiology, Ecology and Biochemistry*. Elsevier, pp. 505–537. https://doi.org/10.1016/B978-0-12-415955-6.00017-7

Parton, W. J., Morgan, J., Kelly, R., Ojima, D. 2000. Modeling soil C responses to environmental change in grassland systems. In: Follett, R. F., Kimble, J. M., Lal, R. (Eds.), *The Potential of U.S. Grazing Lands to Sequester Carbon and Mitigate the Greenhouse Effect*. CRC Press, pp. 371–398. https://doi.org/10.1201/9781420032468.ch15

Parton, W. J., Ojima, D. S., Schimel, D. S. 1994. Environmental change in grasslands: Assessment using models. *Climate Change*, 28, 111–141. https://doi.org/10.1007/BF01094103

Parton, W. J., Scurlock, J. M. O., Ojima, D. S., Gilmanov, T. G., Scholes, R. J., Schimel, D. S., Kirchner, T., Menaut, J.-C., Seastedt, T., Garcia Moya, E., Kamnalrut, A., Kinyamario, J. I. 1993. Observations and modeling of biomass and soil organic matter dynamics for the grassland biome worldwide. *Global Biogeochemical Cycles*, 7, 785–809. https://doi.org/10.1029/93GB02042

Peinetti, H. R., Fredrickson, E. L., Peters, D. P. C., Cibils, A. F., Roacho-Estrada, J. O., Laliberte, A. S. 2011. Foraging behavior of heritage versus recently introduced herbivores on desert landscapes of the American Southwest. *Ecosphere*, 2, 1–14. https://doi.org/10.1890/ES11-00021.1

Peters, D. P. C., Bestelmeyer, B. T., Herrick, J. E., Fredrickson, E. L., Monger, H. C., Havstad, K. M. 2006. Disentangling complex landscapes: New insights into arid and semiarid system dynamics. *BioScience*, 56, 491–501.

Peters, D. P. C., Havstad, K. M., Archer, S. R., Sala, O. E. 2015. Beyond desertification: New paradigms for dryland landscapes. *Frontiers in Ecology and the Environment*, 13(1), 4–12.

Peters, D. P. C., Havstad, K. M., Cushing, J., Tweedie, C., Fuentes, O., Villanueva-Rosales, N. 2014a. Harnessing the power of big data: Infusing the scientific method with machine learning to transform ecology. *Ecosphere*, 5 (6), art. no. A67.

Peters, D. P. C., Loescher, H. W., Sanclements, M. D., Havstad, K. M. 2014b. Taking the pulse of a continent: Expanding site-based research infrastructure for regional- to continental-scale ecology (2014). *Ecosphere*, 5 (3), 1–23, https://doi.org/10.1890/ES13-00295.1

Petrie, M. D., Collins, S. L., Swann, A. M., Ford, P. L., Litvak, M. E. 2015. Grassland to shrubland state transitions enhance carbon sequestration in the northern Chihuahuan Desert. *Global Change Biology*, 21, 1226–1235. https://doi.org/10.1111/gcb.12743

Pietrasiak, N., Drenovsky, R. E., Santiago, L. S., Graham, R. C. 2014. Biogeomorphology of a Mojave Desert Landscape – configurations and feedbacks of abiotic and biotic land surfaces during landform evolution. *Geomorphology*, 206, 23–36.

Pietrasiak, N., Johansen, J. R., Drenovsky, R. E. 2011a. Geologic composition influences distribution of microbiotic crusts in the Mojave and Colorado Deserts at the regional scale. *Soil Biology and Biochemistry*, 43, 967–974.

Pietrasiak, N., Johansen, J. R., La Doux, T., Graham, R. C. 2011b. Spatial distribution and comparison of disturbance impacts to microbiotic soil crust in the Little San Bernardino Mountains of Joshua Tree National Park, California. *Western North American Naturalist*, 71, 539–552.

Pietrasiak, N., Regus, J. U., Johansen, J. R., Lam, D., Sachs, J. L., Santiago, L. S. 2013. Biological soil crust community types differ in key ecological functions. *Soil Biology and Biochemistry*, 65, 168–171.

Plaza, C., Zaccone, C., Sawicka, K., Méndez, A. M., Tarquis, A., Gascó, G., Heuvelink, G. B. M., Schuur, E. A. G., Maestre, F. T. 2018. Soil resources and element stocks in drylands to face global issues. *Scientific Reports*, 8. https://doi.org/10.1038/s41598-018-32229-0

Polley, H. W., Briske, D. D., Morgan, J. A., Wolter, K., Bailey, D. W., Brown, J. R. 2013. Climate change and North American rangelands: Trends, projections, and implications. *Rangeland Ecology and Management*, 66, 493–511. https://doi.org/10.2111/REM-D-12-00068.1

Pombubpa, N., Pietrasiak, N., De Ley P. Stajich, J. E. 2020. Insights into drylands biocrust microbiome: Geography, soil depth, and crust type affect biocrust microbial communities and networks in Mojave Desert, USA. *FEMS Microbial Ecology*, 96, fiaa125.

Power, M. E., Tilman, D., Estes, J. A., Menge, B. A., Bond, W. J., Mills, L. S., Daily, G., Castilla, J. C., Lubchenco, J., Paine, R. T. 1996. Challenges in the quest for keystones: Identifying keystone species is difficult-but essential to understanding how loss of species will affect ecosystems. *BioScience,* 46, 609–620. https://doi.org/10.2307/1312990

Rachal, D. M., Okin, G. S., Alexander, C., Herrick, J. E., Peters, D. P. C. 2015. Modifying landscape connectivity by reducing wind driven sediment redistribution Northern Chihuahuan Desert, USA. *Aeolian Research*, 17, 129–137. https://doi.org/10.1016/j.aeolia.2015.03.003

Raich, J. W., Schlesinger, W. H. 1992. The global carbon dioxide flux in soil respiration and its relationship to vegetation and climate. *Tellus* (44B), 81–99.

Ravi, S., Breshears, D. D., Huxman, T. E., D'Odorico, P. 2010. Land degradation in drylands: Interactions among hydrologic-aeolian erosion and vegetation dynamics. *Geomorphology,* 116, 236–245.

Raymond, P. A., Bauer, J. E. 2001. Riverine export of aged terrestrial organic matter to the North Atlantic Ocean. *Nature*, 409, 497–500.

Rivera-Aguilar, V., Montejano, G., Rodríguez-Zaragoza, S., Durán-Díaz, A. 2006. Distribution and composition of cyanobacteria, mosses and lichens of the biological soil crusts of the Tehuacán Valley, Puebla, México. *Journal of Arid Environments*, 67, 208–225.

Rook, A. J., Dumont, B., Isselstein, J., Osoro, K., WallisDeVries, M. F., Parente, G., Mills, J. 2004. Matching type of livestock to desired biodiversity outcomes in pastures – A review. *Biological Conservation*, 119, 137–150. https://doi.org/10.1016/j.biocon.2003.11.010

Rossi, F., De Philippis, R. 2015. Role of cyanobacterial exopolysaccharides in phototrophic biofilms and in complex microbial mats. *Life*, 5, 1218–1238. doi:10.3390/life5021218

Rossi, F., Mugnai, G., De Philippis, R. 2022. Cyanobacterial biocrust induction: A comprehensive review on a soil rehabilitation-effective biotechnology. *Geoderma*, 415, 115766. https://doi.org/10.1016/j.geoderma.2022.115766

Rotz, C. A. 2018. Modeling greenhouse gas emissions from dairy farms. *Journal. Of Dairy Science*, 101, 6675–6690. https://doi.org/10.3168/jds.2017-13272

Rotz, C. A., Asem-Hiablie, S., Place, S., Thoma, G. 2019. Environmental footprints of beef cattle production in the United States. *Agricultural Systems*, 169, 1–13.

Rowley, M. C., Grand, S., Verrecchia, É. P. 2018. Calcium-mediated stabilisation of soil organic carbon. *Biogeochemistry*, 137, 27–49. https://doi.org/10.1007/s10533-017-0410-1

Rumpel, C., Kögel-Knabner, I. 2011. Deep soil organic matter—A key but poorly understood component of terrestrial C cycle. *Plant Soil*, 338, 143–158.

Russell, M. L., Bailey, D. W., Thomas, M. G., Witmore, B. K. 2012. Grazing distribution and diet quality of angus, brangus, and brahman cows in the chihuahuan desert. *Rangeland Ecology and Management*, 65, 371–381. https://doi.org/10.2111/REM-D-11-00042.1

San José, J. J., Bravo, C. R. 1991. CO_2 exchange in soil algal crusts occurring in the trachypogon savannas of the Orinoco Llanos, Venezuela. *Plant and Soil*, 135, 233–244.

Sancho, L. G., Belnap, J., Cplesie, C., Raggio, J., Weber, B. 2016. Carbon budgets of biological soil crusts at micro-, meso-, and global scales. In: Weber, B., Büdel, B., Belnap, J. (Eds.), *Biological Soil Crusts: An Organizing Principle in Drylands*. Springer International Publishing, pp. 287–304 (Ecological Studies).

Sanderman, J., Berhe, A. A. 2017. The soil carbon erosion paradox. *Nature Climate Change*, 7, 317–319.

Sanderman, J., Chappell, A. 2013. Uncertainty in soil carbon accounting due to unrecognized soil erosion. *Global Change Biology*, 19, 264–272.

Satoh, K., Hirai, M., Nishio, J., Yamaji, T., Kashino, Y., Kioke, H. 2002. Recovery of photosynthetic systems during rewetting is quite rapid in a terrestrial cyanobacterium, Nostoc commune. *Plant and Cell Physiology*, 43, 170–176.

Scasta, J. D., Lalman, D. L., Henderson, L. 2016. Drought mitigation for grazing operations: Matching the animal to the environment. *Rangelands*, 38, 204–210. https://doi.org/10.1016/j.rala.2016.06.006

Schlesinger, W. H. 1985. The formation of caliche in soils of the Mojave Desert, California. *Geochimica Et Cosmochimica Acta,* 49, 57–66.

Schlesinger, W. H. 2002. Inorganic carbon and the global cycle. In: Lal, R. (Ed.), *Encyclopedia of Soil Science*. Marcel Dekker, pp. 27–42.

Schlesinger, W. H. 2017. An evaluation of abiotic carbon sinks in deserts. *Global Change Biology* 23, 25–27.

Schuurman, G. W., Cole, D. N., Cravens, A. E., Covington, S., Crausbay, S. D., Hoffman, C. H., Lawrence, D. J., Magness, D. R., Morton, J. M., Nelson, E. A., O'Malley, R. 2022. Navigating ecological transformation: Resist–accept–direct as a path to a new resource management paradigm. *BioScience*, 72, 16–29. https://doi.org/10.1093/biosci/biab067

Secretariat of the Convention on Biological Diversity, Global Mechanism of the United Nations Convention to Combat Desertification and OSLO consortium (SCBD). 2013. Valuing the biodiversity of dry and sub-humid lands. Technical Series No. 71. Secretariat of the Convention on Biological Diversity, Montreal, 94 pp.

Shao, Y. 2004. Simplification of a dust emission scheme and comparison with data. *Journal of Geophysical Research,* 109, D10202, doi:10.1029/2003JD004372

Shao, Y., Wyrwoll, K.-H., Chappell, A., Huang, J., Lin, Z., McTainsh, G. H., Mikami, M., Tanaka, T. Y., Wang, X., Yoon, S. 2011. Dust cycle: An emerging core theme in Earth system science. *Aeolian Research*, 2, 181–204.

Shibata, M., Terada, F. 2010. Factors affecting methane production and mitigation in ruminants. *Animal Science Journal*, 81, 2–10. https://doi.org/10.1111/j.1740-0929.2009.00687.x

Six, J., Bossuyt, H., Degryze, S., Denef, K. 2004. A history of research on the link between (micro)aggregates, soil biota, and soil organic matter dynamics. *Soil and Tillage Research*, 79, 7–31.

Skjemstad, J. O., Clarke, P., Taylor, J. A., Oades, J. M., Mcclure, S. G. 1996. The chemistry and nature of protected carbon in soil. *Australian Journal of Soil Research*, 34, 251–271.

Spaeth, K. E., Pierson, F. B., Herrick, J. E., Shaver, P. L., Pyke, D. A., Pellant, M., Thompson, D., Dayton, B. 2003. New proposed national resources inventory protocols on nonfederal rangelands. *Journal of Soil and Water Conservation*, 58, 18A–21A.

Spiegal, S., Estell, R. E., Cibils, A. F., James, D. K., Peinetti, H. R., Browning, D. M., Romig, K. B., Gonzalez, A. L., Lyons, A. J., Bestelmeyer, B. T. 2019. Seasonal divergence of landscape use by heritage and conventional cattle on desert rangeland. *Rangeland Ecology and Management*, 72, 590–601. https://doi.org/10.1016/j.rama.2019.02.008

Sterk, G., Herrmann, L., Bationo, A. 1996. Wind-blown nutrient transport and productivity changes in southwest Niger. *Land Degradation and Development*, 7, 325–335.

Stolpe, N., Muñoz, C., Zagal, E., Ovalle, C. 2008. Modeling soil carbon storage in the "Espinal" agroecosystem of central Chile. *Arid Land Research and Management*, 22, 148–158. https://doi.org/10.1080/15324980801958042

Stovall, M., Ganguli, A. G., Faist, A., Schallner, J. W., Yu, Q., Pietrasiak, N. 2022. Can biological soil crusts still be prominent landscape components in rangelands? A case study from New Mexico, USA. *Geoderma*, 410, 115658.

Svejcar, L. N., Kildisheva, O. A. 2017. The age of restoration: Challenges presented by dryland systems. *Plant Ecology*, 218, 1–6. Springer Netherlands. https://doi.org/10.1007/s11258-016-0694-6

Svejcar, T., Angell, R., Bradford, J. A., Dugas, W., Emmerich, W., Frank, A. B., Gilmanov, T., Haferkamp, M., Johnson, D. A., Mayeux, H., Mielnick, P., Morgan, J., Saliendra, N. Z., Schuman, G. E., Sims, L., Snyder, K. 2008. Carbon fluxes on North American Rangelands. *Rangeland Ecology and Management*, 61, 465–474.

Tamm, A., Caesar, J., Kunz, N., Colesie, C., Reichenberger, H., Weber, B. 2018. Ecophysiological properties of three biological soil crust types and their photoautotrophs from the Succulent Karoo, South Africa. *Plant Soil*, 429, 127–146.

Tfaily, M. M., Hess, N. J., Koyama, A., Evans, R. D. 2018. Elevated [CO_2] changes soil organic matter composition and substrate diversity in an arid ecosystem. *Geoderma*, 330, 1–8. https://doi.org/10.1016/j.geoderma.2018.05.025

Thomas, M. G., Mohamed, A. H., Sawalhah, M. N., Holechek, J. L., Bailey, D. W., Hawkes, J. M., Luna-Nevarez, P., Molinar, F., Khumalo, G. 2015. Long-term forage and cow-calf performance and economic considerations of two stocking levels on chihuahuan desert rangeland. *Rangeland Ecology and Management*, 68, 158–165. https://doi.org/10.1016/j.rama.2015.01.003

Throop, H. L., Archer, S. R., Monger, H. C., Waltman, S. 2012. When bulk density methods matter: Implications for estimating soil organic carbon pools in rocky soils. *Journal of Arid Environments*, 77, 66–71. https://doi.org/10.1016/j.jaridenv.2011.08.020

Throop, H. L., Belnap, J. 2019. Connectivity dynamics in dryland litter cycles: Moving decomposition beyond spatial stasis. *Bioscience*, 69, 602–614.

Throop, H. L., Lajtha, K. J. 2018. Spatial and temporal changes in ecosystem carbon pools following juniper encroachment and removal. *Biogeochemistry*, 140, 373–388.

Torell, G., Torell, L. A., Gonzalez, A. L., Cibils, A. F., Estell, R. E., Diaz, J., Anderson, D. M., Rotz, C. A., McIntosh, M. M., Spiegal, S. 2023. Economics of Raramuri Criollo and British crossbred cattle production in the Chihuahuan Desert: Effects of foraging distribution and finishing strategy. *Journal of Arid Environments*, 213, 104922. https://doi.org/10.1016/j.jaridenv.2022.104922

Trumper, K., Ravilious, C., Dickson, B. 2008. Carbon in drylands: Desertification, climate change and carbon finance. Prepared on behalf of UNEP by UNEP-WCMC.

Umstatter, C. 2011. The evolution of virtual fences: A review. *Computers and Electronics in Agriculture*, 75, 10–22. https://doi.org/10.1016/j.compag.2010.10.005

Van Oost, K., Quine, T. A., Govers, G., De Gryze, S., Six, J., Harden, J. W., Ritchie, J. C., McCarty, G. W., Heckrath, G., Kosmas, C., Giraldez, J. V., Marques da Silva, J. R., Merckx, R. 2007. The impact of agricultural soil erosion on the global carbon cycle. *Science*, 318, 626–629.

Van Oost, K., Six, J. 2022. The soil carbon erosion paradox reconciled, Biogeosciences Discuss. [preprint], https://doi.org/10.5194/bg-2022-1, in review.

Vetter, S. 2005. Rangelands at equilibrium and non-equilibrium: Recent developments in the debate. *Journal of Arid Environments*, 62, 321–341. https://doi.org/10.1016/j.jaridenv.2004.11.015

Wand, S. J. E., Midgley, G. F., Jones, M. H., Curtis, P. S. 1999. Responses of wild C4 and C3 grass (Poaceae) species to elevated atmospheric CO_2 concentration: A meta-analytic test of current theories and perceptions. *Global Change Biology*, 5, 723–741. https://doi.org/10.1046/j.1365-2486.1999.00265.x

Wang, B., Waters, C., Orgill, S., Cowie, A., Clark, A., Li Liu, D., Simpson, M., McGowen, I., Sides, T., 2018. Estimating soil organic carbon stocks using different modelling techniques in the semi-arid rangelands of eastern Australia. *Ecological Indicators*, 88, 425–438. https://doi.org/10.1016/j.ecolind.2018.01.049

Wang, G., Li, J., Ravi, S., Dukes, D., Howell, B. G., Sankey, J. B. 2019. Post-fire redistribution of soil carbon and nitrogen at a grassland–shrubland ecotone. *Ecosystems*, 22, 174–188.

Wang, J., Monger, C., Wang, X., Serena, M., Leinauer, B. 2016. Carbon sequestration in response to grassland-shrubland-turfgrass conversions and a test for carbonate biomineralization in desert soils, New Mexico, USA. *Soil Science Society of America Journal*, 80, 1591–1603. https://doi.org/10.2136/sssaj2016.03.0061

Webb, N. P., Chappell, A., Strong, C. L., Marx, S. K., McTainsh, G. H. 2012b. The significance of carbon-enriched dust for global carbon accounting. *Global Change Biology*, 18, 3275–3278.

Webb, N. P., Herrick, J. E., Duniway, M. C. 2014. Ecological site-based assessments of wind and water erosion: Informing accelerated soil erosion management in rangelands. *Ecological Applications*, 24, 1405–1420.

Webb, N. P., LeGrand, S. L., Cooper, B. F., Courtright, E. M., Edwards, B. L., Felt, C., Van Zee, J. W., Ziegler, N. P. 2021. Size distribution of mineral dust emissions from sparsely vegetated and supply-limited dryland soils. *Journal of Geophysical Research: Atmospheres*, 126, e2021JD035478.

Webb, N. P., Stokes, C. J., Scanlan, J. C., 2012a. Interacting effects of vegetation, soils and management on the sensitivity of Australian savanna rangelands to climate change. *Climatic Change*, 112, 925–943.

Webb, N. P., Strong, C. L., Chappell, A., Marx, S. K., McTainsh, G. H. 2013. Soil organic carbon enrichment of dust emissions: Magnitude, mechanisms and its implications for the carbon cycle. *Earth Surface Processes and Landforms*, 38, 1662–1671.

Weber, B., Belnap, J., Büdel, B. 2016. Synthesis on biological soil crust research. In: Weber, B., Büdel, B., Belnap, J. (Eds.), *Biological Soil Crusts: An Organizing Principle in Drylands*. Springer International Publishing, pp. 527–534 (Ecological Studies).

West, N. E., Stark, J. M., Johnson, D. W., Abrams, M. M., Ross Wight, J., Heggem, D., Peck, S. 1994. Effects of climatic change on the edaphic features of arid and semiarid lands of Western North America. *Arid Soil Res. Rehabil.* 8, 307–351. https://doi.org/10.1080/15324989409381408

Whitford, W. G. 2010. Keystone arthropods as webmasters in desert ecosystems. In: Coleman, D. C., Hendrix, P. F. (Eds.), *Invertebrates as Webmasters in Ecosystems*. CABI Publishing, pp. 25–41. https://doi.org/10.1079/9780851993942.0025

Williams, A. J., Buck, B. J., Soukup, D. A., Merkler, D. J. 2013. Geomorphic controls on biological soil crust distribution: A conceptual model from the Mojave Desert (USA). *Geomorphology*, 195, 99–109.

Williams, A. P., Cook, B. I., Smerdon, J. E. 2022. Rapid intensification of the emerging southwestern North American megadrought in 2020–2021. *Nature Climate Change*, 12, 232–234. https://doi.org/10.1038/s41558-022-01290-z

Williams, C. J., Pierson, F. B., Spaeth, K. E., Brown, J. R., Al-Hamdan, O. Z., Weltz, M. A., Nearing, M. A., Herrick, J. E., Boll, J., Robichaud, P. R., Goodrich, D. C., Heilman, P., Guertin, D. P., Hernandez, M., Wei, H., Hardegree, S. P., Strand, E. K., Bates, J. D., Metz, L. J., Nichols, M. H. 2016. Incorporating hydrologic data and ecohydrologic relationships into ecological site descriptions. *Rangeland Ecology and Management*, 69, 4–19.

Willig, M. R., Presley, S. J. 2018. Biodiversity and disturbance. *Encyclopedia of the Anthropocene*, 1–5, 45–51. https://doi.org/10.1016/B978-0-12-809665-9.09813-X

Winder, J. A., Walker, D. K., Bailey, C. C. 1996. Effect of breed on botanical composition of cattle diets on Chihuahuan desert range. *Journal of Range Management*, 49, 209–214.

Wu, L., Lei, Y., Lan, S., Hu, C. 2017. Photosynthetic recovery and acclimation to excess light intensity in the rehydrated lichen soil crusts. *PLoS ONE*, 12(3), e0172537

Wuebbles, D. J., K. Hayhoe. 2002. Atmospheric methane and global change. *Earth-Science Reviews*, 57, 177–210.

Yahdjian, L., Sala, O. E., Havstad, K. M. 2015. Rangeland ecosystem services: Shifting focus from supply to reconciling supply and demand. *Frontiers in Ecology and the Environment*, 13, 44–51.

Zhang, B., Zhang, Y., Li, X., Zhang, Y. 2018. Successional changes of fungal communities along the biocrust development stages. *Biology and Fertility of Soils*, 54, 285–294. doi: 10.1007/s00374-017-1259-0

Zhang, Y., Lavallee, J. M., Robertson, A. D., Even, R., Ogle, S. M., Paustian, K., Cotrufo, M. F. 2021. Simulating measurable ecosystem carbon and nitrogen dynamics with the mechanistically defined MEMS 2.0 model. *Biogeosciences*, 18, 3147–3171. https://doi.org/10.5194/bg-18-3147-2021

Zhao, M. S., Heinsch, F. A., Nemani, R. R., Running, S. W. 2005. Improvements of the MODIS terrestrial gross and net primary production global data set. *Remote Sensing Environment*, 95, 164–176.

Zilverberg, C. J., Williams, J., Jones, C., Harmoney, K., Angerer, J., Metz, L. J., Fox, W. 2017. Process-based simulation of prairie growth. *Ecological Modelling*. 351, 24–35. https://doi.org/10.1016/j.ecolmodel.2017.02.004

Zuberer, D. A., Hallmark, C. T., Wilding, L. P. 1996. Processes and forms of carbon sequestration in calcareous soils of arid and semi-arid regions. Project Report, NRCS Global Change Program. Texas A&M University, College Station, TX.

4 Crop Nutrition Management for Semiarid Areas of Sub-Saharan Africa with Increasingly Variable Climate

Charles Wortmann[1], Aliou Faye[2], Maman Garba[3], Idriss Serme[4], and Zachary P. Stewart[5]

[1]Department of Agronomy and Horticulture, University of Nebraska-Lincoln, Lincoln NE 68506. cwortmann2@unl.edu;

[2]ISRA – Centre of Excellence on Dry Cereals and Associated Crops (CERAAS), BP 3320 Thies Senegal. aliouselbe11@gmail.com

[3]Département Gestion des Ressources Naturelles (DGRN), Institut National de la Recherche Agronomique du Niger: Niamey, Niamey, NE. maman_garba@yahoo.fr;

[4]Intitut de l'Environnement et de Recherches Agricoles, 04 B.P.8645 Ouagadougou 04, Burkina Faso; sermeidriss@yahoo.fr; 00226 70232198;

[5]Center for Agriculture-Led Growth, Bureau for Resilience and Food Security, United States Agency for International Development, Washington, DC 20004

CONTENTS

4.1 Background	84
4.2 Conditioning Soil and Fields for Sustainable Production Increases	85
4.2.1 Reduced Tillage	85
4.2.2 Barriers to Runoff	86
4.2.3 Perennials in Rotation or Permanently with Annual Crops	87
4.2.3.1 Trees and Shrubs	88
4.2.3.2 Perennial Grass	91
4.3 Fertilizer Use Optimization	92
4.3.1 Optimum Fertilizer Rates	92
4.3.2 Fertilizer Use Synergisms and Targeting	95
4.4 Adapting Crop Nutrition Management for More Challenging Rainfall Distribution	98
4.5 Translating Science into Action	99
4.6 Conclusions	101
References	101

DOI: 10.1201/b22954-4

4.1 BACKGROUND

The semiarid areas of sub-Saharan Africa (SSA) include the Sahel and most of the Sudan savanna extending from southern Senegal to South Sudan and Ethiopia; parts of the Rift Valley; southern Somalia; some coastal and upland areas of Kenya; central Tanzania; southern parts of Angola, Zambia, Zimbabwe, Mozambique, and Madagascar: northeastern Namibia; most of Botswana; and northeastern South Africa. While population statistics are not reported by ecozones, >200 million people live in semiarid areas of SSA. Some of these areas have very high population growth rates, such as 3.8% in 2020 for Niger (World Population Review). The food security and economy of these populations are heavily dependent on semiarid crop production. Production increases in SSA have been attributed to increased cropland area, while the average yield increase has been about 1% yr^{-1} compared with 5% yr^{-1} in Asia (Henao and Baanante, 2006).

Soil water deficits are frequent in semiarid areas due to low annual rainfall relative to potential evapotranspiration. These low ratios are often associated with sandy soil with low soil organic matter content and very low plant available water holding capacity, such as with Arenosols in the Sahel, Somalia, Namibia, and Botswana. The greatest yield loss associated with soil water deficits in semiarid areas is generally due to inadequate water during grain-fill (Wortmann et al., 2009; Wildemeersch et al., 2015a) although poor crop establishment and mid-season stress are also important. Rainfall is distributed across >6 months in two growing seasons near the Equator rather than in 3–4 months with monomodal rainfall distribution and one growing season at >10° latitude from the Equator. Less monthly rainfall during a cropping season adds to the occurrence of soil water deficits.

The expectation of higher temperatures and intensification of rainfall with heavier rainfall events and longer dry spells due to global warming will greatly affect future crop production for many semiarid areas. Panthou et al. (2018), for example, analyzed weather data from several decades and 121 Sahelian sites and found that the rate of warming for the Sahel, especially for Senegal, has been greater than the global average rate, with more frequent occurrence of relatively hot years since 2000. They also found that since 1975 the number of high rainfall events and duration of dry spells has increased, especially in eastern Mali. The resulting increase in potential evapotranspiration likely with less rainfall infiltration and more runoff and erosion implies more frequent occurrence and longer duration of soil water deficits. Similar evidence is provided by Sylla et al. (2016) and UNHCR (2022). Other predictions indicate that the average onset of the wet season will be delayed while the cessation will be earlier (Saar, 2012). Less well-documented are evidence of similar trends for other semiarid areas of SSA.

Smallholder farmers account for a very high percentage of the production in semiarid SSA. These smallholders typically operate under severe constraints. They are generally severely constrained financially with little capacity for investment, implying that investments need to bring high returns in the short term with little risk. They are heavily reliant on family manual labor and family illnesses often prevent timely implementation of field operations. They commonly have only partial control of their cropland and may have uncontrolled grazing during the dry season and insecurity of long-term investments in land improvement. Their poverty makes them very vulnerable to risk when their few assets can be wiped out in a single bad year or with a single disaster. Recovery from disaster often requires years when successful.

Cropping and farming systems are diverse across these semiarid areas. Integration of ruminant livestock with crop production is common. Cropping systems may include monoculture but more typical are crop rotations and intercropping. Major crops include grain sorghum (*Sorghum* spp.), pearl millet (*Pennisetum glaucum* (L.) R. Br.), sesame (*Sesamum indicum* L.), cowpea (*Vigna unguiculata*), and groundnut (*Arachis hypogaea* L.). Maize (*Zea mays* L.) and rice (*Oryza sativa* L.) are often important in low-lying areas. Also important are teff (*Eragrostis tef*) in Ethiopia and fonio (*Digitaria exilis*) in West Africa. Farming practices typically are with tillage, little input use, and crop residue harvest, and without irrigation. Compared with temperate areas, in tropical areas

Crop Nutrition for Semiarid Africa

insect pest and diseases are typically more constraining to crop productivity and seasonal radiation is less.

Given the climatic, edaphic, economic, and management constraints, crop yields are low for semiarid SSA, but the food produced is important for food security and the overall livelihoods of the farming families, their communities and their countries. Human populations have increased more than production increases. Unfortunately, yields generally have not increased in recent decades and production increases are primarily due to increased cropland area. Food insecurity has been estimated to increase by 43% over 20 yr in Africa due to anticipated growth in population exceeding growth in production (Funk and Brown, 2009). Worsening climate may result in a greater more food insecurity which implies declines in gross domestic product and health.

Farming in SSA has been associated with net nutrient losses due to nutrient harvest, erosion, and little fertilizer use. Stoorvogel et al. (1993) estimated net loss of N, P, and K to average 22 kg, 2.5 kg, and 15 kg, respectively, in 1982–1984, and predicted greater rates of loss in the future. Henao and Baanante (2006) estimated the total of N, P, and K losses from cropland in SSA to be <30 kg ha^{-1} yr^{-1} for one country, 30–60 kg ha^{-1} yr^{-1} for 19 countries, and >60 kg ha^{-1} yr^{-1} for another 19 countries. Net nutrient losses with no fertilizer applied were greater with maize and soybean production in eastern Uganda than with other crops or pasture, but nutrient losses from sloping cropland could be reduced by 50% through use of living barriers or micro-catchments to reduce erosion (Wortmann and Kaizzi, 1998).

Nutrient management for semiarid areas of SSA with increasingly variable climate needs to go well beyond good fertilizer and manure use, which must be well integrated with optimal agronomic practices which will vary by socioeconomic and biophysical situation. Reduced tillage such as with conservation agriculture, physical water entrapment to reduce runoff, and soil conditioning for better water infiltration and resistance to erosion such as with well-managed perennials in the cropping system are addressed below. Fertilizer use, with and without manure application, is addressed for maximizing profit per hectare as well as profit from financially constrained affordable investments, that is, profit:cost ratios (PCR).

4.2 CONDITIONING SOIL AND FIELDS FOR SUSTAINABLE PRODUCTION INCREASES

Important to future crop nutrition management in semiarid areas, likely with more intense rainfall events and longer periods of soil water deficits, will be the conditioning of soils and fields for greater resilience accompanied by reduced soil erosion and runoff. Greater soil aggregate stability for improved rainwater infiltration and resistance to water and wind erosion will be needed. Part of this resilience will be achieved through more vigorous crop growth due to major changes in agronomic management. Improved ground cover by living plants and plant residues together with reduced tillage will be important for trapping sediment and reducing erosion. Land application of diverse organic materials, hereafter referred to as manure, will continue to be important; these materials are already mostly used but there may be opportunities for enhanced agronomic and economic synergy with other practices. The interaction of fertilizer use optimization with soil conditioning practices, with the expectation of more difficult rainfall patterns, is also addressed below.

4.2.1 REDUCED TILLAGE

Reduced tillage on its own or integrated with other practices such as rotation and leaving much crop residue in the field, as in conservation agriculture, can improve water infiltration and reduce evaporation for productivity, economic, and environmental benefits (Kaizzi et al., 2007; Serme et al., 2015; Nansamba et al., 2016; Pittelkow et al., 2017). It must be well targeted in consideration of climate, soil, and management variables, with greater benefit expected with hot and semiarid areas compared with cooler and more humid areas, and with consistent use over time, especially where soils are prone to crusting (Ikpe et al., 1999; Liben et al., 2017, 2018; Laborde et al., 2020; Mupangwa et al., 2020). The benefits of reduced runoff and erosion with low tillage and conservation agriculture may increase with more erratic weather. The value of crop residue retention in fields has been recognized by several researchers (e.g., Mason et al., 2014; Coulibaly et al., 2020), but the residues are commonly harvested for livestock feeding and other competing needs. The benefits of crop rotation are often known by farmers and others, but economics often favors cereal production (Mason et al., 2014). The benefits of rotation can be at least partly achieved by intercropping which is often preferred by smallholders.

4.2.2 Barriers to Runoff

There are localized traditions and recent adoption of several management options designed to prevent runoff and erosion, such as tie-ridging, basin-farming, *zai*, stone lines, grass bands, and semi-circular bunds. Tie-ridging or furrow-diking perpendicular to the slope is primarily for row-crop production, with planting on ridges of maybe 0.15-m height or in the furrows (Jones and Clark, 1987; Sanders et al., 1990; van Duivenbooden et al., 2000; Gusha, 2002) (Photo 4.1).

The ridges are separated by approximately 0.75 m or more, with the furrows tied at intervals of >2m. Tie-ridging has had great impact on yield where rainfall infiltration rate is slow, such as with silt loam Entisols prone to crusting and Vertisols that have slow infiltration once the surface soil is wet (Brhane et al., 2006; Brhane and Wortmann, 2008; Mesfin et al., 2009). The soils of the latter studies included Vertisols, Andosols, and Entosols, with ranges of clay to sandy loam for soil texture

PHOTO 4.1 Tie-ridging can be highly effective in preventing runoff with soils of low permeability, even with low slope. Photo credit: Charles Wortmann.

PHOTO 4.2 A basin tillage practice known as *zai* in western Africa has been effective in preventing runoff and increasing productivity. Photo credit: Vara Prasad.

and from 0.02 to 0.04 m m^{-1} for slope. There is often a strong synergistic effect of tie-ridging with fertilizer use (Sanders et al., 1990).

Basin-farming or -tillage is commonly used for row-crop production with basins that might be 0.75 m × 1.45 m in size, with the placement of plant residues, manure, fertilizer, and seed in the basins (Jones and Stewart, 1990). The basins prevent runoff and erosion. A variation of basin tillage is the traditional practice of *zai* or micro-basin farming in the Sudan savanna of West Africa for prevention of runoff (Photo 4.2).

Zai pits are 0.2–0.3 m wide and 0.1–0.2 m deep (Fatondji et al., 2001). Yields have been much increased with *zai* (Fatondji et al., 2006; Mason et al., 2015; Wildemeersch, 2015b). The use of semicircular bunds, sometimes accompanied with rock barriers, is another form of basin farming (Reij et al., 2009) (Photo 4.3). The bunds partly encircle flat bottoms that are typically >1 m in width.

These various types of basins are sometimes maintained with no tillage except for reforming such as at weeding. Landscape position is important for tie-ridging and basin-farming as too much runoff entering a field or too much slope may result in runoff breaking the furrows or bunds with more erosion and crop destruction than might have occurred with flat surface farming. In some cases, fields need barriers or diversions to prevent erosive runoff from entering the field.

4.2.3 Perennials in Rotation or Permanently with Annual Crops

Well-managed perennial trees, shrubs, and grasses can be a means for maintaining and improving soil productivity and annual crop response to applied nutrients, especially when the perennials have economic value. The benefits of such perennials in cropping systems are often recognized by farmers for improving annual crop productivity while yielding valuable products. Local policies that restrict farmer control of the land may constrain inclusion of perennials in annual cropping systems (Boffa, 1999). Of greatest concern is the common practice of uncontrolled grazing during the dry season that prevents the successful establishment of trees and shrubs, and good management of perennial grass ley.

PHOTO 4.3 Semicircular bunds trap runoff in semiarid areas with marginal soils. The upper photo shows the half-moon formation with compost recently applied. The second photo shows pearl millet growth in protected area. Photo credits: Zachary Stewart.

4.2.3.1 Trees and Shrubs

Trees provide benefits beyond improving soil productivity (Mafongoya et al., 2006; Sileshi and Mafongoya, 2006; Akinnifesi et al., 2008), such as through production of fodder, fruits, timber, and fuel wood (Garrity et al., 2010). In some cases, the value of tree production exceeds the value of annual crop. The soil productivity gain may be through a combination of erosion control with sediment trapping (Boffa, 1999; Nomaou et al., 2015), improved soil aggregation and water infiltration (Chirwa et al., 2007), improved soil microbiology (Faye et al., 2009), biological nitrogen fixation and nutrient

PHOTO 4.4 Parkland farming such as with annual crops grown in stands of faidherbia trees (upper; Photo credit: Zachary Stewart) or a mix of trees can greatly increase annual crop yield while yielding tree products. Photo credit: Serme Idriss.

cycling (Barnes and Fagg, 2003), manure excretion by livestock seeking shade or feeding under the trees, pest and weed suppression (Sileshi and Mafongoya, 2006), soil organic C (Akinnifesi et al., 2007) and soil inorganic C sequestration (Lal and Kimble, 2000), and hydraulic lift (Bogie et al., 2018).

Trees may be included in cropping systems by intercropping with rows of trees alternated with strips of annual crops with regular pruning or coppicing of the trees to prevent excessive height

PHOTO 4.5 Pearl millet growth is much greater near faidherbia trees compared with further away. Photo credit: Zachary Stewart.

and width (Sileshi and Mafongoya, 2006). Intercropping may also be by dispersal of single trees in the annual crop fields as is often practiced with faidherbia (*Faidherbia albida* (Delile) A.Chev.) in "parkland" farming for semiarid areas (Garrity et al., 2010) (Photo 4.4).

Another option is to rotate two or more years of tree fallow with periods of annual crop production (Mafongoya et al., 2006; Akinnifesi et al., 2009). The intercropping and rotation options may be more appropriate for more humid areas, while parkland agriculture is more common in semiarid areas. Crop growth is commonly suppressed in the vicinity of trees in more humid conditions and with better soil. However, annual crop growth near trees in semiarid "parklands" is commonly much better than growth further from the trees likely due to a combination of benefits addressed above (Saka et al., 1994).

Especially effective for improving crop yields is faidherbia in semiarid areas. It is an N-fixing species indigenous to Africa that is especially suited to parkland agriculture in semiarid areas because of its "reverse phenology." It sheds leaves during the wet season and produces leaves during the dry season (Barnes and Fagg, 2003) (Photo 4.5).

Therefore, even though the trees may be large with wide canopies, there is little shading of the annual crops and trees provide shade and fodder during the dry season. Fodders from faidherbia leaves and pods are greatly valued in livestock feeding and sold as a relatively high-priced feed supplement in Niamey. Other tree species including several *Acacia* and *Combretum* species, shea (*Vitellaria paradoxa*), and tamarind (*Tamarindus indica*) are also valuable in parkland agriculture, but shading may constrain crop growth under the tree canopy (Ouattara et al., 2017).

Crop growth near shrubs is also commonly better than further from the shrub in some semiarid areas. *Guiera senegalensis* is a common indigenous species in western Africa of great interest (Dossa et al., 2012, 2013). It has a hydraulic lifting and redistribution property by which water is brought up through a taproot with excretion at night providing water to neighboring crop plants (Nomaou et al., 2015; Bogie et al., 2018) (Photo 4.6).

The shrubs are often browsed, and branches are periodically harvested for wood. Other shrub species, including leguminous *Philiostigma reticulatum* (DC.) Hochst., also enhance crop productivity while providing dry season fodder and wood.

PHOTO 4.6 Pearl millet growth near the shrub *Guiera senegalensis* is much greater compared to further from the shrub. The shrub can take up deep soil water and release it to nearby plants. Photo credit: Zachary Stewart.

4.2.3.2 Perennial Grass

Perennial grass barriers that are perpendicular to slopes or the prevailing wind of the dry season may improve water infiltration while reducing runoff and wind and water erosion (Black and Siddoway, 1976). Such barriers of elephant grass (*Pennisetum purpureum*) or *Brachiaria* spp. are common in humid crop-dairy farms for fodder production and erosion control (Odero-Waitituh, 2017).

Inclusion of well-managed perennial grass ley in rotation with periods of annual crop production has great potential for semiarid areas. Wortmann et al. (2021) determined that rotation of periods of perennial grass ley with periods of annual crop production is likely to increase annual crop yields sufficiently to partly compensate for less total grain production over time compared with continuous annual cropping. In addition, the fodder produced may be of greater value than the grain produced from a similar land area. The ley rotation is expected to reduce erosion and increase rainwater infiltration, trap sediment, and improve soil fertility and annual crop response to fertilizer. Expected additional benefits from ley rotation are increased farm profitability, reduced economic risk, reduced labor ha^{-1}, better distribution of labor demand, and increased biodiversity. A major obstacle to such ley rotations is inadequate grazing control.

The ley needs to be well managed for good grass establishment with good above- and belowground growth in contrast to natural fallow which relies on natural regeneration and is severely and continuously overgrazed, resulting in little above- and belowground growth and little soil improvement (Wortmann et al., 2021). Grass establishment can be slow in semiarid areas but andropogon (*Andropogon guyanus* Knuth) is easily established from rooted tillers, including by intercropping with cowpea or early harvested pearl millet (Bowden, 1963). Andropogon can be highly productive in semiarid areas, including with sandy soil (Stephens, 1960; Diatta et al., 1997). Andropogon forms reedy growth which can be prevented to get high-quality fodder by harvesting at mid-season and again during the dry season when soil water is depleted (Bowden, 1963).

The grass can be in wide strips and alternated with similar strips for annual crops placed perpendicular to the slope or to prevailing dry season winds to prevent runoff and erosion by water and wind. The stripes can be gradually shifted over time, such as by converting part of the strip width each year. For example, a 15-m-wide strip in a 5-yr rotation may have 3 m yr^{-1} converted to cropland each year while providing planting material to add 3 m of grass to the other side of the ley strip.

PHOTO 4.7 Net returns to investment in fertilizer use can be high at low rates, such as with micro-dosing, for some nutrients applied to some crops (Photo credit: Vara Prasad).

Sediment trapping, increased water infiltration attributed to increased formation of earthworm casts (Wilkinson, 1975), increased soil organic C (Jones, 1971) which likely resulted in a small but valuable increase in plant available soil water holding capacity (Minasny and Mcbratney, 2018), and deep nutrient cycling (Wortmann et al., 2021) during the ley stage contribute to increased productivity. The soil improvements can, however, be short-lived with much tillage and termination of the grass, and annual crop management should be with little or no tillage. Cowpea or another legume crop might follow the ley to minimize problems of N immobilization associated with the large amounts of above- and belowground plant biomass returning to the soil with ley termination. Annual crop response to fertilizer-N can be much greater with ley rotation compared with continuous annual crop production (Stephens, 1960).

Much of the research with ley rotations in Africa was conducted decades ago but the results are relevant when interpreted in consideration of more recent findings in the Americas and Europe (Wortmann et al., 2021). Farmer acceptance of replacing annual crops, even though of very low productivity, with well-managed perennial grass ley is a concern. However, in many places, especially in and along roads to urban and peri-urban areas, there is high market demand for fodder (Gomma et al., 2017; Samireddypalle et al., 2017). This demand can make perennial grass production highly profitable in addition to the increased productivity during the annual crop phase.

4.3 FERTILIZER USE OPTIMIZATION

4.3.1 Optimum Fertilizer Rates

Increased fertilizer use is needed for sustainable intensification of food crop production in the semiarid areas of SSA, but little fertilizer is used due to financial and other constraints. Crop yields could be much increased with more fertilizer use but under current economic conditions, profit-maximizing yield is about one-quarter of the biophysically possible yield (Bonilla-Cedrez et al.,

Crop Nutrition for Semiarid Africa

2021). Perceptions of high profitability at low risk are major drivers for farmer adoption of new practices (Wortmann et al., 2020). When finance is adequate for fertilizer use, farmers might strive to maximize profit ha^{-1} from fertilizer use ($ ha^{-1}). However, when fertilizer use is financially constrained as it is with most smallholder farmers, profit earned on the affordable investment in fertilizer (PCR, $ $$^{-1}$) is most important. The profit potential of fertilizer use varies greatly for different crop–nutrient–rate choices as has been often reported (e.g., Kaizzi et al., 2017; Maman et al., 2017a; Ouattara et al., 2017; Liben et al., 2020b).

Fertilizer use rates for profit optimization are expected to be low in semiarid areas compared with more productive situations. Microdosing of fertilizer use has been promoted where fertilizer is point-applied near plants at a blanket rate expected to give some yield increase and high PCR (Bagayoko et al., 2011) (Photo 4.7).

An approach to optimization of profit from fertilizer use that considers crop–nutrient response functions determined from many trials, the fertilizer costs, the crop values, and farmer's financial capacity for fertilizer use was applied by the 13-country research network called Optimization of Fertilizer Use for Africa (Kaizzi et al., 2017).

Much research appropriate for the optimization of fertilizer use in SSA has been conducted. Wortmann (2020) published a data set consisting of >6,100 geo-referenced response functions determined from field trials conducted in SSA. The data originated from >250 sources which are cited in the data set. The trials were conducted on land considered common for crop production in the area and were not conducted on cropland with unusually severe edaphic constraints such as unusually shallow depth, high sand and gravel content, steep slope, or Al toxicity. The trials determined responses to N, P, and K individually at a range of rates sufficient to calculate response functions. Expecting that in most cases, N was more limiting than P and P more limiting than K, response to P was determined with N uniformly applied for non-legumes, and to K with N and P uniformly applied for all crops. Management was according to researchers' choice of practices and fertilizer-N was typically split-applied. Early application of fertilizer was by different means and times, with the most common being broadcast application before tillage, or band or point application at planting or after emergence. All cases had some tillage and no irrigation. The geo-referenced yield response results were standardized as curvilinear to plateau response functions according to the following equation:

$$\text{Yield} = (\text{Mg ha}^{-1}) = a - bc^r \qquad (\text{Eq. } 4.1)$$

where a was the yield at the plateau for the nutrient application, b was the maximum increase in yield due to the nutrient application, c (<1.0) was a determinant of the response curve shape, and r was the elemental nutrient application rate (kg ha^{-1}) (Jansen et al., 2013; Kaizzi et al., 2017).

Data from sites with annual precipitation to annual potential evaporation ratios of <0.4 were analyzed to determine typical responses for semiarid areas in SSA using the methodology of Wortmann and Stewart (2021) (Figure 4.1). The analyses were for cowpea, groundnut, pearl millet, maize, irrigated rice, and sorghum, and for responses to fertilizer-N, -P, and -K. The experiments were conducted after 1985. The number of experiments included in the analysis ranged from 255 for N applied to maize to 19 for K applied to pearl millet with a median of 60 experiments per crop–nutrient combination (Table 4.1). The median responses were used to calculate the overall crop–nutrient response functions for semiarid areas. Only the yield responses to nutrient application were evaluated here rather than the actual yield so that a and b from the above asymptotic equation were equal.

These yield response functions were used to calculate the economically optimal nutrient rates (EOR), that is, the rate of maximum profit ha^{-1}, and the expected mean net returns to fertilizer use ($ ha^{-1}) for each crop–nutrient response at EOR. Crop prices and fertilizer use costs (US$ kg^{-1}), which in reality vary greatly across time and location, used in the analyses were as follows: 0.22 for maize,

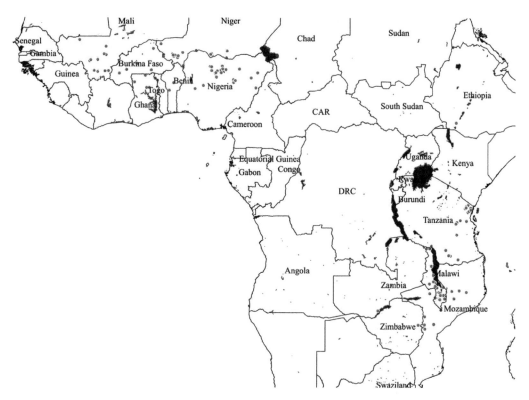

FIGURE 4.1 Maps showing the semiarid research sites that were sources of results used in an analysis of food crop responses to applied nutrients for sub-Saharan Africa.

0.25 for sorghum and pearl millet, 0.50 for rice, 0.55 for unshelled groundnuts, and 0.30 for cowpea; and 1.00 for N, 2.20 for P, and 0.90 for K. The PCR for crop–nutrient applications at 100% and 50% EOR were calculated as the difference in partial returns to fertilizer use minus the fertilizer use cost with the difference (net returns to fertilizer use) divided by the fertilizer use cost.

The median yield responses relative to the mean yield with no fertilizer were 16% for N, 11% for P, and 15% for K (Table 4.1; Figure 4.2). Relative responses were least with legumes and greatest with pearl millet for N and P and maize for K. There was little additional response to rates of >60 kg ha^{-1} N, >25 kg ha^{-1} P, and >30 kg ha^{-1} K for cereals. Half of the total response achieved with less than 20 kg ha^{-1} N, 10 kg ha^{-1} P, and 10 kg ha^{-1} K, and with lower rates for legumes. The standard errors of the mean for the overall maximum yield response are not reported here but were often high showing the variability in response. As percent of the mean maximum yield response, the standard errors of the mean ranged from 7% for maize P to 126% for cowpea P, with overall medians of 17% for N, 30% for P, and 28% for K.

The EOR in the semiarid areas ranged from 12 kg ha^{-1} N for cowpea to 51 kg ha^{-1} N for lowland rice, 0 kg ha^{-1} P for groundnut to 17 kg ha^{-1} P for maize, and 12 kg ha^{-1} K for groundnut to 34 kg ha^{-1} K for rice (Table 4.2; Figure 4.2). Net returns to fertilizer use were especially high with applications of N, P, and K to rice; P and K to maize; and K to sorghum (Figure 4.3; Table 4.2). The profit potential was intermediate with some application of N to maize; N and K to cowpea; K to groundnut; and N to pearl millet. There was relatively low profit potential with application of P to cowpea, N to groundnut, P and K to pearl millet, and N and P to sorghum. With equal land area for each crop, the mean PCR was 2.2 $ $^{-1}$ at 100% EOR and 4.3 $ $^{-1}$ at 50% EOR (Table 4.2).

Therefore, the potential PCR is high with fertilizer use decisions that optimize choice among crop–nutrient–rate options in consideration of the response functions, fertilizer use costs, commodity

Crop Nutrition for Semiarid Africa

TABLE 4.1
Mean Yields of Crops in Semiarid Areas of Sub-Saharan Africa with No Fertilizer Applied and the Mean Yield Response (YR) Function Coefficients with YR = a − bcʳ Where a = b (Mg ha⁻¹) is the Maximum Response to the Applied Nutrient, c is a Curvature Coefficient, and r is the Rate of N, P, or K (kg ha⁻¹). Yields Were on an Air-Dried Basis. n = the Number of Georeferenced Functions Included in the Determination of the Crop–Nutrient Functions

Crop	Yield (Mg ha⁻¹), no fertilizer	--------- N ---------			------- P -------			-------- K ---------		
		a=b	c	n	a=b	c	n	a=b	c	n
Cowpea	0.67	0.098	0.825	50	0.057	0.836	59	0.100	0.853	23
Groundnut	0.84	0.072	0.902	65	0.031	0.894	61	0.068	0.863	22
Maize	2.32	0.550	0.943	255	0.584	0.933	190	0.631	0.940	77
Pearl millet	0.90	0.289	0.942	97	0.241	0.913	70	0.153	0.929	19
Rice	3.23	0.587	0.951	53	0.368	0.865	54	0.746	0.952	39
Sorghum	1.48	0.177	0.921	91	0.166	0.887	82	0.241	0.824	34

TABLE 4.2
The Economically Optimal Rates (EOR) for N, P, and K Application for Several Crops and the Profit to Cost Ratios by Crop–Nutrient in Semiarid Areas of Sub-Saharan Africa

Crop	N	P	K	N	P	K
	Economically optimal rate (EOR), kg ha⁻¹			Profit to cost ratio at EOR		
Cowpea	12	5	14	2.44	0.52	2.54
Groundnut	14	0	12	1.06		1.86
Maize	33	17	15	1.94	1.85	7.16
Pearl millet	24	10	15	1.18	0.64	0.89
Rice	51	16	34	3.15	0.47	8.95
Sorghum	16	7	13	0.93	0.45	3.73
Average				1.78	0.78	4.19

values, and money available to the farmer for investment in fertilizer use. Fertilizer use decision tools using linear programming for maximizing PCR for affordable investment have been developed in Excel, cell phone applications, and paper formats (Kaizzi et al., 2017). The validity of response functions does not extend to cropland with edaphic constraints that are unusually severe relative to local cropland conditions as such constraints are likely to prevent response to fertilizer application.

4.3.2 FERTILIZER USE SYNERGISMS AND TARGETING

The grain and stover yield response to fertilizer application for cereal crops in semiarid areas is often greater with improved practices, such as with tied ridging (Sanders et al., 1990; van Duivenbooden et al., 2000), increased crop residue retention (Bationo et al., 1993), annual crop rotation (Bationo and Ntare, 2000), tree fallow or perennial grass ley in rotation with annual cropping (Stephens, 1960; Akinnifesi et al., 2009), and manure application (Baidu-Forson and Bationo, 1992). Garba et al. (2018a) also found a synergism of manure plus fertilizer application for sandy soil in the Sahel but an

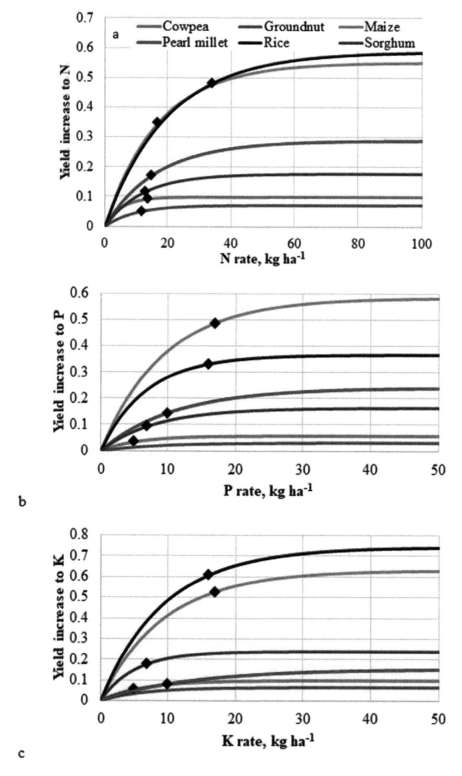

FIGURE 4.2 Grain yield increase (Mg ha^{-1}) due to fertilizer application of N (a), P (b), and K (c), with economical optimal rates indicated by black diamonds, for semiarid areas of sub-Saharan Africa.

Crop Nutrition for Semiarid Africa

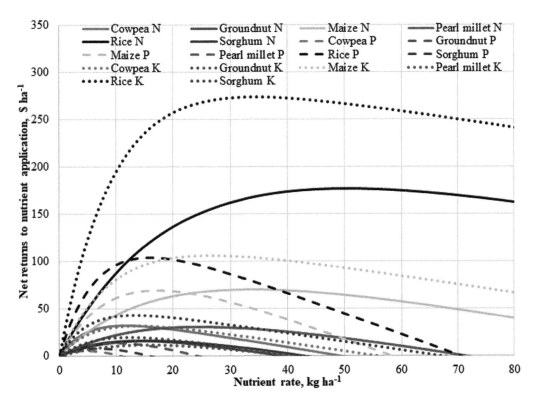

FIGURE 4.3 Net returns to use of fertilizer nutrients for semiarid areas of sub-Saharan Africa.

additive effect for loamy sand or sandy loam soil in the Sudan savanna zone with the additive effect seemingly mostly due to a soil amendment effect of the manure. The economic responses to fertilizer application were greater with intercropping than with sole crop and the base has been established for determining optimal fertilizer rates for intercropping using the sole crop response functions of the intercropped non-legume crop (Maman et al., 2017b, d, c; Ndungu-Magiroi et al., 2017; Maman et al., 2018). Positive N × P interactions can occur but more common are merely additive effects as found in the above-cited studies. The effects of water conservation practices, residue retention, and the effects of perennials in cropping systems on the EOR have not been adequately determined with a need for more research. The effects of manure application, manure type, the residual effect of manure, annual crop rotation, green manure crops, and agroforestry on fertilizer EOR have been estimated for several countries such as by Maman et al. (2017a), Ouattara et al. (2017), and Kibunja et al. (2017), but the information basis needs to be strengthened using more research results. Fertilizer rates are not commonly adjusted for tillage practices, but there is inconsistency in research results for the fertilizer-N rate × tillage interaction (Pittelkow et al., 2015; Liben et al., 2020a).

Optimization of fertilizer rates requires good targeting to technology extrapolation domains (Liben et al., 2020b; GYGA, 2022) or recommendation domains (Wortmann et al., 2017). This is indicated by the variation in EOR determined for sorghum in semiarid recommendation domains such as in Burkina Faso, Ethiopia, Kenya, Niger, and Tanzania (Demissie and Bekele, 2017; Kibunja et al., 2017; Maman et al., 2017a; Ouattara et al., 2017; Senkoro et al., 2017). The analysis presented above was with data across all semiarid recommendation domains to illustrate the value of crop-nutrient response functions for determining the most profitable fertilizer nutrient rates for farmers' financial ability to use fertilizer. Location specific recommendations were determined from data of relatively narrow ranges of inference (Wortmann et al., 2017).

Site-specific fertilizer use within fields or between fields within a recommendation domain may present an opportunity for improved overall response to fertilizer. Even with tailored fertilizer recommendations based on detailed soil mapping (EthioSIS, 2022), 20–80% of the sites had no response to applied fertilizer, depending on the region in Ethiopia, although response was associated with landscape position (Amede, 2018). Soil test information for nutrient availability has not been predictive of response to applied fertilizer nutrients (Kihara et al., 2016; Garba et al., 2018b). Within-field variation due to differences in soil sand content and soil rooting depth for effects on available water holding capacity, proximity to trees or shrubs, and topographic position can also guide site-specific application. There is a need for fertilizer use decisions that use context-specific, user-sourced information to account for field-level heterogeneity (Herrick et al., 2022) as demonstrated by Amede and Diallo (2022). In-season variation in canopy growth and color may indicate variation in sidedress N EOR such as with vigorous growth together with symptoms expressing N deficiency often indicating potential for profitability with a relatively high N rate.

In-season decisions, such as in consideration of plant density, can be important to fertilizer use optimization in semiarid areas (Bastos et al., 2022). This is especially important considering the weak financial status of most smallholder farmers who need to have a good chance of high PCR from fertilizer use within <1 yr. In contrast, more financially able farmers can wait for crop response in later years to the residual effects of unused fertilizer nutrients, especially if nutrient sorption is relatively low as it is for most soils in semiarid areas of SSA, with exceptions such as high P sorption with calcareous Vertisols (Mamo and Wortmann, 2009). Basal fertilizer application in semiarid areas is commonly delayed until the crop is well emerged realizing the risk of failed establishment (e.g. Maman et al., 2017a). The risk of failed establishment is even greater with dry soil planting even though this can be a valuable practice for taking full advantage of the wet season (Liben et al., 2015) and may become more important with future climate. Delayed application gives the opportunity to not apply fertilizer or to adjust rates depending on crop condition. The decision of side-dress application of fertilizer-N also needs to be dependent on the condition of the crop.

4.4 ADAPTING CROP NUTRITION MANAGEMENT FOR MORE CHALLENGING RAINFALL DISTRIBUTION

As mentioned above, rainfall distribution in at least some semiarid areas of SSA is likely to change for more intense rainfall events with increased potential for leaching, runoff, and erosion. Increased mean time between rainfall events with more frequent and longer periods of soil water deficit is expected. This is complicated by the expectation that farmers in these areas will be financially constrained in input use and highly vulnerable to risk. These factors will further challenge annual crop production and may drive an increase in perennials in cropping systems, especially if complemented by supportive land tenure policies. Annual crop production will increasingly need greater use of rainwater entrapment practices if not greater water infiltration capacity, such as with conservation agriculture or perennials in cropping systems.

More information complementary to local knowledge is needed for fine-tuning the targeting and local optimization of practices for water entrapment, conservation agriculture, and perennials in cropping systems. Experiential learning with smallholders who have a tradition with a practice is needed as a starting point for developing a basis for targeting and optimizing practices for diverse semiarid areas. For example, suitability of water entrapment practices is likely to be slope limited. As reported above, tie-ridging was highly effective at increasing yield for slopes of 0.02–0.06 m m^{-1} for a wide range of soil texture classes, but can the range be extended for some soil types? Large tree and shrub densities have been suggested above but are these always optimal and what are the constraints to achieving the desired densities? Optimization of perennial grass ley in cropping system has questions of species choice, establishment method, strip-cropping widths, duration of the rotated periods of ley and annual cropping, conversion of ley back to cropland with no or little

Crop Nutrition for Semiarid Africa

tillage, the optimal sequence of the following annual crops, and fertilizer use for the ley and the annual crops (Wortmann and Stewar 2021).

Available crop–nutrient response research results mostly predate the impact of global warming on the distribution of rainfall and the research was not conducted with water entrapment practices or with perennials in the cropping system. More information is needed to determine how annual crop responses to applied nutrients are affected by major changes in climate and agronomic practices. There is a lack of compelling evidence for having different nutrient management strategies with untilled compared to tilled soils (Liben et al., 2020a). Water entrapment is expected to increase yields and nutrient uptake even if it does not enhance nutrient availability and has increased leaching of nutrients. Inclusion of perennials in the cropping system is expected to increase crop yield and uptake of nutrients, including deep nutrients, and with increased biological N fixation. All of these practices are expected to enhance annual crop root growth which should favor nutrient uptake.

The research to fine-tune and better target the integration of practices for semiarid areas broadly should build on experiential learning with farmers who are experienced with a practice, systematically making and interpreting observations on the agronomics and socio-economics of the practice. Field experimentation will be needed to get additional information, such as to determine and interpret the interactions of crop–nutrient rate with alternative practices to provide the basis to adapt current crop–nutrient response functions for other agronomic practices and for future scenarios. The results of experiential learning and field experimentation need to support adaptation and calibration of crop growth simulation models for optimizing the integration of improved agronomic practices and fertilizer use decisions with future climate scenarios to extend the range of inference for current and newly acquired information. Functionally significant variables will gradually be identified and improved, with user input and verification, for continually improving the capacity for targeting and scaling of the integration of water conservation, soil improvement, and fertilizer use integration for resource-constrained farmers in diverse semiarid areas.

4.5 TRANSLATING SCIENCE INTO ACTION

The "Improvement of smallholder farming systems in Africa" was addressed in a symposium during the 2018 International Annual Meeting of the American Society of Agronomy. Much of the following has greater elaboration in the published results of this symposium (Wortmann et al., 2020). Most emphasized was the need for adequate infrastructure and services. The adequate infrastructure includes adequate roads and transport, markets, input supply, and financial services. The services include strong advisory service.

As discussed above, crop management for semiarid areas with increasingly variable climate will require integration of diverse practices, each well-targeted to a farmer's context. This implies more knowledge and information-intensive decision-making for the choice of best agronomic practices than what is currently common. While most farmers tend to be responsive to improved profit opportunities, the opportunities often will not be immediately obvious. National policies are needed to enable strong advisory services. These advisors will need to be capable of much more than spouting "universal" solutions or recommendations. They will need to work with smallholders, individually and in groups, to build on their knowledge of their multi-disciplinary situation and consider the best choice of practices. There is a need to better link researchers and other innovation generators to these advisors for turning science into farm-level action while enabling bidirectional learning and iterative adaptation.

The advisors must be equipped with appropriate decision-making tools such as for choice of tillage practices, the integration of perennials in cropping systems, and fertilizer use optimization. These tools must maintain precision across often heterogeneous farms and landscapes, while also being applicable at scale. Such approaches often require combining geospatially relevant information (top-down) paired with user-sourced site information (bottom-up). Research needs are addressed above and the resulting information and improvements in decision tools must flow quickly to advisors and farmers for fine-tuning of management practices, targeting, and decision-making. The advisors may be government, private sector, or non-government staff, but working with inter-institutional consensus to avoid conflicting messages. There are numerous factors to be considered in the evaluation of an innovation as demonstrated in Figure 2 of Wortmann et al. (2020) and Figure 2 of Wortmann et al. (2021). Therefore, they will need technical support from specialists to achieve positive spin-offs and avoid negative effects on family life and the environment, including agricultural, social, economic, health, nutrition, and other support. Ongoing learning with farmers will be important. Solid progress with some farmers needs to be recognized by later adopters for demand-driven extension with eventual achievement of spontaneous adoption. Often such innovations in advisory services can occur without complete reform of the service, but employment of and strong support to very capable and motivated advisors will be essential.

Much improvement can be achieved by farmers individually on their farms, such as profit optimization from fertilizer use or change in tillage practice. However, other essential improvements will often require community commitment and action, such as for the introduction of perennials into cropping systems which is feasible only if dry season grazing is controlled. Livestock are often a valuable component of semiarid farming systems in SSA and dry season grazing of croplands following harvest is a common practice. Often, the grazing is uncontrolled with all cropland open to herds of ruminant livestock, some from far away. Such uncontrolled grazing makes establishment and maintenance of trees, shrubs, and perennial grass very difficult if not impossible. Basins and other rainfall retainment structures are damaged. Complete removal of crop residues and repeated traffic of herds across fields is damaging to soil aggregation (Photo 4.8). Some communities may be able to control grazing of cropland but more often communities will need the support of national policies.

Fertilizer use subsidies can be valuable for short-term production increases and may strengthen the farmer's financial ability for fertilizer use in future years. When a policy of subsidizing fertilizer and hybrid seed varieties use was implemented in Malawi, maize production was greatly increased

PHOTO 4.8 Uncontrolled grazing often leaves field bare of crop residue (upper) and soil aggregates pulverized (lower). Photo credit: Zachary Stewart.

but subsidy in one year did not result in increased input use or production in subsequent years (Ricker-Gilbert and Jayne, 2016). This lack of enduring effect may have been due to some level of fertilizer and hybrid seed use prior to the subsidization. The enduring effect may be greater in situations of very little prior use of such inputs as apparently was the case in Mozambique (Carter et al., 2014).

Practices needed for improved resiliency and efficient production with increasingly variable climate are expected to increase soil organic C through sequestration. Payment for C sequestration credits is likely to motivate adoption (Manlay et al., 2020).

There is a need for sustained investment in coordination hubs that better link innovation generators to farmer advisors to better turn science into action. Clear prioritized agendas are needed with alignment between donors and policymakers with international, national, and more localized institutions and sectors involved in research, outreach, and supply of inputs and services. Such African-led coordination efforts may be improving such as through the African Union Commission's Fertilizer and Soil Health Summit Action Plan.

4.6 CONCLUSIONS

Crop nutrition management for semiarid areas of SSA with increasingly variable climate will require much change in agronomic practices even to maintain current yield levels with system resilience, while potential for profitable yield increases with resilience is possible. Maintenance of adequate soil water supply during the growing season is critical. The optimal practice will vary with field situations. It may require *in situ* rain water entrapment such as with some version of basin farming. It may require improved rain water infiltration independent of or together with entrapment, such as with integration of trees, shrubs, or perennial grass, including hydraulic lifting species, in annual crop production systems. Such practices will often need integration with low tillage, crop residue retention or manure use, and crop rotation or intercropping. Once the frequency and duration of soil water deficits are reduced, increased yield and response to fertilizer use will occur. Financially constrained smallholders will need high returns on use of inputs with little risk of failed returns, and crop–nutrient–rate choices according to the financial ability will be critical. Using experiential learning with farmers, field experimentation, and improved simulation models, functionally significant variables will need to be determined and applied for "leapfrogging" efficient targeting and optimization of the integration of water conservation, soil improvement, and fertilizer use integration for resource-constrained farmers in diverse semiarid areas. This integration is expected to have synergistic effects resulting in higher yields and profits compared with using any of these alone. Weak farmer control of their cropland will often obstruct the integration of the needed practices, especially uncontrolled dry season grazing which often prevents successful basin farming, crop residual retention, and establishment and maintenance of perennials in cropping systems. National and international efforts, such as those of the Africa Fertilizer and Soil Health Summits (IFDC, 2022), need to bring together national and international stakeholders, including donors, in addressing the research–farmer–market continuum, and the implementation of land tenure policy reforms. Improvements of rural infrastructure and creation of farmers services centers are of high priority, as is increased support to research. Local communities and farmers need more control over their cropland. Fertilizer use subsidizes and C sequestration credits can be useful when well-targeted.

REFERENCES

Akinnifesi, F.K., P.W. Chirwa, O.C. Ajayi, G. Sileshi, P. Matakala, F.R. Kwesiga, R. Harawa, and W. Makumba. 2008. Contributions of agroforestry research to livelihood of smallholder farmers in Southern Africa: 1. Taking stock of the adaptation, adoption and impact of fertilizer tree options. *Agricultural Journal* 3: 58–75.

Akinnifesi, F.K., W. Makumba, G. Sileshi, O.C. Ajayi, and D. Mweta. 2007. Synergistic effect of inorganic N and P fertilizers and organic inputs from *Gliricidia sepium* on productivity of intercropped maize in Southern Malawi. *Plant Soil* 294: 203–217.

Akinnifesi, F.K., G. Sileshi, S. Franzel, O.C. Ajayi, R. Harawa, W. Makumba, S. Chakeredza, S.A. Mng'omba, J. de Wolf, and J. Chianu. 2009. On-farm assessment of legume fallows and other fertility management options used by smallholder farmers in southern Malawi. *Agricultural Journal* 4: 260–271.

Amede, T. 2018. Feeding degraded soils in Ethiopia to feed the people and the environment. https://cgsp ace.cgiar.org/bitstream/handle/10568/91676/wheat_icrisat_2018.pdf?sequence=1andisAllowed=y. Accessed Aug 2022.

Amede, T. and A. Diallo. 2022. Enhancing agronomic efficiency of fertilizers in Sub-Saharan Africa: evidence from the field. *Growing Africa* 18.

Bagayoko, M., N. Maman, S. Pale, and S. Sirifi. 2011. Microdose and N and P fertilizer application rates for pearl millet in West Africa. *African Journal of Agricultural Science* 6: 1141–1150.

Baidu-Forson, J. and A. Bationo. 1992. An economic evaluation of a long-term experiment on phosphorus and manure amendments to sandy Sahelian soils: using a stochastic dominance model. *Fertilizer Research* 33: 193–202.

Barnes, R.D. and C.W. Fagg. 2003. Faidherbia albida. Monograph and Annotated Bibliography. Tropical Forestry Papers No 41, Oxford Forestry Institute, Oxford, UK. 281 pp.

Bastos, L.M., A. Faye, Z.P. Stewart, T.M. Akplo, D. Min, P.V. Prasad, and I.A. Ciampitti. 2022. Variety and management selection to optimize pearl millet yield and profit in Senegal. *European Journal of Agronomy* 139. Doi: 10.1016/j.eja.2022.126565.

Bationo, A., C.B. Christianson, and M.C. Klaij. 1993. The effect of crop residue and fertilizer use on pearl millet yields in Niger. *Fertilizer Research* 34: 251–258.

Bationo, A. and B. Ntare. 2000. Rotation and nitrogen fertilizer effect on pearl millet, cowpea and groundnut yield and soil chemical properties in a sandy soil in the semi-arid tropics, West Africa. *The Journal of Agricultural Science* 134: 277–284.

Black, A.L. and F.H. Siddoway. 1976. Dryland cropping sequences with a tall wheat-grass barrier. *Journal of Soil and Water Conservation* 31: 101–105.

Boffa, J.M. 1999. Agroforestry parklands in sub-Saharan Africa. FAO Conservation Guide 34, Food and Agriculture Organization, Rome. 254 pp.

Bogie, N.A., R. Bayala, I. Diedhiou, M.H. Conklin, M.L. Fogel, R.P. Dick, and T.R. Ghezzehei. 2018. Hydraulic redistribution by native sahelian shrubs: bioirrigation to resist in-season drought. *Frontiers in Environmental Science: Agroecology* 6.

Bonilla-Cedrez, C., J. Chamberlin, and R.J. Hijmans. 2021. Fertilizer and grain prices constrain food production in sub-Saharan Africa. *Nature Food* 2: 766–772.

Bowden, B.N. 1963. Studies on *Andropogon gayanus* Kunth. I. The use of *Andropogon gayanus* in agriculture. *Empire Journal of Experimental Agriculture* 31: 267–273.

Brhane, G., C.S. Wortmann, M. Mamo, H. Gebrekidan, and A. Belay. 2006. Micro-basin tillage for grain sorghum production in semi-arid areas of northern Ethiopia. *Agronomy Journal* 98: 124–128.

Brhane, G. and C.S. Wortmann. 2008. Tie-ridge tillage for high altitude pulse production in northern Ethiopia. *Agronomy Journal* 100: 447–453. Doi: 10.2134/agronj2007.0159

Carter, M.R., R. Laajaj, and D. Yang. 2014. Subsidies and persistent technology adoption: field experimental evidence from Mozambique. National Bureau of Economic Research Working Paper No. 20465.

Chirwa, P.W., C.K. Ong, J. Maghembe, and C.R. Black. 2007. Soil water dynamics in intercropping systems containing *Gliricidia sepium*, pigeon pea and maize in southern Malawi. *Agroforestry Systems* 69: 29–43.

Coulibaly, A., M. Bagayoko, S. Traore, and S.C. Mason. 2000. Effect of crop residue management and cropping system on pearl millet and cowpea yield. *African Crop Science Journal* 8: 1–8.

Demissie, N. and I. Bekele. 2017. Optimizing fertilizer use within an integrated soil fertility management framework in Ethiopia. In: C.S. Wortmann and K. Sones (Eds.), *Fertilizer Use Optimization in sub-Saharan Africa*. CABI, London, UK, pp. 52–66. ISBN 9781786392046. doi: 10.1079/9781786392046.0052

Diatta, A., A. Bodian and D. Babene. 1997. Production fourragère des graminées *Andropogon gayanus Kunth* et *Panicum maximum Jacq. CV. CI* utilisées en substitution de la jachère en Haute Casamance au Sénégal. In: C. Floret and R. Pontanier (Eds.), *Jachère et maintien de la fertilité*. CORAF Europe, pp. 127–131.

Dossa, E.L., I. Diedhiou, M. Khouma, and M.D. Sene. 2012. Crop productivity and nutrient dynamics in a shrub (Guiera senegalensis)-based farming system of the Sahel. *Agronomy Journal* 104: 1255–1264. doi: 10.2134/agronj2011.0399

Dossa, E.L., I. Diedhiou, M. Khouma, M.D. Sene, A.N. Badiane, S.A.N. Samba, K.B. Assigbetse, S. Sall, A. Lufla, F. Kizito, R.P. Dick, and I. Saxena. 2013. Crop productivity and nutrient dynamics in a shrub-based farming system of the Sahel. *Agronomy Journal* 105: 1237–1246. doi: 10.2134/agronj2012.0432

van Duivenbooden, N., M. Pala, C. Studer, C.L. Bielders, and D.J. Beukes. 2000. Cropping systems and crop complementary in dryland agriculture to increase soil water use efficiency: a review. *Netherlands Journal of Agricultural Science* 48: 213–236.

EthioSIS. 2022. The Ethiopian Soil Information System (EthioSIS). Retrieved from www.ata.gov.et/programs/highlighted-deliverables/ethiosis/; Aug 2022.

Fatondji, D., C. Martius, C.L. Bielders, P.L.G. Vlek, A. Bationo, and B. Gerard. 2006. Effect of planting technique and amendment type on pearl millet yield, nutrient uptake, and water use on degraded land in Niger. *Nutrient Cycling in Agroecosystems* 76: 203–217.

Fatondji, D., C. Martius, and P. Vlek. 2001. Zaï – A traditional technique for land rehabilitation in Niger. *ZEFnews* 8: 1–2. (Zentrum für Entwicklungsforschung, Universität Bonn, Bonn, Germany.

Faye, A., T. Krasova-Wade, M. Thiao, J. Thioulouse, M. Neyra, Y. Prin, A. Galiana, I. Ndoye, B. Dreyfus, and R. Duponnois. 2009. Controlled ectomycorrhization of an exotic legume tree species *Acacia holosericea* affects the structure of root nodule bacteria community and their symbiotic effectiveness on *Faidherbia albida*, a native sahelian Acacia. *Soil Biology and Biochemistry* 41: 1245–1252.

Funk, C.C., and M.E. Brown. 2009. Declining global per capita agricultural production and warming oceans threaten food security. *Food Security* 1: 271–289.

Garba, M., I. Serme, N. Maman, K. Ouattara, A. Gonda, C.S. Wortmann, and S.C. Mason. 2018a. Crop response to manure plus fertilizer in Burkina Faso and Niger. *Nutrient Cycling in Agroecosystems* 111: 175–188. doi: 10.1007/s10705-018-9921-y

Garba, M., I. Serme, and C.S. Wortmann. 2018b. Crop yield response to fertilizer relative to soil properties in Sub-Saharan Africa. *Soil Science Society of America Journal* 82: 862–870. doi: 10.2136/sssaj2018.02.0066

Garrity, D.P., F.K. Akinnifesi, O.C. Ajayi, S.G. Weldesemayat, J.G. Mowo, A. Kalinganire, M. Larwanou, and J. Bayala. 2010. Evergreen Agriculture: a robust approach to sustainable food security in Africa. *Food Security* 2: 197–214.

Gomma, A.D., I. Chaibou, I., M.N. Banoin, and E. Schlecht. 2017. Forages trade and nutritive value in urban centers in Niger: Maradi and Niamey cities cases. *International Journal of Innovation and Applied Studies* 21: 508–521.

Gusha, A.C. 2002. Effects of tillage on soil micro relief, surface depression storage and soil water storage. *Soil Tillage Research* 76: 105–114.

GYGA. 2022. Global Yield Gap Atlas. Retrieved from www.yieldgap.org/cz-ted; Aug 2022.

Henao, J., and C. Baanante. 2006. Agricultural production and soil nutrient mining in Africa: Implications for resource conservation and policy development. International Fertilizer Development Center, Muscle Shoals AL. https://vtechworks.lib.vt.edu/bitstream/handle/10919/68832/4566_Henao2006_Ag_production_nutrient_mining_.pdf. Accessed Sep 2022.

Herrick, J.E., J. Maynard, B. Bestelmeyer, A. Ganguli, J. Glover, K. Johnson, D. Kimiti, J. Neff, G. Peacock, J. Peters, S. Salley, K. Shepherd, P. Shaver, Z. Stewart, and R. van den Bosch. 2022. Simple guidelines for deciding when soil variability does–and doesn't–matter for rangeland management and restoration. www.kalro.org/igc-irc2021congresskenya/wp-content/uploads/2022/02/Simple-Guidelines-for-Deciding-When-Soil-Variability-Does.pdf. Accessed Aug. 2022.

IFDC. 2022. Africa Fertilizer and Soil Health Summit. https://ifdc.org/africa-fertilizer-summit-ii/. Accessed Aug 2022.

Ikpe, F.N., J.M. Powell, N.O. Isirimah, T.A.T. Wahua, and E.M. Ngodigha. 1999. Effects of primary tillage and soil amendment practices on pearl millet yield and nutrient uptake in the Sahel of West Africa. *Experimental Agriculture* 35: 437–448.

Jansen, J., C.S. Wortmann, M.C. Stockton, and K.C. Kaizzi. 2013. Maximizing net returns to financially constrained fertilizer use. *Agronomy Journal* 105: 573–578. doi: 10.2134/agronj2012.0413

Jones, M.J. 1971. The maintenance of soil organic matter under continuous cultivation at Samaru, Nigeria. *Journal of Agricultural Science Cambridge* 77: 473–482. doi: 10.1017/ S0021859600064558

Jones, O.R., and R.N. Clark. 1987. Effects of furrow dikes on water conservation and dryland crop yields. *Soil Science Society of America Journal* 51: 1307–1314.

Jones, O.R., and B.A. Stewart. 1990. Basin tillage. *Soil Tillage Research* 18: 249–265.

Kaizzi, C.K., J. Byalebeka, C.S. Wortmann, and M. Mamo. 2007. Low input approaches for soil fertility management in semi-arid eastern Uganda. *Agronomy Journal* 99: 847–853. doi: 10.2134/agronj2006.0238

Kaizzi, C.K., B.M. Mohammed, and N. Maman. 2017. Fertilizer use optimization: principles and approach. In: C.S. Wortmann and K. Sones (Eds.), *Fertilizer Use Optimization in Sub-Saharan Africa.* CABI, London, UK, pp. 9–19. ISBN 9781786392046. doi: 10.1079/9781786392046.0009

Kibunja, C.N., K.W. Ndungu-Magiroi, D.K. Wamae, T.J. Mwangi, L. Nafuma, M.N. Koech, J. Ademba, and E.M. Kitonyo. 2017. Optimizing fertilizer use within the context of integrated soil fertility management in Kenya. In: Charles S. Wortmann and Keith Sones (Eds.), *Fertilizer Use Optimization in sub-Saharan Africa* CABI, London, UK, pp. 82–99. ISBN 9781786392046. doi: 10.1079/9781786392046.0028

Kihara, J., G. Nziguheba, S. Zingore, A. Coulibaly, A. Esilaba, V. Kabambef, S. Njoroge, C. Palm, and J. Huising. 2016. Understanding variability in crop response to fertilizer and amendments in sub-Saharan Africa. *Agriculture, Ecosystems, and Environment* 229: 1–12. doi: 10.1016/j.agee.2016.05.012

Laborde, J., C. Wortmann, H. Blanco-Canqui, G. Baigorria, and J. Lindquist. 2020. Identifying the drivers and predicting the outcome of conservation agriculture globally. *Agricultural Systems* 170. doi: 10.1016/ j.agsy.2019.102692

Lal, R., and J.M. Kimble. 2000. Pedogenic carbonates and the global carbon cycle. In: R. Lal, J.M. Kimble, H. Eswaran, and B.A. Stewart (Eds.), *Global Climate Change and Pedogenic Carbonates.* U.S. Department of Energy Office of Scientific and Technical Information, Oak Ridge, TN, pp. 1–14.

Liben, F., S.J. Hassan, B.T. Weyesa, C.S. Wortmann, H.K. Kim, M.S. Kidane, G.G. Yeda, and B. Beshir. 2017. Conservation agriculture for maize and bean production in the Central Rift Valley of Ethiopia. *Agronomy Journal* 109: 2988–2997. doi:10.2134/agronj2017.02.0072

Liben, F., B. Tadesse, Y.T. Tola, C.S. Wortmann, H.K. Kim, and W. Mupangwa. 2018. Conservation agriculture effects on crop productivity and soil properties in Ethiopia. *Agronomy Journal* 110: 758–767. doi: 10.2134/agronj2017.07.0384

Liben, F., K. Tesfaye, and C.S. Wortmann. 2015. Dry soil planting of sorghum for Vertisols of Ethiopia. *Agronomy Journal* 10: 469–474. doi: 10.2134/agronj14.0621

Liben, F., D. Wegary, C. Wortmann, Z. Stewart, H. Yang, J. Lindquist, T. Tadesse, and H.K. Liben, F., C. Wortmann, and A. Tirfessa. 2020a. Geospatial modeling of conservation tillage and nitrogen timing effects on yield and soil properties. *Agricultural Systems* 177: 1–10. doi: 10.1016/j.agsy.2019.102720

Liben, F., C. Wortmann, H. Yang, T. Tadesse, Z.P. Stewart, D. Wegary, and W. Mupangwa. 2020b. Nitrogen response functions targeted to technology extrapolation domains in Ethiopia using CERES-Maize. *Agronomy Journal.* 113: 436–450. doi: 10.1002/agj2.20439

Mafongoya, P.L., E. Kuntashula, and G. Sileshi. 2006. Managing soil fertility and nutrient cycles through fertilizer trees in southern Africa. In: N. Uphoff, A.S. Ball, E. Fernes, H. Herren, O. Husson, M. Liang, C. Palm, J. Pretty, P. Sanchez, N. Sanginga, and J. Thies (Eds.), *Biological Approaches to Sustainable Soil Systems,* CRC Press, Boca Raton, FL, pp. 273–289.

Maman, N., G. Abdoul, M. Garba, and C. Wortmann. 2018. Sesame sole crop and intercrop response to fertilizer in semi-arid Niger. *Agronomy Journal* 111: 2069–2074. doi: 10.2134/agronj2018.12.0756

Maman, N., M.K. Dicko, G. Abdou, Z. Kouyaté, and C. Wortmann. 2017b. Pearl millet and cowpea intercrop response to applied nutrients in West Africa. *Agronomy Journal* 109: 2333–2342. doi: 10.2134/ agronj2017.03.0139

Maman, N., M.K. Dicko, G. Abdou, and C. Wortmann. 2017c. Sorghum and groundnut sole and intercrop nutrient response in semi-arid West Africa. *Agronomy Journal* 109: 2907–2917. doi: 10.2134/ agronj2017.02.0120

Maman, N., M. Garba, and C.S. Wortmann. 2017a. Optimizing fertilizer use within the context of integrated soil fertility management in Niger. In: C.S. Wortmann and K. Sones (Eds.), *Fertilizer Use Optimization in sub-Saharan Africa.* CABI, London, UK, pp. 136–147. ISBN 9781786392046. doi: 10.1079/ 9781786392046.0136

Maman, N., L. Traoré, M. Garba, M.K. Dicko, A. Gonda, and C.S. Wortmann. 2017d. Maize sole crop and intercrop response to fertilizer in Mali and Niger. *Agronomy Journal* 110: 728–736. doi: 10.2134/agronj2017.06.0329

Mamo, M. and C.S. Wortmann, 2009. Phosphorus sorption as affected by soil properties and termite activity in eastern and southern Africa. *Soil Science Society of America Journal* 73: 2170–2176. doi: 10.2136/sssaj2007.0373

Manlay, R.J., G.T. Freschet, L. Abbadie, B. Barbier, J.L. Chotte, et al. 2020. Séques- tration du carbone et usage durable des savanes ouest-africaines: synergie ou antagonisme?. In: Carbone des sols en Afrique. Impacts des usages des sols et des pratiques agricoles. Tiphaine Chevallier, Tantely Razafimbelo, Lydie Chapuis-Lardy, and Michel Brossard. IRD Éditions/FAO, pp. 241–254. Collec- tion: Synthèses, 9782709928366. hal-02974833

Mason, S.C., N. Maman, and S. Palé. 2015. Pearl millet production practices in semi-arid West Africa: A review. *Experimental Agriculture* 51: 501–521. doi: 10.1017/S0014479714000441

Mason, S.C., K. Ouattara, S.J.B. Taonda, S. Palé, A. Sohoro, and D. Kaboré. 2014. Soil and cropping system research in semi-arid West Africa as related to the potential for conservation agriculture. *International Journal of Agricultural Sustainability* 13: 1–15. doi: 10.1080/14735903.2014.945319

Mesfin, T., G.B. Tesfahunegn, C.S. Wortmann, O. Nikus, and M. Mamo. 2009. Tie-ridging and fertilizer use for sorghum production in semi-arid Ethiopia. *Nutrient Cycling in Agroecosystems* 85: 87–94. doi: 10.1007/s10705-009-9250

Minasny, B. and A.B. Mcbratney. 2018. Limited effect of organic matter on soil available water capacity. *European Journal of Soil Science* 16: 39–47. doi: 10.1111/ejss.12475

Mupangwa, W., I. Nyagumbo, F. Liben, L. Chipindu, P. Craufurd, and S. Mkuhlani. 2021. Maize yields from rotation and intercropping systems with different legumes under conservation agriculture in contrasting agro-ecologies. *Agriculture, Ecosystems and Environment* 306. doi: 10.1016/j.agee.2020.107170

Nansamba, A., K.C. Kaizzi, A.B. Twaha, P. Ebanyat, and C.S. Wortmann. 2016. Grain sorghum response to reduced tillage, rotation, and soil fertility management in Uganda. *Agronomy Journal* 108: 1–10. doi: 10.2134/agronj2015.0608

Ndungu-Magiroi, K.W., C.S. Wortmann, C. Kibunja, C. Senkoro, T.J.K. Mwangi, D. Wamae, M. Kifuko-Koech, and J. Msakyi. 2017. Maize bean intercrop response to nutrient application relative to maize sole crop response. *Nutrient Cycling in Agroecosystems* 109:17–27. doi: 10.1007/s10705-017-9862-x

Nomaou, D.L., G. Yadji, T. Abdourahamane, T.A. Didier, A.M. Nassirou, and A.J.M. Karimou. 2015. Effet des touffes de Guiera senegalensis (J.F. Gmel) sur la fertilité des sols dans la région de Maradi (Niger). *Journal of Applied Bioscience.* doi: 10.4314/jab.v94i1.8

Odero-Waitituh, J.A. 2017. Smallholder dairy production in Kenya: A review. Livestock Research for Rural Development 29. http://lrrd.cipav.org.co/lrrd29/7/atiw29139. Accessed Aug 2022.

Ouattara, K., I. Serme, A.A, Bandaogo, S. Ouedraogo, A. Sohoro, Z. Gnankambary, S. Youl, P. Yaka, T. Pare. 2017. Optimizing fertilizer use within an integrated soil fertility management framework in Burkina Faso. In: C.S. Wortmann and K. Sones (Eds.), *Fertilizer Use Optimization in sub-Saharan Africa.* CABI, London, UK, Ch. 4, pp. 40–51. ISBN 9781786392046. doi: 10.1079/9781786392046.0040

Panthou, G., T. Lebel, T. Vischel, G. Quantin, Y. Sane, A. Ba, O. Ndiaye, A. Diongue-Niang, and M. Diopkane. 2018. Rainfall intensification in tropical semi-arid regions: The Sahelian case. *Environmental Research Letters* 13. doi: 10.1088/1748-9326/aac334/aac334

Pittelkow, C.M., B.A. Linquist, M.E. Lundy, X. Liang, K.J. van Groenigen, J. Lee, N. va Gestel, J. Six, R.T. Venterea, and C. van Kessel. 2015. When does no-till yield more? A global meta-analysis. *Field Crops Research* 183: 156–168. doi: 10.1016/j.fcr.2015.07.020

Reij, C., G. Tappan, and M. Smale. 2009. Agroenvironmental transformation in the Sahel: another kind of "Green Revolution". IFPRI Discussion Paper 00914. Washington DC: International Food Policy Research Institute.

Ricker-Gilbert, J., and T.S. Jayne. 2016. Estimating the enduring effects of fertiliser subsidies on commercial fertiliser demand and maize production: Panel data evidence from Malawi. *Journal of Agricultural Economics.* doi: 10.1111/1477-9552.12161

Saka, A.R., W.T. Bunderson, O.A. Itimu, H.S.K. Phombeya, and Y. Mbekeani. 1994. The effects of Acacia albida on soils and maize grain yields under smallholder farm conditions in Malawi. *Forest Ecology and Management* 64: 217–230.

Samireddypalle, A., O. Boukar, E. Grings, C. Fatokun, K. Prasad, R. Devulapalli, I. Okike, and M. Blümmel. 2017. Cowpea and groundnut haulms fodder trading and its lessons for multidimensional cowpea improvement for mixed crop livestock systems in West Africa. *Frontiers in Plant Science* 8. doi: 10.3389/fpls. 2017.00030

Sanders, J.H., J.G. Nagy, and S. Ramaswamy. 1990. Developing new agricultural technologies for the Sahelian countries: The Burkina Faso case. *Economic Development and Cultural Change* 39: 1–22.

Sarr, B. 2012. Present and future climate change in the semi-arid region of West Africa: A crucial input for practical adaptation in agriculture. *Atmospheric Science Letters* 13: 108–112. doi: 10.1002/asl.368

Senkoro, C.J., G.J. Ley, A.E. Marandu, C. Wortmann, M. Mzimbiri, J. Msaky, R. Umbwe and S.D. Lyimo. 2017. Optimizing fertilizer use within the context of integrated soil fertility management in Tanzania. In: C.S. Wortmann and K. Sones (Eds.), *Fertilizer Use Optimization in sub-Saharan Africa*. CABI, London, UK, pp. 176–192. ISBN 9781786392046. doi: 10.1079/9781786392046.0025

Serme, I., K. Ouattara, V. Logah, J.B. Taonda, S. Pale, C. Quansah, and C.R. Abaidoo. 2015. Impact of tillage and fertility management options on selected soil physical properties and sorghum yield. *International Journal of Biological and Chemical Sciences* 9. doi: 10.4314/ijbcs.v9i3.2

Sileshi, G. and P.L. Mafongoya. 2006. Long-term effect of legume-improved fallows on soil invertebrates and maize yield in eastern Zambia. *Agriculture, Ecosystems and Environment* 115: 69–78.

Stephens, D. 1960. Three rotation experiments with grass fallows and fertilizers. *Empire Journal of Experimental Agriculture* 28: 165–178.

Stoorvogel, J.J., E.M.A. Smaling, and B.H. Janssen. 1993. Calculating soil nutrient balances in Africa at different scales. *Fertilizer Research* 35: 227–235.

Sylla, M.B., M.N. Pinghouinde, P. Gibba, I. Kebe, and N.A.B. Klutse. 2016. Climate Change over West Africa: Recent Trends and Future Projections. In: J.A. Yaro and J. Hesselberg (Eds.), *Adaptation to Climate Change and Variability in Rural West Africa*. Springer International Publishing, Switzerland, Ch 3, pp. 25–40. doi: 10.1007/978-3-319-31499-0_3

UNHCR. 2022. Climate risk profile-Sahel. United Nations High Commissioner for Refugees. www.unhcr.org/61a49df44.pdf. Accessed Aug 2022

Wildemeersch, J.C.J., M. Garba, M. Sabiou, D. Fatondji, and W.M. Cornelis. 2015a. Agricultural drought trends and mitigation in Tillaberí, Niger. *Soil Science and Plant Nutrition* 61: 414–425. doi: 10.1080/00380768.2014.999642

Wildemeersch, J.C.J., M. Garba, M. Sabiou, S. Sleutel, and W. Cornelis. 2015b. The effect of water and soil conservation (WSC) on the soil chemical, biological, and physical quality of a plinthosol in Niger. *Land Degradation and Development* 26: 773–783.

Wilkinson, G.E. 1975. Effect of grass fallow rotations on the infiltration rate of water into a savannah zone soil of northern Nigeria. *Tropical Agriculture* 52: 97–103.

Wortmann, C.S., T. Amede, M. Bekunda, C. Kome, P. Masikati, K. Ndungu-Magiroi, S. Snapp, Z. Stewart, M.E. Westgate, and Z. Zida. 2020. Improvement of smallholder farming systems in Africa. *Agronomy Journal* 112: 5325–5333. doi: 10.1002/agj2.20363

Wortmann, C.S., A. Bilgo, K. Kaizzi, F. Liben, M. Garba, N. Maman, I. Serme, and Z. Stewart. 2021. Perennial grass ley rotations with annual crops in tropical Africa. *Agronomy Journal* 113: 4510–4526. doi: 10.1002/agj2.20634

Wortmann, C.S. and C.K. Kaizzi. 1998. Nutrient balances and expected effects of alternative practices in farming systems of Uganda. In: E.M.A. Smaling (Ed.), *Nutrient Balances as Indicators of Productivity and Sustainability in sub-Saharan African Agriculture*. Agricultural Ecosystems and Environment. Wageningen University and Research, The Netherlands, 71: 115–130.

Wortmann, C.S., M. Mamo, C. Mburu, E. Letayo, G. Abebe, K.C. Kayuki, M. Chisi, M. Mativavarira, S. Xerinda, and T. Ndacyayisenga. 2009. Atlas of sorghum (Sorghum bicolor (L.) Moench) production in eastern and southern Africa. University of Nebraska-Lincoln, Lincoln, NE. https://digitalcommons.unl.edu/cgi/viewcontent.cgi?article=1001andcontext=intsormilpubs

Wortmann, C.S., M. Milner, and G.B. Tesfahunegn. 2017. Spatial analysis for optimization of fertilizer use. In: C.S. Wortmann and K. Sones (Eds.), *Fertilizer Use Optimization in sub-Saharan Africa*. CABI, London, UK, pp. 20–24. ISBN 9781786392046. doi: 10.1079/9781786392046.0020

Wortmann, C.S. and Z.P. Stewart. 2021. Nutrient management for sustainable food crop intensification in African tropical savannas. *Agronomy Journal* 113. doi: 10.1002/agj2.20851

5 Soil–Plant–Water– Environment Interaction in Dryland Agriculture

Sushil Thapa[1]*, Qingwu Xue[2], and Rajan Ghimire[3]
[1]Department of Agriculture, University of Central Missouri, Warrensburg, MO 64093, USA
[2]Texas A&M AgriLife Research and Extension Center, Amarillo, TX 79106, USA
[3]Agricultural Science Center, New Mexico State University, Clovis, NM 88101, USA
*Correspondence: sthapa@ucmo.edu

CONTENTS

5.1 Introduction 108
5.2 Agriculture in the US Great Plains 110
5.3 Crop Response to Soil Moisture 111
5.4 Crop Response to Temperature 114
5.5 Crop Response to Alternative Planting Geometry 115
5.6 Increasing Water Use Efficiency 118
5.7 Using Cover Crops in Dryland Agriculture 120
5.8 Conclusions 121
References 122

5.1 INTRODUCTION

Water scarcity and drought are the major constraints for agricultural production in many parts of the world (Badr, El-Tohamy, and Zaghloul, 2012; Huang, Pray, and Rozelle, 2002; Rosegrant and Cline, 2003), including the US Great Plains, where dryland farming is commonly practiced. Extending from Mexico to Canada, the Great Plains Region covers the central midsection of the US and is divided into the northern plains (part of Montana, Nebraska, North Dakota, South Dakota, and Wyoming) and the southern plains (part of Colorado, Kansas, Oklahoma, New Mexico, and Texas) (Figure 5.1). The area experiences some of the coldest and hottest temperatures and sharp changes in precipitation gradient from east to west. For example, the eastern margin in Nebraska receives 625 mm precipitation annually, and the western margin in Montana receives less than 375 mm. Similarly, the southern plains receive 380–640 mm and the northern plains receive 300–375 mm. More rainfall occurs in the summer months than in winter, except in some of the northwestern parts of the Great Plains (Robinson and John, 2022). Multiple climatic abnormalities and weather hazards are quite common, including severe thunderstorms, winter storms, rapid temperature fluctuations, droughts, tornadoes, and even hurricanes in the far southeast section (Karl, Melillo, and Peterson, 2009). Most parts of the region are dryland farming areas characterized by growing season precipitation of 200–300 mm only (Weinheimer, Johnson, and Mitchell, 2013). High solar radiation, temperature, wind speed, and vapor pressure deficit (VPD) lead to high evaporative demand, especially in the southern Great Plains area (Stewart and Burnett, 1987).

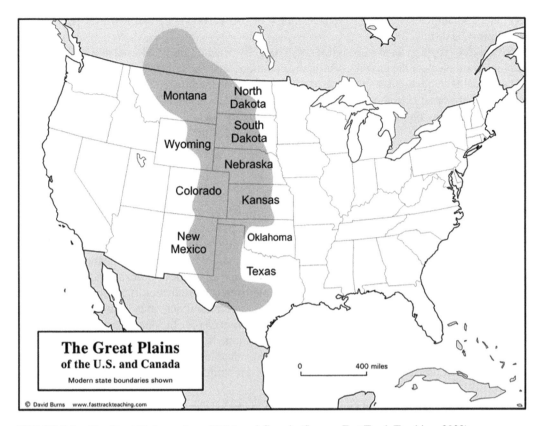

FIGURE 5.1 The Great Plains region of USA and Canada (Source: Fast Track Teaching, 2022).

Ogallala Aquifer, which spans about 450,000 km² (173,746 mi²) of South Dakota, Wyoming, Colorado, Nebraska, Kansas, Oklahoma, Texas, and New Mexico, is one of the largest freshwater aquifers in the world and is the primary source of water for crop production in the Great Plains region (Colaizzi et al., 2008). It has been estimated that more than 90% of the water withdrawals from the aquifer are used for agricultural irrigation (Colaizzi et al., 2008). Availability of irrigation tremendously affected the rural economies of the Ogallala Aquifer regions (Leatherman et al., 2004; Guerrero et al., 2010) because it increased the land production values by more than $12 billion annually (Hornbeck and Keskin, 2014). However, due to low precipitation, the natural recharge rate in the central and southern regions of the Ogallala Aquifer is very low, ~11.00 mm yr^{-1} (Scanlon et al., 2010). As a result, the aquifer has been rapidly depleting over the years, considerably affecting crop yields and farm profitability (McGuire, 2004; Roberts, Male, and Toombs, 2007; Colaizzi et al., 2008).

Dryland farming is common in areas of the Great Plains where irrigation water is not available. Although dryland farming and rainfed farming are used interchangeably, they are not the same (Stewart and Thapa, 2016). Dryland farming is generally practiced in areas where lack of moisture limits crop and/or pasture production to some part of the year (Stewart et al., 2006), whereas rainfed farming involves growing crops without irrigation. As such, dryland farming can be a part of rainfed agriculture. Dryland farming areas in North America span from the Canadian Prairies and the US Great Plains to the Pacific Northwest and Southwest of the US and parts of the intermountain regions (Cannell and Dregne, 1983). Dryland agriculture may constitute the world's largest biome and is indispensable for food production, which covers approximately 47% of the Earth's land surface (UNEP, 1997; Schimel, 2010). It supports the life of many microorganisms

110   Soil and Drought

and plant and animal species and produces much of the world's grain and livestock (Koohafkan and Stewart, 2008).

Successful cropping in water-limited areas largely depends on plant-available soil water at planting to supplement the growing season precipitation. Since both the stored water on soil profile and seasonal precipitation are meager in the US Great Plains, crop yields in the area are generally low and highly variable from year to year. The objective of this chapter is to review and summarize the existing studies that are focused on understanding different aspects of soil, crop, water, and environment and their interaction for improving yield and water use efficiency (WUE) in the US Great Plains and beyond.

5.2 AGRICULTURE IN THE US GREAT PLAINS

Agriculture is the major land use in the Great Plains, where over 80% of the area is designated for cropland, pastureland, and rangeland (Shafer et al., 2014). The agricultural sector generates a total market value of about $92 billion, approximately equally divided between crop and livestock sectors (USDA ERS, 2012). In the northern Great Plains, crop production is dominated by pasture, alfalfa (*Medicago sativa* L.), wheat (*Triticum aestivum* L.), barley (*Hordeum vulgare* L.), corn (*Zea mays* L.), and soybeans (*Glycine max* L.), while livestock production is concentrated on beef cattle along with some sheep, dairy cows, and pigs. Agricultural activities in the southern Great Plains are focused on wheat, corn, grain sorghum (*Sorghum bicolor* L. Moench), cotton (*Gossipum hirsutum* L.), and intensive livestock production in feedlots. In the water-limited areas of the Great Plains, where irrigation is not available or land is not suitable for cash crop production, livestock grazing centered on pastureland or rangelands is the predominant operation (Collins, Mikha, and Brown, 2012).

Dryland areas in the US Great Plains usually experience harsh weather conditions. Predominantly low seasonal precipitation has a major impact on crops and soil health. For example, the biological, chemical, and physical properties of soils can be affected by prolonged drought. The impact of drought is usually severe in dryland farming areas because precipitation amounts in drylands are considerably less than the potential evapotranspiration (PET) (FAO, 2004; Stewart and Peterson, 2015). To a large extent, crop yields are determined by the seasonal evapotranspiration (ET). However, seasonal precipitation in most parts of the Great Plains is not adequate to meet the crop ET requirement, resulting in poor yields. Therefore, one of the most effective ways to utilize available water in dryland areas is to alter the balance between evaporation and transpiration (Cooper et al., 1987). This is done to increase plant transpiration and decrease evaporation from the soil surface as much as possible.

The soils in the Great Plains vary with rainfall and natural grass cover. In the more humid region with taller and heavier grass cover, deep and black soils are found that contain much organic matter. Areas with less moisture have lighter and shallower soils with less organic matter. Most of the soils accumulate carbonates in the profile of about 1.0 m deep layer, translocated there by infiltrating water. As the climate gets progressively drier, the depth of carbonates increases from east to west (Robinson and John, 2022). Low soil moisture and high temperatures can directly affect the decomposition of soil organic matter (SOM) because microorganisms need an optimal temperature, moisture, and soil pH to break down SOM successfully. In dryland soils, microbial activity is limited by moisture availability and low levels of SOM. Low SOM and low biological activity led to the gradual degradation of soils over the years. To improve SOM storage, increase soil-water retention, and reduce erosion, conservation tillage is practiced in some areas. No-till covers the soil with crop residues and creates a blanketing effect on the soil surface. Further, the no-till protects soil surface against erosion, traps soil moisture, and keeps soil temperatures cooler during summer and warmer during winter compared to the fields with conventional tillage. Practices like summer fallowing, growing crops every two years, and using cover crops are also followed in some areas with mixed results.

Grain sorghum, corn, and wheat are the major cereals grown under irrigated as well as dryland conditions in the Great Plains. Sorghum is a drought-tolerant and water-use-efficient crop grown in semiarid tropical and subtropical environments (Blum, 2004; Rooney, 2004) because of its ability to remain dormant under adverse conditions and resume the growth when soil moisture conditions improve (Bennett, Tucker, and Maunder, 1990; Unger, 1994). Sorghum produces ears over a longer period of time because tillers develop continuously for several weeks. Consequently, short periods of drought do not seriously damage pollination and fertilization. Drought tolerance in grain sorghum is also associated with a dense and prolific root system capable of extracting soil water deep in the soil profile (Wright and Smith, 1983; Singh and Singh, 1995). Despite its ability to tolerate drought, a lack of water during the reproduction and grain-filling stages results in a major yield loss. For example, Craufurd, Flower, and Peacock (1993) found that water stress during booting and flowering resulted in as high as 85% yield loss in grain sorghum.

Corn is a major irrigated summer crop in the southern Plains. The crop is widely used as animal feed and a biofuel (ethanol) crop. Although corn plants are relatively tolerant to water stress during the vegetative growth and grain ripening stages (Doorenbos and Kassam, 1979), water deficit is one of the major threats limiting grain yield (Lobell et al., 2008). As a short-day plant, corn does not respond to photoperiod until the end of the juvenile stage, during which only vegetative growth occurs. When the inductive stage starts, it becomes sensitive to photoperiod. For instance, increasing photoperiod delays tassel initiation in corn (Ellis et al., 1992).

Wheat, the most widely cultivated crop globally, is a major crop in the Great Plains for grain and forage production. Over the decades, wheat evolved into a plant that can tolerate cold temperatures, survive under different soil conditions, and mature even with limited soil moisture availability (Smith, 1995). Therefore, wheat is grown over a wide range of moisture and temperature conditions from southern to northern plains. Generally, wheat is a photoperiod-sensitive crop, but some cultivars in hot and dry areas are photoperiod insensitive. Such non-photoperiodic cultivars would likely initiate reproductive growth too early in some northern regions and therefore be more susceptible to frost damage (Smith, 1995).

Cotton is also a major crop in the southern Great Plains, where there is a sufficient frost-free growing season of at least 180–200 days, ample sunlight, and enough growing degree days (GDD) accumulated during the growing season. Barley, oat (*Avena sativa* L.), and sunflower (*Helianthus annuus* L.) are cultivated in some dryland areas of the northern Great Plains. Sunflower has a deep rooting system capable of extracting water and nutrients to 3 m in the soil profile, contributing to its adaptation to dryland environments (Jones and Johnson, 1983).

5.3 CROP RESPONSE TO SOIL MOISTURE

Conserving some soil water and making it available to the plants, especially during the reproduction and grain filling, is one of the most challenging aspects of crop production in the dryland farming areas. Stewart and Burnett (1987) stated that dryland farming emphasizes water conservation in every practice throughout the year. Soil water between field capacity (FC) and permanent wilting point (PWP) is available to the growing plants (Figure 5.2). When the soil water reaches the wilting point, plants cannot extract water, but the remaining water in the soil can evaporate. To return the soil back to FC, first, the amount of water lost below the wilting point must be added, and then the plant available water (PAW) (Stewart and Peterson, 2015). The PAW in soil controls the physical environment of plant roots and affects soil chemical and biological conditions (Asgarzadeh et al., 2014). Figure 5.2 shows the plant available soil water (PASW) for three corn seasons in the Texas High Plains. At the early growth stage (~140 days of year), PASW was similar among the three years, but at later growth stage (~200 days of year and thereafter), PASW was the highest in 2010 and lowest in 2011. The difference in PASW was largely due to the variation in rainfall among the seasons, which was 224 mm in 2010 compared to 85 mm in 2011 (Thapa et al., 2020b).

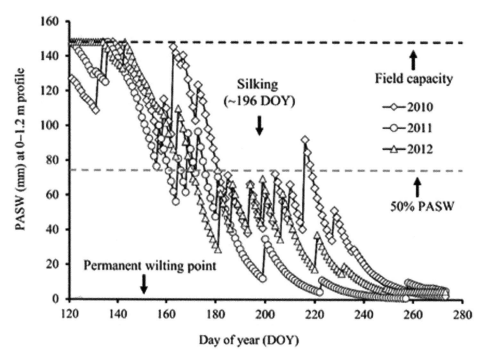

FIGURE 5.2 Plant available soil water (PASW) in the 0- to 1.2-m soil profile for corn in the Texas High Plains. (Source: Thapa et al., 2020b).

Corn is sensitive to water deficit conditions and other environmental stresses, especially during reproduction (Bryant et al., 1992; Otegui, Andrade, Suero, 1995; Thapa et al., 2021). Rainfall distribution for the 3-week period during tasseling greatly influences corn yield (Brown et al., 1985). During tasseling, water deficit will have more effect on the timing of silk emergence than on tassel development and pollen shed. Because of the hot and dry environmental conditions, the tassel may develop and shed pollen before the ear and silk formation has been completed, resulting in poor pollination. For example, according to Kansas State University Extension (2007), one day of water stress in corn within a week soon after silking can result in up to 8% yield loss (Figure 5.3). Thapa et al. (2021) reported grain-filling as the most critical period for water management in corn because transpiration efficiency (TE) was highest when corn plants were harvested at the mid-grain-filling stage, and TE decreased thereafter. They also found that the drought-tolerant hybrid had higher TE than the conventional hybrid. Reduced kernel number, kernel weight, and root growth as an effect of water stress in corn were reported by various scholars (Claassen and Shaw, 1970; Newell and Wilhelm, 1987). Studies also have suggested that corn yield is a linear function of seasonal ET, and the effect of water stress depends on the severity of water stress affecting seasonal ET (Stone, 2003; Klocke et al., 2004; Payero et al., 2006).

Although both corn and grain sorghum follow the C_4 photosynthetic pathway and are grown under similar agronomic conditions, grain sorghum is more drought tolerant than corn (Blum, 2004; Rooney, 2004). Therefore, many growers in water-limited areas prefer sorghum as an alternative crop to corn (Berenguer and Faci, 2001). However, a minimum amount of water is required to establish crop for grain production. This minimal level varies with the environmental demand, that is, 150–180 mm of PAW in the southern Plains (Hanks, 1974). Stone et al. (1996) conducted a study over 14 years in Tribune, Kansas, to establish the yield versus water application relationships in corn and grain sorghum and found that as the total irrigation amount increased from 100 to 400 mm, corn

Soil–Plant–Water–Environment Interaction in Dryland Agriculture

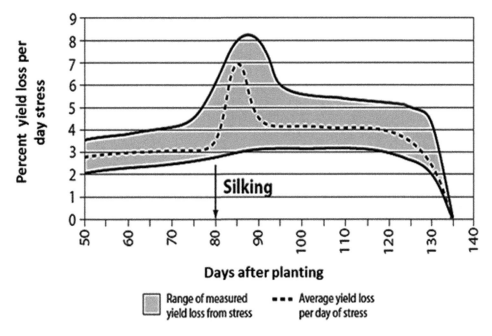

FIGURE 5.3 Corn yield loss due to one day of moisture stress after silking. (Source: Kansas State University Extension, 2007).

out-yielded sorghum at total irrigation amounts of 345 mm and above. They suggested that if grain production is the major objective, grain sorghum is better than corn at less than 206 mm of irrigation, whereas corn is a better choice than sorghum at more than 206 mm of irrigation.

A wheat plant is not highly drought resistant and cannot survive long periods of drought. However, it adjusts to moisture stress by producing smaller cells, resulting in the shorter height, smaller leaves, and reduced stomatal opening. Therefore, the potential yield of wheat can only be obtained under well-watered conditions (Evans, 1975). Under drought stress, grain number per ear and grain weight are the most important factors determining yield. In drylands, maximum yields cannot be realized as there is a linear relationship between the yield of a cultivar of wheat and the severity of drought (Fischer and Maurer, 1978). Water stress in wheat at the booting stage (1–2 weeks prior to anthesis) causes the greatest reduction in yield (Day and Intalap, 1970; Fischer and Slatyer, 1973). Further, water stress at flower formation results in reduced grain number, while drought at later stages results in reduced kernel size because the number of grains has already been established (Asana, Saini, and Ray, 1958). Under the Great Plains environment, good crop yields are obtained in most years when the soil is wet to a depth of 0.9 m at the time of seeding (Army and Hide, 1959). In the Texas High Plains, Thapa et al. (2017a) found that winter wheat yield was largely determined by the soil water extraction (SWE). In a season with a high yield (3548 kg ha^{-1} in 2016), net SWE occurred in the 0.0–2.4-m deep profile. In contrast, in a season with a very low yield (650 kg ha^{-1} in 2011), net SWE was limited to the upper 1.2 m. They found that more recent cultivars (TAM 110, TAM 111, and TAM 112) were able to extract more water from deeper soil profile (particularly between jointing and maturity) and had higher ET, biomass, and yield than the older cultivar (TAM 105). This result was evident during the historic dry seasons of 2011 and 2012. Although irrigation helps, in addition to adding water at critical growth stages, ensuring early plant growth could promote root development for extracting soil water from the deeper profile later in the season and maximize crop production with limited irrigation (Thapa et al., 2017a).

5.4 CROP RESPONSE TO TEMPERATURE

Different environmental factors, such as solar radiation, temperature, heat, and water stress, influence crop production in dry areas. Specifically, temperature and heat are directly related to water stress in crops because fluctuations in temperature can lead to many problems in crop physiology. Crops grow and develop within a certain threshold of temperatures. A base temperature (T_{base}) is the minimum temperature required for the crop to grow and develop. For example, the base temperature for summer crops such as corn and sorghum is 10 °C, while it is 4.5 °C for winter wheat. Increasing temperatures can be a concern when growing crops because higher temperatures will lead to higher ET rates, limiting the available water for crop growth and development. High heat also affects the process of photosynthesis in plants. Under severe water stress, crops will close their stomata to help slow or prevent water loss from transpiration. This hinders the photosynthesis process, reduces carbon uptake, and ultimately decreases yields (Leng and Hall, 2019). Several studies have shown that increased temperature coupled with drought can reduce crop yields by as much as 50% (Lamaoui, 2018).

Most plants tolerate normal temperature fluctuations. However, at the reproductive stage, high temperatures can affect pollen viability, fertilization, and grain or fruit formation (Hatfield and Prueger, 2015). When the temperature increased from 30 °C to 35 °C, the potential kernel growth rate and final kernel size were reduced in corn, significantly decreasing the grain yield (Jones, Ouattar, and Crookston, 1984). To adjust the heat or temperature stress, plants undergo physiological and biochemical changes which include long-term evolutionary phenological and morphological adaptations and short-term avoidance or acclimation mechanisms such as changing transpirational cooling, leaf orientation, or alteration of membrane lipid compositions. Similarly, plants growing in a hot climate avoid heat stress by reducing the absorption of solar radiation using their small hairs (tomentose) and protective waxy covering (Hasanuzzaman et al., 2013). High temperature and low humidity induce leaf water stress when the uptake of water through the root system is inadequate to cope with high transpiration rates (Grange and Hand, 1987). Sinclair and Bennett (1998) found increased evaporative demand as well as crop transpiration with the increasing atmospheric VPD. An increase in VPD from 1 to 1.8 kPa determines the major reduction in plant growth in several crops because an increasing VPD can cause an inhibition of photosynthesis (Bunce, 1984). For example, as the VPD increased, the photosynthetic CO_2 exchange rates decreased during the afternoon as compared to the morning (Singh et al., 1987; Hirasawa and Hsiao, 1999). Because of the low vapor pressure in the arid regions, the VPD is high, hence, the higher loss of water through stomata. Gholipoor et al. (2010) found substantial intra-specific variability in the sensitivity of stomatal response to changes in VPD (1.6–2.7 kPa) for 17 sorghum genotypes in Manhattan, Kansas. This trait (stomatal response to VPD) helps to limit the transpiration rate to a constant value when the VPD is high and contributes to increasing grain yield (Sinclair et al., 2005). The C_4 plants tolerate higher temperatures and can be grown in the warmer seasons of the year. Their higher TE can result in higher yields than C_3 plants with the same rainfall and irrigation. So, C_4 crops are preferable in drylands for producing biomass. To cope with drought and high-temperature stress, plants have evolved several adaptive strategies (Borrell et al., 2006; Araus et al., 2008). Conservative water use early in the growing season is one of those adaptations that make water available for the plants later in the growing season (Sinclair, Hammer, and van Oosterom, 2005; Messina et al., 2011).

As the temperature and water stress increase, plant canopy temperature (CT) also increases. CT is one of the many physiological traits that may help identify heat or drought-tolerant cultivars. Under high solar radiation and drought conditions, stomatal conductance decreases when soil moisture is not adequate to keep up with evaporative demands, increasing CT (Urban et al., 2007). Canopy temperature depression (CTD) is expressed as the difference between air temperature and canopy temperature (CTD $= T_{air} - T_{canopy}$) (Jackson et al., 1981; Balota et al., 2008). A winter wheat field study conducted at Bushland, Texas, showed a genotypic variation for CTD, and there was a significant

Soil–Plant–Water–Environment Interaction in Dryland Agriculture

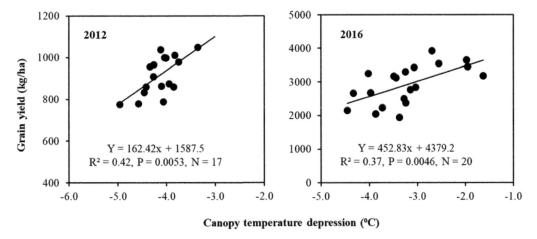

FIGURE 5.4 The relationship between grain yield and canopy temperature depression (CTD) in the 2012 season and 2016 season. Daytime (10.00 to 18.00 h) canopy temperature was measured every 5 minutes using IRT crop sensors during the grain filling period. Canopy temperature was measured, and grain yield was determined from three replications for each genotype. Therefore, each data point is the average of three replications. The 2012 season indicates dry season (seasonal precipitation = 60 mm), while the 2016 season indicates a relatively wet season (seasonal precipitation = 315 mm). (Source: Adapted from Thapa et al., 2018a).

positive linear relationship between grain yield and daytime CTD (Figure 5.4) (Thapa et al., 2018a). Amani, Fischer, and Reynolds (1996), Fischer et al. (1998), and Lopes and Reynolds (2010) also found a significant positive correlation between CTD and grain yield in spring wheat.

5.5 CROP RESPONSE TO ALTERNATIVE PLANTING GEOMETRY

Growers continually search for agronomic methods that help to increase crop yields and reduce the cost of production, or a combination of both. Manipulating planting geometry has a long history. Over the years, reducing plant population and modifying plant or row spacing have been adopted by growers for better utilization of available soil moisture under water-limited conditions (Larson and Vanderlip, 1994; Stewart, 2009). The availability of water and nutrients often determine the selection of planting geometry for a specific area and crop (Thapa et al., 2017b). Brown (1985) discussed the way the Native American tribes used to grow corn in the desert areas of New Mexico and Arizona in the US. The corn was planted 10–12 plants per hill and about 2 m apart. The idea was to reduce the desiccation of foliage, anthers, and silk, allowing normal fertilization in the hot and dry environment. Weatherwax (1954) explained the methods of growing corn in clumps in his book *Indian Corn in Old America*. When Weatherwax asked the growers about the clumps, one of them answered that they had tried various methods and clump planting yielded more corn (Stewart and Thapa, 2016).

In the conventional evenly spaced planting (ESP), growers use different row-spacing and plant populations to improve crop yields. Narrow row-spacing in corn and some other crops suppresses weed growth due to the smothering effect and increases radiation use efficiency compared with wider row-spacing (Dwyer, Tollenear, and Stewart, 1991; Tollenaar and Aguilera, 1992). However, under the conditions of limited water supply, growers usually increase the row spacing or decrease the plant population to ensure that available soil water meets the crop demand, especially late in the season. Bean and Gerik (2005) reported higher yields of corn planted in 1.0 m rows than that planted in 0.76 and 0.51 m rows in the Texas Panhandle. Some alternative planting geometries such as clump, cluster, and skip-row are discussed below (Figures 5.5 and 5.6).

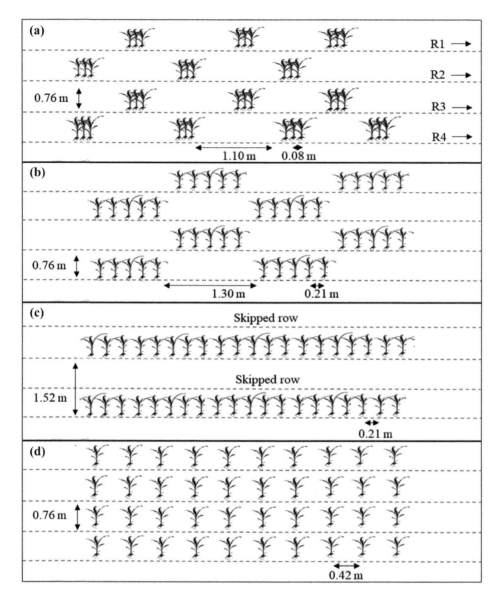

FIGURE 5.5 An illustration of the arrangements of clump (A), cluster (B), skip-row (C), and standard evenly spaced plating (ESP; D) geometries. (Source: Adapted from Thapa et al., 2018b).

Planting three to five plants close together in clumps or clusters is based on the rationale that growing plants in clumps will increase plant competition, resulting in less use of water, nutrients, and sunlight (Stewart, 2009; Thapa, Xue, and Stewart, 2020a). There will be less vegetative growth in clump planting, mainly because of less tillering. This will leave more water for the plants during the reproduction and grain filling periods and result in higher grain yields (Stewart, 2006). Clump planting of sorghum and corn has been shown to decrease CT and the number of tillers per plant and increase grain yields and increase harvest index (HI) in the central and southern Great Plains (Bandaru et al., 2006; Mohammed, Blaser, and Stewart, 2012; Kapanigowda et al., 2010; Thapa et al., 2018b, 2020a). Bandaru et al. (2006) compared clump (three plants clump^{-1}) versus ESP in the Texas Panhandle for 3 years and in Tribune, Kansas, for 1 year under dryland conditions. Results showed that planting grain sorghum in clumps reduced the number of tillers and increased

FIGURE 5.6 Clump geometry (three plants close together; left picture) and cluster geometry (five plants close together; right picture). Picture credit: B. A. Stewart, West Texas A&M University.

grain yield by as much as 100% when yields were below the 3.0 Mg ha^{-1} range, and there was a slight decrease when yields exceeded 6 Mg ha^{-1}. Kapanigowda et al. (2010) evaluated corn planted in three plants per clump versus ESP in the Texas Panhandle under different water regimes. They reported that clumps produced significantly fewer tillers but a greater grain yield and HI than ESPs. Mohammed, Blaser, and Stewart (2012) evaluated dryland corn geometries in the Texas Panhandle and found that clumps reduced LAI at the vegetative growth stage by 5–14% and increased HI by 5–10% than ESPs. Thapa et al. (2016, 2018b) planted corn in ESP, clump, and cluster geometries at different locations of the Texas High Plains, analyzed thermal images taken at the hottest part of the day (2:00–3:00 p.m. CST) at silking and grain-filling stages. For all locations and measurement dates, plants in ESPs had warmer CTs ranging from 0.7 to 1.2 °C higher than those in clumps and clusters (Figure 5.7). Thapa et al. (2016, 2017b) measured air temperature and relative humidity

FIGURE 5.7 Corn images taken by a thermal infrared camera (A and B) and by a digital camera (C and D). In the left image, A and C represent clump and B and D represent standard ESP geometries. In the right image, A and C represent cluster and B and D represent ESP geometries. Grey patches in thermal pictures are the soil surfaces filtered out for temperature measurement. The vertical color bar indicates the CT, which changes from dark red to light yellow as the temperature increases. The mean CT difference was about 2°C for the left and 3°C for the right image (Source: Thapa et al., 2016, 2018b).

every 5 min within sorghum and corn plant canopies for different planting geometries. The VPDs for clump and ESP treatments did not differ during the night hours. However, as the day progressed, clumps consistently had lower VPDs than those for ESPs at different growth stages.

Growers in dryland areas of the US Great Plains also practice skip-row configuration (one row planted, one row skipped). This method will conserve some water in the skipped area for use by plants during the flowering and grain filling periods (McLean et al., 2003; Routley et al., 2003). Water stress in the late season can have bad effects on crop yield and HI since early crop senescence limits grain development and the production of harvestable yield (Sinclair and Weiss, 2010). Skip-row planting is expected to be the most effective method where the soil has a high water-holding capacity and can conserve significant amounts of water for use by plants (Abunyewa et al., 2010). Vigil et al. (2008) reported that skip-row configurations in corn and grain sorghum offered an average of 0.38 Mg ha^{-1} grain yield advantage over conventional ESP when studied across 11 site years in Colorado and Kansas. Lyon et al. (2009) conducted 23 field trials across the central Great Plains, where skip-row planting patterns resulted in increased grain yields compared to the ESP. They suggested skip-row planting if yields are expected to fall between 4.71 and 6.27 Mg ha^{-1} and ESPs for areas with yield potentials of greater than 6.27 Mg ha^{-1}. The yield advantage of skip-row planting, especially at low yield levels, was also reported by Pavlista et al. (2010) in the Nebraska Panhandle. Thapa et al. (2018b) implemented ESP, clump, cluster, and skip-row geometries in a commercial production field with large plots (>1,000 m^2 plot^{-1}) in Texas High Plains and found positive yield response of alternative geometries. Musick and Dusek (1982) conducted skip-row graded furrow irrigation studies at Bushland, Texas, in corn and found that skip-row irrigation reduced average water intake from 130 to 60 mm or 46% of every row irrigation compared with ESP. Nielsen, Vigil, and Henry (2018) grew grain sorghum for 3 years at Akron, Colorado, in three planting configurations – ESP, skip-row, and double skip-row – and found that skip-row planting used a greater amount of water during the second half of the growing season. This result supports the idea of conserving some soil water by alternative planting geometries for the later growth stages. Yield advantage of alternative geometries, especially skip-row, has been reported in other parts of the world, such as in Australia (Hammer et al., 2014; Whish et al., 2005; Routley et al., 2003; Spackman et al., 2001; Collins et al., 2006).

5.6 INCREASING WATER USE EFFICIENCY

Soil cover, minimum tillage, appropriate planting methods, plant population, and planting geometry help optimize WUE and crop yields in water-limited areas. Although crop yields have increased in most dryland farming regions over the years, data show that yields may have plateaued during the 1980s. Among many factors, reducing water loss by evaporation and runoff and making more water available for the plants are strategies that contribute to yield under dryland conditions (Stewart and Thapa, 2016). Sinclair and Weiss (2010) reported that a C_4 grass growing in an average transpiration environment of 2 KPa VPD has a transpiration rate of approximately 220 units of water for each unit of dry biomass production. A C_3 species growing in the same environment will transpire about 330 units of water for each unit of dry matter produced. Using field data from the Great Plains area, Stewart and Peterson (2015) estimated the transpiration rates of corn and grain sorghum to be generally between 225 and 275 g of water per g of biomass production.

There is a close relationship between plant available water in the soil profile during the reproductive period and total biomass production. Therefore, plants with high WUE are suggested for dryland production (Blum, 2009). The term "water use efficiency" has been used to describe harvestable products per unit of water used in ET called grain water use efficiency (WUEg) (Stewart and Peterson, 2015). WUE is also expressed as units of total aboveground dry matter produced per unit of water consumed in ET called biomass water use efficiency (WUEb) (Begg and Turner,

Soil–Plant–Water–Environment Interaction in Dryland Agriculture 119

1976; Jensen et al., 1981). Passioura (1996) described grain yield as a partial function of WUE and suggested the following equation.

$$Y = WUE \times WU \times HI \qquad \text{(Eq. 5.1)}$$

Or,

$$Y = (B/WU) \times WU \times HI \qquad \text{(Eq. 5.2)}$$

This turns to the basic equation as given below (Donald and Hamblin, 1976).

$$Y = B \times HI \qquad \text{(Eq. 5.3)}$$

where Y is the grain yield (g DM m^{-2}), WUE is water use efficiency (g DM/g H$_2$O), WU is water use (g H$_2$O m^{-2}), HI is harvest index (g Y/ g B), DM is dry matter (g), ET is evapotranspiration (g H$_2$O m^{-2}), and B is the aboveground biomass (g DM).

Sparse or erratic rainfall distribution in space and time coupled with water loss through runoff, deep percolation (below the root zone), and evaporation from the soil surface are the factors that result in low WUE (Mando, 1997). The WUE values can vary with crop type, a portion of the crop harvested, and weather conditions. One of the major environmental factors influencing the WUE of a crop is atmospheric humidity. A lowering of the VPD of the atmosphere around a leaf will proportionally increase the transpiration rate of the leaf. Thus, if some compensatory closure of stomata does not occur, a decrease in atmospheric humidity will decrease the crop WUE. Other environmental factors influencing WUE are ambient temperature, solar radiation, and soil water availability. The level of CO$_2$ in the atmosphere also affects the efficiency of water use by plants, where WUE increases with the increase in CO$_2$ concentration (Turner and Burch, 1983). WUE in the semiarid US Great Plains is primarily limited by soil water deficit. For example, in the southern Plains, compared to rainfed production, limited irrigation increased WUE in winter wheat from 0.40 kg m^{-3} to 0.94 kg m^{-3} in the historic dry season of 2012 (Thapa 2017a, 2019a). Reduced grain yield and WUE in winter wheat under severe water stress were also reported by Xue et al. (2014) and Musick et al. (1994) at the same location. This was mainly due to the harsh environmental conditions such as drought and high temperature (heat) stress, which are the two major environmental factors limiting crop growth and yield in dryland farming areas (Prasad and Staggenborg, 2008). In addition to soil water deficit, crop WUE is also affected by the genotype. For example, Thapa et al. (2019b) reported a significantly different WUE among six winter wheat genotypes in the Texas High Plains.

Figure 5.8 shows a graphical model of the water balance for a wide range of climatic conditions (Ponce, 1995). This model was based on the range of climates in the Sertao (meaning semiarid) region of Brazil, and it demonstrates the use of water over a range of climatic zones. Although this diagram is conceptual and does not show the amount of percolation water, it demonstrates the fact that only a small proportion of annual precipitation in arid and semiarid regions is used for ET and that a large proportion is lost as evaporation. Wani et al. (2012) stated that in arid areas, as little as 10% of precipitation is used for transpiration, while the remaining 90% can be lost as evaporation. This estimate supports the concept illustrated in Figure 5.8.

Summer fallowing and growing crops every other year are common practices for conserving precipitation in the soil and increasing WUE in most areas of Great Plains. The conserved soil water can help in meeting crop demand, especially at the time of seed germination and early stand crop establishment. Although soil water may not be sufficient to fully meet crop requirements for

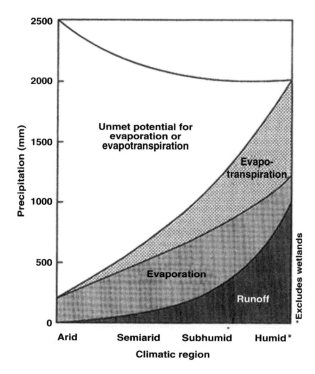

FIGURE 5.8 A graphical model of the water balance for a wide range of climatic conditions. (Source: Ponce, 1995).

yield maximization, it can help ensure a certain yield and protect growers from complete yield loss (Li et al., 2012). Ridge (1986) found that summer fallow enhanced yield stability without reducing the overall productivity of the cropping systems. However, summer fallowing is not free of critics (Peterson, Wesfall, and Cole, 1993; Unger, 2001). According to Li et al. (2012), during the fallow period, the bare soil surface was subjected to severe wind and water erosion in dryland areas of China, and this is also common in the central and southern Great Plains. Further, fallowing decreased SOM content, especially when used with intensive tillage (Power, 1990; Rasmussen and Collins, 1991). There was also a considerable loss of large amounts of water through evaporation (Zhang and Wu, 1994; Peterson, Westfall, and Hansen, 2012). In such situations, mulching materials like straw or crop residue on the soil surface may help preventing excessive water loss through evaporation. This means the mulch holds the water until it decomposes. As the mulch decomposes into the soil, it adds organic matter to the soil and increases the water-holding capacity, which allows the soil to remain moist in times of limited moisture availability. Overall, covered soils can hold a higher amount of water even at the time of low rainfall and increasing environmental temperature compared to uncovered soils, which eventually helps to increase the WUE.

5.7 USING COVER CROPS IN DRYLAND AGRICULTURE

Cover crops are increasingly adopted across the US to improve soil health, weed suppression, and WUE. However, cover cropping is adopted with caution in southern Great Plains where PET exceeds 400% of the annual precipitation, thereby considerably affecting crop production (Hansen et al., 2012). Studies from central and southern Great Plains have shown that cover cropping or continuous cropping in drylands can deplete soil water for the succeeding crops and consequently reduce crop yields (Aiken et al., 2013; Miller, Engel, and Holmes, 2006). Standing cover crops deplete soil water through transpiration, while they increase water availability for the following crop by improving infiltration and soil water storage (Unger and Vigil, 1998). Specifically, cover crops increase soil cover between the harvest of the previous crop and the planting of a subsequent crop in a crop rotation (Holman et al., 2018; Nielsen et al., 2015a). By replacing fallow in crop-fallow systems, cover crops can increase microbial activity, soil carbon sequestration, soil health, and productivity of cropping systems (Thapa et al., 2022). Specifically, in dryland cropping systems, cover cropping with reduced- or no-tillage can increase soil organic carbon (SOC) accumulation (Thapa et al., 2019c), improve soil structure and aggregation, and reduce soil susceptibility to wind and water erosion (Blanco-Canqui and Ruis, 2020).

In the semiarid Great Plains, wind erosion is the most significant concern in the late winter and early spring (Allmaras, 1983). The presence of cover crops or their residues will provide adequate protection for the soil surface against wind erosion. Cover crops also reduce water erosion and increase available soil water, water-holding capacity, and soil water storage (Blanco-Canqui et al., 2013). Because fallowing is proved to be inefficient for precipitation storage, cover crops can help improve the precipitation use efficiency in the semiarid Great Plains (Nielsen et al., 2015a; Nielsen et al., 2015b; Stewart and Peterson, 2015). Using cover crops to partially replace the fallow can increase precipitation use efficiency because cover crops can benefit from soil moisture for transpiration that would otherwise be lost through evaporation alone (Stewart and Liang, 2015; Holman, Assefa, and Obour, 2021). The increased residue input from cover crops and the depletion of soil water compared to fallow could increase precipitation storage efficiency or fallow efficiency between cover crop termination and next crop planting (Nielsen et al., 2015a; Paye et al., 2022).

Challenges associated with limited and erratic precipitation in the Great Plains region have restricted cover crop adoption in dryland production systems (Holman et al., 2018; Holman, Assefa, and Obour, 2021; Schlegel and Havlin, 1997). Such challenges include difficulties in cover crop establishment, highly variable cover crop biomass production, incompatibility with some conventional residual herbicides, and limited soil moisture availability for subsequent crops (Ghimire et al., 2019). Limited precipitation and soil moisture following the harvest of primary crops make cover crops a significant challenge for dryland producers. For example, soil moisture is often low following wheat harvest in June (Holman et al., 2018). Therefore, following wheat, a successful cover crop establishment will depend on timely rainfall in July and August. Spring planted cover crops following corn or grain sorghum harvest in late fall will often have enough soil moisture for germination but need timely rainfall for good establishment and biomass production (Holman, Assefa, and Obour, 2021).

5.8 CONCLUSIONS

The Dust Bowl that happened in the US and Canada during the 1930s is considered as the worst ecological disaster in North America and it was largely associated with poor farming practices in dryland areas. Since then, some of the degraded cropland was returned to grassland/rangeland and drastic changes were made in tillage and other management practices to reduce soil erosion. Land cover using crop residues or cover crops, organic matter management, use of drought-tolerant or better-adapted cultivars, conservation tillage (no-till), improved soil and water management practices, and other technological innovations have increased production stability in dryland agriculture during the past few decades. However, this stability is relative, and its sustainability is always a major concern. Reducing runoff and storing precipitation water in the soil profile for later use by the crop, minimizing soil evaporation, and ensuring some soil water available to plants during the later growth stage (reproduction and grain filling) are recommended by various scholars for successful dryland production. Research results also show that using improved cultivars that can extract water from the deeper soil profile and maintain cooler canopy temperature has improved WUE and minimized the yield loss due to water stress and other harsh climatic conditions in the semiarid Great Plains. Furthermore, manipulation of planting geometry (clump, cluster, and skip-row) and cropping systems diversification have shown some encouraging results in drylands and should be adapted in the fullest extent feasible. These practices will help to maintain or even enhance the soil physical, chemical, and biological properties, thereby minimizing water and wind erosion and maximizing WUE. Therefore, improved knowledge of soil–plant–water–environment interactions, adoption of best or proven management practices, and producer education on soil and water management need to be further strengthened for sustainable crop production in the semiarid US Great Plains and beyond.

REFERENCES

Abunyewa, A. A., R. B. Ferguson, C. S. Wortmann, D. J. Lyon, S. C. Mason, and R. N. Klein. 2010. Skip-row and plant population effects on sorghum grain yield. *Agronomy Journal* 102: 296–302.

Aiken, R. M., D. M. O'Brien, B. L. Olson, and L. Murray. 2013. Replacing fallow with continuous cropping reduces crop water productivity of semiarid wheat. *Agronomy Journal* 105: 199–207. doi: 10.2134/agronj2012.0165

Allmaras, R. R. 1983. Soil conservation: Using climate, soils, topography, and adapted crops information to select conserving practices. In: H. E. Dregne and W. Willis (Eds.), *Dryland Agriculture*. 1st ed. American Society of Agronomy, Madison, WI, pp. 140–154.

Amani, I. R., R. A. Fischer, and M. P. Reynolds. 1996. Canopy temperature depression association with yield of irrigated spring wheat cultivars in a hot climate. *Journal of Agronomy and Crop Science* 176: 119–129. http://dx.doi.org/10.1111/j.1439-037X.1996.tb00454.x

Araus, J. L., G. A. Slafer, C. Royo, and M. D. Serret. 2008. Breeding for yield potential and stress adaptation in cereals. *Critical Reviews in Plant Sciences* 27: 377–412.

Army, T. J., and J. C. Hide. 1959. Effects of green manure crops on dryland wheat production in the Great Plains are of Montana. *Agronomy Journal* 51: 196–198.

Asana, R. D., A. D. Saini, and D. Ray. 1958. Studies in physiological analysis of yield III. The rate of grain development in wheat in relation to photosynthetic surface and soil moisture. *Plant Physiology* 11: 655–665.

Asgarzadeh, H., M. R. Mosaddeghi, A. R. Dexter, A. A. Mahboubi, and M. R. Neyshabouri. 2014. Determination of soil available water for plants: Consistency between laboratory and field measurements. *Geoderma* 226–227: 8–20.

Badr, M. A., W. A. El-Tohamy, and A. M. Zaghloul. 2012. Yield and water use efficiency of potato grown under different irrigation and nitrogen levels in an arid region. *Agricultural Water Management* 110: 9–15.

Balota, M., W. A. Payne, S. R. Evett, and T. R. Peters. 2008. Morphological and physiological traits associated with canopy temperature depression in three closely related wheat lines. *Crop Science* 48: 1897–1910.

Bandaru, V., B. A. Stewart, R. L. Baumhardt, S. Ambati, C. A. Robinson, and A. Schlegel. 2006. Growing dryland grain sorghum in clumps to reduce vegetative growth and increase yield. *Agronomy Journal* 98: 1109–1120.

Bean, B., and T. Gerik. 2005. Evaluating corn row spacing and plant population in the Texas Panhandle. Result demonstration report from TAES-Amarillo. http://amarillo.tamu.edu/files/2010/11/Evaluationofcorn.pdf (accessed March 23, 2022).

Begg, J. E., and N. C. Turner. 1976. Crop water deficits. *Advances in Agronomy* 28: 161–217.

Bennett, W. F., B. B. Tucker, and A. B. Maunder. 1990. *Modern Grain Sorghum Production*, 1st ed. Iowa State University Press, Ames, IA.

Berenguer, M. J., and J. M. Faci. 2001. Sorghum (*Sorghum bicolor* L. Moench) yield compensation processes under different plant densities and variable water supply. *European Journal of Agronomy* 15: 43–55.

Blanco-Canqui, H., J. D. Holman, A. J. Schlegel, J. Tatarko, and T. M. Shaver. 2013. Replacing fallow with cover crops in a semiarid soil: Effects on soil properties. *Soil Science Society of America Journal* 77(3): 1026. doi: 10.2136/sssaj2013.01.0006

Blanco-Canqui, H., and S. J. Ruis. 2020. Cover crops impact on soil physical properties: A review. *Soil Science Society of America Journal* 84(5): 1527–1576.

Blum, A. 2004. Sorghum physiology. In: H. T. Nguyen and A. Blum (Eds.), *Physiology and Biotechnology Integration for Plant Breeding*. Marcel Dekker, NY, pp. 141–223.

Blum, A. 2009. Effective use of water (EUW) and not water-use efficiency (WUE) is the target of crop yield improvement under drought stress. *Field Crops Research* 112: 119–123. doi: 10.1016/j.fcr.2009.03.009

Borrell, A., D. Jordan, J. Mullet, B. Henzell, and G. Hammer. 2006. Drought adaptation in sorghum. In: J. M. Ribaut (Ed.), *Drought Adaptation in Cereals*. The Haworth Press, Binghamton, NY, pp. 335–399.

Brown, W. L. 1985. New technology related to water policy: Plants. In: W. R. Jordan (Ed.), *Water and Water Policy in World Food Supplies: Proceedings of the Conference on Water and Water Policy in World Food Supplies*. Texas A&M University Press, College Station, TX, pp. 37–41.

Brown, W. L., M. S. Zuber, L. L. Darrah, and D. V. Glover. 1985. *Origin, Adaptation, and Types of Corn. Cooperative Extension Service*. Iowa State University, Ames, IA.

Bryant, K. J., V. W. Benson, J. R. Kiniry, J. R. Williams, and R. D. Lacewell. 1992. Simulating corn yield response to irrigation timings: Validation of the Epic model. *Journal of Production Agriculture* 5: 237–242.

Bunce, J. A. 1984. Effects of humidity on photosynthesis. *Journal of Experimental Botany* 35: 1245–1251.

Cannell, G. H., and H. E. Dregne. 1983. Regional setting. In: H. E. Dregne and W. O. Willis (Eds.), *Dryland Agriculture*. ASA, CSSA and SSSA, Madison, WI, pp. 3–17.

Claassen, M. M., and R. H. Shaw. 1970. Water deficit effects on corn. II. Grain components. *Agronomy Journal* 62: 652–655.

Colaizzi, P. D., P. H. Gowda, T. H. Marek, and D. O. Porter. 2008. Irrigation in the Texas High Plains: A brief history and potential reductions in demand. *Irrigation and Drainage* 58: 257–274. doi: 10.1002/ird.418

Collins, H. P., M. M. Mikha, and T. T. Brown. 2012. Agricultural management and soil carbon dynamics: Western U.S. croplands. In: M. A. Liebig, A. J. Franzluebbers, and R. F. Follett (Eds.), *Managing Agricultural Greenhouse Gases*. Elsevier, New York, NY, pp. 59–78.

Collins, R., S. Buck, D. Reid, and G. Spackman. 2006. Manipulating row spacing to improve yield reliability of grain sorghum in Central Queensland. In N. Turner and T. Acuna (Eds.), *Proceedings of the 13th Australian Agronomy Conference*, Perth, Western Australia. Australian Society of Agronomy. http://agronomyaustraliaproceedings.org/images/sampledata/2006/concurrent/systems/4522_collinsr.pdf (accessed April 10, 2022).

Cooper, P. J. M., P. J. Gregory, J. D. H. Keatinge, and S. C. Brown. 1987. Effects of fertilizer, variety and location on barley production under rainfed conditions in Northern Syria. II. Soil water dynamics and crop water use. *Field Crops Research* 16: 67–84.

Craufurd, P. Q., D. J. Flower, and J. M. Peacock. 1993. Effect of heat and drought stress on sorghum (*Sorghum bilcolor*). I. Panicle development and leaf appearance. *Experimental Agriculture* 29: 61–76.

Donald, C. M., and J. Hamblin. 1976. The biological yield and harvest index of cereals as agronomic and plant breeding criteria. *Advances in Agronomy* 28: 361–405.

Doorenbos, J., and A. K. Kassam. 1979. Yield response to water. Irrigation and Drainage Paper 33. Food and Agriculture Organization of the United Nations, Rome, Italy.

Dwyer, I. M., M. Tollenear, and D. W. Stewart. 1991. Changes in plant density dependence of leaf photosynthesis of maize hybrids. *Canadian Journal of Plant Science* 71: 1–11.

Ellis, R. H., R. J. Summerfield, G. O. Edmeades, and E. H. Roberts. 1992. Photoperiod, temperature and the interval from sowing to tassel initiation in diverse cultivars of maize. *Crop Science* 32: 1225–1232.

Evans, L. T. (Ed.) 1975. *Crop Physiology: Some Case Histories*. Cambridge University Press, London.

FAO. 2004. Carbon sequestration in dryland soils. Food and Agriculture Organization of the United Nations, Rome, Italy.

Fischer, R. A., and R. Maurer. 1978. Drought resistance in spring wheat cultivars: 1. Grain yield response. *Australian Journal of Agricultural Research* 29: 897–912.

Fischer, R. A., D. Rees, K. D. Sayre, Z. M. Lu, A. G. Condon, and A. L. Saavedra. 1998. Wheat yield progress associated with higher stomatal conductance and photosynthetic rate and cooler canopies. *Crop Science* 38: 1467–1475.

Fischer, R. A., and R. O. Slatyer (Eds.). 1973. Plant response to climatic factors. United Nations Educational, Scientific, and Cultural Organization, Paris.

Ghimire, R., B. Ghimire, A.O. Mesbah, U.M. Sainju, and O.J. Idowu. 2019. Soil health response of cover crops in winter wheat-fallow system. *Agronomy Journal* 111: 2108–2115.

Gholipoor, M., P. V. Vara Prasad, R. N. Mutava, and T. R. Sinclair. 2010. Genetic variability of transpiration response to vapor pressure deficit among sorghum genotypes. *Field Crops Research* 119: 85–90.

Grange, R. I., and D. W. Hand. 1987. A review of the effects of atmospheric humidity on the growth of horticultural crops. *Journal of Horticultural Science* 62(2): 125–134.

Guerrero, B. L., S. H. Amosson, T. H. Marek, and J. W. Johnson. 2010. Economic evaluation of wind energy as an alternative to natural gas powered irrigation. *Journal of Agricultural and Applied Economics* 42: 277–287.

Hammer, G. L., G. McLean, S. Chapman, B. Zheng, A. Doherty, M. T. Harrison, E. van Oosterom, and D. Jordan. 2014. Crop design for specific adaptation in variable dryland production environments. *Crop and Pasture Science* 65: 614–626. https://doi.org/10.1071/CP14088

Hanks, R. J. 1974. Model for predicting plant yield as influenced by water use. *Agronomy Journal* 66: 660–665.

Hansen, N. C., B. L. Allen, R. L. Baumhardt, and D. J. Lyon. 2012. Research achievements and adoption of no-till, dryland cropping in the semi-arid U.S. Great Plains. *Field Crops Research,* 132: 196–203. doi: 10.1016/j.fcr.2012.02.021

Hasanuzzaman, M., K. Nahar, M. M. Alam, R. Roychowdhury, and M. Fujita. 2013. Physiological, biochemical, and molecular mechanisms of heat stress tolerance in plants. *International Journal of Molecular Science,* 14(5): 9654–9684. doi: 10.3390/ijms14059643

Hatfield, J. L., and J. H. Prueger. 2015. Temperature extremes: Effect on plant growth and development. *Weather and Climate Extremes,* 10: 4–10. doi: 10.1016/j.wace.2015.08.001

Hirasawa, T., and T. C. Hsiao. 1999. Some characteristics of reduced leaf photosynthesis at midday in maize growing in the field. *Field Crop Research* 62: 53–62.

Holman, J. D., K. Arnet, J. Dille, S. Maxwell, A. Obour, T. Roberts, K. Roozeboom, and A. Schlegel. 2018. Can cover or forage crops replace fallow in the semiarid central great plains? *Crop Science* 58(2): 932–944. doi: 10.2135/cropsci2017.05.0324

Holman, J. D., Y. Assefa, and A. K. Obour. 2021. Cover-crop water use and productivity in the high plains wheat-fallow crop rotation. *Crop Science* 61: 1374–1385. https://doi.org/10.1002/csc2.20365

Hornbeck, R., and P. Keskin. 2014. The historically evolving impact of the Ogallala Aquifer: Agricultural adaptation to groundwater and drought. *American Economic Journal: Applied Economics* 6: 190–219.

Huang, J., C. Pray, and S. Rozelle. 2002. Enhancing the crops to feed the poor. *Nature* 418: 678–684.

Jackson, R. D., S. B. Idso, R. J. Reginato, and P. J. Pinter. 1981. Canopy temperature as a crop water-stress indicator. *Water Resources Research* 17: 1133–1138.

Jensen, M. E., D. S. Harrison, H. C. Korven, and F. E. Robinson. 1981. The role of irrigation in food and fiber production. In: M. E. Jensen (Ed.), *Design and Operation of Farm Irrigation Systems.* ASAE Monograph No. 3, pp. 15–41.

Jones, O. R., and W. C. Johnson. 1983. Cropping practices: Southern Great Plains. In: H. E. Dregne and W. O. Willis (Eds.), *Dryland Agriculture.* Agronomy Monograph No. 23. ASA, CSSA, SSSA, Madison, WI, pp. 365–385.

Jones, R. J., S. Ouattar, and R. K. Crookston. 1984. Thermal environment during endosperm cell division and grain filling in maize: effects on kernel growth and development in vitro. Crop Science 24: 133–137.

Kansas State University Extension. 2007. *Corn Production Handbook.* Agricultural Experiment Station and Cooperative Extension Service, Kansas State University, Manhattan, Kansas.

Kapanigowda, M., B. A. Stewart, T. A. Howell, H. Kadasrivenkata, and R. L. Baumhardt. 2010. Growing maize in clumps as a strategy for marginal climatic conditions. *Field Crops Research* 118: 115–125.

Karl, T. R., J. M. Melillo, and T. C. Peterson (Eds.). 2009. Global climate change impacts in the United States. Cambridge University Press, New York, NY, p. 196. http://downloads.globalchange.gov/usimpacts/pdfs/climate-impacts-report.pdf. (accessed October 20, 2016).

Klocke, N. L., J. P. Schneekloth, S. Melvin, R. T. Clark, and J. O. Payero. 2004. Field scale limited irrigation scenarios for water policy strategies. *Applied Engineering in Agriculture* 20: 623–631.

Koohafkan, P., and B. A. Stewart. 2008. Water and cereals in drylands. The Food and Agriculture Organization of the United Nations. Rome, Italy.

Lamaoui, M., M. Jemo, R. Datla, and F. Bekkaoui. 2018. Heat and drought stresses in crops and approaches for their mitigation. Frontiers in Chemistry, 6. https://doi.org/10.3389/fchem.2018.00026

Larson, E. J., and R. L. Vanderlip. 1994. Grain sorghum yield response to nonuniform stand reductions. *Agronomy Journal* 86: 475–477.

Leatherman, J. C., H. A. Cader, and L. E. Bloomquist. 2004. When the well runs dry: The value of irrigation to the western Kansas economy. *Kansas Policy Review* 26: 7–20.

Leng, G., and J. Hall. 2019. Crop yield sensitivity of global major agricultural countries to droughts and the projected changes in the future. *The Science of the Total Environment*, 654, 811–821. https://doi.org/10.1016/j.scitotenv.2018.10.434

Li, S. X., Z. H. Wang, S. Q. Li, Y. J. Gao, and X. H. Tiana. 2012. Effect of plastic sheet mulch, wheat straw mulch, and maize growth on water loss by evaporation in dryland areas of China. *Agricultural Water Management* 116: 39–49.

Lobell, D. B., M. B. Burke, C. Tebaldi, M. D. Mastrandrea, W. P. Falcon, and R. L. Naylor. 2008. Prioritizing climate change adaptation needs for food security in 2030. *Science* 319: 607–610.

Lopes, M. S., and M. P. Reynolds. 2010. Partitioning of assimilates to deeper roots is associated with cooler canopies and increased yield under drought in wheat. *Functional Plant Biology* 37: 147–156.

Lyon, D. J., A. D. Pavlista, G. W. Hergert, R. N. Klein, C. A. Shapiro, S. Knezevic, S. C. Mason, L. A. Nelson, D. D. Baltensperger, R. W. Elmore, M. F. Vigil, A. J. Schlegel, B. L. Olson, and R. M. Aiken. 2009. Skip-row planting patterns stabilize corn grain yields in the central Great Plains. Plant Management Network. www.agronext.iastate.edu/corn/contact/roger_elmore/docs/skip.pdf (accessed March 28, 2016).

Mando, A. 1997. The impact of termites and mulch on the water balance of crusted Sahelian soil. *Soil Technology* 11(2): 121–138.

McGuire, V. L. 2004. Water-level changes in the High Plains Aquifer, predevelopment to 2002, 1980–2002, and 2001–2002. Fact Sheet 2004–3026. U.S. Geological Survey, Lincoln, Nebraska.

McLean, G., J. Whish, R. Routley, I. Broad, and G. Hammer. 2003. The effect of row configuration on yield reliability in grain sorghum: II. Modeling the effects of row configuration. pp. 0–4. In: *Proceedings of the 11th Australian Agronomy Conference (Feb. 2–6, 2003)*. Geelong, Victoria, Australia. www.regional.org.au/au/asa/2003/c/9/mclean.htm (accessed January 27, 2016).

Messina, C. D., D. Podlich, Z. S. Dong, M. Samples, and M. Cooper. 2011. Yield-trait performance landscapes: From theory to application in breeding maize for drought tolerance. *Experimental Botany* 62: 855–868.

Miller, P. R., R. E. Engel, and J. A. Holmes. 2006. Cropping sequence effect of pea and pea management on spring wheat in the Northern Great Plains. *Agronomy Journal* 98: 1610–1619. doi: 10.2134/agronj2005.0302

Mohammed, S., B. C. Blaser, and B. A. Stewart. 2012. Planting geometry and plant population affect dryland maize grain yield and harvest index. *Journal of Crop Improvement* 26(1): 130–139.

Musick, J. T., and D. A. Dusek. 1980. Irrigated corn yield response to water. *Transactions of the American Society of Agricultural Engineering* 23: 92–98.

Musick, J. T., O. R. Jones, B. A. Stewart, and D. A. Dusek. 1994. Water-yield relationship for irrigated and dryland wheat in the U.S. Southern Plains. *Agronomy Journal* 86: 980–986.

Newell, R. L., and W. W. Wilhelm. 1987. Conservation tillage and irrigation effects on corn root development. *Agronomy Journal* 79: 160–165.

Nielsen, D. C., D. J. Lyon, G. W. Hergert, R. K. Higgins, and J. D. Holman. 2015b. Cover crop biomass production and water use in the Central Great Plains. Agronomy Journal 107(6): 2047–2058. doi: 10.2134/agronj15.0186

Nielsen, D. C., D. J. Lyon, R. K. Higgins, G. W. Hergert, J. D. Holman, and M. F. Vigil. 2015a. Cover crop effect on subsequent wheat yield in the central Great Plains. *Agronomy Journal* 108(1): 243–256. doi: 10.2134/agronj2015.0372

Nielsen, D. C., M. F. Vigil, and W. B. Henry. 2018. Skip row planting configuration shifts grain sorghum water use under dry conditions. *Field Crops Research* 223: 66–74.

Otegui, M. E., F. H. Andrade, and E. E. Suero. 1995. Growth, water use, and kernel abortion of maize subjected to drought at silking. *Field Crops Research* 40: 87–94.

Passioura, J. B. 1996. Drought and drought tolerance. *Plant Growth Regulation* 20: 79–83.

Pavlista, A. D., D. J. Lyon, D. D. Baltensperger, and G. W. Hergert. 2010. Yield components as affected by planting dryland maize in a double-skip row pattern. *Journal of Crop Improvement*, 24: 131–141. https://doi.org/10.1080/15427520903565307

Paye, W. S., R. Ghimire, P. Acharya, A. Nilahyane, and A. Mesbah. 2022. Cover crop water use and corn silage production in semiarid irrigated conditions. *Agricultural Water Management* 260: 107275, doi: 10.1016/j.agwat.2021.107275

Payero, J. O., N. L. Klocke, J. P. Schneekloth, and D. R. Davison. 2006. Comparison of irrigation strategies for surface-irrigated corn in West Central Nebraska. *Irrigation Science* 24: 257–265.

Peterson, G. A., D. G. Wesfall, and C. V. Cole. 1993. Agro-ecosystem approach to soil and crop management research. *Soil Science Society of America Journal* 57(5): 1354–1360.

Peterson, G. A., D. G. Westfall, and N. C. Hansen. 2012. Enhancing precipitation use efficiency in the world's dryland agroecosystems. In: R. Lal and B. A. Stewart (Eds.), *Advancement in Soil Science – Soil Water and Agronomic Productivity*. CRC Press, Boca Raton, FL, pp. 455–476.

Ponce, V. M. 1995. Management of droughts and floods in the semiarid Brazilian Northeast – the case for conservation. *Journal of Soil and Water Conservation* 50(5): 422–431.

Power, J. F. 1990. Fertility management and nutrient cycling. *Advances in Soil Science* 13: 131–149.

Prasad, P. V. V., S. Staggenborg, and Z. Ristic. 2008. Impacts of drought and/or heat stress on physiological, developmental, growth, and yield processes of crop plants. *Advances in Agricultural Systems Modeling Series 1*. ASA, CSSA, SSSA, 677 S. Segoe Rd., Madison, WI 53711, USA.

Rasmussen, P. E., and H. P. Collins. 1991. Long-term impacts of tillage, fertilization, and crop residues on soil organic matter in temperate semi-arid regions. *Advances in Agronomy* 45: 93–134.

Ridge, P. E. 1986. A review of long fallows for dryland wheat production in southern Australia. *Journal of Australian Institute of Agricultural Science* 52: 37–44.

Roberts, M. G., T. D. Male, and T. P. Toombs. 2007. Potential impacts of biofuel expansion on natural resources; a case study of the Ogallala aquifer region, Environmental Defense Report. Environmental Defense, NY.

Robinson, E. B., and D. L. John. 2022. Great Plains. *Encyclopedia Britannica*, 10 May 2022. www.britannica.com/place/Great-Plains (accessed April 7, 2022).

Rooney, W. 2004. Sorghum improvement-integrating traditional and new technology to produce improved genotypes. *Advances in Agronomy* 83: 37–109.

Rosegrant, M., and S. Cline. 2003. Global food security: Challenges and policies. *Science* 302: 1917–1919.

Routley, R., I. Broad, G. McLean, J. Whish, and G. Hammer. 2003. The effect of row configuration on yield reliability in grain sorghum: I. Yield, water use efficiency and soil water extraction. *Agron. Conf. Australian Soc. of Agron.*, Gosford, Australia. http://era.deedi.qld.gov.au/427/1/Routley EffectConfiguration1-SEC.pdf (accessed May 15, 2022).

Scanlon, B. R., R. C. Reedy, J. B. Gates, and P. H. Gowda. 2010. Impact of agroecosystems on groundwater resources in the central high plains, USA. *Agriculture, Ecosystem and Environment* 139(4): 700–713.

Schimel, D. S. 2010. Drylands in the earth system. *Science* 327: 418–419.

Schlegel, A. J., and J. L. Havlin. 1997. Green fallow for the central Great Plains. *Agronomy Journal* 89(5): 762–767. doi: 10.2134/agronj1997.00021962008900050009x

Shafer, M., D. Ojima, J. M. Antle, D. Kluck, R. A. McPherson, S. Petersen, B. Scanlon, and K. Sherman. 2014. Ch. 19: Great Plains. In: J. M. Melillo, Terese (T. C.) Richmond, and G. W. Yohe (Eds.), *Climate Change Impacts in the United States: The Third National Climate Assessment*, U.S. Global Change Research Program, pp. 441–461. doi: 10.7930/J0D798BC

Sinclair, T. R., and J. M. Bennett. 1998. Water. In: T. R. Sinclair and F. P. Gardner (Eds.), *Principles of Ecology in Plant Production*. CABI, Wallingford, UK, pp. 103–120.

Sinclair, T. R., G. L. Hammer, and E. J. Van Oosterom. 2005. Potential yield and water-use efficiency benefits in sorghum from limited maximum transpiration rate. *Functional Plant Biology* 32: 945–952.

Sinclair, T. R., and A. Weiss. 2010. *Principles of Ecology in Plant Production*. 2nd ed. CABI, Cambridge, MA.

Singh, B. R., and D. P. Singh. 1995. Agronomic and physiological responses of sorghum maize and pearl millet to irrigation. *Field Crops Research* 42: 57–67.

Singh, D. P., D. B. Peters, P. Singh, and M. Singh. 1987. Diurnal patterns of canopy photosynthesis, evapotranspiration and water use efficiency in chickpea (*Citer arietinum* L.) under field conditions. *Photosynthesis Research* 11: 61–69.

Smith, C. W. 1995. *Crop Production*. John Wiley & Sons Inc., NY.

Spackman, G. B., K. J. McCosker, A. J. Farquharson, and M. J. Conway. 2001. Innovative management of grain sorghum in Central Queensland. In: *Proceedings of the 10th Australian Agronomy Conference*. Geelong, Victoria, Australia. www.agronomyaustraliaproceedings.org/images/sampledata/2001/1/a/spackman.pdf (accessed May 12, 2022).

Stewart, B. A. 2006. Growing dryland crops in clumps: What are the benefits? In: *Proceedings of the 28th Annual Southern Conservation Systems Conference*, Amarillo, Texas, pp. 47–56.

Stewart, B. A. 2009. Manipulating tillage to increase stored soil water and manipulating plant geometry to increase water-use efficiency in dry land areas. *Journal of Crop Improvement* 23: 71–82.

Stewart, B. A., and E. Burnett. 1987. Water conservation technology in rainfed and dryland agriculture. In: W. R. Jordan (Ed.), *Water and Water Policy in World Food Supplies*. Texas A&M University Press, pp. 355–359.

Stewart, B. A., and W. L. Liang. 2015. Strategies for increasing the capture, storage, and utilization of precipitation in semiarid regions. *Journal of Integrative Agriculture* 14(8): 1500–1510. doi: 10.1016/S2095-3119(15)61096-6

Stewart, B. A., and G. A. Peterson. 2015. Managing green water in dryland agriculture. *Agronomy Journal* 107: 1544–1553.

Stewart, B. A., and S. Thapa. 2016. Dryland farming: Concept, origin, and brief history. In: M. Farooq and H. M. Siddique (Eds.), *Innovations in Dryland Agriculture*. Springer International Publishers, pp. 3–29. https://doi.org/10.1007/978-3-319-47928-6_1

Stone, L. R. 2003. Crop water use requirements and water use efficiencies. *In: Proceedings of the 15th Annual Central Plains Irrigation Conference and Exposition*. Colby, Kansas, pp. 127–133.

Thapa, S., B. A. Stewart, E. Ashiadey, Q. Xue, B. C. Blaser, and R. Shrestha. 2021. Transpiration efficiency of corn hybrids at different growth stages. *Journal of Crop Improvement* 36: 389–399. doi: 10.1080/15427528.2021.1972376

Thapa, S., Q. Xue, K. E. Jessup, J. C. Rudd, S. Liu, R. N. Devkota, and J. Baker. 2018a. Canopy temperature depression at grain filling correlates to winter wheat yield in the U.S. Southern High Plains. *Field Crops Research* 217: 11–19. doi: 10.1016/j.fcr.2017.12.005

Thapa, S., B. A. Stewart, Q. Xue, M. B. Rhoades, B. Angira, and J. Reznik. 2018b. Canopy temperature, grain yield, and harvest index in corn as affected by planting geometry in a semi-arid environment. *Field Crops Research* 227: 110–118. doi: 10.1016/j.fcr.2018.08.009

Thapa, S., Q. Xue, K. E. Jessup, J. C. Rudd, S. Liu, G. P. Pradhan, R. N. Devkota, and J. Baker. 2017a. More recent wheat cultivars extract more water from greater soil profile depths to increase yield in the Texas High Plains. *Agronomy Journal* 109: 1–10. doi: 10.2134/agronj2017.02.0064

Thapa, S., B. A. Stewart, Q. Xue, and Y. Chen. 2017b. Manipulating plant geometry to improve microclimate, grain yield, and harvest index in grain sorghum. *PLoS One* 12(3): e0173511. doi: 10.1371/journal.pone.0173511

Thapa, S., B. A. Stewart, Q. Xue, P. Pokhrel, T. Barkley, and M. Bhandari. 2016. Growing corn in clumps reduces canopy temperature and improves microclimate. *Journal of Crop Improvement* 30: 614–631. doi: 10.1080/15427528.2016.1217445

Thapa, S., Q. Xue, and B. A. Stewart. 2020a. Alternative planting geometries reduce production risk in corn and sorghum in water-limited environments. *Agronomy Journal* 112(4): 3322–3334. https://doi.org/10.1002/agj2.20347

Thapa, S., Q. Xue, T. Marek, W. Xu, D. Porter, and K. Jessup. 2020b. Corn production under restricted irrigation in the Texas High Plains. *Agronomy Journal* 112(2): 1190–1200. https://doi.org/10.1002/agj2.20003

Thapa, S., Q. Xue, K. E. Jessup, J. C. Rudd, S. Liu, R. N. Devkota, and J. Baker. 2019a. Soil water extraction and use by winter wheat cultivars under limited irrigation in a semi-arid environment. *Journal of Arid Environment* 174: 104046. doi: 10.1016/j.jaridenv.2019.104046

Thapa, S., Q. Xue, K. E. Jessup, J. C. Rudd, S. Liu, T. H. Marek, R. N. Devkota, J. A. Baker, S. Baker. 2019b. Yield determination in winter wheat under different water regimes. *Field Crops Research* 233: 80–87. doi: 10.1016/j.fcr.2018.12.018

Thapa, V. R., R. Ghimire, V. Acosta-Martinez, and M. Marsalis. 2019c. Conservation systems for positive net ecosystem carbon balance in semiarid drylands. *Agroecosystems, Geosciences and Environment* 2: 190022.

Thapa V. R., R. Ghimire, D. VanLeeuwen, V. Acosta-Martinez, and M. K. Shukla. 2022. Response of soil organic matter to cover cropping in water-limited environments. *Geoderma*. https://doi.org/10.1016/j.geoderma.2021.115497

Tollenaar, M., and A. Aguilera. 1992. Radiation use efficiency of an old and new maize hybrid. *Agronomy Journal* 84: 536–541.

Turner, N. C., and G. J. Burch. 1983. The role of water in plants. In: I. D. Teare, and M. M. Peet (Eds.), *Crop-water Relations*. John Wiley & Sons, Inc., NY.

UNEP. 1997. *World Atlas of Desertification*. 2nd ed. United Nations Environment Programme, Nairobi, Kenya.

Unger, P. W. 1994. Tillage effects on dryland wheat and sorghum production in the southern Great Plains. *Agronomy Journal* 86: 310–314.

Unger, P. W. 2001. Total carbon, aggregation, bulk density, and penetration resistance of cropland and nearby grassland soils. In: R. Lal (Ed.), *Soil Carbon Sequestration and the Greenhouse Effect*. SSSA, Madison, WI, pp. 77–92.

Unger, P. W., and M. F. Vigil. 1998. Cover crop effects on soil water relationships. *Journal of Soil and Water Conservation* 53: 200–207.

Urban, O., D. Janous, M. Acosta, R. Czerny, I. Markova, M. Navratil, M. Pavelka, R. Pokorny, M. Sprtova, R. Zhang, V. Spunda, J. Grace, and M. V. Marek. 2007. Ecophysiological controls over the net ecosystem exchange of mountain spruce stand. Comparison of the response in direct vs. diffuse solar radiation. *Global Change Biology* 13: 157–168.

USDA ERS. 2012. *Atlas of Rural and Small-Town America*. Washington, DC: U.S. Department of Agriculture, Economic Research Service. www.ers.usda.gov/data-products/atlas-of-rural-and-small-town-america. aspx. (accessed March 17, 2022).

Vigil, M. F., B. Henry, F. J. Calderon, D. Poss, D. C. Nielsen, J. G. Benjamin, and R. Klein. 2008. A use of skip-row planting as a strategy for drought mitigation in the west-central Great Plains. In: A. Schlegel (Ed.), *Proceedings Great Plains Soil Fertility Conference* (12): 101–106. Great Plains Soil Fertility Conference, Monticello, IL.

Wani, A. P., K. K. Garg, A. K. Singh, and J. Rockström. 2012. Sustainable management of scarce water resources in tropical rainfed agriculture. In: R. Lal and B. A. Stewart (Eds.), *Soil Water and Agronomic Productivity, Advances in Soil Science*. CRC Press, Boca Raton, FL, pp. 347–408.

Weatherwax, P. 1954. *Indian Corn in Old America*. Macmillan, New York, NY.

Weinheimer, J., P. Johnson, and D. Mitchell. 2013. Texas High Plains initiative for strategic and innovative irrigation management and conservation. *Journal of Contemporary Water Research & Education* 151: 43–49.

Whish, J., G. Butler, M. Castor, S. Cawthray, I. Broad, P. Carberry, G. Hammer, G. McLean, R. Routley, and S. Yeates. 2005. Modelling the effects of row configuration on sorghum yield reliability in north-eastern Australia. *Australian Journal of Agricultural Research*, 56: 11–23. https://doi.org/10.1071/AR04128

Wright, G. C., and R. C. G. Smith. 1983. Differences between two grain sorghum genotypes in adaptation to draught stress. II. Root water uptake and water use. *Australian Journal of Agricultural Research* 34(6): 627–636.

Xue, Q., J. C. Rudd, S. Liu, K. E. Jessup, R. N., Devkota, and J. R. Mahan. 2014. Yield determination and water use efficiency of wheat under water-limited conditions in the U.S. Southern High Plains. *Crop Science* 54: 34–47.

Zhang, Z. X., and J. L. Wu. 1994. Study on the techniques for conservation of water during summer fallow period in precipitation deficient year. In: W. Q. Chen, and N. Q. Xin (Eds.), *Strategies for Comprehensive Development of Agriculture in North China*. China's Agri. Sci. and Tech. Press, Beijing, pp. 195–198.

6 Managing Soil and Water Resources by Tillage, Crop Rotation, and Cover Cropping

Paul Bradley DeLaune[1], Katie Lynn Lewis[2], and Joseph Alan Burke[2]
[1]Texas A&M AgriLife Research, Vernon, TX 76384, USA
[2]Texas A&M AgriLife Research, Lubbock, TX 79403, USA

CONTENTS

6.1 Introduction 129
6.2 Conservation Tillage 131
6.3 Crop Rotation 132
6.4 Cover Crops 133
6.5 Impact of Conservation Agriculture on Soil and Water Resources 134
     6.5.1 Soil Properties 134
     6.5.2 Erosion 138
     6.5.3 Soil Water and Water Use Efficiency 140
     6.5.4 Integrated Crop–Livestock Systems 145
6.6 Further Advancement of Conservation Agriculture 147
References 147

6.1 INTRODUCTION

Recent catastrophic events related to drought conditions have highlighted climate change and water scarcity. These events also raise awareness of the potential impact on the global food chain and agriculture in general. Drought is not just the absence of rain; it is fueled by land degradation and the climate crisis. Drought has affected millions of people and resulted in significant economic losses over the past century throughout the world, expanding from Africa to Europe to the United States (US) and to Asia (Taylor et al., 2017; Guha-Sapir et al., 2021; NOAA-NCEI, 2021). A great portion of the world's food is produced in semi-arid and arid regions. Providing sufficient food, feed, and fuel for an increasing global population without degrading natural resources has been identified as a significant agricultural challenge of the 21st century (Rockstrom et al., 2009; Tilman et al., 2002). Irrigation has accounted for much of the production in these environments. With a growing population, there is an ever-increasing need for not only food but also the very water that is critical to producing much of the world's food. Competing entities for water resources continue to heighten and pressure policymakers to guide decisions on how water is allocated. In some instances, reduced water allocation for agriculture or declining aquifer levels have led traditionally irrigated agricultural areas to reduce irrigation inputs or transition to rainfed agricultural environments. In such instances, years of low-quality water use or low residue cropping systems can lead to degraded soil resources.

Soil degradation has been estimated to have decreased soil ecosystem services by 60% between 1950 and 2010 (Lal, 2015; Leon et al., 2014). Soil health and soil quality have recognized renewed

DOI: 10.1201/b22954-6

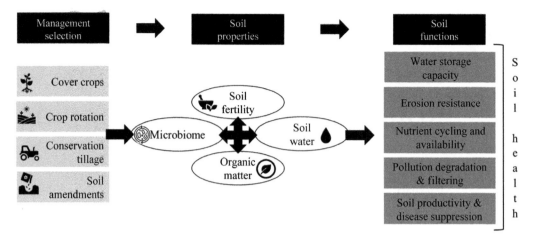

FIGURE 6.1 A conceptual model of the impact of management strategies on microbiome, soil organic matter, macroaggregates, and nutrient cycling that will benefit several soil functions and overall soil health (Adapted from Cano et al., 2018).

importance due to strong links to human health and livelihoods (Abrahams, 2002; Lal, 2009; Brevik and Sauer, 2015; Wall et al., 2015; Steffan et al., 2018). Lal (2015) noted that once the process of soil degradation is set in motion, often by land misuse and soil mismanagement along with extractive farming, it feeds on itself in an ever-increasing downward spiral. Improper use of tillage often initiates degradation of soil physical properties and a subsequent decline in resilience and quality of soil and the environment (Lal, 1993, 2015; Busscher and Bauer, 2004; Holland, 2004; Ramzan et al., 2019). Such processes are particularly severe in rain-limited environments, where rainfall variability exacerbates crop failure and resource degradation (Tittonell et al., 2012).

Soil quality can be maintained or restored through sustainable management, which often begins with some form of conservation agriculture (Wall et al., 2015; Lal, 2015). Over the last decade or more, several terms have been used to describe what basically circles back to soil and water conservation. Whether soil quality, soil health, regenerative agriculture, climate smart agriculture practices, resilient agriculture, or any other relatable term, the ultimate goal is to restore degraded soil resources and reverse the downward spiral that is often set in motion by human beings. Conservation approaches such as conservation tillage, crop rotation, and cover cropping are often cited as soil health-promoting practices that can restore soil resources and subsequently result in systems that are resilient to drought conditions. Selection of these management practices has been reported to affect soil properties, subsequent soil function, and ultimately overall soil health (Figure 6.1). The Food and Agriculture Organization (FAO) describes conservation agriculture as a concept in support of sustainable land management, environmental protection and climate change adaptation and mitigation (FAO, 2022). The FAO promotes the adoption of the following conservation agriculture principles that are universally applicable in all agricultural landscapes and cropping systems:

a. Minimum mechanical soil disturbance (i.e., no/zero tillage through direct seed and/or fertilizer placement to reduce soil erosion and preserve organic matter);
b. Permanent soil organic cover of at least 30% with crop residue and/or cover crops to maintain a protective layer of vegetation on the soil surface to suppress weeds, protect the soil from the impact of extreme weather patterns, help preserve soil moisture, and avoid compaction of the soil; and
c. Species diversification through varied crop sequences and associations involving at least three different crop species through which, if well-designed, will promote good soil structure, foster

Managing Soil and Water Resources 131

a diverse range of soil flora and fauna that contributes to nutrient cycling and improved plant nutrition and helps to prevent pest and diseases.

Building soil organic carbon (SOC) is a chief goal of conservation agriculture adoption, as SOC is often viewed as the cornerstone to improved soil function and subsequent soil and water conservation. Soil organic matter (SOM) is a key factor contributing to the water-holding capacity of soil, up to 10,800 liters more water per hectare (ha) can be retained with a 1% increase in SOM (Libohova et al., 2018). In this chapter, we will review these promoted conservation agriculture approaches and their potential to conserve fragile soil and water resources.

6.2 CONSERVATION TILLAGE

Conservation tillage is a broadly defined practice that includes no-tillage, strip tillage, and mulch tillage systems (Claassen et al., 2018). Often conservation tillage practices are coupled with other practices such as crop rotations and cover cropping where reduced soil disturbance results in at least 30% of residue remaining on the soil surface. Stewart (2007) explained the expansion of cultivation from humid to semiarid regions of the US Great Plains in the late 1800s and early 1900s where the greatest expansion largely occurred during a decade with higher-than-average annual precipitation, which when coupled with more than adequate plant nutrients coming from the decomposition of SOM from newly plowed grassland soils resulted in high wheat yields and prosperous farming. Of course, this decade was followed by much drier years and the infamous era known as the "Dust Bowl," one of the greatest ecological disasters ever caused by human activities. Thereafter, the concept of stubble mulch tillage was developed to combat soil erosion and further advancement in machinery and herbicide technologies led to expanded adoption of conservation tillage (Derpsch, 2004).

Conservation agriculture has rapidly grown in recent years. Kassam et al. (2019) reported conservation agriculture increased to 180 million ha in 2015/2016 compared to 106 million ha in 2008/09 and only 2.8 million ha in 1973/74. Globally, North America and South America represent nearly 75% of the 12.5% cropland under conservation agriculture (Kassam et al., 2019; Table 6.1). Across the US, approximately 72% of cropland reported using conservation tillage in 2017 compared to 62% in 2012 (NASS, 2019). Most of the corn (65% in 2016), wheat (67% in 2017), and soybean (70% in 2012) were produced using conservation tillage, while only 40% of cotton used conservation tillage in 2015 (Claassen et al., 2018). The reduction in tillage associated with these systems has historically been viewed as a method to reduce soil erosion by enhanced aggregation, increase SOC, enhance water capture and storage, and provide resilience and yield stability to crop production (Kladivko et al., 1986; Mikha et al., 2015; Zuber et al., 2015; Claassen et al., 2018).

Tillage management practices influence biological, chemical, and physical soil properties, all of which contribute to the health of soil. Water dynamics are a major component of soil health and are directly influenced by physical soil properties and processes such as aggregate formation and stability and by biological processes including SOM. Tillage is one of the primary management decisions that can have a direct impact on soil physical processes including aggregate formation and water flow and storage (Mikha et al., 2015). In semiarid environments where water is the greatest limiting resource, practices resulting in improved water and soil conservation should be encouraged. The type of tillage and the number of tillage operations can have a great effect on soil disturbance, residue management, and soil properties. Moldboard tillage may bury approximately 95% of the residue with one operation compared to 90% with disk tillage, 50% with chisel plows, and about 25% with under cutters (sweeps 50–75 cm wide) (Stewert, 2007). Stewart (2007) noted that even the sweep plows that are often used during long fallow periods may bury most of the crop residues because there are usually three or more tillage operations during the fallow period when tillage

TABLE 6.1
Cropland Area under Conservation Agriculture (M ha) by Region in 2015/16; as % of Global Total Cropland, and as % of Cropland of Each Region

Region	Conservation Agriculture Cropland Area	% of Global Conservation Agriculture Cropland Area	% of Cropland Area in the Region
South America	69.90	38.7	63.2
North America	63.18	35.0	28.1
Australia & New Zealand	22.67	12.6	45.5
Asia	13.93	7.7	4.1
Russia & Ukraine	5.70	3.2	3.6
Europe	3.56	2.0	5.0
Africa	1.51	0.8	1.1
Global Total	**180.44**	**100**	**12.5**

Source: Adapted from Kassam et al. (2019).

is used exclusively for weed control. Reduced tillage without ground cover may not achieve the greatest benefits related to soil and water conservation.

6.3 CROP ROTATION

Crop rotation is the changing of crop species grown in a particular field from year to year, an ancient practice that dates back to pre-BC times. In contrast, monocropping is planting and growing the same species in the same field year after year. Some media and/or literature define monoculture as growing a single crop at one time annually, even though a different species may be grown in the second year. For this chapter, a corn crop grown in year 1 followed by a soybean crop in the following year is regarded as a crop rotation, often referred to as a short-cycle or simple crop rotation. Within the US, crop rotation has been shown to vary by crop and region. While wheat was produced in monoculture settings on only 25% of planted areas, monoculture wheat consisted of as much as 86% of planted areas in certain states. Cotton was identified as the crop grown mostly in monoculture systems, with 60% of total planted areas in monoculture and as high as 92% of those planted in some states (Padgitt et al., 2000). Increasing crop diversity through crop rotations can enhance crop productivity as well as soil and environmental quality (Tautges et al., 2016); thus, diversity of crop rotations is an important factor in the capability of a crop rotation to provide improved soil function.

The value of crop rotation has long been noted, as Magdof and Van Es (2009) begin the chapter entitled "Crop Rotations" in *Building Soils for Better Crops: Sustainable Soil Management* with a quote attributed to Henry Snider from 1896:

> with methods of farming in which grasses form an important part of the rotation, especially after those that leave a large residue of roots and culms, the decline of the productive power is much slower than when crops like wheat, cotton, or potatoes, which leave little residue on the soil, are grown continuously.

Rotations have historically been driven by the need to build sufficient fertility through the crop sequence and to maintain control of weeds, pests, and diseases (Knox et al., 2011). Several campaigns to restore soil health call for land stewards to mimic nature, which usually lends itself to a highly diverse system. For example, most US farms in the early 1900s produced a variety of crops and animal species together on the same farm in complementary ways (MacDonald et al., 2013). Perhaps this was recognized by ancient growers, often growing crops that best fit local climate and

Managing Soil and Water Resources

could feed their local population. Wahlqvist and Meei-Shyuan (2007) noted a remarkable resilience and ingenuity of people and their food systems, but monoculture and lack of diversity encourage food system failure.

Thus, one may question how the agriculture community departed from this proven practice. Industrialization of agriculture is often viewed as the catalyst to the loss of crop diversity as farm mechanization, increased use and reliance of pesticides and fertilizers, farm specialization, government subsidies, and improved transportation have induced a spatial concentration of crop types (Wang et al., 2021; Crosslet et al., 2020; MacDonald et al., 2013). In addition, expansion of simple crop rotations has increased due to crop price increases, increased food and industrial uses, economic and world trade benefits, and genetic improvements (Karlen et al., 2004; Wand and Chowdhury, 2020). For example, cropland areas devoted to small grains, sunflowers, and flaxseed decreased by over 50% from 1980 to 2012 to grow corn and soybeans on the US Great Plains (Smart et al., 2021). These factors not only affect how cropland is managed, but also have resulted in conversion of grassland to cropland. It has been estimated that 2.3 million ha of grassland was converted to cropland during the high crop price period of 2006–2015 in mostly the US Northern Great Plains along with a 5.3 million ha decrease in CRP lands during the same period throughout the US Great Plains (Smart et al., 2021; USDA-FSA, 2019). Within the US, crop diversity in 2012 was significantly lower than any other census period from 1978 to 2012 (Aguilar et al., 2015). Lin (2011) stated that a major challenge for the implementation of diversified agricultural systems for farmers is finding the appropriate balance of diversification within the farm system to satisfy both production and protection values.

6.4 COVER CROPS

To increase the ecosystem services benefits of semiarid cropping systems, the inclusion of cover cropping is rapidly becoming a popular option. Cover crops can be loosely defined as terminated crops grown between cash crops, in the absence of a normal cash crop rotation, or intercropped (Reves, 1994). At a basic level, cover crops can eliminate a traditional fallow period in most cropping systems that leave soil bare or exposed, especially in conventionally tilled cropping systems. The US Department of Agriculture further defines cover crops as "Crops, including grasses, legumes, and forbs, for seasonal cover and other conservation purposes." Cover crops are primarily used for erosion control, soil health improvement, and water quality improvement. A cover crop managed and terminated according to these guidelines is not considered a "crop" for crop insurance purposes. The cover crop may be terminated by natural causes, such as frost, or intentionally terminated through chemical application, crimping, rolling, tillage, or cutting (USDA-NRCS, 2014a, 2014b). They have been used extensively in different regions of the world to increase SOM and nitrogen (N) availability (green manures), decrease soil susceptibility to erosion (cover crops), and minimize nitrate leaching following cash crop harvest (catch crops) (Koudahe et al., 2022). Studies have indicated that cover crops can also help control noxious weeds, minimize pesticide losses, and increase drought resilience (Myers et al., 2019). Despite these potential benefits, cover crop adoption in the US varies greatly by region, with most of their use in temperate and sub-humid regions (Blanco-Canqui et al., 2015).

In 2017, farmers planted 6.2 million ha of cover crops compared to 4.2 million ha in 2012 (Wallander et al., 2021). However, cover crop adoption varies greatly in the US, with a majority of the cover cropped acres occurring in the eastern and southeastern US; however, in semiarid regions of West Texas, cover crop adoption rates are greater in cotton monocultures than corn and soybeans in more temperate regions like the Midwest (Wallander et al., 2021). One potential reason for the increased adoption of cover crops in cotton production is because of the limited amount of biomass produced by cotton. Cover crop adoption accounts for approximately 8% of the Southern Great Plains region of Texas and Oklahoma (Figure 6.2) which accounts for nearly 40% of US cotton

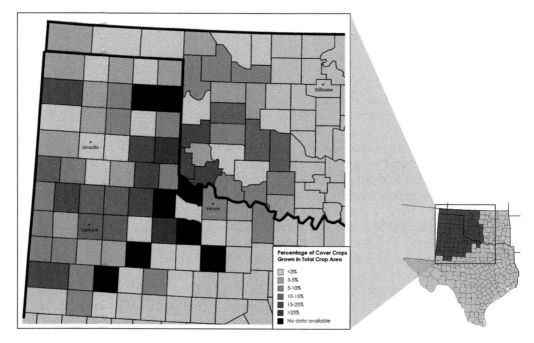

FIGURE 6.2 Cover crop adoption rates in the semi-arid Texas and Oklahoma regions of the U.S. Southern Great Plains. Data source: 2017 U.S. Census of Agriculture (NASS, 2018).

production (NASS, 2018; Burke et al., 2021). While the United States Department of Agriculture – Natural Resources Conservation Service (USDA-NRCS) has not currently defined a goal for cover crop adoption in the US, conservation organizations have called for 40.5 million ha of cover crop adoption by 2025 (Hamilton et al., 2017).

The benefits and consequences of cover crop adoption depend on many factors, including cover crop species selection, environmental conditions, and producer knowledge, which impact the adoption rate of cover crops in the US. The most commonly grown cover crops in the US are grasses and legumes followed by non-legume broadleaves and brassicas (Haramoto, 2019). The choice of cover crop significantly influences the ecosystem benefits rendered from it, which is compounded by producer attitudes and knowledge (Blanco-Canqui et al., 2015; Halvorson et al., 2019; Nichols et al., 2021). Dabney et al. (2001) noted that the adoption of cover crops in cropping systems is based on the perceived balance between advantages and disadvantages. In semiarid environments, reduction in soil moisture by cover crops is a potential disadvantage (Dabney et al., 2001; Balkcom et al., 2007). Water use by cover crops has been shown to greatly reduce yields of subsequent crops in semiarid regions (Unger and Vigil, 1998; Reicosky and Forcella, 1998; Nielsen et al., 2016; Holman et al., 2018). Crop yield is often used as a major criterion to determine if a particular management practice provides an advantage or disadvantage. Hence, striking a balance between residue management, soil water use, and crop yield must be carefully weighed in water-limited environments prior to cover crop adoption.

6.5 IMPACT OF CONSERVATION AGRICULTURE ON SOIL AND WATER RESOURCES

6.5.1 Soil Properties

Enhanced aggregate formation and stability because of decreased tillage intensity can increase SOC storage by providing protection from microbial oxidation and soil erosion (Balesdent et al.,

2000; Zotarelli et al., 2005; Razafimbelo et al., 2008; Mikha et al., 2015; Zhao et al., 2017). Also important to chemical composition and biological activity, SOC improves nutrient cycling and holding capacity of a soil (Sadeghpour et al., 2016). Reduced tillage and no-till (NT) have been reported to reduce net carbon dioxide (CO_2) losses from the soil, thereby increasing potential carbon storage (Lopez-Fando and Pardo, 2011; Zuber et al., 2015; McDonald et al., 2019). Due to the positive changes associated with an increase in SOC, agricultural productivity is often enhanced; however, in semiarid environments yield benefits are not consistently reported (Zuber et al., 2015; Lewis et al., 2018). Aggregation is the arrangement of sand, silt, and clay particles around SOM and through particle associations (Oades, 1995). Aggregates exist as macro- and microaggregates which give physical structure to soil, reducing compaction and providing protection to SOM. Larger macroaggregates are more stable and better able to resist wind and water erosion. The area between aggregates provides pore space for retention and exchange of water and air. High-intensity tillage practices disturb aggregates, decreasing aggregate stability and size, which not only has a negative impact on water dynamics and soil erosion but also results in a net loss of C into the atmosphere as microorganisms decompose SOM releasing C as CO_2.

Greater macroaggregates and aggregate stability have been reported with decreased tillage intensity and no-tillage systems versus conventional tillage management practices (Hontoria et al., 2016; Weidhuner et al., 2021). In a more tropical region, Zotarelli et al. (2005) reported mean weight diameter to be 0.5 mm greater with no-tillage than conventional tillage. Conventional tillage decreased the proportion of the largest macroaggregate class (>2 mm) by approximately 10% compared to no-tillage. Occasional tillage is often needed for weed control purposes, but as reported by Weidhuner et al. (2021), periodic tillage in a mostly NT system can be as detrimental to soil structure as practicing yearly reduced tillage, if not more. Reverting to tillage after 12 years of NT significantly increased the bulk density by 19%, reduced the total porosity by 21% and soil water content by 28% compared to NT with cover crops after 3 years of tillage in the southern Great Plains (Mubvumba et al., 2022). Converting from NT after 6 years to conventional till via disking increased the total sediment loss by 78% in a winter wheat system (DeLaune and Sij, 2012).

Crop rotation coupled with conservation tillage often will enhance soil properties. Crop rotation and soil tillage have been described as the most critical soil practices affecting soil physical and hydrological characteristics (Hill, 1990; Katsvairo et al., 2002). Diversification of crop species grown in crop sequence or association and minimum mechanical soil disturbance are the interlinked principles of conservation agriculture (FAO, 2010). Crop rotation has been shown to improve soil stability, soil structure, aggregation, SOC, and water infiltration by alternating crops with differing rooting depths (Karlen et al., 2006; Janvier et al., 2007). However, the effects of crop rotation on soil properties are not consistent among studies and may be site specific, which may be attributed to variations in management operations, crop and soil types, antecedent conditions, inherent soil properties, duration of study, and environmental settings (Blanco-Canqui et al., 2004; Ferreras et al., 2000; Lal and Van Doren, 1990). SOC is important for soil fertility and function through subsequent effects on physical, biological, and chemical properties. There has been an increasing interest in estimating SOC changes in light of climate change, as these changes affect the rate of accumulation of atmospheric CO_2 (Janzen, 2004).

All crop rotations are not created equal – in other words, selection of crops in the rotation can have varying effects on ecosystem services. For example, inclusion of soybean within a crop rotation has been shown to cause soil structure degradation, erosion problems, lower SOC, and higher bulk density (Bathke and Blake, 1984; Jagadamma et al., 2008; Coulter et al., 2009). Kirschenmann (2002) concluded that cropping systems that specialize in one or two crops provide minimal or no plant diversity to a system and ultimately lead to biological and physical soil property degradation and in many instances to soil chemical property degradation. The corn–soybean rotation is common in the US Midwest and northern Great Plains, as corn can utilize the N remaining from the legume crop. A global data analysis indicated that enhancing rotation complexity can sequester an average of 20 ±12 g C myr^{-1}, excluding a change from continuous corn to a corn–soybean rotation which may not result in a significant accumulation of SOC (West and Post, 2002). Dold et al. (2017) found that corn had a positive C balance, while soybean showed a negative balance.

Diversifying crop rotations can enhance important soil properties to build more resilient systems. Alhameid et al. (2019) demonstrated the benefits of expanding a simple corn–soybean rotation to a 4-year rotation that included corn–soybean–wheat–oat. Analysis of soil samples collected during the corn phase indicated that a 4-year rotation under NT significantly decreased soil penetration resistance and increased water-filled micropores compared to 2- and 3-year crop rotations under NT management and 2-, 3-, and 4-year crop rotations under conventional till management. During both the soybean and corn phases, the NT 4-year crop rotation also significantly increased soil water content at varying soil water matric potentials compared to all other crop rotation and tillage management scenarios and significantly increased water infiltration rate compared to all other scenarios other than a 3-year crop rotation under NT management. After 24 years of these rotations, the 4-year crop rotation increased SOC, C and N fractions, and soil aggregation compared to a corn–soybean rotation (Maiga et al., 2019). Varvel (2006) evaluated monoculture and crop rotation options for corn, soybean, and sorghum in the western US Corn Belt and found that rotation increased SOC levels after the first 8 years. However, after switching to a greater tillage depth, decreases in SOC values for the following 10 years were much greater in monoculture and 2-year cropping systems compared to 4-year cropping systems that included an oat/clover mixture in a soybean–sorghum–corn rotation. Adding legumes via an oat/clover mixture or alfalfa for 1–2 years in 4-year rotations has been shown to be important to maintain SOC levels (Varvel, 2006; Russel et al., 2005). A meta-analysis of 122 crop rotation studies concluded that adding one or more crops in rotation to a monoculture increased soil C by 3.6% and microbial biomass C by 20.7%, while increasing total soil N by 5.3% and microbial biomass N by 26.1% (McDaniel et al., 2014). McDaniel et al. (2014) also found that rotations did not increase soil C compared to corn monocultures, but they did increase soil C when compared to soybeans (11%), sorghum (7.9%), and wheat (2.9%). In addition, the meta-analysis showed that adding cover crops to a rotation increases total C by 8.5% and total N by 12.8%. Bowles et al. (2020) reported based on 347 site-years of yield data from 11 experiments that rotation diversification increased corn yields during adverse weather, including droughts.

Semiarid environments can present greater challenges as the main feature of a semiarid region is frequent drought caused by the prolonged absence of rain (Barbosa et al., 2015; Marengo et al., 2016). In semiarid regions, such as the US Great Plains, simplified crop rotations with extended fallow periods often lead to the deterioration of soil properties (Shaver et al., 2003; Benjamin et al., 2007). The wheat–fallow system has been a dominant farming system in the US Great Plains due to perceived improved soil water conditions following fallow. Haas et al. (1957) reported C losses of over 50% (C concentration basis) at many Great Plains sites after 30±40 years of cultivation in a wheat–fallow system with conventional tillage. Improvements in the ability to capture and retain precipitation in the soil during the non-crop periods of the cropping cycle through the advent of reduced- and NT systems made it possible to reduce the frequency of fallow and intensify cropping systems relative to wheat–fallow systems (Peterson et al., 1998). Peterson et al. (1996) documented that with NT, intensified cropping is possible from Texas to North Dakota in the Great Plains region. Peterson et al. (1998) concluded that adoption of residue management systems, such as reduced- or NT systems, coupled with cropping intensification relative to wheat–fallow, should allow soils to store C in the organic pool. Continuous cropping, with no summer fallow period, increased the SOC content of the surface 2.5 cm of soil by 39% relative to wheat–fallow and the increases in SOC were directly linked to increased crop residue biomass returned to the soil (Shaver et al., 2002). Blanco-Canqui et al. (2010) also concluded that intensification of cropping systems such as continuous wheat coupled with NT improved soil physical properties and increased SOC concentration. In this study, continuous wheat and continuous sorghum systems were compared to wheat–sorghum–fallow, wheat–fallow, and sorghum–fallow systems. Continuous wheat resulted in the lowest values of bulk density, cone index, shear strength, and the greatest increases in wet aggregate stability, macroaggregation, aggregate water repellency, total and effective porosity, cumulative water infiltration, soil water retention at low suctions, and SOC concentrations (Blanco-Canqui et al., 2010). In Saskatchewan, Canada, SOC stocks in a continuous wheat system were 7–32% higher than those in the fallow–wheat rotations between 1986 and 2011 in the 0–7.5-cm soil layer (Maillard et al., 2018). They also determined that the replacement of the fallow phase by pulse crops offered promising potential to rebuild SOC stock after fallow–wheat in combination with a reduction in tillage. Acosta-Martinez et al. (2007) reported that shifting from a typical wheat–fallow rotation to a corn–millet–wheat or corn–fallow–wheat rotation for 15 years did not lead to significant differences in SOC, but a shift toward greater soil fungal populations and an increase in C and P enzyme activities were also noted. Based on a study of 96 NT dryland fields in the Great Plains, reduced frequency of summer fallow led to increases in SOC, soil aggregate stability, and microbial biomass (Rosenzweig et al., 2018). Liu et al. (2020) found that pulse species had no significant effect on SOC in the 0–15 cm depth after 8 years of pulse wheat rotations. However, it was noted that inclusion of pulse crops in rotation actually increased SOC over time, contrasting concerns that low C input from pulse crops residue may degrade soil quality. In the UK, diversifying wheat systems mitigated the effect of drought through increased water availability and lower canopy temperatures (Degani et al., 2019). These results were observed although the diverse rotation had a negative impact on soil C due to an increase in tillage. Hence, the authors concluded that a combination of management approaches such as conservation tillage and crop diversification may increase soil function.

Like crop rotations and reduced tillage, cover crops have also been shown to increase soil physical, chemical, and biological characteristics (Blanco-Canqui et al., 2015). Of the most commonly measured soil physical characteristics, aggregate stability showed the greatest increase following the adoption of cover crops in numerous studies (McVay et al., 1989; Wagger and Denton, 1989; Hermawan and Bomke, 1997; Sainju et al., 2003; Liu et al., 2005; Villamil et al., 2006; Blanco-Canqui et al., 2011; Steele et al., 2012; Hubbard et al., 2013; Acuna and Villamil et al., 2014). Cover crops can increase soil aggregation by protecting the soil surface from the physical disturbance of precipitation events and increasing SOC through plant biomass and stimulation of the soil microbial community (Blanco-Canqui et al., 2015). Other physical characteristics like bulk density were also beneficially impacted by cover crop adoption (Wagger and Denton, 1989; Villamil et al., 2006; Blanco-Canqui et al., 2011; Steele et al., 2012; Hubbard et al., 2013); however, the soil water status at sampling can significantly impact the differences in penetration resistance of semiarid soils following cover crop use (DeLaune et al., 2019). Soil temperature is also significantly impacted by cover crop adoption where cover crop residues decrease soil temperature in the summer by shading the soil surface (Kahimba et al., 2008; Blanco-Canqui et al., 2011) and increase soil temperature in the winter through increased microbial activity (Kahimba et al., 2008). Grasses typically have a greater impact on soil temperature because they produce greater amounts of herbage mass compared to legumes and brassicas (Teasdale, 1993).

Cover crops can have a profound impact on soil biological parameters. Globally, cover crops increase soil microbial abundance, activity, and diversity relative to fallow cropping systems, but the benefit is dependent on climate, termination timing and method, and tillage practices (Kim et al., 2020). The greatest increases in soil biological parameters from cover crops stem from the fresh carbon inputs to the soil microbiome during traditional fallow periods (Holman et al., 2021). Additionally, the increases in soil biological function increase the SOC sequestration potential of cropping systems, which is a primary concern with the looming threat of climate change (Blanco-Canqui et al., 2013, 2015). Improvement in biological activity following cover crops has also been shown to decrease the soil pH and increase the soil nutrient cycling potential (Lewis et al., 2018; Burke et al., 2019, 2022).

6.5.2 Erosion

Reducing soil erosion was a major factor in the adoption of NT agriculture within the US. Over 15 years, Owens et al. (2002) found that NT reduced sediment loss by more than 99% compared with a chisel-plow/disk system. In northeastern US, Clausen et al. (1996) reported that reduced tillage decreased sediment losses from paired watersheds by 99% compared with conventional tillage. In Europe, Jordan et al. (2000) reported a 68% reduction in sediment when conservation tillage was implemented. This reduction in soil loss has generally been attributed to the effect of increased residue on the soil surface and improved soil aggregate stability. NT technologies have a great potential to increase organic matter content of the soil and sequester carbon while building and maintaining good soil structure and health compared to intensive tillage systems that does exactly the opposite (Kassam et al., 2009; FAO, 2008; Friedrich et al., 2009). A large proportion of SOC content is concentrated near the soil surface and therefore is highly vulnerable to mineralization processes associated with soil erosion (Lal et al., 1999). Thaler et al. (2021) estimated that topsoil erosion reduces crop yield and causes an annual profit loss of \$2.8 billion on average across the US Corn Belt alone. Globally, the average rate of soil erosion is 30 Mg ha^{-1} year^{-1}, resulting in substantial amounts of cropland becoming unproductive annually – approximately 10 million ha (Gachene et al., 2020).

Gantzer et al. (1991) estimated that essentially all of the topsoil was lost from continuous corn plots within 100 years compared to about half of the topsoil lost from a 6-year rotation of corn–oat–wheat–clover–timothy–timothy in a system with annual fall tillage. As interest in NT for erosion control increased, combining crop rotation to conservation tillage became more evident as a sustainable management practice in many studies. For example, NT alone in a soybean/wheat succession in Brazil was not enough to control soil erosion, even during low-intensity events (Deuschle et al., 2019). However, crop rotation intensification of the NT system with increased phytomass production by crops reduced runoff soil losses by up to 84%. Hunt et al. (2019) found that the addition of oat and alfalfa to a 2-year corn–soybean rotation reduced sediment loading by 60%. Furthermore, they found that adding at least one additional crop phase to a simple crop rotation resulted in a 36–39% reduction in N runoff. Soil loss via water erosion is also a major concern in arid and semiarid regions that are characterized by frequent occurrence of droughts followed by intense rainy seasons (Fengrui et al., 2000; Mechesha et al., 2012). In semiarid northwest China, shifting from a continuous winter wheat–fallow system to a system that cultivates winter wheat followed by a 2–3-month fallow crop in one year and a summer crop cultivation in the next is expected to greatly reduce soil erosion (Fengrui et al., 2000). This system allows the soil to be covered during easily eroded rainy periods but lies bare 6 months every 2 years, where most of this 6-month period is winter when soil evaporation is low and the danger of erosion is also minimal. As evident in these aforementioned studies, conservation tillage accompanied with proper residue management and timing and placement of the correct crop is paramount to achieve the greatest reduction in soil erosion via water.

As monoculture or simplified crop rotation systems can compromise ecosystem services, wind erosion becomes paramount during drought conditions. Annual economic loss in China due to dust and sandstorms has been reported to be as high as \$6.3 billion, which is 16% of the global economic loss caused by sand and dust storms. Covering 20% of the soil surface has been estimated to reduce soil losses via wind erosion by 57%, and a 50% cover reduces soil losses by 95% compared to soils with no cover (Fryrear, 1985). Wang et al. (2002) noted that altering crop rotation options to include corn or millet that offer greater top growth and wheat with high-density seeding rates in pea and potato rotations will significantly reduce wind erosion in semiarid regions of northwest China. Feng et al. (2011) found that shifting from winter wheat to spring wheat rotations resulted in wetter soils in the spring and retained more surface residue in the late summer that reduces the risk of wind erosion in the US Pacific Northwest. Spring rotations consisted of non-erodible aggregates ranging from 67% to 74% compared to 39% to 50% for winter rotations. Soils comprised of less than 40% non-erodible aggregates are at a serious risk to wind erosion (Campbell et al., 1993). Schnarr et al. (2021) concluded that given the postharvest amount of wheat residue cover and its persistence over time, wheat-based rotations may provide more enduring wind erosion protection for the soil compared to rotations based in summer annual crops, such as corn or forages, especially on highly wind erosion-prone soils. In semiarid northwestern China, NT with the most intensive crop rotation significantly decreased the amount of sand transported near the field surface at high wind velocities and the rate of soil wind erosion, thereby reducing the negative environmental impacts of crop production (Yang et al., 2020).

As previously discussed, cover crops can increase the physical characteristics of soil which can decrease its susceptibility to wind and water erosion (Kaspar et al., 2001). Wendt and Burwell (1985) demonstrated the potential hazard of removing crop residue on soil loss, where annual soil loss from NT corn grown for silage (crop residue harvested from the field) was 22 Mg ha^{-1} compared to 0.6 Mg ha^{-1} for NT corn grown for grain (crop residue retained). However, addition of a wheat or rye cover crop in the NT silage system reduced the annual soil loss from 22 Mg ha^{-1} to 0.9 Mg ha^{-1}. Similarly, Zhu et al. (1989) reported that annual soil loss was decreased 87% by common chickweed, 95% by Canada bluegrass, and 96% by downy brome cover crops compared to no cover crops in a NT soybean system. Blanco-Canqui et al. (2013) reported that sediment loss was 3.7 times lower from triticale and spring pea cover crops compared to fallow in a NT wheat system.

Within semiarid regions, cover crops have historically been implemented to reduce wind erosion (Unger and Vigil, 1998; Hansen et al., 2012; Blanco-Canqui et al., 2013). In semiarid ecoregions, wind erosion is the primary environmental concern where up to 18 Mg ha^{-1} yr^{-1} of soil can be eroded (Hansen et al., 2012). The Texas High Plains exemplifies the benefits of cover crop adoption to reduce wind erodibility. Bilbro et al. (1994) showed that a rye cover crop planted during the traditional winter fallow period of sorghum could significantly reduce the wind erosion potential in northwestern Texas. While wind erosion potentials have decreased in the 21st century, the potential for wind erosion on the US Great Plains still averages more than 4.7 Mg ha^{-1} annually (Van Pelt et al., 2013). Proper cover crop management coupled with reduced or no-tillage and crop rotations can decrease this value more (Blanco-Canqui and Wortmann, 2017). Blanco-Canqui et al. (2013) recommended that cover crop growth and/or termination should be near to times when water and wind erosion events are most likely to occur, as benefits are rapidly lost with time after termination in semiarid environments. A terminated wheat cover crop provides protection of cotton seedlings without compromising crop yield or irrigation water use efficiency (WUE) in the Texas Rolling Plains (Figure 6.3; DeLaune et al., 2020). Additional research is needed to better understand how to

FIGURE 6.3 Cotton emerging in a: (A) conventionally tilled system, (B) no-till system, and (C) no-till system with a chemically terminated winter wheat cover crop (Photo Credit: Paul DeLaune).

maximize the benefits of cover crops in semiarid dryland production systems to minimize the potential yield losses from limited precipitation in the Great Plains.

6.5.3 Soil Water and Water Use Efficiency

Water shortage is a serious constraint in crop production and is predicted to be an increasing problem in coming years due to changing climate scenarios (IPCC, 2014). Levia et al. (2020) warned that current global trends in land-use change toward monocultures are risking a more homogeneous terrestrial water cycle and urged policymakers to design forests and agricultural systems that embrace differences among plant species. Converting natural vegetative systems to potentially monoculture-based cropping systems has the potential to degrade many aspects affecting the water cycle, such as increased soil erosion, loss of organic matter, reservoir siltation, water table drawdown and decreased infiltration (Figure 6.4; Levia et al., 2020). Agriculture is the largest use of human consumptive water use, so finding ways to sustain irrigated agriculture while conserving water for other critical needs is essential (Rehkamp et al., 2021).

In semiarid environments, conservation tillage can reduce evaporative losses and conserve groundwater and surface water resources while enhancing soil stability and reducing wind erosion. NT is a practice that can increase surface residue and subsequently improve water holding capacity, soil water storage, and WUE (McVay et al., 2006; Schwartz et al., 2010; Baumhardt et al., 2012). Because of limited rainfall, the majority of irrigated cropland is in the arid, western US (NASS, 2019). Approximately 23 million ha (57 million acres) of US cropland reported receiving irrigation in 2017 (Figure 6.5). According to the 2017 Census of Agriculture, farms with some form of irrigation accounted for more than 54% of the total value of US crop sales, while irrigated land accounted for less than 20% of harvested cropland. Irrigation reduces year-to-year crop yield variability by

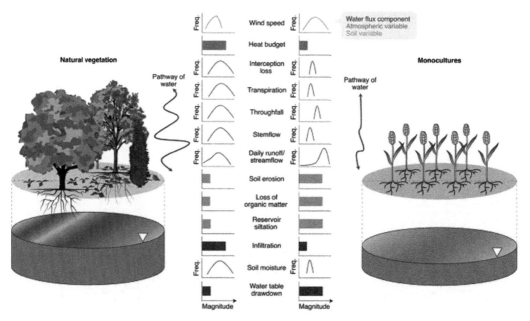

FIGURE 6.4 Conceptual diagrams showing the differences in the water cycle of natural vegetation and monocultures. Plots show the frequency distribution (lines) and relative magnitude (bars) of the hydrological fluxes (blue), atmospheric variables (pink), and soil processes (yellow). No specific scales are shown; magnitudes depend on region and specific vegetation. The vertical dotted yellow lines indicate the distance to the water table (white triangle) (Adapted from Levia et al., 2020).

Managing Soil and Water Resources 141

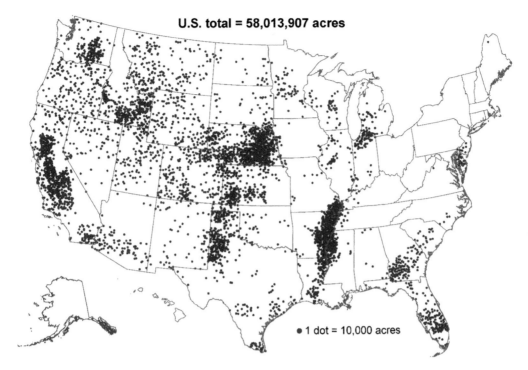

FIGURE 6.5 U.S. acres of irrigated land in 2017 by county. (Source: USDA, Economic Research Service using data from USDA, National Agricultural Statistics Service, 2017 Census of Agriculture).

about 41% in the US (Kukal and Irmak, 2020). However, global water storage has been depleted in all major irrigation areas over the past two decades (Gleick et al., 2014). The Ogallala Aquifer is the largest freshwater aquifer in the world that underlies 450,660 km^2 across the US Great Plains. Deines et al. (2020) concluded that 24% of currently irrigated land in the Ogallala Aquifer may be unable to support irrigated agriculture by 2100 and that 13% of these areas are not suitable for dryland agriculture or pasture use. Thus, developing management strategies to efficiently use irrigation water and capture precipitation and soil water in rainfed environments is crucial to avoid degradation of soil and water resources. In southern regions of the US Great Plains, shifting from full irrigation toward limited irrigation or dryland production may be replaced by higher residue forage crops (e.g., sorghum and perennial grasslands) to promote soil health C inputs and soil cover (Cano et al., 2018). Colaizzi et al. (2009) noted that conservation tillage in the Texas High Plains has been shown to increase available soil moisture by 25 mm or more at the time of planting compared with conventional tillage. Assuming a 25 mm reduction in irrigation water applied, each 10% increase in irrigated area under conservation tillage would result in a 1% reduction in irrigation demand.

Crop selection can alter irrigation needs as well, as application of irrigation water at 60–85% crop ET replacement or skipping early vegetative stage irrigation results in maximizing WUE in C$_3$ crops such as cotton, wheat, and barley or C$_4$ crops such as corn and sugarcane (Attia et al., 2016; Howell, 2001; Karam et al., 2003; Mansour et al., 2017; Singh et al., 2007). Yu et al. (2022) recommended limiting the planting of crops with high water consumption and replacing with drought-resistant varieties to sustain production and water shortages in semiarid environments of China. In this region, crops such as potato, corn, cotton, and other crops were recommended as potential rotational options as these crops are drought-tolerant, require less water, and can make full use of favorable conditions (Deng et al., 1988). Crop variety changes (long- to short-season grain corn and grain sorghum), change in high-water use crops (grain corn) to lower-water use crops (grain sorghum, soybean, and cotton), and conversion of marginally irrigated lands to dryland production (cotton, grain sorghum, and winter wheat) offered the greatest water savings of 20%, but all had very negative economic impacts when shifting from full irrigation to dryland systems in the Texas High Plains (Amosson et al., 2005). Marek et al. (2017) simulated total water use and irrigation demand for continuous corn, continuous cotton, and rotations of corn–cotton, corn–wheat, sorghum–cotton, and cotton–wheat in the Texas High Plains. These simulations indicated that annual rotations required at least 18.6% less total water and 42.3% less irrigation than continuous corn. Colaizzi et al. (2009) concluded that groundwater withdrawals in the Texas High Plains could be reduced by 8% by converting half of the irrigated corn area to cotton. In areas with decreasing or limited irrigation capacity, crop selection and splitting irrigations between crops at or of different growth stages can increase WUE. Saseendran et al. (2008) found that when available water for irrigation was limited to 100 mm, irrigating 50% of the area with 200 mm of water at 20:80 split irrigations between the vegetative and reproductive stages of corn produced greater yield than irrigating 100% of the area with 100 mm water. Producers in Texas with limited well capacities have found success irrigating half of a center pivot planted to a crop of greater water demand while planting the other half to a crop with a lesser water demand or leaving the other half under dryland. Another example of a conservation measure in the Texas Rolling Plains utilizes a corn–peanut rotation to add much needed residue (strip-till corn) to a high soil disturbance peanut system, where the peak water demand for corn precedes the peak water demand for peanut. This allows for greater irrigation rates for each crop at critical growth periods under this split field approach.

Miller et al. (2002) reported improved WUE and nutrient uptake under diversified cropping compared with monocropping. Bordovsky et al. (1994) found that the greatest cotton yield and soil water increases (20–25%) were obtained when NT was combined with a cotton–wheat–fallow rotation compared to continuous cotton. Baumhardt et al. (2012) concluded that the more intensive continuous wheat cropping sequence maintained higher overall infiltration than a wheat–sorghum–fallow rotation, which was attributed to more stable soil aggregates. Baumhardt et al. (2012) suggested that more intensive rotations with residue-retaining cultural practices improve fallow efficiency through increased rain infiltration. Peterson and Westfall (2004) ultimately concluded that if fallow is to be used in a cropping system, one should make every attempt to avoid the periods when precipitation storage is likely to be grossly inefficient. The greatest precipitation use efficiency will be achieved if non-crop periods can be decreased in length and if one can grow a crop during the time when the precipitation is being received. In the central Great Plains, WUE of winter wheat was improved by 18–56% by including a broadleaf crop in the grass-based rotation (Tanaka et al., 2005). In semiarid northeastern Spain, barley under rotation with rapeseed or vetch performed better than barley under monoculture in yield and WUE (Alvaro-Fuentes et al., 2009). However, there were no differences between continuous wheat systems and wheat in rotation. As rapeseed and vetch failed in 80 and 35% of the growing seasons, it was noted that other crops more adapted to limited water conditions should be considered as rotational crops (Alvaro-Fuentes et al., 2009). In the Texas High Plains, using stubble mulch and NT residue management compared with disc tillage during all rotation phases increased the 4-year study mean soil water storage from 4% to 16% of the mean fallow precipitation (Baumhardt et al., 2013). Despite greater runoff and drainage, plant available soil water for the 1.8-m profile at wheat and sorghum planting averaged 194 mm for NT compared with 166 mm for stubble mulch tillage because of reduced evaporation under a rainfed system in the Texas High Plains (Baumhardt et al., 2017). Cropping systems management should consider the quality and persistence of crop residue coupled with conservation tillage to capture precipitation and conserve stored soil water.

Diversifying crop rotations based on local environmental conditions or expected climate change could be a successful adaptive strategy to improve cropping system resiliency. Li et al. (2019) used the term "robust agriculture" to emphasize the ability to maintain a desired level of agricultural outputs in the presence of disturbance (Figure 6.6). Li et al. (2019) considered pulse crops in place of typical

Managing Soil and Water Resources 143

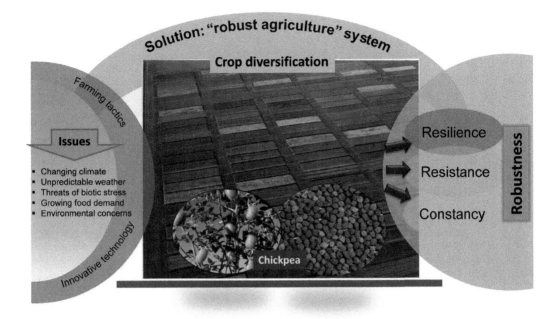

FIGURE 6.6 A schematic illustration showing that a "robust agriculture" system can help alleviate the major issues and challenges in agriculture. Chickpea is used as an example to demonstrate that a robust agricultural system can be developed through resilience (from various perturbations), resistance to biotic stresses, and constancy in crop yield and profitability across disturbances (Adapted from Li et al., 2019).

small grain-fallow systems that have been previously discussed. Gan et al. (2015) hypothesized that pulses could be used to intensify wheat–fallow systems due to the shallow rooting depth of pulse plants that would allow water use from only the upper 0.6 m and leaving water in deeper layers for following deeper-rooted crops; better WUE of pulse crops compared with oilseed crops; and a shorter growing season of pulse crops that results in a longer postharvest period during which soil water can accumulate prior to planting subsequent crops the following spring. Gan et al. (2015) found that approximately 79% of the precipitation during the growing season was not conserved by a summer-fallow practice. However, the main difference between the summer-fallow systems and the diversified system with pulses in terms of water use occurs during the summer period of the cropping sequence, where 79% of the un-conserved rainwater by the summer-fallow system was utilized for grain production through the adoption of the alternative, non-summer fallowing pulse systems. In this scenario, diversifying cropping systems with pulses provided an effective alternative to a traditional small grain-fallow cropping system in a semi-arid rainfed area. Furthermore, the system proposed by Li et al. (2019) estimated a 14% advantage in system robustness for the most diversified systems.

Cover crops can have positive (Daigh et al., 2014; Burke et al., 2021), negative (Unger and Vigil, 1998; Nielsen and Vigil, 2005; Nielsen et al., 2015), or neutral (Holman et al., 2012; Burgess et al., 2014) effects on soil water availability for the subsequent cash crop. In semiarid west Texas cotton production, no-tillage integrated with single and mixed species cover crops significantly decreased profile soil water prior to cover crop termination; however, following termination, there was significantly greater profile soil water with the conservation practices compared to conventionally grown cotton (Figure 6.7). In addition to increases in soil water during the cotton growing season, the conservation systems generally increased cotton WUE but resulted in a decrease in cotton lint yield compared to the conventional system (Burke et al., 2021). The increase in water availability is

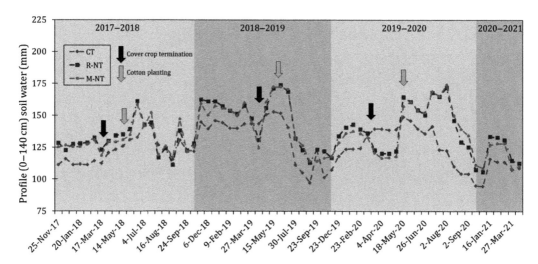

FIGURE 6.7 Profile soil water content from November 2017 to April 2021 at the Agricultural Complex for Advanced Research and Extension Systems in Lamesa, TX, USA. Conventional tillage winter fallow, no-tillage rye cover, and no-tillage mixed species cover are represented as CT, R-NT, and M-NT, respectively (Adapted from Burke et al., 2021, 2022).

thought to be offset by the increase in early-season N immobilization from soil microbes decomposing the cover crop residues. In addition to increased lint yield, NT with a wheat cover crop also significantly increased irrigation WUE compared with strip-tillage and conventional tillage in a semiarid continuous cropping system under subsurface drip irrigation (DeLaune et al., 2020). Over a 6-year average, irrigation WUE was 11% greater for NT with a wheat cover crop and 9% greater for NT (no cover crop) compared with conventional tillage. Baumhardt et al. (2013) reported that disc tilled cotton lint yields irrigated at 5 mm d^{-1} were typically less than NT and stubble-mulch tillage cotton irrigated at 2.5 mm d^{-1}. This higher lint yield associated with NT were attributed to reduced evaporation and greater infiltration with NT residue from previous wheat cover crops. Continuous cotton systems return little residue to the soil surface (Bouquet et al., 2004; Nouri et al., 2019). Hence, implementing a cover crop could enhance soil function of NT. Research has shown that a wheat or rye cover crop significantly improved water infiltration early in the growing season of continuous cotton systems in the Southern Great Plains (DeLaune et al., 2016; DeLaune et al., 2019).

Little is known about the impact of cover crops on soil water in semiarid dryland cropping systems (Blanco-Canqui et al., 2015). Reducing this knowledge gap may increase the adoption of cover crops in semiarid cropping systems. While water use by cover crops may be deemed a disadvantage, particularly in semiarid dryland systems, they can increase infiltration and precipitation capture efficiency (Keisling et al., 1994; Blanco-Canqui et al., 2011; 2015; Nielsen et al., 2016). In a southern Great Plains dryland winter wheat system, stored soil water was 22–26% lower due to cover crops at the time of termination, but the positive change in stored soil water from cover crop termination to wheat planting was two to four times greater for cover crop treatments compared to non-cover crop treatments (Mubvumba et al., 2021). In the same system, cover crops enhanced a 12-year NT system as evident in significant improvements in permanganate oxidizable carbon (19–32%), large macroaggregates (37–51%), mean weight diameter (22–31%), bulk density (8–13%), total porosity (10–18%), and measured soil water content (11–14%) (Mubvumba et al., 2022). Thus, greater infiltration and water capture after termination has the potential to make up for the loss in soil water that cover crops used. To offset potential concerns with cover crop water use, timely termination of cover crops has been considered important to prevent excessive water use or allow time to

Managing Soil and Water Resources

capture precipitation and recharge the soil profile (Unger and Vigil, 1998). In NT systems that have low residue input and long fallow periods, cover crops could offset the soil degradation that occurs under such circumstances.

6.5.4 INTEGRATED CROP–LIVESTOCK SYSTEMS

Integrated crop–livestock systems have received renewed interest in recent years to promote soil health. Roesch-McNally et al. (2018) found that producers with livestock were more likely to use extended crop rotations. In addition, Wang et al. (2021) found that producers who integrated livestock in their cropping system were 23.8% more likely to adopt diverse crop rotation practices. Historically small and diverse farming operations in the US and Europe have been replaced with larger specialized row crop and/or animal feeding operations, with only an estimated 15% of current European farms operating some sort of mixed crop–livestock system (Singer et al., 2009; Peyraud et al., 2014). Franzluebbers and Gastal (2019) hypothesized that pastures (i.e., any form of perennial grass/legume mixture as a fodder for ruminant animals or as feedstock for biofuel production) in rotation with crops managed with some form of conservation tillage on the majority of agricultural lands around the world will lead to greater SOC and total N. In semiarid west Texas, cotton rotated with grazed wheat, rye, and Old World bluestem used 25% less irrigation water, required 36% less N fertilizer, and had fewer other chemical inputs than continuous monoculture cotton (Allen et al., 2012). In a region of Argentina receiving 870 mm annual precipitation, conventionally tilled annual cropping for 7 years decreased SOC 4.4 g kg^{-1} but increased to original levels after 3–4 years under pasture (Studdert et al., 1997). Similarly, crop–pasture rotations with NT maintained or increased the original SOC content while also producing the same low erosion rate of natural pastures in Uruguay (Garcia-Prechac, 2004). Diaz-Zorita et al. (2002) inferred that the inclusion of pastures in rotation with annual crops under NT systems is more of an economic than a soil management decision because both pastures and NT row cropping maintain the productivity of the soils in the sub-humid and semiarid Pampas of Argentina. Franzluebbers and Gastal (2019) recommend that 3 years of forage could be followed by 3–7 years of cropping using some form of conservation tillage with minimal soil disturbance during the transition and during cropping.

Pasture-cropping, the planting of an annual crop into an established perennial forage system, has received interest due to the potential as a multiple source of income via grazing and grain production while providing soil health benefits. In southwestern Australia, WUE for grain production was lower in pasture-cropped plots; however, this was offset by pasture production, so that over a full 12-month period, water-use efficiency for biomass production was generally greater for the pasture-cropped plots than for either the pasture or crop monocultures (Ward et al., 2014). In semiarid environments, residue management should be considered to ensure soil resources are maintained and protected while others have recommended more research for pasture-cropping capabilities in water-limited environments. Nielsen et al. (2015, 2016) proposed that grazing of cover crops may be considered to provide an economic benefit of cover crops, although removal of biomass would have to be carefully considered in light of the increased potential for soil loss by wind erosion. Although flash grazing resulted in significant removal of cover crops biomass, no significant negative effects were observed from this practice regarding soil water storage under dryland conditions in the Texas Rolling Plains (Mubvumba et al., 2021). Holman et al. (2018) concluded that the best way for producers to potentially improve soil properties over time and be profitable is to grow a forage crop and not a summer cover crop in the semi-arid Central Great Plains region. A survey in Nebraska indicated that livestock grazing and use as a secondary forage in a rotation accounted for 33.6% of the reasoning for planting a cover crop on an agricultural property (Jansen et al., 2019). Thus, incorporating livestock into cropping systems may enhance the adoption of forages or cover crops as alternative forms of conservation agriculture.

FIGURE 6.8 Pictures of dust storms (a) from Hansford County, TX in 1935; (b) Terry County, TX in 2013; and (c) Lubbock County, TX in 2021 (Photo Credits: (a) Arthur Rothstein, (b) Lisa Lewis, and (c) Joseph Burke).

6.6 FURTHER ADVANCEMENT OF CONSERVATION AGRICULTURE

Management practices such as conservation tillage, crop rotation, and cover cropping often proved significant benefits to reverse the course of soil degradation. However, these practices are not always effective if not used to fit the correct landscape or environment. These practices are often more beneficial when used in conjunction with one another. A "stacked" approach should be encouraged and should spread the risks of producing the world's food supply during times of harsh environmental conditions or an ever-changing climate. Although crop rotation in some form is often practiced, the definition set forth by the FAO that recommends the use of at least three crops is often not implemented and these crop rotations are often economically driven. While North America has among the greatest land area under some form of conservation agriculture compared to other regions, only about 28% of the total acres in this region are under conservation agriculture practices. Similar images from the infamous "Dust Bowl" era are periodically observed even today (Figure 6.8), indicating the continued need for the implementation of conservation agriculture approaches.

Ultimately, social challenges and barriers of adoption of conservation practices must be addressed. Available survey responses seem to indicate that producers are willing to adopt conservation practices; however, economic concerns must be considered and met. As Roesch-McNally et al. (2018) summarized, integrating diversity into current cropping systems proves difficult with continued technological trajectory toward greater specialization, field- and landscape-scale homogeneity rather than heterogeneity, and the need for major cross-scale policy solutions. Rosa et al. (2020) recently projected that irrigated agriculture could be expanded by 70–350 million ha that could feed an additional 300 million to 1.4 billion people. In contrast, freshwater limitations in some regions could require the reversion of 20–60 million ha from irrigated to rain-fed management by the end of the century due to climate change (Elliot et al., 2014). With expansion of irrigated agriculture and potential intensification of these systems coupled with reverting from irrigated to dryland agriculture that may not be suitable for rainfed farming, lessons learned from the past must be considered to ensure that further degradation of soil and water resources does not continue to spiral in the wrong direction. With educated decisions, conservation practices such as conservation tillage, crop rotation and diversity, and cover cropping can successfully be employed to protect and restore soil and water resources and overall soil and food security.

REFERENCES

Abrahams, P. 2002. Soils: Their implications to human health. *Sci. Total Environ.* 291:1–32.

Acosta-Martínez, V., M. Mikha, M.F. Vigil. 2007. Microbial communities and enzyme activities in soils Ander alternative crop rotations compared to wheat-fallow for the Central Great Plains. *Appl. Soil Ecol.* 37:41–52.

Acuna, J.C.M., and M.B. Villamil. 2014. Short-term effects of cover crops and compaction on soil properties and soybean production in Illinois. *Agron. J.* 106:860–870. doi.org/10.2134/agronj13.0370

Aguilar J., G.G. Gramig, J.R. Hendrickson, D.W. Archer, F. Forcella, and M.A. Liebig. 2015. Crop species diversity changes in the United States: 1978–2012. *PLoS ONE* 10(8):e0136580. doi:10.1371/journal.pone.0136580

Alhameid, A., J. Singh, U. Sekaran, E. Ozlu, S. Kumar, and S. Singh. 2019. Crop rotational diversity impacts soil physical and hydrological properties under long-term no- and conventional-till soils. *Soil Res.* doi:10.1071/SR18192

Allen, V.G., C.P. Brown, R. Kellison, P. Green, C.J. Zilverberg, P. Johnson, J. Weinheimer, T. Wheeler, E. Segarra, V. Acosta-Martinez, T.M. Zobeck, and J.C. Conkwright. 2012. Integrating cotton and beef production in the Texas Southern High Plains: I. Water use and measures of productivity. *Agron. J.* 104:1625–1642.

Alvaro-Fuentes, J., J. Lampurlanes, and C. Cantero-Martinez. 2009. Alternative crop rotations under Mediterranean no-tillage conditions: Biomass, grain yield, and water-use efficiency. *Agron. J.* 101:1227–1233.

Amosson, S.H., L.K. Almas, F. Bretz, D. Gaskins, B. Guerrero, D. Jones, T.H. Marek, L. New, and N. Simpson. 2005. Water Management Strategies for Reducing Irrigation Demands in Region A. Report to the Panhandle Water Planning Group, Amarillo, TX, January. www.panhandlewater.org

Attia, A., N. Rajan, Q. Xue, S. Nair, A. Ibrahim, and D. Hays. 2016. Application of DSSAT CERES-Wheat model to simulate winter wheat response to irrigation management in the Texas high plains. *Agric. Water Manag.* 165:50–60.

Balesdent, J., C. Chenu, and M. Balabane. 2000. Relationship of soil organic matter dynamics to physical protection and tillage. *Soil Till. Res.* 53:215–230.

Balkcom, K., H. Schomberg, W. Reeves, A. Clark, L. Baumhardt, H. Collins, J. Delgado, S. Duiker, T. Kaspar, and J. Mitchell. 2007. Managing cover crops in conservation tillage systems In: Clark A (ed) Managing Cover Crops Profitably. 3rd ed. Beltsville, MD. United Book Press, Inc. 2007. 44–61. Print.

Barbosa, H. A., T.V. Lakshmi Kumar, and L.R.M Silva. 2015. Recent trends in vegetation dynamics in the South America and their relationship to rainfall. *Nat. Haz.* 77:883–899. doi:10.1007/s11069-015-1635-8

Bathke, G.R., and G. Blake. 1984. Effects of soybeans on soil properties related to soil erodibility. *Soil Science Society of America Journal* 48:1398–1401. doi:10.2136/sssaj1984.03615995004800060040x

Baumhardt, R.L., G.L. Johnson, and R.C. Schwartz. 2012. Crop rotation effects on simulated rain infiltration and sediment transport. *Soil Sci. Soc. Amer. J.* 76:1370–1378.

Baumhardt, R.L., R.C. Schwartz, T. Howell, S.R. Evett, and P. Colaizzi. 2013. Residue management effects on water use and yield of deficit irrigated corn. *Agron. J.* 105:1035–1044.

Baumhardt, R.L., R.C. Schwartz, O.R. Jones, B.R. Scanlon, R.C. Reedy, and G.W Marek. 2017. Long-term conventional and no-tillage effects on field hydrology and yields of a dryland crop rotation. *Soil Sci. Soc. Amer. J.* 81:200–209.

Benjamin, J.G., M. Mikha, D.C. Nielsen, M.F. Vigil, F. Calderon, and W.B. Henry. 2007. Cropping intensity effects on physical properties of a no-till silt loam. *Soil Sci. Soc. Am. J.* 71:1160–1165.

Bilbro, J.D., B.L. Harris, and O.R. Jones. 1994. Erosion control with sparse residue. p. 30-32. In: B.A. Stewart and W.C. Moldenhauer (eds.), *Crop Residue Management to Reduce Erosion and Improve Soil Quality.* Conservation Research Report No. 37, Agricultural Research Service, United States Department Agriculture, Washington, D.C.

Blanco-Canqui, H., C.J. Gantzer, S.H. Anderson, and E.E. Alberts. 2004. Tillage and crop influences on physical properties for an epiaqualf. *Soil Sci. Soc. Am. J.* 68 (2):567–576.

Blanco-Canqui, H., J.D. Holman, A.J. Schlegel, J. Tatarko, and T. Shaver. 2013. Replacing fallow with cover crops in a semiarid soil: Effects on soil properties. *Soil Sci. Soc. Am. J.* 77:1026–1034. doi:10.2136/sssaj2013.01.0006

Blanco-Canqui, H., M.M. Mikha, D.R. Pressley, and M.M. Claassen. 2011. Addition of cover crops enhances no-till potential for improving soil physical properties. *Soil Sci. Soc. Amer. J.* 75:1471–1482. doi:10.2136/sssaj2010.0430

Blanco-Canqui, H., T.M. Shaver, J.L. Lindquist, C.A. Shapiro, R.W. Elmore, C.A. Frances, and G.W. Hergert. 2015. Cover crops and ecosystem services: Insights from studies in temperate soils. *Agron. J.* 107(6): 2449–2474.

Blanco-Canqui, H., L.R. Stone, A.J. Schlegel, J.G. Benjamin, M.F. Vigil, and P.W. Stahlman. 2010. Continuous cropping systems reduce near-surface maximum compaction in no-till soils. *Agron. J.* 102:1217–1225. doi:10.2134/agronj2010.0113

Blanco-Canqui, H., and C. Wortmann. 2017. Crop residue removal and soil erosion by wind. *J. Soil Water Conser.* 72(5):97A–104A.

Boquet, D.J., R.L. Hutchinson, and G.A. Breitenbeck. 2004. Long-term tillage cover crop, and nitrogen rate effects on cotton: Yield and fiber properties. *Agron. J.* 96:1436–1442.

Bordovsky, J.P., W.M. Lyle, and J.W. Keeling. 1994. Crop rotation and tillage effects on soil water and cotton yield. *Agron. J.* 86:1–6.

Bowles, T.M., M. Mooshammer, Y. Socolar, F. Calderon, M.A. Cavigelli, S.W. Culman, W. Deen, C.F. Drury, A.G. y Garcia, A.C.M. Gaudin, W.S. Harkcom, R.M. Lehman, S.L. Osborne, G.P. Robertson, J. Salerno, M.R. Schmer, J. Strock, and A.S. Grandy. 2020. Long-term evidence shows that crop-rotation diversification increases agricultural resilience to adverse growing conditions in North America. *One Earth* 2 (3):284–293.

Brevik, E.C., and T.J. Sauer. 2015. The past, present, and future of soils and human health studies. *Soil* 1:35–46. doi:10.5194/soil-1-35-2015

Burgess, M., P. Miller, C. Jones, and A. Bekkerman. 2014. Tillage of cover crops affects soil water, nitrogen, and wheat yield components. *Agron. J.* 106: 1497–1508.

Burke, J.A., K.L. Lewis, P.B. DeLaune, C.J. Cobos, and J.W. Keeling. 2022. Soil water dynamics and cotton production following cover crop use in a semi-arid ecoregion. *Agronomy* 12:1306.

Burke, J.A., K.L. Lewis, G.L. Ritchie, P.B. DeLaune, J.W. Keeling, V. Acosta-Martinez, J. Moore-Kucera, and T. McLendon. 2021. Net positive soil water content following cover crops with no tillage in irrigated semi-arid cotton production. *Soil Till. Res.* 208:104869.

Burke, J.A., K.L. Lewis, G.L. Ritchie, J. Moore-Kucera, P.B. DeLaune, and J.W. Keeling. 2019. Temporal variability of soil carbon and nitrogen in cotton production on the Texas High Plains. *Agron. J.* 111(5): 2218–2225.

Busscher, W.J., and P.J. Bauer. 2004. Soil strength, cotton root growth and lint yield in a Southeastern USA Coastal loamy sand. *Soil Till. Res.* 74:151–159.

Campbell, C.A., A.P. Moulin, D. Curtin, G.P. Lafond, and L. Townley-Smith. 1993. Soil aggregation as influenced by cultural practices in Saskatchewan. 1. Black Chernozemic soils. *Can. J. Soil Sci.* 73, 579–595.

Cano, A., A. Núñez, V. Acosta-Martinez, M. Schipanski, R. Ghimire, C. Rice, and C. West. 2018. Current knowledge and future research directions to link soil health and water conservation in the Ogallala Aquifer region. *Geoderma* 328:109–118.

Claassen, R., M. Bowman, J. McFadden, D. Smith, and S. Wallander. 2018. Tillage intensity and conservation cropping in the United States. EIB-197. September 2018. Washington, DC: USDA-ERS.

Clausen, J.C., W.E. Jokela, F.I. Potter III, and J.W. Williams. 1996. Paired watershed comparison of tillage effects on runoff, sediment, and pesticide losses. *J. Environ. Qual.* 25:1000–1007.

Colaizzi, P.D., P.H. Gowda, T.H. Marek, and D.O. Porter. 2009. Irrigation in the Texas High Plains: A brief history and potential reductions in demand. *Irrig. Drain.* 58:257–274.

Coulter, J.A., E.D. Nafziger, and M.M. Wander. 2009. Soil organic matter response to cropping system and nitrogen fertilization. *Agronomy Journal* 101:592–599. doi:10.2134/agronj2008.0152x

Crossley, M.S., K.D. Burke, S.D. Schoville, and V.C. Radeloff. 2020. Recent collapse of crop belts and declining diversity of US agriculture since 1840. *Global Change Biol.* 27(1):151–164.

Dabney, S.M., J.A. Delgado, and D.W. Reeves. 2001. Using winter cover crops to improve soil and water quality. *Commun. Soil Sci. Plant Anal.* 32:1221–1250.

Daigh, A.L., M.J. Helmers, E. Kladivko, X. Zhou, R. Goeken, and J. Cavdini. 2014. Soil water during the drought of 2012 as affected by rye cover crops in fields in Iowa and Indiana. *J. Soil Water Consrv.* 69:564–573.

Degani, E., S.G. Leigh, H.M. Barber, H.E. Jones, M. Lukac, P. Sutton, and S.G. Potts. 2019. Crop rotations in a climate change scenario: Short-term effects of crop diversity on resilience and ecosystem service provision under drought. *Agric. Ecosyst. Environ.* 285:106625.

Deines, J.M., M.E. Schipanski, B. Golden, S.C. Zipper, S. Nozari, C. Rottler, B. Gerttero, and V. Sharda. 2020. Transitions from irrigated to dryland agriculture in the Ogallala Aquifer: Land use suitability and regional economic impacts. *Agric. Water Manag.* 233:106061.

DeLaune, P.B., P. Mubvumba, S. Ale, and E. Kimura. 2020. Impact of no-till, cover crop, and irrigation on cotton yield. *Ag. Water Manage.* doi:10.1016/j.agwat.2020.106038

DeLaune, P., P. Mubvumba, and K. Lewis. 2016. Influence of management practices on soil health in cotton cropping systems. In: *Proc. 2016 Beltwide Cotton Conference.* New Orleans, LA. 4–6 Jan. 2016. [CD].

DeLaune, P.B., P. Mubvumba, K.L. Lewis, and J.W. Keeling. 2019. Rye cover crop impacts soil properties in a long-term cotton system. *Soil Sci. Soc. Am. J.* 83(5): 1451–1458.

DeLaune, P.B., and J.W. Sij. 2012. Impact of tillage on runoff in long term no-till wheat systems. *Soil Tillage Res.* 124:32–35.

Deng, Z.Y., Q. Zhang, J.Y.. Pu, D.X. Liu, H. Guo, Q.F. Wang, H. Zhao, and H.L. Wang. 2008. The impact of climate warming on crop planting and production in northwestern China. *Acta. Ecol. Sin.* 28(8):3760–3768.

Derpsch, F. 2004. History of crop production, with and without tillage. *Leading Edge* 3:150–154. www.notill.org/sites/default/files/history-of-crop-production-with-without-tillage-derpsch.pdf

Deuschle, D., J.P.G. Minella, T. de A.N. Hörbe, A.L. Londero, and F.J.A. Schneider. 2019. Erosion and hydrological response in no-tillage subjected to crop rotation intensification in southern Brazil. *Geoderma* 340 :157–163.

Díaz-Zorita, M., G.A. Duarte, and J.H. Grove. 2002. A review of no-till systems and soil management for sustainable crop production in the subhumid and semiarid Pampas of Argentina. *Soil Till. Res.* 65(1):1–18.

Dold, C., H. Büyükcangaz, W. Rondinelli, J. Prueger, T. Sauer, and J. Hatfield. 2017. Long-term carbon uptake of agro-ecosystems in the Midwest. *Agric. For. Meteorol.* 232:128–140.

Elliot, J., D. Deryng, C. Muller, K. Frieler, M. Konzmann, D. Gerten, M. Glotter et al. 2014. Constraints and potentials of future irrigation water availability on agricultural production under climate change. *Proc. Natl. Acad. Sci. U.S.A.* 111 :3239–3244.

FAO. 2010. What is conservation? Available at www.fao.org/conservation-agriculture/ overview/what-is-conservation-agriculture/en/

FAO. 2022. Conservation agriculture. Available at www.fao.org/3/cb8350en/cb8350en.pdf

Feng, G., B. Sharratt, and F. Young. 2011. Soil properties governing soil erosion affecting cropping systems in the U.S. Pacific Northwest. *Soil Till. Res.* 111:168–174.

Fengrui, L., Z. Songling, and G.T. Geballe. 2000. Water use patterns and agronomic performance for some cropping systems with and without fallow crops in a semi-arid environment of northwest China. *Agric. Ecosyst. Environ.* 79:129–142.

Ferreras, L.A., J.L. Costa, F.O. Garcia, and C. Pecorari. 2000. Effect of no-tillage on some soil physical properties of a structural degraded Petrocalcic Paleudoll of the southern "Pampa" of Argentina. *Soil Till. Res.* 54:31–39.

Franzluebbers, A.J., and F. Gastal. 2019. Building agricultural resilience with conservation pasture-crop rotations. In: G. Lemaire, P.C.D.F. Carvalho, S. Kronberg, and S. Recous (Eds.), *Agroecosystem Diversity: Reconciling Contemporary Agriculture and Environmental Quality.* Elsevier, Academic Press, pp. 109–121.

Friedrich, T. and A.H. Kassam. 2009. Adoption of conservation agriculture technologies: constraints and opportunities. In: Proceedings of the 4th World Congress on Conservation Agriculture4–7 February 2009: New Delhi, India.

Fryrear, D.W. 1985. Soil cover and wind erosion. *Transactions American Society Agric. Engin.* 28:781–784.

Gachene, C.K.K., S.O. Nyawade, and N.N. Karanja. 2020. Soil and water conservation: An overview. In: W. Leal Filho, A.M. Azul, L. Brandli, P.G. Özuyar, and T. Wall (Eds.), *Zero Hunger. Encyclopedia of the UN Sustainable Development Goals.* Springer, Cham. doi:10.1007/978-3-319-95675-6_91

Gan, Y., C. Hamel, J.T. O'Donovan, H. Cutforth, R.P. Zentner, C.A. Campbell, Y. Niu, and L. Poppy. 2015. Diversifying crop rotation with pulses enhances system productivity. *Sci. Reports* 5:14625.

Gantzer, C.J., S.H. Anderson, A.L. Thompson, and J.R. Brown. 1991. Evaluation of soil loss after 100 years of soil and crop management. *Agron. J.* 83:74–77.

García-Prechac, F., O. Ernst, G. Siri-Prieto, and J.A. Terra. 2004. Integrating no-till into crop-pasture rotations in Uruguay. *Soil and Tillage Research* 77:1–13.

Gleick, P.H., N. Ajami, J. Christian-Smith, H. Cooley, K. Donnelly, J. Fulton, M.L. Ha, M. Heberger, E. Moore, et al. 2014. *The World's Water: The Biennial Report on Freshwater Resources,* Volume 8; Island Press, Washington, DC, USA.

Guha-Sapir, D. R. Below, and P. Hoyois. 2021. EM-DAT: The CRED/OFDA International Disaster Database. www.emdat.be

Haas, H.J., C.E. Evans, and E.F. Miles, 1957. Nitrogen and carbon changes in Great Plains soils as influenced by cropping and soil treatments. *USDA Tech. Bull.* 1164, US Government Printing Office, Washington, DC.

Halvorson, J.J., D.W. Archer, M.A. Liebig, K.M. Yeater, and D.L. Tanaka. 2019. Impacts of intensified cropping systems on soil water use by spring wheat. *Soil Sci. Soc. Am. J.* 83: 1188–1199.

Hamilton, A.V., D. A. Mortensen, and M.K. Allen. 2017. The state of the cover crop nation and how to set realistic goals for the popular conservation practice. *J. Soil Water Conserv.* 72(5): 111A–115A.

Hansen, N.C., B.L. Allen, R.L. Baumhardt, and D.J. Lyon. 2012. Research achievements and adoption of no-till, dryland cropping in the semi-arid US Great Plains. *Field Crops Res.* 132:196–203.

Haramoto, E.R. 2019. Species, seeding rate, and planting method influence cover crop services prior to soybean. *Agron. J.* 111:1068–1078.

Hermawan, B., and A.A. Bomke. 1997. Effects of winter cover crops and successive spring tillage on soil aggregation. *Soil Tillage Res.* 44:109–120. doi:https://doi.org/10.1016/S0167-1987(97)00043-3

Hill, R. 1990. Long-term conventional and no-tillage effects on selected soil physical properties. *Soil Science Society of America Journal* 54:161–166. doi:10.2136/sssaj1990.03615995005400010025x

Holland, J.M. 2004. The environmental consequences of adopting conservation tillage in Europe: reviewing the evidence. *Agric. Ecosyt. Environ.* 103:1–25.

Holman, J.D., K. Arnet, J. Dille, S. Maxwell, A. Obour, T. Roberts, K. Roozeboom, and A. Schlegel. 2018. Can cover or forage crops replace fallow in the semiarid Central Great Plains? *Crop Sci.* 58(2):932–944

Holman, J.D., Y. Assefa, and A. Obour. 2021. Cover-cropwateruse and productivity in the High Plains wheat-fallow crop rotation. *Crop Sci.* 61:1374–1385. doi.org/10.1002/csc2.20365

Holman, J.D., T. Dulmer, T. Roberts, and S. Maxwell. 2012. Fallow replacement crop effects of wheat yield. Rep. Progr. 1070. Kansas State Univ. Coop. Ext. Serv., Manhattan.

Hontoria, C., C. Gómez-Paccard, I. Mariscal-Sancho, M. Benito, J. Pérez, and R. Espejo. 2016. Aggregate size distribution and associated organic C and N under different tillage systems and Ca-amendment in a degraded Ultisol. *Soil Till. Res.* 160:42–52. doi:10.1016/j.still.2016.01.003.

Howell, T.A., 2001. Enhancing water use efficiency in irrigated agriculture. *Agron. J.* 93:281–289.

Hubbard, R.K., T.C. Strickland, and S. Phatak. 2013. Effects of cover crop systems on soil physical properties and carbon/nitrogen relationships in the Coastal Plain of southeastern USA. *Soil Till. Res.* 126:276–283. doi:https://doi.org/10.1016/j.still.2012.07.009

Hunt, N.D., J.D. Hill, and M. Liebman. 2019. Cropping system diversity effects on nutrient discharge, soil erosion, and agronomic performance. *Environ. Sci. Technol.* 53:1344–1352.

IPCC. 2014. Climate Change 2014: Synthesis Report. Contribution of Working Groups I, II and III to the Fifth Assessment Report of the Intergovernmental Panel on Climate Change. IPCC, Geneva, Switzerland, p. 151.

Jagadamma, S., R. Lal, R.G. Hoeft, E.D. Nafziger, and E.A. Adee. 2008. Nitrogen fertilization and cropping system impacts on soil properties and their relationship to crop yield in the central Corn Belt, USA. *Soil Till. Res.* 98: 120–129. doi:10.1016/j.still.2007.10.008

Jansen, J., J. Stokes, and J. Parsons. 2019. Cover crop utilization across Nebraska and implications for cropland lease arrangements in 2019. *Cornhusker Economics.* 1009. https://digitalcommons.unl.edu/agecon_cor nhusker/1009

Janvier, C., F. Villeneuve, C. Alabouvette, V. Edel-Hermann, T. Mateille, and C. Steinberg. 2007. Soil health through soil disease suppression. Which strategy from descriptors to indicators? *Soil Bio. Biochem.* 39:1–23. doi:10.1016/j.soilbio.2006.07.001

Janzen, H.H., 2004. Carbon cycling in earth systems—a soil science perspective. *Agric. Ecosyst. Environ.* 104(3):399–417.

Jordan, V. W. L., A. R. Leake, and S. Ogilvy. 2000. Agronomic and environmental implications of soil management practices in integrated farming systems. *Aspects of Applied Biology* 62:61–66.

Kahimba, F.C., R. Sri Ranjan, J. Froese, M. Entz, and R. Nason. 2008. Cover crop effects on infiltration, soil temperature and soil moisture distribution in the Canadian prairies. *Appl. Eng. Agric.* 24:321–333. https://doi.org/10.13031/2013.24502

Karam, F., J. Breidy, C. Stephan, and J. Rouphael, 2003. Evapotranspiration, yield and water use efficiency of drip irrigated corn in the Bekaa valley of Lebanon. *Agric. Water Manag.* 63:125–137.

Karlen, D.L., E.G. Hurley, S.S. Andrews, C.A. Cambardella, D.W. Meek, M.D. Duffy, and A.P. Mallarino. 2006. Crop rotation effects on soil quality at three northern corn/soybean belt locations. *Agronomy Journal* 98:484–495. doi:10.2134/agronj2005.0098

Kaspar, T.C., J.K. Radke, and J.M. Laflen. 2001. Small grain cover crops and wheel traffic effects on infiltration, runoff, and erosion. *J. Soil Water Conserv.* 56 (2): 160–164.

Kassam, A., T. Friedrich, and R. Derpsch. 2019. Global spread of conservation agriculture. *Internat. J. Environ. Stud.* 76:29–51. doi:10.1080/00207233.2018.1494927

Katsvairo T., W.J. Cox, and H. Van Es. 2002. Tillage and rotation effects on soil physical characteristics. *Agronomy Journal* 94: 299–304. doi:10.2134/agronj2002.0299

Keisling, T.C., H.D. Scott, B.A. Waddle, W. Williams, and R.E. Frans. 1994. Winter cover crops influence on cotton yield and selected soil properties. *Commun. Soil Sci. Plt. Anal.* 25(19–20):3087–3100.

Kim, N., M.C. Zabaloy, K. Guan, and M.B. Villamil. 2020. Do cover crops benefit soil microbiome? A meta-analysis of current research. *Soil Biol. Biochem.* 142:107701.

Kirschenmann, F.L., 2002. Research on the Farm: Challenges for the 21st Century. Excerpts from Keynote Address: ISU Northeast Research Farm Silver Anniversary Field Day, June 26, 2001. Iowa State University Research and Demonstration Farms Progress Reports, 2001(1).

Kladviko, E.J., D.R. Griffith, and J.V. Mannering. 1986. Conservation tillage effects on soil properties and yield of corn and soybeans in Indiana. *Soil Till. Res.* 8:277–287.

Knox, O.G.G., A.R. Leake, R.L. Walker, A.C. Edwards, and C.A. Watson. 2011. Revisiting the multiple benefits of historical crop rotations within contemporary UK agricultural systems. *J. Sustain. Agric.* 35:169–179.

Koudahe, K., S. Allen, and K. Djaman, 2022. Critical review of impact of cover crops on soil properties. *Int. Soil Water Conserv. Res.* 10:343–354.

Kukal, M.S., and S. Irmak. 2020. Impact of irrigation on interannual variability in United States agricultural productivity. *Agric. Water Manag.* 234:106141.

Lal, R. 1993. Tillage effects on soil degradation, soil resilience, soil quality, and sustainability. *Soil Till. Res.* 27:1–8.

Lal, R. 2009. Soil degradation as a reason for inadequate human nutrition. *Food Sec.* 1:45–57.

Lal, R. 2015. Restoring soil quality to mitigate soil degradation. *Sustainability* 7:5875–5895. doi:10.3390/su7055875

Lal, R., R.F. Follett, J. Kimble, and C.V. Cole. 1999. Managing U.S. cropland to sequester carbon in soil. *J. Soil Water Cosnerv.* 54:374–381.

Lal, R., and D.M. Van Doren. 1990. Influence of 25 years of continuous corn production by three tillage methods on water infiltration for two soils in Ohio. *Soil Till. Res.* 16:71–84.

Leon, J., and N. Osorio. 2014. Role of litter turnover in soil quality in tropical degraded lands of Colombia. *Sci. World J.* Article ID 693981. doi:10.1155/2014/693981

Levia, D.F., I.F. Creed, D.M. Hannah, K. Nanko, E.W. Boyer, D.E. Carlyle0Moses, M. van de Giesen, D. Grasso, A.J. Guswa, J.E. Hudson, S.A. Hudson, et al. 2020. Homogenization of the terrestrial water cycle. *Nature Geosci.* 13:656–660.

Lewis, K.L., J.A. Burke, W.S. Keeling, D.M. McCallister, P.B. DeLaune, and J.W. Keeling. 2018. Soil benefits and yield limitations of cover crop use in Texas High Plains cotton. *Agron. J.* 110(4):1616–1623.

Li, J., L. Huang, J. Zhang, J.A. Coulter, L. Li, and Y. Gan. 2019. Diverifying crop rotation improves system robustness. *Agron. Sustain. Develop.* 39:38 doi:10.1007/s13593-019-0584-0

Libohova, Z., C. Seybold, D. Wysocki, S. Wills, P. Schoeneberger, C. Williams, D. Lindbo, D. Stott, and P.R. Owens. 2018. Reevaluating the effects of soil organic matter and other properties on available water-holding capacity using the National Cooperative Soil Survey Characterization Database. *J. Soil Water Conserv.* 73(4):411–421. doi:10.2489/jswc.73.4.411

Lin, B.B. 2011. Resilience in agriculture through crop diversification: adaptive management for environmental change. *Bioscience.* 61:183–193.

Liu, A.G., B.L. Ma, and A.A. Bomke. 2005. Effects of cover crops on soil aggregate stability, total organic carbon, and polysaccharides. *Soil Sci. Soc. Am. J.* 69:2041–2048. doi.org/10.2136/sssaj2005.0032

Liu, K., M. Bandara, C. Hamel, J.D. Knight, and Y. Gan. 2020. Intensifying crop rotations with pulse crops enhances system productivity and soil organic carbon in semi-arid environments. *Field Crops Res.* 248:107657.

Lopez-Fando, C., and M.T. Pardo. 2011. Soil carbon storage and stratification under different tillage systems in a semi-arid region. *Soil Till. Res.* 111:224–230. doi:10.1016/j.still.2010.10.011

MacDonald, J.M., P. Korb, and R.A. Hoppe. 2013. Farm size and the organization of U.S. crop farming, ERR-152. U.S. Department of Agriculture, Economic Research Service.

Magdoff, F., and H. van Es. 2009. Building soils for better crops: sustainable soil management. Sustainable Agriculture Research and Education Program.

Maiga, A., A. Alhameid, S. Singh, A. Polat, J. Singh, S. Kumar, and S. Osborne. 2019. Responses of soil organic carbon, aggregate stability, carbon and nitrogen fractions to 15 and 24 years of no-till diversified crop rotations. *Soil Res.* doi:10.1071/SR18068

Maillard, É., B.G. McConkey, M.St. Luce, D.A. Angers, and J. Fan. 2018. Crop rotation, tillage system, and precipitation regime effects on soil carbon stocks over 1 to 30 years in Saskatchewan, Canada. *Soil Till. Res.* 177:97–104.

Mansour, E., I.M. Abdul-Hamid, T. Yasin, N. Qabil, and A. Attia. 2017. Identifying drought tolerant genotypes of barley and their responses to various irrigation levels in a Mediterranean environment. *Agric. Water Manag.* 194:58–67.

Marek, G.W., P.H. Gowda, T.H. Marek, D.O. Porter, R.L. Baumhardt, and D.K. Brauer. 2017. Modeling long-term water use of irrigated cropping rotations in the Texas High Plains using SWAT. *Irrig. Sci.* 35:111–123.

Marengo, J. A., R.R. Torres, and L.M. Alves. 2016. Drought in Northeast Brazil – past, present, and future. *Theor. App. Clim.* 20:1–12. doi:10.1007/s00704-016-1840-8

McDaniel, M.D., L.K. Tiemann, and A.S. Grandy. 2014. Does agricultural crop diversity enhance soil microbial biomass and organic matter dynamics? A meta-analysis. *Ecol. Appl.* 24: 560–570.

McDonald, M.D., K.L. Lewis, G.L. Ritchie, P.B. DeLaune, K.D. Casey, and L.C. Slaughter. 2019. Carbon dioxide mitigation potential of conservation agriculture in a semi-arid agriculture region. *AIMS Agri. Food.* 4:206–222 doi:10.3934/agrfood.2019.1.206

McVay, K.A., J.A. Budde, K. Fabrizzi, M.M. Mikha, C.W. Rice, A.J. Schlegel, D.E. Peterson, D.W. Sweeney, and C. Thompson. 2006. Management effects on soil physical properties in long-term tillage studies in Kansas. *Soil Sci. Soc. Amer. J.*, 70(2):434–438.

McVay, K.A., D.E. Radcliffe, and W.L. Hargrove. 1989. Winter legume effects on soil properties and nitrogen fertilizer requirements. *Soil Sci. Soc. Am. J.* 53: 1856–1862. https://doi.org/10.2136/sssaj1989.03615995005300060040x

Meshesha, D.T., A. Tsunekawa, M. Tsubo, and N. Haregeweyn. 2012. Dynamics and hotspots of soil erosion and management scenarios of the Central Rift Valley of Ethiopia. *Inter. J. Sediment Res.* 27:84–99.

Mikha, M.M., G.W. Hergert, J.G. Benjamin, J.D. Jabro, and R.A. Nielsen. 2015. Long-term manure impacts on soil aggregates and aggregate-associated Carbon and Nitrogen. *Soil Sci. Soc. Am. J.* 79:626–636.

Miller, P.R., B.G. McConkey, G.W. Clayton, S.A. Brandt, J.A. Staricka, A.M. Johnston, G.P. Lafond, B.G. Schatz, D.D. Baltensperger, and K.E. Neill. 2002. Pulse crop adaptation in the Northern Great Plains. *Agronomy Journal* 94:261–272. doi:10.2134/agronj2002.0261Mubvumba, P., P.B. DeLaune, and F. Hons. 2021. Soil water dynamics under a warm-season cover crop mixture in continuous wheat. *Soil Till. Res.* 206:104283 doi:10.1016/j.still.2020.104823

Mubvumba, P., P.B. DeLaune, and F.M. Hons. 2022. Enhancing long-term no-till wheat systems with cover crops and flash grazing. 2022. *Soil Sec.* doi: 10.1016/j.soisec.2022.100067

Myers, R., A. Weber, and S. Tellatin. 2019. Cover crop economics: Opportunities to improve your bottom line in row crops. *SARE Technical Bulletin*. June.

National Agricultural Statistics Service. 2018. Crops county data [Online]. Available at http://quickstats.nass.usda.gov/?source_desc=CENSUS (verified 23 Mar. 2021).

National Agricultural Statistics Service. 2019. 2017 Census of Agriculture, United States Department of Agriculture. Volume 1, Part 51, AC-17-A-51.

Nichols, V.A., E.B. Moore, S. Gailans, T.C. Kaspar, and M. Liebman. 2021. Site-specific effects of winter wheat cover crops on soil water storage. *Agrosyst. Geosci. Environ.* 5: e20238.

Nielsen, D.C., D.J. Lyon, G.W. Hergert, R.K. Higgins, F.J. Calderon, and M.F. Vigil. 2015. Cover crop mixtures do not use water differently than single-species plantings. *Agron. J.* 107: 1025–1038.

Nielsen, D.C., D.J. Lyon, R.K. Higgins, G.W. Hergert, J.D. Holman, and M.F. Vigil. 2016. Cover crop effect on subsequent wheat yield in the Central Great Plains. Agron. J. 108, 243–256.

Nielson, D.C. and M.F. Vigil. 2005. Legume green fallow effect on soil water content at wheat planting and wheat yield. *Agron. J.* 97: 684–689.

NOAA-NCEI. (2021). U.S. Billion-Dollar Weather and Climate Disasters. www.ncdc.noaa.gov/billions/, doi: 10.25921/stkw-7w73

Nouri, A., J. Lee, X. Yin, D.D. Tyler, and A.M. Saxton. 2019. Thirty-four year of no-tillage and cover crops improve soil quality and increase cotton yield in Alfisols, Southeastern US. *Geoderma* 337:998–1008.

Oades, J.M., 1995. Organic matter: chemical and physical fractions. In: R.D.B. Lefroy, G.J. Blair, and E.T. Craswell (Eds.), *Soil Organic Matter Management for Sustainable Agriculture*. ACIAR, Canberra, pp. 135–139.

Owens, L.B., R.W. Malone, D.L. Hothem, G.C. Starr, and R. Lal. 2002. Sediment carbon concentration and transport from small watersheds under various conservation tillage. *Soil Till. Res.* 67:65–73.

Padgitt, M., D. Newton, R. Penn, and C. Sandretto. 2000. Production practices for major U.S. agriculture, 1990–1997. *Statistical Bulletin* No. 969.

Peterson, G.A., A.D. Halvorson, J.L. Havlin, J.L., O.R. Jones, D.G. Lyon, and D.L. Tanaka. 1998. Reduced tillage and increasing cropping intensity in the Great Plains conserve soil carbon. *Soil Till. Res.* 47:207–218.

Peterson, G.A., A.J. Schlegel, D.L. Tanaka, and O.R. Jones. 1996. Precipitation use efficiency as impacted by cropping and tillage systems. *J. Prod. Agric.* 9: 80–186.

Peterson, G.A. and D.G. Westfall. 2004. Managing precipitation use in sustainable dryland agroecosystems. *Annals Appl. Biology* 144(2):127–138.

Peyraud, J.L., M. Taboada, and J.L. Delaby. 2014. Integrated crop and livestock systems in Western Europe and South America: A review. *European J. Agron.* 57:41–42.

Ramzan, S., A. Pervez, M.A. Wani, J. Jeelani, I. Ashraf, R. Rasool, M.A. Bhat, and M. Maqbool. 2019. Soil health: Looking for the effect of tillage on soil physical health. *Intern. J. Chem. Stud.* 7:1731–1736.

Razafimbelo, T.M., A. Albrecht, R. Oliver, T. Chevallier, L. Chapuis-Lardy, and C. Feller. 2008. Aggregate associated-C and physical protection in a tropical clayey soil under Malagasy conventional and no-tillage systems. *Soil Till. Res.* 98(2): 140–149.

Reeves, D., 1994. Cover crops and rotations. In: J.L. Hatfield and B.A. Steward (Eds.), *Advances in Soil Science: Crop Residue Management.* CRC Press, Inc., Boca Raton, FL, pp. 125–172.

Rehkamp, S., P. Canning, and C. Birney. 2021. Tracking the U.S. Domestic Food Supply Chain's Freshwater Use over Time. U.S. Department of Agriculture, Economic Research Service: Washington, DC, USA, ERR-288.

Reicosky, D.C., and F. Forcella. 1998. Cover crop and soil quality interactions in agroecosystems. *J. Soil Water Conserv.* 53:224–229.

Rockstrom, J., W. Steffen, K. Noone, A. Persson, F.S. Chapin, E.F. Lambin, T.M. Lenton, M. Scheffer, C. Folke, H.J. Schellnhuber, B. Nykvist, C.A. deWit, T. Hughes, S. van der Leeuw, H. Rodhe, S. Sorlin, P.K. Snyder, R. Costanza, U. Svedin, M. Falkenmark, L. Karlberg, R.W. Corell, V.J. Fabry, J. Hansen, B. Walker, D. Liverman, K. Richardson, P. Crutzen, and J.A. Foley. 2009. A safe operating space for humanity. *Nature* 461(7263):472–475.

Roesch-McNally, G.E., J.G. Arbuckle, and T.C. Tyndall. 2018. Barriers to implementing climate resilient agricultural strategies: The case of crop diversification in the U.S. Corn Belt. *Global Environ. Change* 48:206–215.

Rosa, L., D.D. Chiarelli, M. Sangiorgio, A.A. Beltran-Peña, M.C. Rulli, P. D'Odorico, and I. Fung. 2020. Potential for sustainable irrigation expansion in a 3 °C warmer climate. *PNAS* 117(47):29526–29534.

Rosenzweig, S., S.J. Fonte, and M.E. Schipanski, 2018. Intensifying rotations increases soil carbon, fungi, and aggregation in semi-arid agroecosystems. *Agric. Ecosyst. Environ.* 258:14–22.

Russel, A.E., D.A. Laird, T.B. Parkin, and A.P. Mallarino. 2005. Impact of nitrogen fertilization and cropping system on carbon sequestration in midwestern mollisols. *Soil Sci. Soc. Amer. J.* 69:413–422.

Sadeghpour, A., Q.M. Ketterings, F. Vermeylen, G.S. Godwin, and K.J. Czymmek. 2016. Soil properties under Nitrogen- vs. Phosphorus-based manure compost management of corn. *Soil Sci. Soc. Am. J.* 80(5):1272–1282.

Sainju, U.M., W.F. Whitehead, and B.P. Singh. 2003. Cover crops and nitrogen fertilization effects on soil aggregation and carbon and nitrogen pools. *Can. J. Soil Sci.* 83:155–165. https://doi.org/10.4141/S02-056

Saseendran, S.A., L.R. Ahuja, D.C. Nielsen, T.J. Trout, and L. Ma. 2008. Use of crop simulation to evaluate limited irrigation management options for corn in a semiarid environment. *Water Resources Res.* 44. W00E02. doi:10.1029/2007WR006181

Schnarr, C., M. Schipanski, and J. Tatarko. 2021. Crop residue cover dynamics for wind erosion control in a dryland, no-till system. *J. Soil Water Conserv.* 77(3):221–229.

Schwartz, R.C., R.L. Baumhardt, and S.R. Evett. 2010. Tillage effects on soil water redistribution and bare soil evaporation throughout a season. *Soil Till. Res.* 110:221–229. doi:10.1016/j.still.2010.07.015

Shaver, T.M., G.A. Peterson, L.R. Ahuja, D.G. Westfall, L.A. Sherrod, and G. Dunn, 2002. Surface soil physical properties after 12 years of dryland no-till management. *Soil Sci. Soc. Am. J.* 66:1296–1303.

Shaver, T.M., G.A. Peterson, and L.A. Sherrod. 2003. Cropping intensification in dryland systems improves soil physical properties: Regression relationships. *Geoderma* 116:149–164. doi:10.1016/S0016-7061(03)00099-5

Singer, J.W., A.J. Franzluebbers, and D.L. Karlen. 2009. Grass-based farming systems: soil conservation and environmental quality. In: W.F. Wedin, and S.L. Fales (Eds.), *Grassland: Quietness and Strength for a New American Agriculture. Am. Soc. Agron., Crop Sci. Soc. Am., Soil Sci. Soc. Am*, Madison, WI, pp. 121e136.

Singh, P.N., S.K. Shukla, and V.K. Bhatnagar. 2007. Optimizing soil moisture regime to increase water use efficiency of sugarcane (Saccharum spp. hybrid complex) in subtropical India. *Agric. Water Manag.* 90:95–100.

Smart, A.J., D. Redfearn, R. Mithcell, T. Wang, C. Zilverberg, P.J. Bauman, J.D. Derner, J. Walker, and C. Wright. 2021. Forum: Integration of crop-livestock systems: an opportunity to protect grasslands from conversion to cropland in the U.S. Great Plains. *Rangeland Ecol. Manage.* 78:250–256.

Steele, M.K., F.J. Coale, and R.L. Hill. 2012. Winter annual cover crop impacts on no-till soil physical properties and organic matter. *Soil Sci. Soc. Am. J.* 76:2164–2173. https://doi.org/10.2136/sssaj2012.0008

Steffan, J.J., E.C. Brevik, L.C. Burgess, and A. Cerda. 2018. The effect of soil on human health: an overview. *Euro. J. Soil Sci.* 69:159–171. doi:10.111/ejss.12451

Stewart, B.A. 2007. Experience with conservation agriculture in semi-arid regions of the USA. In: Proceedings of the International Workshop on Conservation Agriculture for Sustainable Land Management to Improve the Livelihood of People in Dry Areas, 7-9 May 2007, Damascus, Syria. pp. 141–155.

Studdert, G.A., H.E. Echeverria, and E.M. Casanovas. 1997. Crop–pasture rotation for sustaining the quality and productivity of a Typic Argiudoll. *Soil Sci. Soc. Am. J.* 61:1466–1472.

Tanaka, D.L., R.L. Anderson, and S.C. Rao. 2005. Crop sequencing to improve use of precipitation and synergize crop growth. *Agron. J.* 97:385–390.

Tautges, N.E., T.S. Sullivan, C.L. Reardon, and I.C. Burke. 2016. Soil microbial diversity and activity linked to crop yield and quality in a dryland organic wheat production system. *Applied Soil Ecology* 108:258–268.

Taylor, C.M., D. Belušić, F. Guichard, D.J. Parker, T. Vischel, O. Bock, P.P. Harris, S. Janicot, C. Klein, and G. Panthou. 2017. Frequency of extreme Sahelian storms tripled since 1982 in satellite observations. *Nature.* 544 (7651):475–478. https://doi.org/10.1038/nature220699

Teasdale, J.R., and C.L Mohler. 1993. Light transmittance, soil temperature, and soil moisture under residue of hairy vetch and rye. *Agron J.* 85:673–680.

Thaler, E.A., I.J. Larsen, and Q. Yu. 2021. The extent of soil loss across the U.S. Corn Belt. *Proc. Natl. Acad. Sci. Unit. States Am.* 118, e1922375118.doi.org/10.1073/pnas.1922375118

Tilman, D., K.G. Cassman, P.A. Matson, R. Naylor, and S. Polasky. 2002. Agricultural sustainability and intensive production practices. *Nature* 418 (6898):671–677.

Tittonell, P., E. Scopel, N. Andrieu, H. Posthumus, P. Mapfumo, M. Corbeels, G.E. van Halsema, R. Lahmar, S. Lugandu, J. Rakotoarisoa, F. Mtambanengwe, B. Pound, R. Chikowo, K. Naudin, B. Triomphe, and S. Mkomwa. 2012. Agroecology-based aggradation-conservation agriculture (ABACO): Targeting innovations to combat soil degradation and food insecurity in semi-arid Africa. *Field Crops Res.* 132:168–174.

Unger, P.W., and M.F. Vigil. 1998. Cover crop effects on soil water relationships. *J. Soil Water Conserv.* 53:200–207.

USDA-FSA. 2019. Conservation Reserve Program Statistics. U.S. Department of Agriculture, Farm Service Agency, Washington, DC. Available at: www.fsa.usda.gov/programs-and-services/conservation-progr ams/reports-and-statistics/conservation-reserve-program-statistics/index

U.S. Department of Agriculture, Natural Resources Conservation Service. 2014a. NRCS Cover Crop Termination Guidelines.

U.S. Department of Agriculture, Natural Resources Conservation Service. 2014b. Revised Cover Crop Termination Guidelines. *NRCS National Bulletin* NB 450-15-1 TCH.

Van Pelt, R.S., M.C. Baddock, T.M. Zobeck, V. Acosta-Martinez, A.J. Schlegel, and M.F. Vigil. 2013. Field wind tunnel testing of two silt loam soils on the North American Central High Plains. *Aeolian Res.* 10:53–59.

Varvel, G.E. 2006. Soil organic carbon changes in diversified rotations of the Western Corn Belt. *Soil Sci. Soc. Amer. J.* 70(2):426–433.

Villamil, M.B., G.A. Bollero, R.G. Darmody, F.W. Simmons, and D.G. Bullock. 2006. No-till corn/soybean systems including winter cover crops: Effects on soil properties. *Soil Sci. Soc. Am. J.* 70: 1936–1944. doi.org/10.2136/sssaj2005.0350

Wagger, M.G., and H.P. Denton. 1989. Influence of cover crop and wheel traffic on soil physical properties in continuous no-till corn. *Soil Sci. Soc. Am. J.* 53:1206–1210. doi.org/10.2136/sssaj1989.036159950053 00040036x

Wahlqvist, M.L, and L. Meei-Shyuan. 2007. Regional food culture and development. *Asia Pac. J. Clin. Nutr.* 16:2–7.

Wall, D.H., U.N. Nielsen, and J. Six. 2015. Soil biodiversity and human health. *Nature* 528:69–76. doi:10.1038/ nature15744

Wallander, S., D. Smith, M. Bowman, and R. Claassen. 2021. Cover crop trends, programs, and practices in the United States. US Department of Agriculture, Economic Research Service, EIB 222. chrome-www.ers.usda.gov/webdocs/publications/100551/eib-222.pdf?v=8242.7

Wang, E., W.L. Harman, J.R. Williams, and C. Xu. 2002. Simulated effects of crop rotations and residue management on wind erosion in Wichuan, West-Central Inner Mongolia, China. *J. Environ. Qual.* 31:1240–1247.

Wang, T., and I. Chowdhury. 2020. Crop diversity reduced in South Dakota. South Dakota State University Extension Article available at: https://extension.sdstate.edu/crop-diversity-reduced-south-dakota

Wang, T., H. Jin, Y. Fan, O. Obembe, and D. Li. 2021. Farmers' adoption and perceived benefits of diversified crop rotations in the margins of the US Corn Belt. *J. Environ. Manag.* 293: 112903. doi.org/10.1016/j.jenvman.2021.112903

Ward, P.R., R.A. Lawes, and D. Ferris. 2014. Soil-water dynamics in a pasture-cropping system. *Crop Past. Sci.* 65:1016–1021.

Weidhuner, A., A. Hanauer, R. Krausz, S.J. Crittenden, K. Gage, and A. Sadeghpour. 2021. Tillage impacts on soil aggregation and aggregate-associated carbon and nitrogen after 49 years. *Soil Till. Res.* 208: 104878. doi.org/10.1016/j.still.2020.104878

Wendt, R.C., and R.E. Burwell. 1985. Runoff and soil losses from conventional, reduced, and no-till corn. *J. Soil Water Conserv.* 40:450–454.

West, T.O., and W.M. Post. 2002. Soil organic carbon sequestration rates by tillage and crop rotation: A global data analysis. *Soil Sci. Soc. Amer. J.* 66(6):1930–1946.

Yang, C., Y. Geng, X.Z. Fu, J.A. Coulter, and Q. Chai. 2020. The effects of wind erosion depending on cropping system and tillage method in a semi-arid region. *Agron.* 20, 732. doi:10.3390/agronomy10050732

Yu, T., L. Mahe, Y. Li, X. Wei, X. Deng, and D. Zhang. 2022. Benefits of crop rotation on climate resilience and its prospects in China. *Agronomy* 12(2):436. doi.org/10.3390/agronomy12020436.

Zhao, J., S. Chen, R. Hu, and Y. Li. 2017. Aggregate stability and size distribution of red soils under different land uses integrally regulated by soil organic matter, and iron and aluminum oxides. *Soil Till. Res.* 167:73–79.

Zotarelli, L., B.J.R. Alves, S. Urquiaga, E. Torres, H.P. dos Santos, K. Paustian, and R.M. Boddey. 2005. Impact of tillage and crop rotation on aggregate-associated carbon in two Oxisols. *Soil Sci. Soc. Am. J.* 69(2):482–491.

Zuber, S.M., G.D. Behnke, E.D. Nafziger and M.B. Villamil. 2015. Crop rotation and tillage effects on soil physical and chemical properties in Illinois. *Agron. J.* 107(3):971–978.

7 Managing Drought Stress in Agro-Ecosystems of Latin America and the Caribbean Region

Stoécio Malta Ferreira Maia[1], Carlos de Oliveira Galvão[2], Thalita Fernanda Abbruzzini[3], Julio Campo[4], José Antonio Marengo Orsini[5], Carlos Eduardo Pellegrino Cerri[6], and Teogenes Senna de Oliveira[7]

[1]Federal Institute of Education, Science and Technology of Alagoas, Brazil;
[2]Department of Civil Engineering, Federal University of Campina Grande, Brazil;
[3]Institute of Geology, Department of Environmental and Soil Sciences, National Autonomous University of Mexico, Mexico;
[4]Terrestrial Biogeochemistry and Climate Laboratory, Institute of Ecology, National Autonomous University of Mexico, Mexico;
[5]Center for Monitoring and Early Warning of Natural Disasters – Cemaden, Brazil;
[6]Soil Science Department, Luiz de Queiroz College of Agriculture, University of São Paulo, Brazil;
[7]Soils Department, Federal University of Viçosa, Brazil;

CONTENTS

7.1 General Considerations on LAC 158
7.2 Climate Changes in the Drylands of LAC..160
    7.2.1 The Present...160
    7.2.2 The Future 161
7.3 Water Security in the LAC 162
    7.3.1 Overview 162
    7.3.2 The Emergency of Adaptation 162
    7.3.3 Adaptation Measures 162
    7.3.4 Governance and Financing..163
7.4 Managing Drought Stress in Agroecosystems..163
    7.4.1 Brazil...165
    7.4.2 Argentina..170
    7.4.3 Mexico 170
        7.4.3.1 Evidence from Crop Systems in Central Mexico 172
        7.4.3.2 Evidence from Livestock Systems in the South-Southeast
             of Mexico 173

DOI: 10.1201/b22954-7

7.5   Conclusions 173
References...175

7.1   GENERAL CONSIDERATIONS ON LAC

According to the United Nations, Latin America and the Caribbean (LAC) encompasses 33 countries (21 in Latin America and 12 in the Caribbean), and 15 dependences of other territories. These countries comprise an area of 21.95 million (M) km^2 (approximately 15% of the world's total area) (Mahlknecht et al., 2020) located south of the United States of America, starting with Mexico in North America and spreading through Central America and the Caribbean down into the southernmost tip of South America. Nowadays, the LAC population is approximately 658 M, being one of the most urbanized regions worldwide, since eight out of ten citizens live in cities (WB, 2022).

The LAC is probably the most diverse region on the planet as the climate of the region ranges from the hot and humid Amazon River basin to the dry and desert-like conditions of northern Mexico and northern Chile (Mahlknecht et al., 2020). Besides, LAC is home to about half of the world's remaining tropical forests, with major biomes including the Amazon, along with drier forests/woodlands and savannahs, such as the Gran Chaco (spanning parts of Argentina, Bolivia, Brazil, and Paraguay), the Cerrado (in Brazil), and the Chiquitano (in Bolivia and Brazil). These forests not only store close to half of the biomass carbon of all tropical forests (Pendrill and Persson, 2017), but many have also been identified as prioritized areas for global biodiversity conservation (Brooks et al., 2006).

Yet, this region has a high aptitude for agriculture and livestock, contributing to 7% of the global agricultural added value and with a relatively large share of the global volume of specific products, such as coffee (58%), soybeans (52%), sugar (29%), beef (26%), poultry (22%), and maize (13%) (ECLAC, 2017; Mahlknecht et al., 2020). Grain production in the region has increased on average by 3.5% per year since 1990, and overall, regional production is largely driven by Brazil, Argentina, and Mexico, which together account for 80% of the total. However, despite its suitability and extensive areas under agriculture, the contribution of the farm sector to the gross domestic product (GDP) decreased from 13.0% to 5.6%, and likewise the share of agriculture in total employment decreased from 19.5% to 14.2% between 1995 and 2015 (OECD/IEA, 2022).

Obviously, there are areas of LAC which have high levels of productivity. However, compared with other regions in terms of yields, LAC still has potential for growth in agricultural terms, since its production yields remain low when compared with those of the main production regions worldwide (FAO, 2015). These lower levels of productivity are mainly due to socioeconomic aspects, since LAC is among the most unequal regions in the world regarding income, with a Gini coefficient of about 0.5 (FAO, 2018), and approximately 30% of the total population has been classified as poor (ECLAC, 2017), and the socioeconomic limitations, in turn, occur mainly in the drylands of LAC.

While well known for rainforests, two-thirds of the LAC region has a semiarid to an arid climate, namely, central-north Mexico, northeast Brazil, central-south Argentina, central-northern Chile, and some parts of Bolivia and Peru (Figure 7.1) (Mahlknecht et al., 2020). According to FAO (2022), drylands are characterized by a scarcity of water, which affects both natural and managed ecosystems and constrains the production of livestock as well as crops, wood, forage, and other plants and affects the delivery of environmental services. For millennia, drylands have been shaped by a combination of low precipitation, droughts, and heat waves, as well as human activities such as fire use, livestock grazing, the collection of wood and non-wood forest products, and soil cultivation. In this context, LAC, like other regions of the planet (especially the African continent), must face a series of challenges to manage its agroecosystems, with emphasis on water scarcity and droughts, and these phenomena that have always been present are already becoming more frequent and intense due to the effects of global climate change. Droughts vary in duration, severity, spatial coverage,

Managing Drought Stress

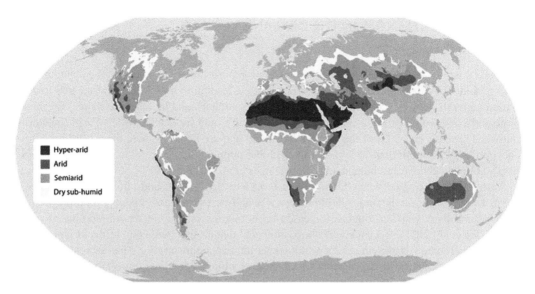

FIGURE 7.1 The world's drylands and subtypes. Prepared using spatial data from UNEP-WCMC (2007). Source: FAO (2022a).

and impacts caused by the water deficit characteristic of specific event. Droughts are generally classified as meteorological, agricultural, hydrological, and socioeconomic. The "meteorological drought" is defined as the degree of dryness in a specific region, often measured by a deviation from normal or climatological precipitation (Wilhite, 2000). "Agricultural drought" is the result of deficient soil moisture induced by deficiency in water availability for crop or plant growth leading to biomass reduction. "Hydrological drought" is associated with a reduction of mean level water in surface and subsurface water resources like lakes, reservoirs, aquifers, and streams. It can occur for an extended period of time, maybe some days, a season, a year, or even several years. Finally, the "socioeconomic drought" is associated with impacts on human activities, including direct and indirect impacts on agricultural production and other economic activities and national and regional GDP (Wilhite, 2000; Mahlknecht et al., 2020).

Regardless of the type, droughts always cause stress and negatively impact agricultural production. For instance, drought-induced production losses in Brazil were observed in 2012 and 2016, with a production decline of 62.4% and 48.2% for beans, 46.6% and 42.3% for maize, and 34.3% and 15.6% for cassava, respectively (Carvalho et al., 2020). Economic losses in Brazil in 2020 were US$3 billion (Aon Benfield UCL Hazard Research Centre, 2020). Similar losses were observed in other Latin American countries such as Argentina, El Salvador, and Guatemala, which experienced severe losses in agriculture totaling about US$6 billion due to droughts in 2018 (Aon Benfield UCL Hazard Research Centre, 2019). Drought-related global economic losses from 2000 to 2020 have been estimated to be US$ 438 billion (Aon Benfield UCL Hazard Research Centre, 2020).

The management of stresses caused by drought in agriculture can occur in different ways, such as genetic improvement of crops, plant resistance mechanisms (escape, avoidance, and tolerance mechanisms), and the use of best management practices (Seleiman et al., 2021; Adeyemi et al., 2020). Despite the countless possibilities of coexistence with droughts, the focus in this chapter is on the management of agroecosystems, especially in LAC drylands, seeking to highlight management practices and systems that have the potential to accumulate carbon in soil and improve storage and the use efficiency of water. Additionally, the chapter also addresses aspects related to climate change and water security in LAC, because these are matters directly linked to managing drought stress in agroecosystems.

7.2 CLIMATE CHANGES IN THE DRYLANDS OF LAC

7.2.1 THE PRESENT

Drylands are places of water scarcity, where rainfall may be limited or may only be abundant for a short period. They experience high mean temperatures, leading to high rates of water loss to evaporation and transpiration. Drylands are found on all continents, and include grasslands, savannahs, shrublands, and woodlands; cover over 40% of the earth's land surface; provide 44% of the world's cultivated systems and 50% of the world's livestock; and are home to more than 2 billion people. They are home to more than 38% of the total global population and are one of the most sensitive areas to climate change and human activities (Davies et al., 2012).

Aridity, and associated water scarcity, is a long-term hydrologic and climatic condition, with pervasive influences on dynamics in human society and terrestrial ecosystems (e.g., more frequent hydrologic and ecosystem droughts, increased economic losses, and decreased land-based carbon sink) (Chai et al., 2021). Recent studies have indicated that aridity, defined in terms of atmospheric supply (precipitation) and demand for water (potential evapotranspiration), has increased globally during recent decades and is projected to increase significantly in the future (Koutroulis, 2019). Natural variability (e.g., Pacific Decadal Oscillation and El Niño–Southern Oscillation) modulates large-scale aridity via atmosphere–ocean feedbacks and atmospheric teleconnections. However, the natural variability alone cannot explain the increase in aridification over long timescales. Overall, anthropogenic greenhouse gas (GHG) emissions can enhance global precipitation and modify stomatal conductance and plant water use, likely relieving the aridity stress (Bonfils et al., 2020).

Drylands, as defined by the United Nations Environment Programme (UNEP) Aridity Index (AI) (UNEP AI), are primarily distributed in middle and low latitudes (Figure 7.1). Major hyperarid (AI < 0.05) areas are located over the central and northern Sahara Desert, the Rub al Khali Desert, and Arab Plateau in the eastern and northwestern Arabian Peninsula, and the Taklimakan Desert in northwestern China. Arid ($0.05 \leq$ AI <0.2) areas are located over the southern Sahara Desert, the Kalahari Desert in southwestern Africa, western and central Arabian Peninsula, Central Asia, southwestern United States, eastern Patagonia, and much of Australia. Semiarid ($0.2 \leq$ AI <0.5) and dry subhumid ($0.5 \leq$ AI <0.65) regions mainly occur in the western United States, the west coast of South America, Central and East Asia, southern Africa, and a large portion of Australia outside of the central desert regions (Huang et al., 2016a). In LAC drylands, as in the rest of the world, anthropogenic activities are dominant drivers of global environmental change and desertification. Central and South America are highly exposed, vulnerable, and strongly impacted by climate change, a situation amplified by inequality, poverty, population growth and high population density, land use change, particularly deforestation with the consequent biodiversity loss, soil degradation, and high dependence of national and local economies on natural resources to generate ecosystem services.

Drylands in South America represent approximately 31% of the continent's total land area and 8.7% of the global drylands (FAO, 2019). Following Ganem et al. (2022) and using the UNEP AI, four dryland subtypes are recognized in South America, covering an estimated 5.1 M km². Semiarid is the most common dryland subtype (46%), followed by the dry subhumid (41%), arid (8%), and hyperarid (5%) zones. The semiarid zone represents 14% of South America, with its highest country-level proportions found in Argentina (38%) and Brazil (9%). The most critical water-stressed subtypes are found in arid areas largely distributed over Argentina (70%) and the hyperarid regions predominantly in Chile (over 60%) and Peru (35%).

Drylands in South America are primarily characterized by savannas, shrublands, woodlands, and grasslands, generally grouped into biomes. The Caatinga is the driest forest on the continent, consisting of xeric shrubland and thorn woodland. The Llanos (plains) del Orinoco is the second largest ecosystem of northern South America, encompassing parts of Venezuela and Colombia. The subtropical shrubland in central Argentina, known as Espinal, is considered one of the most vulnerable

Managing Drought Stress

ecoregions. Concerning drylands that fall within the hyperarid classification, the Atacama Desert of northern Chile and the Mexican Sonoran Desert are broadly acknowledged as the driest globally. The drylands region shared by the United States and Mexico currently faces multiple sustainability challenges at the intersection of the human and natural systems.

Warming and drying conditions threaten surface water and groundwater availability, disrupt land- and marine-based livelihood systems, and challenge the sustainability of human settlements. These biophysical challenges are exacerbated by a highly mobile and dynamic population, volatile economic and policy conditions, increased exposure to extreme events, and urbanization on marginal, vulnerable lands (NAS, 2018).

The rural drylands of LAC are perceived as areas of sparse vegetation with a population dedicated to unprofitable economic activities (Ocampo-Melgar et al., 2022). Nickl et al. (2020) evaluated whether drylands have expanded and become increasingly arid across 33 Latin American countries using the UNEP AI from 1960 to 2017. Their results show that, with some exceptions, most LAC countries have experienced aridification from the 1961–1990 period to the 1991–2017 period, and the trend is for this pattern to continue.

In Brazil, the drylands are mostly concentrated in the Northeast region (N.E.), between $2.5°$ S and $16.1°$ S, and between $34.8°$ W and $46°$ W, with an area of about 1.54 M km^2, representing 18.3% of Brazilian territory. It includes Maranhão, Piauí, Ceará, Rio Grande do Norte, Paraíba, Pernambuco, Alagoas, Sergipe, and Bahia states. The region that is affected regularly by droughts in N.E. was referred to as the Polígono das Secas (Drought Polygon) and included an area of about 1.64 M km^2. One sub-region within this Drought Polygon is called Sertão or the semiarid region in N.E., with an area of 0.91 M km^2 (Marengo and Bernasconi, 2015).

7.2.2 THE FUTURE

Extremes of climate variability and climate change in drylands strongly influence the economy and society, especially in developing countries (Huang et al., 2016a, 2017). Warming and drying trends occur in the historical period, and both are projected to persist into the future. Thus, these two factors may be linked through dynamics and climate. Huang et al. (2016b) projected dryland area under representative concentration pathways (RCPs): RCP8.5 and RCP4.5 from CMIP5 models, with an increase of 23% and 11%, respectively, relative to 1961–1990 baseline, equaling 56% and 50%, respectively, of the total land surface. Such an expansion of drylands would lead to reduced carbon sequestration and enhanced regional warming, resulting in warming trends over the present drylands that are double those over humid regions. The increasing aridity, enhanced warming, and rapidly growing human population will exacerbate the risk of land degradation and desertification soon in the drylands of developing countries, where 78% of dryland expansion and 50% of the population growth will occur under RCP8.5. For LAC, for 2071–2100 relative to 1961–1990, and using the UNEP AI, the RCP8.5 scenario shows an increase of arid and semiarid conditions in central Mexico and isolated regions of southern Peru and Central America, Northeast Brazil (in agreement with Marengo et al. 2021 and Marengo and Bernasconi 2015) and in Northern Argentina. These climatic fluctuations may be most pronounced in the poorest regions with high levels of chronic undernourishment and a great degree of instability. Food price fluctuations already represent a risk to vulnerable populations that are expected to increase with climate change.

Projections by Chai et al. (2021) identified in CMIP6 simulations consistently drying trends using the UNEP AI over the drylands in central North America, and central South America under both SSP3 – 7.0 and SSP5 – 8.5 scenarios. The authors explained that such AI decreases driven by elevated GHG emissions were mainly caused by the increase in air temperature, implying that the land precipitation increase could not keep pace with the growing evaporative demand associated with GHG-dominated warming. However, the model-projected trends of land aridity in the 21st century may contain significant uncertainty.

The largest expansion of drylands has occurred in semiarid regions since the early 1960s. Future aridity changes suggest more pressing environmental challenges that will likely face society (Huang et al. 2017). Dryland expansion will lead to reduced carbon sequestration and enhanced regional warming. The increasing aridity, enhanced warming, and rapidly growing population will exacerbate the risk of land degradation and desertification in the near future in developing countries.

7.3 WATER SECURITY IN THE LAC

7.3.1 OVERVIEW

As shown above, drylands in the LAC already experience water scarcity, land degradation, and desertification. Projections of climate change point out to future exacerbation of such conditions. The projected increase in temperatures will also increase the amount of water evapotranspiration and plant consumption. Both present and future situations demonstrate the need to manage drought stress so that water and food security are maintained, through consistent and efficient agricultural production. Recent data show that agricultural and livestock activities in the LAC drylands are responsible for most of the water consumption, compared to industrial and human ones. Even though water scarcity in the region has already limited the scope of irrigated agriculture, which will make it difficult to manage water stress in the region. (Ocampo-Melgar et al., 2022).

Water availability in the LAC drylands is mostly seasonally distributed and, very often, present high interannual variability. Then, regularization of water availability over time is necessary to provide water security, both for human consumption and agricultural and livestock production. This regularization requires surface reservoirs for water storage during the rainy months, for usage during the dry periods, or wells for exploiting groundwater, when it is available. Water regularization can reduce water scarcity and supply water for irrigation, reducing the risk of rainfed agriculture and making irrigated agriculture viable.

On the other hand, high interannual variability of the rainfall and hydrological regimes implies uncertainty about water availability and security, even where river discharges are regularized, and groundwater is available. Periods of low river flows and/or groundwater levels may last for several years during prolonged droughts, when reservoirs and aquifers may reach exhaustion.

Then, adaptive planning of agroecosystems is mandatory in such regions for achieving water security and should consider water and soil availability and management, water use efficiency (WUE) and demands, unconventional water sources, and drought stress management.

7.3.2 THE EMERGENCY OF ADAPTATION

Uncertainties in water security and agricultural production, due to climate and hydrological variability and change, adapt such agroecosystems a necessity, in the short-, medium-, and long terms. IPCC (2022) highlights two important aspects regarding adaptation: climate, ecosystems (including biodiversity), and human society should be considered as coupled systems and adaptation measures should be "effective, feasible and conform to principles of justice." The same report warns that some systems are reaching their limits to adaptation. It emphasizes that "smallholder farmers in Central and South America have reached soft limits" to adaptation, and "inequity and poverty also constrain adaptation, leading to soft limits and resulting in disproportionate exposure and impacts for most vulnerable groups" (IPCC, 2022).

7.3.3 ADAPTATION MEASURES

Opportunities for short- to long-term water management can enhance water security. Among them, the chapter highlights the following, relevant to agricultural production in LAC drylands: climate and hydrological services, water allocation, infrastructure management, water and soil conservation, and unconventional water sources.

Climate and hydrological services are active in LAC for decades, providing information on weather, climate, and drought monitoring and forecasting (Coelho et al., 2006). If inclusive of diverse users and providers, they may effectively inform appropriate and efficient water and land use, and infrastructure management (IPCC, 2022). Mexican and Brazilian drought monitors, among others, map drought occurrence and severity almost in real time (Lobato-Sánchez, 2016; Escada et al., 2021). Seasonal forecasting of rainfall, runoff, reservoir storage, and aquifer recharge are consolidated operational services in some regions of LAC (Clarkson et al., 2022). Such services have been used to plan and implement drought preparedness measures. One of such measures is the annual water allocation process in the Brazilian semiarid region (De Nys et al., 2017). During these processes, users and managers negotiate the quantity of water to be allocated to human consumption, and agricultural and industrial uses, considering current water availability and sectorial demands. Water security and drought disaster contingency plans are other important instruments for resilience building.

Unconventional water sources, such as small shallow aquifers, rainwater harvesting, and wastewater reuse, are valuable in drylands, particularly where integrated to soil conservation measures and appropriate crop planning (Laura et al., 2020; Espíndola et al., 2020; Rêgo et al., 2022).

7.3.4 Governance and Financing

Water governance in LAC is institutionally organized in most countries, but fragmented, lacking improvements in policy coherence, integrity, transparency, stakeholder involvement, and financing (Neto et al., 2018). The governance of water and territory, and associated policies and management systems, needs stronger integration with other sectoral policies, such as agricultural and energy policies (Mahlknecht et al., 2020). River basin organizations may have a central role in improving governance, but they are not implemented in all countries in LAC, and their maturity varies widely (Akhmouch, 2012).

Securing investments and appropriate governance for adaptation, such as improving drought stress management, infrastructure upgrading, and operational optimization, are necessary for improving water security in the medium to long term. The continuity of "business-as-usual" in technological development and appropriation, sources of funding, and resource allocation will not lead to the desired adaptation goals. Innovation in financing is mandatory to achieve sustainable water security in LAC drylands (Lentini, 2022).

Empowerment of vulnerable smallholder farmers of LAC drylands is a governance challenge in the region, to promote equity in access to water, financing, technology, and other resources. The recognition, by the governmental and other institutional bodies, of such communities and their important role in the regional economies is the first step toward this goal. Many times, the water governance frameworks do not consider such very numerous farmers (Tsuyuguchi et al., 2020; Ocampo-Melgar et al., 2022). There is a need for the implementation of context-specific governance mechanisms targeted at the poor (Akhmouch, 2012).

7.4 MANAGING DROUGHT STRESS IN AGROECOSYSTEMS

The management of stresses caused by drought in agriculture can occur in different ways, such as genetic improvement of crops, cultivated plant resistance mechanisms (which include escape, avoidance, and tolerance mechanisms), seed priming, plant growth regulators, osmoprotectants, fertilization management (application of certain elements [e.g., potassium, selenium, and silicon]), hydrogel, and microorganisms manipulation, as well as management practices (Adeyemi et al., 2020; Seleiman et al., 2021) that include soil and the diversity of cultivated and/or native plants or animals. Many of these possibilities are still not commercially viable or are difficult to adopt in the drylands of LAC, especially in developing countries, which are characterized by the predominance of small farmers, who do not have access to technical assistance, financial resources, or adequate public policies.

Adapted management practices for drylands of LAC involve some fundamental principles, especially for the development of sustainable dryland farming systems, and include, for example, optimization of the fit between crop growth cycle and the available moisture; weed control; optimized plant population density and spatial arrangement of plants; soil fertility management concerning the water regime; control of soil biotic stress factors that inhibit root development; improved forage/livestock/grains integration and rotation; improved soil and water conservation practices and the associated reduced tillage systems; increasing the soil organic matter (SOM) content; and enhancement of crop and livestock diversification (Pérez-Marin et al., 2017; Maia et al., 2019).

All of it is possible when considering the science of Agroecology. The management of drought stress in agroecosystems should be done in a holistic and integrated approach that simultaneously applies ecological and social concepts and principles to the design and management of sustainable agriculture and food systems. It seeks to optimize the interactions between plants, animals, humans, and the environment while also addressing the need for socially equitable food systems within which people can exercise choice over what they eat and how and where it is produced. This is the definition of Agroecology. Recently, at the Second International Symposium on Agroecology, held in April 2018, FAO launched the "10 Elements of Agroecology Framework," which is designated as a guide to one of the ways to promote sustainable agriculture and food systems. All ten elements contribute to living with droughts; however, three elements can be pointed out as the most important: (i) *efficiency* – innovative agroecological practices produce more using fewer external resources, which involves planning and managing diversity to create synergies among different components of the system, thus increasing efficiency and reducing dependence on external inputs; (ii) *resilience* – agroecological systems are more resilient, they have a greater capacity to recover from disturbances including extreme weather events such as drought, floods, or hurricanes, and to resist pest and disease attack; and (iii) *diversity* – a key element in agroecological systems, impacting all others.

Only from a biological perspective, agroecological systems optimize the diversity of species and genetic resources in different ways, such as (i) the agroforestry systems (AGFs) that organize crops, shrubs, livestock, and trees of different heights and shapes at different levels or strata, increasing vertical diversity; (ii) crop–livestock systems which rely on the diversity of local breeds adapted to specific environments; and (iii) intercropping and crop rotation which combines complementary species to increase spatial and temporal diversity.

By planning and managing diversity, agroecological approaches enhance the provisioning of ecosystem services (ESs), including pollination and soil health, upon which agricultural production depends. Diversification can increase productivity and resource-use efficiency by optimizing biomass and water harvesting (FAO, 2022b). Therefore, the key focus of agroecology is that food and production systems that utilize a diversity of crops, animals, and native biodiversity with different spatial and seasonal arrangements can mimic natural regenerative processes which increase carbon storage and reduce their vulnerability to risks associated with droughts, climate change, and natural disasters (UNCCD, 2022). Thus, functional and biological diversification, despite being an ancient approach, has increasingly proved to be the basis for the establishment of more stable and resilient agroecosystems, which is essential to living with droughts and climate change. In fact, the diversification of farming in one way or another influences or drives most of the other principles mentioned above.

In this context, some examples (case studies) of agroecosystems management adopted in Brazil, Mexico, and Argentina are presented below, which have been shown to be sustainable and stable, thus being able to reduce the risks and impacts generated by droughts.

7.4.1 Brazil

In Brazil, droughts are recurrent phenomena that affect the northeast region with greater frequency and intensity, despite other regions in Brazil having also been affected by droughts in recent years. Impacts have been reported, especially those that are affecting major agricultural producers, as in West Central Brazil (Cunha et al., 2019), the metropolitan area of São Paulo in 2014–2015 due to El Niño (Nobre et al., 2016), or the Amazon region by extreme droughts in 1998, 2005, 2010, and 2015–2016 (Marengo et al., 2018, Cunha et al., 2019).

The northeast region of Brazil has been the most affected by droughts as it almost entirely lies in the Brazilian semiarid region (Marengo et al., 2017). The region has the highest proportion of people living in poverty in the country, rainfed agriculture in this region accounts for 95% of farm land, and according to the IPCC AR5 WG2 (IPCC, 2014), northeast Brazil may be the most vulnerable region to droughts (Marengo et al., 2017), affecting agriculture and increasing the risk of fires in those regions. The semiarid region of Brazil is the focus of the case studies presented here, considering the diversity of agricultural systems as the main adapted strategy for the management of stresses caused by drought. Thus, agricultural systems such as agroforestry, intercropping, and agroecology are relevant options for the Brazilian drylands.

AGFs are conservation-effective management practices that integrate trees in the agricultural and/ or livestock systems and combine agricultural, pastoral, and silvicultural practices at the same time and space (Maia et al., 2007). In the semiarid region of Brazil, different AGF arrangements have been evaluated, seeking to understand the impacts of these systems on different variables. A prominent initiative is at Embrapa (Brazilian Agricultural Research Corporation) Ovinos e Caprinos, in the state of Ceará. Different AGFs arrangements (agrosilvopastoral and silvopastoral) have been evaluated under different aspects since 2002, and the results obtained are encouraging.

For example, Maia et al. (2007) and Aguiar et al., 2014 reported that the agrosilvopastoral and silvopastoral systems are efficient in maintaining or even increasing soil C and N stocks when compared to native vegetation (Caatinga). Such results were confirmed by Primo et al. (2022), who simulated AGFs for 100 years (Figure 7.2) using the Century Model observing that both agrosilvopastoral and silvopastoral have significant potential to accumulate C in the soil compared with the traditional uses from the region and even about the native vegetation. The rotation of agrosilvopastoral and silvopastoral, including a fallow of 7 years in the same area, seems also to be a good strategy.

The accumulation of SOM by AGFs results in many soil processes already studied in the same area such as inputting nutrients (Mendes et al., 2013), reduction of soil erosion (de Aguiar et al., 2010), and increasing water retention (Silva et al., 2011), and is effective in maintaining and sometimes improving soil quality conditions at the same level as that under the native vegetation or even better (Fialho et al., 2013). These positive effects are a consequence of the plant diversity maintained by the management done on AGFs, not only on vegetation (Aguiar et al., 2014; Aguiar et al., 2019) but also on soil fauna (Fialho et al., 2021).

Improved/maintained water retention with AGFs is a very interesting condition for managing drought stress what was seen by Silva et al. (2011). Larger values of the least limiting water range (LLWR) of AGFs conditions compared with intensive cultivation or native vegetation (Figure 7.3) were observed by these authors. The LLWR considers the range of soil water content within which plant growth is least limited by water potential, aeration, and mechanical impedance. The critical limits are associated with field capacity (−0.01 MPa), wilting point (−1.5 MPa), aeration (0.10 m^3 m^{-3} = 10%), and penetration resistance (3.5 MPa). LLWR provides measures of the influence of management systems on soil structure, porosity, bulk density, penetration resistance and water retention. These conditions turn the agricultural system much more resilient to droughts events since successive cropping adversely affects the physical properties and even fallow time after cropping could not be sufficient to allow recovery of the physical conditions.

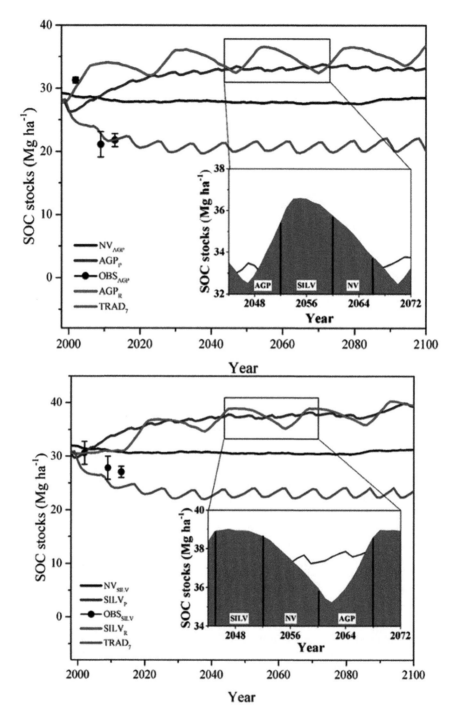

FIGURE 7.2 Soil organic C (SOC) stocks simulated by Century Model and observed (OBS) in the 0–20 cm layer, for traditional management systems with 7 years of fallow (TRAD), agrosilvopastoral system under permanent (AGP$_P$) and in rotation (AGP$_R$) with the silvopastoral system (SILV$_R$) and natural vegetation (NV$_R$) for 100 years simulation. The shaded area highlights a simulation period with the rotation sequence of AGP – SILV – NV and the respective fluctuations of SOC stocks. Source: Primo et al. (2022).

Managing Drought Stress

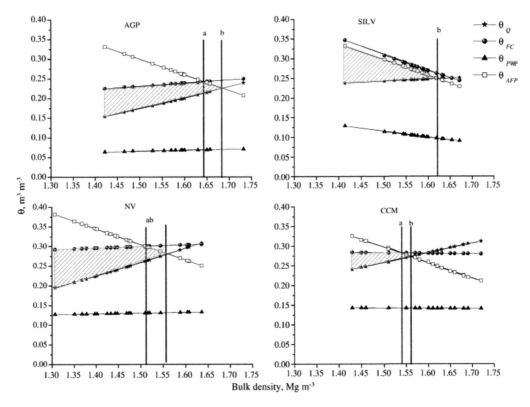

FIGURE 7.3 Water content (θ) at field capacity ($\psi = 0.01$ MPa), permanent wilting point ($\psi = 1.5$ MPa), air-filled porosity of 10% (θ_{AFP}), soil resistance penetration at 3.5 MPa (θ_Q), and the least limiting water range (LLWR) as functions of bulk density for 0–5 cm layer soil samples from a Luvisol under agroforestry, conventional crop management, and natural vegetation located at Sobral – CE, Brazil. AGP: agrosilvopastoral, SILV: silvopastoral, CCM: conventional crop management, NV: natural vegetation. The letter *a* indicates bulk density (BD) when air-filled porosity (θ_{AFP}) becomes the upper limit of LLWR, and *b* indicates critical bulk density (BD$_c$). Source: Silva et al. (2011).

Thus, AGFs for degraded situations due to traditional crop production or abandoned agricultural areas in the Brazilian semiarid region can potentially be applied and should have high priority among the strategies for drylands of LAC.

Intercropping systems, a more simplified agriculture system compared with AGFs reported above, is the simultaneous cultivation of two or more crop species on the same area of land (Campiglia et al., 2015), which can be also effective in dealing with drought. The advantages of intercropping could be associated with higher diet diversity for human consumes, and reduction of risk of crop losses intensifies the use of labor and needs limited external inputs as well as reduced erosion risks (Araújo Girão et al., 2015). Additionally, such agroecosystems provide ecological services such as nutrient recycling and soil C sequestration.

In the Brazilian semiarid region, two initiatives supported by the nongovernmental organization ESPLAR – Centre for Research and Assistance – based on the soil conservation practices application and direct participation of family farmers in the decisions of soil and culture management have produced very relevant results. Both examples are carried out at Ceará State and besides intercropping, other conservation practices are applied, such as contour planting, soil cover with weeds, constant inputs of organic matter, no use of fire, minimum tillage using mainly manual tools, using natural products for fertilization, and ecological control of insects.

The first initiative was at Tauá county, located in the southwest of the State of Ceará, Brazil, where two types of intercropping were evaluated (see more details in Maia et al., 2019): one with bean, sesame, and pigeon pea, and the other with cotton, maize, bean, sesame, and pigeon pea. These systems promoted, respectively, the contribution of 13.7 and 35.9 Mg ha^{-1} year^{-1} of biomass (dry matter), which is much higher than the contribution on average, by the aboveground native vegetation (3.7 Mg ha^{-1} year^{-1}). This is probably the main reason for the observed improvements in soil properties, such as soil density, cation exchange capacity, and maintenance of soil C stocks, concerning areas of native vegetation (Maia et al., 2019). Specifically, on soil C, it was observed that in just 4 years with the intercropping systems, the system with the highest type of intercropping species promoted an increase in the total soil carbon stock (0–50 cm), while the intercropping system with the least diversity kept the stock equal to the levels observed in the native vegetation area (Figure 7.4). The higher the soil C stocks, the better the soil properties, whether physical, chemical, or biological, especially in relation to soil water retention which showed to be better in intercropping areas compared with the traditional agricultural system (Sousa, 2006), as commented for AGFs above.

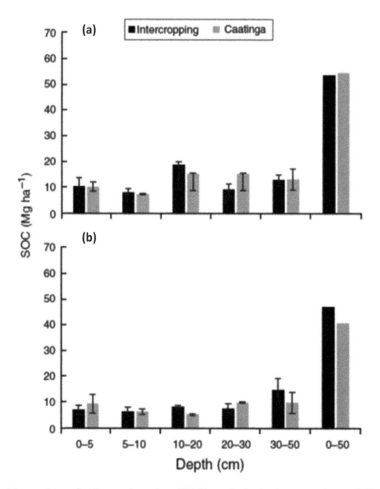

FIGURE 7.4 Mean values of soil organic carbon (SOC) stocks in the intercropping and Caatinga areas of (*a*) Farm 1 and (*b*) Farm 2. The SOC stocks for the 0–50 cm layer was corrected by the equivalent soil mass approach. Source: Maia et al. (2019).

Managing Drought Stress

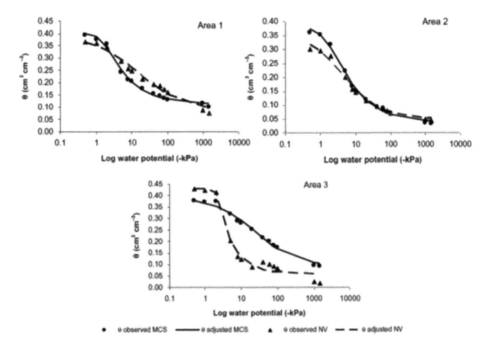

FIGURE 7.5 Water retention curves under multi-cropping systems (MCS) and natural vegetation (NV) of the three studied areas of family farmers from Choró–CE, Brazil. Source: Araújo Girão et al. (2015).

The second initiative, which is on the same basis as the first one, was developed in the Choró county, in the central part of the Ceará State, Brazil, and evaluated an intercropping with cotton, maize, cowpea, and sesame under different arrangements that were cultivated simultaneously and have been replanted every year. Araújo et al. (2013) evaluated the stability of soil aggregates and resistance to penetration and observed that the intercropping systems maintained these properties similar to those observed in native vegetation. Araújo Girão et al. (2015) evaluated soil moisture and water holding capacity and observed that, in general, intercropping areas had higher soil moisture contents, as well as higher water holding capacity (Figure 7.5) compared to native vegetation.

Finally, areas under organic production of Caribbean cherry in the Brazilian semiarid region evaluated the effect of organic and green manure on different soil properties. Xavier et al. (2006) found that such practices resulted in the maintenance and recovery of C and N contents of microbial biomass and light organic matter. Likewise, it was observed that there was an increase in the available-P and most labile P pool, contributing to the P incorporation into a biological cycle (Xavier et al., 2009). The agroecological cultivation of cotton in the semiarid region of the state of Ceará, Brazil (Lima et al., 2007; Almeida et al., 2009), resulted in a greater number of species and individuals of the fauna (macro and meso) of the soil when compared to conventional cultivation. These works show the effects of agroecological management on soil health, which is verified through the organisms present, and corroborate other studies that evaluate the importance of organisms (in particular, microorganisms) to develop protection mechanisms against water stress and desiccation (Frasier et al., 2016a; Seleiman et al., 2021; Braga et al., 2022). In this sense, Braga et al. (2022) evaluated the inoculation of soybean with *Bacillus* sp. and *Paenibacillus* sp. in semiarid soils of the states of Pernambuco and Bahia, Brazilian semiarid region, and observed that there was an increase in shoots and roots dry biomass under water restriction, and pointed out that these microorganisms had great potential for promoting soybean growth as well as mitigating adverse effects imposed by drought.

7.4.2 ARGENTINA

In Latin America, Argentina has the second largest area of dryland, which represents 55% of the country's land area, is responsible for 50% of the agricultural production, 47% of the livestock production, and is home to almost one-third of the country's population (Torres et al., 2015). However, unlike Brazil, the drylands of Argentina are characterized by low temperatures, and are receiving water from snow accumulation over the higher elevations of the Andes, which is an important source of water for the recharge of aquifers and the development of agricultural activities (Rivera et al., 2021). These conditions allow that in the Argentine drylands, it is possible to adopt the no-tillage system using mainly corn, soybean, and wheat crops, but also including cover crops or green manure, such as rye, vetch, sorghum, etc., which does not occur in the Brazilian semiarid region.

Regardless of these differences, studies performed in Argentina also point to the diversity of land use as one of the main ways to make agroecosystems more resilient to droughts. Several studies have shown that practices such as double cropping, intercropping, relay cropping, and fallow, especially under no-till (NT), are efficient in increasing soil carbon stocks, biological activity, and WUE (Farage et al., 2007; Alvarez et al., 2014; Frasier et al., 2016b, 2016a).

For example, Noellemeyer et al. (2013) evaluated 15 years of crop rotation with wheat, corn, sunflower, and soybean under NT and conventional till (CT), and observed that NT improved available water use, yields, and water productivity of all studied crops and contributes to reducing the risk of crop water stress in dryland agriculture. Thus, the authors state that "recommendations for crop management in dryland agriculture should take into account the synergistic effect of good agronomic performance of the crops with better water availability in NT to improve overall water efficiency." In this same sense, Caviglia et al. (2004) found that the sequential and relay wheat–soybean double crops when compared with sole crops in the south-eastern Pampas of Argentina were significantly more efficient in water use, since the fraction of annual precipitation captured by crops varied from 0.26 to 0.51 in sole crops and from 0.53 to 0.71 in double crops. According to Fernandez et al. (2008), in the semiarid central region of Argentina, the probability that rainfall meets crop requirements during growing season is less than 10%, therefore fallowing has been the most important practice to assure water availability during the growing season. These authors evaluated the interactions between residue cover, weed control, soil profile depth, and water storage capacity on fallow efficiency. Overall, they found that weed density was the most important factor that controlled available water content, with the areas without weed control resulting in 58% more water, while the treatment with the highest amount of waste (6,000 kg of dry matter ha^{-1}) presented 26% more water in the soil than the treatment with the lowest amount (2,000 kg of dry matter ha^{-1}). Finally, Frasier et al. (2016b) found that sorghum cultivation combined with cover crops (rye, vetch, and rye–vetch mixture) under NT led to increased stocks of total C as well as the microbial biomass carbon when compared to the monoculture system. Besides, litter cover improved soil moisture to 45–50% water-filled pore space (Figure 7.6) and reduced substantially the soil temperatures (Figure 7.7). These findings support the view that cover crops, specifically legumes, in NT systems can increase soil ecosystem services related to water and carbon storage, habitat for biodiversity, and nutrient availability.

7.4.3 MEXICO

The land area of Mexico covers climatic gradients, from humid to dry climate zones. However, most of Mexico undergoes significant water limitation, since approximately 90% of its surface area experiences an annual deficit of rainfall relative to evaporation demand (Díaz-Padilla et al., 2011), which is particularly prevalent in the extensive months of recurrent droughts (November to May). Moreover, approximately 60% of its surface area falls into arid and semiarid regions (Montaño et al., 2016) where water resources are limited. Correspondingly, plants experience a large period without water availability, except for the moist tropical and temperate zones, where maximal plant

Managing Drought Stress

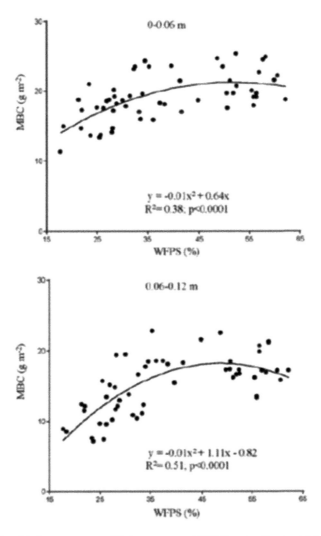

FIGURE 7.6 Relationship between water filled pore space (WFPS) and soil microbial carbon for 0–0.06 m and 0.06–0.12 m depth layers.

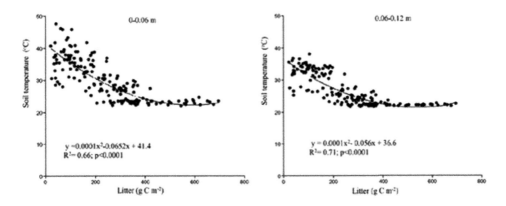

FIGURE 7.7 Relationship between litter carbon and soil temperature at 0–0.06 m and 0.06–0.12 m depth.

growth rates occur (Martínez-Garza et al., 2013). Thus, plant productivity across Mexico is predominantly limited by water, and the rainless period modulates the productivity of both native vegetation (Campo, 2016) and agroecosystems (Figueroa et al., 2020; Abbruzzini et al., 2021). Worrisomely, Mexico, a country where 75% of the agricultural production is rainfed (Conde et al., 2006; Sagarpa et al., 2013), is one of the regions of the world where rainfall is most likely to decrease under climate change (Sánchez et al., 2022).

Substantial precipitation decreases in Mexican drylands have been observed over the past 30 years (Pontifes et al., 2018), and analyses of the climate data of Mexico over the last 30–50 years show consistent patterns of general warming (Magrin et al., 2007). Groisman et al. (2005) found a substantial decrease in precipitation over the Central Plateau of Mexico during the last 30 years. In the same period, the frequency of rain events above 75 mm increased substantially by 110%. Moreover, the extensive land cover changes in this region could have impacted the moisture exchange between the land surface and atmosphere; a quarter of the national agricultural products are cultivated in these drylands, which also accounts for one-third of the national beef production (SAGARPA, 2014). This could be partially responsible for the dramatic escalation of maximum afternoon temperatures recorded during the last 40 years over Central Mexico (Englehart and Douglas, 2005; Pascual et al., 2015). The negative impact of climate on agricultural production was already evident in crop and livestock systems (Stahle et al., 2016; Murray-Tortarolo et al., 2018, 2019).

Climate models predict that, by the end of the 21st century, central Mexico will experience decreases in precipitation amounts of around 25% (Chadwick et al., 2016). In fact, most climate change projections for Mexico report a decrease in rainfall and an increase in mean temperature by mid-century (Murray-Tortarolo et al., 2016), likely leading to longer and more intense droughts. Reductions in rainfall combined with increasing in mean temperature imply lower availability of water and soil moisture, affecting both agricultural and livestock productivities that could put food security under severe threat (Parry et al., 2004; Murray-Tortarolo et al. 2018, 2019).

Despite the limiting conditions given its reliance on seasonal precipitation, rainfed agriculture has been practiced since pre-Columbian times, and it remains today the most important agricultural activity for the majority of subsistence farming in the country (Conde et al., 2006). This type of agriculture has seen rapid growth since the late 1980s (Mendoza-Ponce et al., 2018), due to the expansion of maize and common bean cropping in the tropical arid and semiarid regions of Central Mexico (INEGI, 2019). On the other hand, livestock constitutes the main land use in the country, accounting for 56% of the national territory (Servicio de Información Agroalimentario y Pesquera – SIAP, 2020). Beef production has expanded mainly in the tropical regions of the country (Mendoza-Ponce et al., 2018), encompassing humid, subhumid, and semiarid climates in the southeast of Mexico (Galeana-Pizaña et al., 2018). Together, the Central and South-Southeast regions of Mexico have the greatest agricultural and livestock production in the country (SIAP, 2020), which is predicted to increase by the end of the 21st century (Mendoza-Ponce et al., 2018). The current and expected expansion of rainfed agroecosystems in the tropical regions of the country, together with the decrease in precipitation amount occurring, generate abrupt and large-scale changes in forestlands and alter the regulation of major soil biogeochemical cycles, as was documented for cropping and livestock systems in both regions (i.e., Central and South-Southeast regions).

7.4.3.1 Evidence from Crop Systems in Central Mexico

Regarding the regulation of major soil biogeochemical cycles in rainfed croplands due to climate change, Abbruzzini et al. (2021) assessed metrics of the soil carbon C, N, and P cycles, as well as soil microbial biodiversity, under rainfed maize and common bean cropping systems in arid and semiarid regions (416 and 554 mm of mean annual precipitation) in Central Mexico (see crop management and soil characteristics in Abbruzzini et al., 2021). Decreasing precipitation had a negative effect on soil organic C and total N concentrations of rainfed areas with cultivation (i.e., both maize and common beans). In contrast, differences in soil total P concentrations were nonsignificant

Managing Drought Stress 173

between arid and semiarid regions. An analysis of enzymatic activities showed that both regions are P-limited. The C:N ratios decreased with increasing aridity and reflected N accumulation, suggesting the effects of less soil leaching. Similarly, C:P and N:P ratios showed an abrupt decrease with aridity, indicating that biologically controlled C and N cycles seem to become uncoupled from P biogeochemically controlled cycles in response to increasing aridity, with negative consequences for crop yields. In addition, increased aridity resulted in N depletion in soils under cultivation of maize plants; meanwhile, more efficient N and P cycling were observed in cultivation plots of common beans, showing a striking contrast in the consequences of increased aridity in rainfed agriculture between biological N-fixation species and species without this capacity. Decreasing precipitation also reduced the microbial diversity in soils, suggesting a decrease in soil functional capacity (Wang et al., 2015; Su et al., 2020) with increasing aridity.

7.4.3.2 Evidence from Livestock Systems in the South-Southeast of Mexico

A recent study examined soil C, N, and P dynamics in native forests and different livestock production systems (extensive pasture and intensive silvopastoral systems), across a rainfall gradient from 1611 (humid) to 711 (semiarid) mm of mean annual precipitation, during the dry and rainy seasons in the tropical region of Mexico, to provide an estimate of the sensitivity of land-use change effects and livestock systems to increase in drought (see livestock management and soil characteristics in Figueroa et al., 2020).

The land-use change from native forests to both livestock production systems (i.e., extensive pasture and intensive silvopastoral systems) decreased soil C, N, and P contents, meanwhile increasing the soil basal respiration and the activities of C-acquiring enzymes, which could be related to differences in quantity and quality of C and N inputs through litter between native forests and livestock production systems. Moreover, the activity of P-acquiring enzymes increased significantly in soils under intensive silvopastoral systems compared to soils under natural forests, indicating this intensive livestock system maybe makes P-limitation to soil microbes worse compared to extensive pasture. Finally, no significant changes in soil C, N, and P concentrations, as well as in C and N fluxes, were detected between extensive pasture and intensive silvopastoral systems. Regarding the precipitation regime effects on soils, organic C, total N, total P, and microbial biomass C and N contents were the highest at the driest end of the gradient regardless of the land use. Thus, the expected climate change could have large effects on moisture-sensitive biogeochemical processes, altering future C, N, and P balances in soils of both native forests and agroecosystems, with potentially negative consequences for their sustainability in these tropical regions.

7.5 CONCLUSIONS

Living with droughts has been and will always be a great challenge, especially in drylands, which are regions that naturally have water scarcity. Climate change comes to impose extra challenges, as the trend is for extreme events to become more frequent and more intense, and in the case of LAC drylands, some studies have shown that the aridity index has increased in recent decades, a trend that is confirmed by models that point out that it is the 2071–2100 period relative to 1961–1990 (using the UNEP AI and under RCP8.5 scenario) that there will be an increase of arid and semiarid conditions in central Mexico and isolated regions of southern Peru and Central America, Northeast Brazil, and in Northern Argentina. These climatic fluctuations may be most pronounced in the poorest regions with high levels of chronic undernourishment and a great degree of instability.

Regarding water security, some aspects should be highlighted. First, the need to seek ways to improve the regularity of water availability in LAC drylands, which can be done via surface reservoirs for water storage during the rainy months, for usage during the dry periods, or wells for exploiting groundwater, when it is available. Another point is the adoption of agroecosystems adaptation plans, which should consider water and soil availability and management, WUE and demands, unconventional water sources, and drought stress management. Adaptation planning is urgent, as studies have shown that some systems are reaching their limits to adaptation, mainly for smallholder farmers in Central and South America, which have reached soft limits to adaptation, and inequity and poverty also constrain adaptation, leading to soft limits and resulting in disproportionate exposure and impacts for most vulnerable groups. In this context, climate and hydrological services are essential to contribute to water security in the region, as they can provide appropriate information on efficient water and land use, and infrastructure management which includes, for example, data on the seasonal forecasting of rainfall, runoff, reservoir storage, and aquifer recharge, data that can be used to plan and implement drought preparedness measures. Finally, the water governance in LAC despite being institutionally organized in most countries is fragmented, lacking improvements in policy coherence, integrity, transparency, stakeholder involvement, and financing. Therefore, the governance of water and territory, and associated policies and management systems, needs stronger integration to other sectoral policies, such as agricultural and energy policies. Furthermore, securing investments and appropriate governance for adaptation, such as improving drought stress management, infrastructure upgrading, and operational optimization, are necessary for improving water security in the medium to long term. The continuity of "business-as-usual" in technological development and appropriation, sources of funding, and resource allocation, without the empowerment of vulnerable smallholder farmers of LAC drylands, will not lead to the desired adaptation goals.

Details on the management systems and practices of the case studies can be seen in the works cited, as well as in several others available. However, our proposal is not that such systems and/or techniques are simply copied or replicated, but rather to demonstrate that certain principles can be the basis for the adoption of more sustainable and resilient agroecosystems in the face of droughts and climate change in different drylands of Latin America. These principles are the combination of crop and animal diversity with reduced and conservationist soil tillage. This combination, considering the soil, climate, management, and social specificities of each region, can result in the maintenance or even increase of soil C stocks, which means an increase in soil organic matter with all the benefits associated; an increase in soil fauna, with emphasis on microorganisms, which can contribute to adaptation mechanisms to water restriction in plants; improvement of the physical and hydric conditions of the soils, resulting in greater retention of humidity, greater porosity, and lower soil density; and improvement of soil fertility by providing certain elements, such as potassium, selenium, and silicon, which can play important roles in drought resistance and adaptation.

In addition to the environmental benefits described above, more diversified production systems can also result in better productivity rates. For example, some studies on intercropping have found significant yield advantages of intercropping compared with simple cropping, with a land equivalent ratio of up to 1.34 (Ghaley et al. 2005). Oliveira and de Araújo (2008) found that in the Brazilian semiarid region, the land equivalent ratio in an intercropping with maize, bean, and cottom was 3.8. Otherwise, Knapp and van der Heijden (2018) performed a meta-analysis to assess temporal yield stability (i.e., the variability of yield across years) of organic agriculture and conservation agriculture (no-tillage) versus conventional agriculture and found that organic agriculture has, per unit yield, a significantly lower temporal stability (−15%) compared to conventional agriculture. However, the authors also found that the use of green manure and enhanced fertilization can reduce the yield stability gap between organic and conventional agriculture. It is worth noting that the diversification of production also means greater diversity of income and food supply for producers, helping to reduce exposure to droughts and improve food security in the region.

Despite all the benefits, the adoption of these agricultural management systems or practices still faces several barriers. Its diversity also means greater implementation and management complexity for farmers, in addition to demanding greater investment in some cases (at least in the initial years). In this sense, it is essential to have specialized technical assistance, as well as appropriate financial incentives, which involves the establishment of specific, long-term public policies that consider monitoring and evaluation mechanisms, and that subsidize producers not only in terms of soil and agroecosystem management but also in the use of other technologies mentioned here. Perhaps, a viable path is an integration between governmental agents, such as higher education institutions (universities and institutes etc.), technical assistance companies, nongovernmental organizations, and even private companies. If all these actors act in an integrated manner and on a conservationist basis, the repercussions of the actions are inevitable in the sense of a better coexistence with droughts and climate change.

REFERENCES

Abbruzzini, T.F., Avitia, M., Carrasco-Espinosa, K., Peña, V., Barrón-Sandoval, A., Salazar Cabrera, U.I., Cruz-Ortega, R., Benítez, M., Escalante, A.E., Rosell, J.A., Wegier, A., Campo, J., 2021. Precipitation controls on soil biogeochemical and microbial community composition in rainfed agricultural systems in tropical drylands. *Sustainability* 13, 11848. https://doi.org/10.3390/ su132111848

Adeyemi, O., Keshavarz-Afshar, R., Jahanzad, E., Battaglia, M.L., Luo, Y., Sadeghpour, A., 2020. Effect of wheat cover crop and split nitrogen application on corn yield and nitrogen use efficiency. *Agronomy* 10, 1081. https://doi.org/10.3390/agronomy10081081

Aguiar, M.I.D., Fialho, J.S., Campanha, M.M., Oliveira, T.S., 2014. Carbon sequestration and nutrient reserves under different land use systems. *Rev. Árvore* 38, 81–93. https://doi.org/10.1590/S0100-6762201400 0100008

Aguiar, M.I.D., Fialho, J.S., Campanha, M.M., Oliveira, T.S., 2019. Florística e estrutura vegetal em áreas de Caatinga sob diferentes sistemas de manejo. *Pesqui. Florest. Bras.* 39. https://doi.org/10.4336/2019. pfb.39e201801715

Akhmouch, A., 2012. Water governance in Latin America and the Caribbean: A multi-level approach, OECD Regional Development Working Papers, 2012/04, OECD Publishing, Paris.

Almeida, M.V.R. de, Oliveira, T.S. de, Bezerra, A.M.E., 2009. Biodiversidade em sistemas agroecológicos no município de Choró, CE, Brasil. *Ciênc. Rural* 39, 1080–1087. https://doi.org/10.1590/S0103-847820 09005000047

Alvarez, C., Alvarez, C.R., Costantini, A., Basanta, M., 2014. Carbon and nitrogen sequestration in soils under different management in the semi-arid Pampa (Argentina). *Soil Tillage Res.* 142, 25–31. https://doi.org/ 10.1016/j.still.2014.04.005

Aon Benfield UCL Hazard Research Centre, 2019. Weather, Climate & Catastrophe Insight—2019 Annual Report. https://reliefweb.int/organization/bhrc. (accessed August 2022).

Aon Benfield UCL Hazard Research Centre, 2020. Weather, Climate & Catastrophe Insight—2020 Annual Report. https://reliefweb.int/organization/bhrc. (accessed August 2022).

AR5 Climate Change, 2014. Impacts, Adaptation, and Vulnerability—IPCC, n.d. www.ipcc.ch/report/ar5/wg2/ (accessed 16 September 2022).

Araújo Girão, A.L. de, Oliveira, R.T. de, Ferreira, T.O., Ortiz-Escobar, M.E., Miranda, F.R., Oliveira, T.S. de, 2015. Assessment of soil moisture by family farmers under multi-cropping systems in a semiarid region. *Agroecol. Sustain. Food Syst.* 39, 747–761. https://doi.org/10.1080/21683565.2015.1029602

Araújo Girão, A.L. de, Oliveira, R.T. de, Ferreira, T.O., Romero, R.E., Oliveira, T.S. de, 2013. Evaluation of soil structure using participatory methods in the semiarid region of Brazil. *Rev. Ciênc. Agronômica* 44, 411–418. https://doi.org/10.1590/S1806-66902013000300001

Bonfils, C. J. et al., 2020. Human influence on joint changes in temperature, rainfall and continental aridity. *Nat. Clim. Change* 10, 726–731.

Braga, A.P.A., Cruz, J.M., de Melo, I.S., 2022. Rhizobacteria from Brazilian semiarid biome as growth promoters of soybean (Glycine max L.) under low water availability. *Braz. J. Microbiol.* 53, 873–883. https://doi.org/10.1007/s42770-022-00711-7

Brooks, T.M., Mittermeier, R.A., da Fonseca, G.a.B., Gerlach, J., Hoffmann, M., Lamoreux, J.F., Mittermeier, C.G., Pilgrim, J.D., Rodrigues, A.S.L., 2006. Global biodiversity conservation priorities. *Science* 313, 58–61. https://doi.org/10.1126/science.1127609

Campiglia, E., Mancinelli, R., De Stefanis, E., Pucciarmati, S., Radicetti, E., 2015. The long-term effects of conventional and organic cropping systems, tillage managements and weather conditions on yield and grain quality of durum wheat (Triticum durum Desf.) in the Mediterranean environment of Central Italy. *Field Crops Res.* 176, 34–44. https://doi.org/10.1016/j.fcr.2015.02.021

Campo, J., 2016. Shift from ecosystem P to N limitation at precipitation gradient in tropical dry forests at Yucatan, Mexico. *Environmental Research Letters* 11, 095006. https://doi.org/10.1088/1748-9326/11/9/095006

Carvalho, A.L. de, Santos, D.V., Marengo, J.A., Coutinho, S.M.V., Maia, S.M.F., 2020. Impacts of extreme climate events on Brazilian agricultural production. *Sustentabilidade Em Debate* 11, 197–224. https://doi.org/10.18472/SustDeb.v11n3.2020.33814

Caviglia, O.P., Sadras, V.O., Andrade, F.H., 2004. Intensification of agriculture in the south-eastern Pampas. *Field Crops Res.* 87, 117–129. https://doi.org/10.1016/j.fcr.2003.10.002

Chadwick, R., Good, P., Martin, G., Rowell, D.P., 2016. Large rainfall changes consistently projected over substantial areas of tropical land. *Nature Climate Change* 6, 177–181. https://doi.org/10.1038/nclimate2805

Chai, R., Mao, J., Chen, H. et al., 2021. Human-caused long-term changes in global aridity. *npj Clim Atmos Sci* 4, 65 (2021). https://doi.org/10.1038/s41612-021-00223-5

Clarkson, G., Dorward, P., Poskitt, S., Stern, R.D., Nyirongo, D., Fara, K., Giraldo, D., 2022. Stimulating small-scale farmer innovation and adaptation with Participatory Integrated Climate Services for Agriculture (PICSA): Lessons from successful implementation in Africa, Latin America, the Caribbean and South Asia. *Climate Services*, 26, 100298.

Coelho, C.A.S., Stephenson, D.B., Balmaseda, M., Doblas-Reyes, F.J., Van Oldenborgh, G.J., 2006. Toward an integrated seasonal forecasting system for South America. *Journal of Climate* 19(15), 3704–3721.

Conde, A.C., Ferrer, R., Orozco, S., 2006. Climate change and climate variability impacts on rainfed agricultural activities and possible adaptation measures. A Mexican case study. Atmósfera 19, 181–194. www.revistascca.unam.mx/atm/index.php/atm/article/view/8559

Cunha, A.P.M.A., Zeri, M., Deusdará Leal, K., Costa, L., Cuartas, L.A., Marengo, J.A., Tomasella, J., Vieira, R.M., Barbosa, A.A., Cunningham, C., Cal Garcia, J.V., Broedel, E., Alvalá, R., Ribeiro-Neto, G., 2019. Extreme drought events over Brazil from 2011 to 2019. *Atmosphere* 10, 642. https://doi.org/10.3390/atmos10110642

Davies, J., Poulsen, L., Schulte-Herbrüggen, B., Mackinnon, K., Crawhall, N., Henwood, W.D., Dudley, N., Smith, J., Gudka, M., 2012. Conserving dryland biodiversity. xii +84p, Available from IUCN (International Union for the Conservation of Nature), Nairobi, Kenya.

de Aguiar, M.I., Maia, S.M.F., Xavier, F.A. da S., de Sá Mendonça, E., Filho, J.A.A., de Oliveira, T.S., 2010. Sediment, nutrient and water losses by water erosion under agroforestry systems in the semi-arid region in northeastern Brazil. *Agrofor. Syst.* 79, 277–289. https://doi.org/10.1007/s10457-010-9310-2

De Nys, E., Magalhães, A.R., Engle, N.L., 2017. *Drought in Brazil: Proactive Management and Policy*, CRC Press, Boca Raton, FL.

Díaz-Padilla, G., Sánchez-Cohen, I., Guajardo-Panes, R.A., Del Ángel-Pérez, A.L., Ruíz-Corral, A., Medina-García, G., Ibarra-Castillo, D., 2011. Mapping of the aridity index and its population distribution in Mexico. *Revista Chapingo Serie Ciencias Forestales y Ambientales* 17, 267–275. https://doi.org/10.5154/r.rchscfa.2010.09.069

Donatti, C.I., Harvey, C.A., Martinez-Rodriguez, M.R., Vignola, R., Rodriguez, C.M., 2019. Vulnerability of smallholder farmers to climate change in Central America and Mexico: Current knowledge and research gaps. *Climate and Development* 11, 264–286. https://doi.org/10.1080/17565529.2018.1442796

ECLAC, United Nations. Economic Commission for Latin America and the Caribbean, 2017. Santiago, Chile: Social panorama of Latin America.

Englehart, P.J., Douglas, A.V., 2005. Changing behavior in the diurnal range of surface air temperatures over Mexico. *Geophysical Research Letters* 32, 1–4. https://doi.org/10.1029/2004GL021139

Escada, P., Coelho, C.A.S., Taddei, R., Dessai, S., Cavalcanti, I.F.A., Donato, R., Kayano, M., Martins, E.S.P.R., Miguel, J.C.H., Monteiro, M., Moscati, M.C.L., 2021. Climate services in Brazil: Past, present and future perspectives. *Climate Services* 24, 100276–100279.

FAO – Food and Agriculture Organization of the United Nations, 2015. Regional overview of food insecurity. Santiago, Chile: Latin America and the Caribbean.

FAO – Food, Agriculture Organization of the United Nations, 2018. World food situation. www.fao.org/worldfoodsituation/foodpricesindex/en/ (accessed September 2022).

FAO – Food and Agriculture Organization of the United Nations, 2019. Trees, Forests and Land Use in Drylands: The First Global Assessment: Full Report; FAO Forestry Paper No. 184; FAO: Rome, Italy, 2019; ISBN 978-92-5-131999-4.

FAO – Dryland Forest, 2022a. www.fao.org/dryland-forestry/background/what-are-drylands/en/. (accessed August 2022).

FAO, IFAD, UNICEF, WFP and WHO, 2022b. The State of Food Security and Nutrition in the World. Repurposing food and agricultural policies to make healthy diets more affordable. Rome, FAO. https://doi.org/10.4060/cc0639en

Farage, P., Ardo, J., Olsson, L., Rienzi, E., Ball, A., Pretty, J., 2007. The potential for soil carbon sequestration in three tropical dryland farming systems of Africa and Latin America: A modelling approach. *Soil Tillage Res.* 94, 457–472. https://doi.org/10.1016/j.still.2006.09.006

Fernandez, R., Quiroga, A., Noellemeyer, E., Funaro, D., Montoya, J., Hitzmann, B., Peinemann, N., 2008. A study of the effect of the interaction between site-specific conditions, residue cover and weed control on water storage during fallow. *Agric. Water Manag.* 95, 1028–1040. https://doi.org/10.1016/j.agwat.2008.03.010

Fialho, J.S., de Aguiar, M.I., Magalhães, R.B., de Oliveira, T.S., n.d. Soil quality, resistance and resilience in traditional agricultural and agroforestry ecosystems in Brazil's semiarid region 12.

Fialho, J.S., Primo, A.A., Aguiar, M.I. de, Magalhães, R.B., Maia, L. dos S., Correia, M.E.F., Campanha, M.M., Oliveira, T.S. de, 2021. Pedofauna diversity in traditional and agroforestry systems of the Brazilian semi-arid region. *J. Arid Environ.* 184, 104315. https://doi.org/10.1016/j.jaridenv.2020.104315

Figueroa, D., Ortega-Fernández, P., Abbruzzini, T.F., Rivero-Villlar, A., Galindo, F., Chavez-Vergara, B., Etchevers, J.D., Campo, J., 2020. Effects of land use change from natural forest to livestock on soil C, N and P dynamics along a rainfall gradient in Mexico. *Sustainability* 12, 8656. https://doi.org/10.3390/su12208656.

Frasier, I., Noellemeyer, E., Figuerola, E., Erijman, L., Permingeat, H., Quiroga, A., 2016a. High quality residues from cover crops favor changes in microbial community and enhance C and N sequestration. *Glob. Ecol. Conserv.* 6, 242–256. https://doi.org/10.1016/j.gecco.2016.03.009

Frasier, I., Quiroga, A., Noellemeyer, E., 2016b. Effect of different cover crops on C and N cycling in sorghum NT systems. *Sci. Total Environ.* 562, 628–639. https://doi.org/10.1016/j.scitotenv.2016.04.058

Galeana-Pizaña, J.M., Couturier, S., Monsivais-Huertero, A., 2018. Assessing food security and environmental protection in Mexico with a GIS-based Food Environmental Efficiency index. *Land Use Policy* 76, 442–454. https://doi.org/10.1016/j.landusepol.2018.02.022

Ganem, K.A., Xue, Y., Rodrigues, A.d.A., Franca-Rocha, W., Oliveira, M.T.d., Carvalho, N.S.d., Cayo, E.Y.T., Rosa, M.R., Dutra, A.C., Shimabukuro, Y.E., 2022. Mapping South America's drylands through remote sensing—A review of the methodological trends and current challenges. *Remote Sens* 14, 736. https://doi.org/10.3390/rs14030736

Ghaley, B.B., Hauggaard-Nielsen, H., Jensen, H.H., Jensen, E.S., 2005. Intercropping of wheat and pea as influenced by nitrogen fertilization. *Nutrient Cycling in Agroecosystems* 73(2), 201–212. https://10.1007/s10705-005-2475-9.

Groisman, P.Y., Knight, R.W., Easterling, D.R., Karl, T.R., Hegerl, G.C., Razuvaev, V.N., 2005. Trends in intense precipitation in the climate record. *Journal of Climate* 18, 1326–1350. https://doi.org/10.1175/JCLI3339.1

Huang, J., Ji, M., Xie, Y., Wang, S., He, Y., Ran, J., 2016a. Global semi-arid climate change over last 60 years. *Clim. Dyn.* 46(3–4), 1131–1150, https://doi.org/10.1007/s00382-015-2636-8

Huang, J., Yu, H., Guan, X. et al., 2016b. Accelerated dryland expansion under climate change. *Nature Clim Change* 6, 166–171. https://doi.org/10.1038/nclimate2837

Huang, J. et al., 2017. Dryland climate change: Recent progress and challenges. *Rev. Geophys* 55, 719–778. https://doi.org/10.1002/2016RG000550

Instituto Nacional de Geografía Estadística (INEGI), 2019. Encuesta Nacional Agropecuaria 2019. Accessed at: www.inegi.org.mx/programas/ena/2019/

IPCC, 2014: *Climate Change 2014: Impacts, Adaptation, and Vulnerability. Part A:Global and Sectoral Aspects. Contribution of Working Group II to the Fifth Assessment Report of the Intergovernmental Panel on Climate Change.* C.B. Field,., V.R. Barros, D.J. Dokken, K.J. Mach, M.D. Mastrandrea, T.E. Bilir, M. Chatterjee, K.L. Ebi, Y.O. Estrada, R.C. Genova, B. Girma, E.S. Kissel, A.N. Levy, S. MacCracken, P.R. Mastrandrea, and L.L. White (eds.). Cambridge University Press, Cambridge, UK and New York, NY, USA, 1132 pp. https://www.ipcc.ch/report/ar5/wg2/

IPCC, 2022. Summary for policymakers. In: *Climate Change 2022: Impacts, Adaptation and Vulnerability.* Contribution of Working Group II to the Sixth Assessment Report of the Intergovernmental Panel on Climate Change. Cambridge University Press, Cambridge and New York, pp. 3–33.

Knapp, S., van der Heijden, M.G.A., 2018. A global meta-analysis of yield stability in organic and conservation agriculture. *Nature Communications* 9:3632. https:// 10.1038/s41467-018-05956-1.

Koutroulis, A.G., 2019. Dryland changes under different levels of global warming. *Sci. Total Environ.* 655, 482–511.

Laura, F., Tamara, A., Müller, A., Hiroshan, H., Christina, D., Serena, C., 2020. Selecting sustainable sewage sludge reuse options through a systematic assessment framework: Methodology and case study in Latin America. *Journal of Cleaner Production*, 242, 118389.

Lentini, E., 2022. Building a water security agenda for Latin America and the Caribbean 2030. CAF, Caracas.

Lima, H.V. de, Oliveira, T.S. de, Oliveira, M.M. de, Mendonça, E. de S., Lima, P.J.B.F., 2007. Indicadores de qualidade do solo em sistemas de cultivo orgânico e convencional no semi-árido cearense. *Rev. Bras. Ciênc. Solo* 31, 1085–1098. https://doi.org/10.1590/S0100-06832007000500024

Lobato-Sánchez, R., 2016. El monitor de sequía en México (The drought monitor in Mexico). *Tecnología y Ciencias del Agua* 7(5), 197–211.

Magrin, G., Gay, G.C., Cruz, C.D., Giménez, J.C., Moreno, A.R., Nagy, G.J., Nobre, C., Villamizar, A., 2007. Latin America. In: Parry, M.L, Canziani, O.F., Palutikof, J.P., van der Linden, P.J., Hanson, C.E. (Eds.), *Climate Change 2007: Impacts, Adaptation and Vulnerability.* Contribution of Working Group II to the Fourth Assessment Report of the Intergovernmental Panel on Climate Change. Cambridge University Press, Cambridge, pp. 581–615.

Mahlknecht, J., González-Bravo, R., Loge, F.J., 2020. Water-energy-food security: A nexus perspective of the current situation in Latin America and the Caribbean. *Energy* 194, 116824. https://doi.org/10.1016/j.energy.2019.116824

Maia, S.M.F., Otutumi, A.T., Mendonça, E. de S., Neves, J.C.L., Oliveira, T.S. de, 2019. Combined effect of intercropping and minimum tillage on soil carbon sequestration and organic matter pools in the semiarid region of Brazil. *Soil Res.* 57, 266. https://doi.org/10.1071/SR17336

Maia, S.M.F., Xavier, F.A.S., Oliveira, T.S., Mendonça, E.S., Araújo Filho, J.A., 2007. Organic carbon pools in a Luvisol under agroforestry and conventional farming systems in the semi-arid region of Ceará, Brazil. *Agrofor. Syst.* 71, 127–138. https://doi.org/10.1007/s10457-007-9063-8

Marengo, J.A.; Bernasconi, M., 2015. Regional differences in aridity/drought conditions over Northeast Brazil: present state and future projections. *Climatic Change*, 129, 103–115. https://doi.org/10.1007/s10584-014-1310-1.

Marengo, J.A., Torres, R.R., Alves, L.M., 2017. Drought in northeast Brazil—Past, present, and future. *Theor. Appl. Climatol.* 129, 1189–1200. https://doi.org/10.1007/s00704-016-1840-8

Martínez-Garza, C., Tobon, W., Campo, J., Howe, H.F., 2013. Drought mortality of tree seedlings in an eroded tropical pasture. *Land Degradation and Development* 24, 287–295. https://doi.org/10.1002/ldr.1127

Mendes, M.M. de S., Lacerda, C.F. de, Fernandes, F.É.P., Cavalcante, A.C.R., Oliveira, T.S. de, 2013. Ecophysiology of deciduous plants grown at different densities in the semiarid region of Brazil. *Theor. Exp. Plant Physiol.* 25, 94–105. https://doi.org/10.1590/S2197-00252013000200002

Mendoza-Ponce, A., Corona-Núñez, R., Kraxner, F., Leduc, S., Patrizio, P., 2018. Identifying effects of land use cover changes and climate change on terrestrial ecosystems and carbon stocks in Mexico. *Global Environmental Change* 53 12–23. 10.1016/j.gloenvcha.2018.08.004

Montaño, N.M., Ayala, F., Bullock, S.H., Briones, O., García-Oliva, F., García-Sánchez, R., Maya, Y., Perroni, Y., Siebe C., Tapia-Torres, Y., 2016. Almacenes y flujos de carbono en ecosistemas áridos y semiáridos de México: Síntesis y perspectivas. *Revista Terra Latinoamericana* 34, 39–59. www.terralatinoamericana.org.mx/index.php/terra/article/view/75

Murray-Tortarolo, G.N., Friedlingstein, P., Sitch, S. et al., 2016. The carbon cycle in Mexico: past, present and future of C stocks and fluxes. *Biogeosciences* 13, 223–238. http://dx.doi.org/10.5194/bg-13-223-2016

Murray-Tortarolo, G.N., Jaramillo, V.J., 2019. The impact of extreme weather events on livestock populations: The case of the 2011 drought in Mexico. *Climatic Change* 153, 79–89. https://doi.org/10.1007/s10584-019-02373-1

Murray-Tortarolo, G.N., Jaramilo, V.J., Larsen, J., 2018. Food security and climate change: The case of rainfed maize production in Mexico. *Agricultural and Forest Meteorology* 253–254, 124–131. https://doi.org/10.1016/j.agrformet.2018.02.011

National Academies of Sciences, Engineering, and Medicine, 2018. Advancing Sustainability of U.S.-Mexico Transboundary Drylands: Proceedings of a Workshop. The National Academies Press, Washington, DC. https://doi.org/10.17226/25253

Neto, S., Camkin, J., Fenemor, A., Tan, P.-L., Baptista, J.M., Ribeiro, M.M.R., Schulze, R., Stuart-Hill, S., Spray, C., Elfithri, R., 2018. OECD principles on water governance in practice: An assessment of existing frameworks in Europe, Asia-Pacific, Africa and South America. *Water International*, 43, 60–89.

Nickl, E., Millones, M., Parmentier, B., Lucatello, S., Trejo, A., 2020. Drylands, Aridification, and Land Governance in Latin America: A Regional Geospatial Perspective, Chapter 16. In: Lucatello, S. et al. (Eds.), *Stewardship of Future Drylands and Climate Change in the Global South*. Springer International Publishing.

Nobre, C.A., Marengo, J.A., Seluchi, M.E., Cuartas, A., Alves, L.M., 2016. Some characteristics and impacts of the drought and water crisis in Southeastern Brazil during 2014 and 2015. *Journal of Water Resource and Protection*, 8, 252–262. https://10.4236/jwarp.2016.82022

Noellemeyer, E., Fernández, R., Quiroga, A., 2013. Crop and tillage effects on water productivity of dryland agriculture in Argentina. *Agriculture* 3, 1–11. https://doi.org/10.3390/agriculture3010001

Ocampo-Melgar, A., Lutz-Ley, A., Zúñiga, A., Cerda, C., Goirán, S., 2022. Latin American drylands. Challenges and opportunities for sustainable development. *Metode Science Studies Journal*. University of Valencia. https://doi.org/10.7203/metode.13.21458

OECD/IEA, 2022. Organization for Economic Cooperation and Development/International Energy Agency. IEA Statistics 2014. https://www.iea.org/data-and-statistics (accessed September 2022).

Oliveira, T.S and Araújo, A.L., 2008. Essa terra dá mais legume: construindo a qualidade do solo no Sertão Central do Ceará. *Revista Agriculturas* 5 (3), 1–6. http://aspta.org.br/revista/v5-n3-manejo-sadio-dos-solos/.

Parry, M.L., Rosenzweig, C., Iglesias, A., Livermore, M., Fischer, G., 2004. Effects of climate change on global food production under SRES emissions and socio-economic scenarios. *Global Environmental Change* 14, 53–67. https://doi.org/10.1016/j.gloenvcha.2003.10.008

Pascual, R., Albanil, A., Vazquez, J.L., 2015. Mexico. Regional climates. In: Blunden, J., Arndt, D.S. (Eds.), *State of the Climate in 2014*. *Bulletin of the American Meteorological Society* 96, S172–S173.

Pendrill, F., Persson, U.M., 2017. Combining global land cover datasets to quantify agricultural expansion into forests in Latin America: Limitations and challenges. *PLOS ONE* 12, e0181202. https://doi.org/10.1371/journal.pone.0181202

Pérez-Marin, A.M., Rogé, P., Altieri, M.A., Ulloa Forero, L.F., Silveira, L., Oliveira, V.M., Domingues-Leiva, B.E., 2017. Agroecological and Social Transformations for Coexistence with Semi-Aridity in Brazil. *Sustainability* 9, 990. https://doi:10.3390/su9060990

Pontifes, P.A., García-Meneses, P.M., Gómez-Aíza, L., Monterroso-Rivas, A.I., Caso-Chávez, M., 2018. Land use/land cover change and extreme climatic events in the arid and semi-arid ecoregions of Mexico. *Atmósfera* 31, 355–372. https://doi.org/10.20937/ATM.2018.31.04.04

Primo, A.A., Araújo Neto, R.A., Zeferino, L.B., Fernandes, F.E.P., Araújo Filho, J.A., Cerri, C.E.P., Oliveira, T.S., 2022. Traditional management and permanent or rotation agroforestry systems: A comparative study for C sequestration by century model simulation. *J Environ Manag* (unpublished manuscript).

Rêgo, J.C., Albuquerque, J.P.T., Pontes Filho, J.D.A., Tsuyuguchi, B.B., Souza, T.J., Galvão, C.O., 2022. Sustainable and resilient exploitation of small alluvial aquifers in the Brazilian semi-arid region. In: Re, V., Manzione, R.L., Abiye, T.A., Mukherji, A., MacDonald, A. (Eds.), *Groundwater for Sustainable Livelihoods and Equitable Growth*. CRC Press, Boca Raton, FL, pp. 101–121.

Rivera, J.A., Otta, S., Lauro, C., Zazulie, N., 2021. A decade of hydrological drought in Central-Western Argentina. *Front. Water* 3, 640544. https://doi.org/10.3389/frwa.2021.640544

Sánchez, H.U.R., Fajardo A.L.M, Ortiz, A.D.B., De la Torre, O.V., 2022. The agricultural sector and climate change in Mexico. *Journal of Agriculture and Ecology Research International* 23, 19–44, 2022. https://doi.org/10.9734/JAERI/2022/v23i330222

Secretaría de Agricultura, Ganadería, Desarrollo Rural, Pesca y Alimentación (SAGARPA)., 2013. Programa Sectorial de Desarrollo Agropecuario, Pesquero y Alimentario. Diario Oficial de la Federación, México, 51 pp.

Secretaría de Agricultura, Ganadería, Desarrollo Rural, Pesca y Alimentación (SAGARPA)., 2014. Atlas de las zonas áridas de México. Servicio de Información Agroalimentaria y Pesquera, Secretaria de Agricultura, Ganadería, Desarrollo Rural y Pesca y Alimentación, México, 156 pp.

Seleiman, M.F., Al-Suhaibani, N., Ali, N., Akmal, M., Alotaibi, M., Refay, Y., Dindaroglu, T., Abdul-Wajid, H.H., Battaglia, M.L., 2021. Drought stress impacts on plants and different approaches to alleviate its adverse effects. *Plants* 10, 259. https://doi.org/10.3390/plants10020259

Servicio de Información Agroalimentario y Pesquera (SIAP)., 2020. Panorama Agroalimentario 2020, México, 200 pp.

Silva, G.L., Lima, H.V., Campanha, M.M., Gilkes, R.J., Oliveira, T.S., 2011. Soil physical quality of Luvisols under agroforestry, natural vegetation and conventional crop management systems in the Brazilian semi-arid region. *Geoderma* 167–168, 61–70. https://doi.org/10.1016/j.geoderma.2011.09.009

Sousa, A.F., 2006. Indicadores de sustentabilidade em sistemas agroecológicos por agricultores familiares do semi-árido cearense 104.

Stahle, D.W., Cook, E.R., Burnette, D.J. et al., 2016. The Mexican Drought Atlas: Tree-ring reconstructions of the soil moisture balance during the late pre-Hispanic, colonial, and modern eras. *Quaternary Science Reviews* 149, 34–60. http://dx.doi.org/10.1016/j.quascirev.2016.06.018

Su, X., Su, X., Yang, S., Zhou, G., Ni, M., Wang, C., Qin, H., Zhou, X., Deng, J., 2020. Drought changed soil organic carbon composition and bacterial carbon metabolizing patterns in a subtropical evergreen forest. *Science of the Total Environment* 736, 139568. https://doi.org/10.1016/j.scitotenv.2020.139568

Torres, L., Abraham, E.M., Rubio, C., Barbero-Sierra, C., Ruiz-Pérez, M., 2015. Desertification Research in Argentina. *Land Degrad. Dev.* 26, 433–440. https://doi.org/10.1002/ldr.2392

Tsuyuguchi, B.B., Morgan, E., Rêgo, J.C., Galvão, C.O., 2020. Governance of alluvial aquifers and community participation: a social-ecological systems analysis of the Brazilian semi-arid region. *Hydrogeology Journal*, 02160–02168.

UNCCD, 2022 – United Nations Convention to Combat Desertification, 2022. *The Global Land Outlook*, second edition. UNCCD, Bonn.

Wang, L., Manzoni, S., Ravi, S., Riveros-Iregui, D., Caylor, K., 2015. Dynamic interactions of ecohydrological and biogeochemical processes in water-limited systems. *Ecosphere* 6, 1–27. https://doi.org/10.1890/ES15-00122.1

WB, 2022 – World Bank 2022. World development indicators. https://data.worldbank.org/indicator/SP.POP.TOTL?locations=ZJ (accessed September 2022).

Wilhite, D.A. 2000. Drought as a natural hazard: concepts and definitions. In *Drought: A Global Assessment*, A. Donald and A. Wilhite (eds.). Routledge: New York, NY.

Xavier, F.A. da S., de Oliveira, T.S., Andrade, F.V., de Sá Mendonça, E., 2009. Phosphorus fractionation in a sandy soil under organic agriculture in Northeastern Brazil. *Geoderma* 151, 417–423. https://doi.org/10.1016/j.geoderma.2009.05.007

Xavier, F.A. da S., Maia, S.M.F., Oliveira, T.S. de, Mendonça, E. de S., 2006. Biomassa microbiana e matéria orgânica leve em solos sob sistemas agrícolas orgânico e convencional na Chapada da Ibiapaba – CE. *Rev. Bras. Ciênc. Solo* 30, 247–258. https://doi.org/10.1590/S0100-06832006000200006

8 Conservation Agriculture
Water Use Efficiency in Dryland Agriculture

D.C. Reicosky
Soil Scientist, Emeritus ARS-USDA, Morris, MN
Email: don.reicosky@gmail.com

CONTENTS

8.1 Introduction 182
    8.1.1 Agriculture and Freshwater Use 182
    8.1.2 Climate Change, Environment, and Ecosystem Services 182
    8.1.3 Objective – To Describe and Review Benefits of Conservation Agriculture Increasing Water Use Efficiency in Dryland Agriculture.................................. 182
8.2 Conservation Agriculture as Sustainable Management of Water Resources...... 183
    8.2.1 Efficient Use of Water 183
    8.2.2 Dryland Agriculture 184
8.3 Impact of Carbon Management Practices on Hydrological Processes................ 185
    8.3.1 Soil Carbon Management Terminology... 187
    8.3.2 Soil Structure – Hydrologic Characteristics 188
    8.3.3 Role of Biology in Soil Physical Structure 190
    8.3.4 Tillage Impacts on Soil Structure and Carbon Emissions.................... 192
    8.3.5 Soil Health Metaphor.. 199
    8.3.6 Role of Cover Crops in Balancing Carbon and Water Cycles 200
    8.3.7 Soil Functions and Carbon.. 202
8.4 Conservation Agriculture is Carbon Centric Increasing Water Use Efficiency.... 203
    8.4.1 Water Use Efficiency in Dryland Agriculture 204
    8.4.2 Tillage Impacts on Water Use Efficiency... 204
    8.4.3 Soil Evaporation.. 205
    8.4.4 Evapotranspiration 206
    8.4.5 Infiltration – Compaction Effects... 207
    8.4.6 Water Storage – Pore Space and Bio-pores, and "Sponge Effect" 208
    8.4.7 Runoff/Erosion.. 209
8.5 Conservation Agriculture Benefits 210
    8.5.1 Mitigate Climate Extremes 212
    8.5.2 Enhanced Ecosystem Services.. 213
    8.5.3 Water Quality 214
8.6 Summary and Conclusions... 215
References.. 217

DOI: 10.1201/b22954-8

8.1 INTRODUCTION

8.1.1 AGRICULTURE AND FRESHWATER USE

Agriculture must face the challenges of climate change, which is causing an increase in the severity and frequency of extreme weather events and natural disasters that have the greatest impact on global arid and semiarid regions. Climate change, in combination with the expanding human population, presents a formidable food security challenge. Environmental issues, such as desertification, deforestation, erosion, degradation of water quality, and depletion of water resources related to conventional tillage (CT) farming, are complicating the challenge of food security.

Soil water is key to understanding the land's surface and all the activities that occur there, both seen and unseen. These include agriculture, hydrology, weather, and human health. Simply stated, water is life. Water is of fundamental importance for human well-being, quality of life and socioeconomic development, as well as for healthy biosphere and ecosystems. Since the birth of humanity, all great civilizations viewed water as a symbol and source of life. The distribution of fresh water over the planet is highly uneven; therefore, sustainable planning and management of water is critical. On average, agriculture accounts for 70% of global freshwater withdrawals (FAO, 2017). Water shortages in dryland agriculture with erratic rainfall patterns can have far-reaching effects on food security, nutrition, livelihoods, and other socioeconomic and environmental aspects. With continued population growth, increasing food demand, and pressure on water resources, increased water use efficiency (WUE) in agriculture is crucial. The finite nature of global freshwater has led to considerations of water reuse as a component of sustainable and integrated water resources management (Lazarova and Asano, 2013; Helmecke et al., 2020) with other potential limitations.

8.1.2 CLIMATE CHANGE, ENVIRONMENT, AND ECOSYSTEM SERVICES

When it comes to feeding the world – nearly 10 billion people by 2050 – access to water via the ground or through precipitation is going to be key. Farmers around the world have relied on intensive tillage for the last 10,000 years (Lal et al., 2007; Montgomery, 2007; Reusser et al., 2015). Although this approach has had many benefits, it also has serious problems, notably the resulting susceptibility of soil to erosion and degradation (Delgado et al., 2011). According to Pimentel et al. (1995), about 430 million ha – almost one-third of the global arable land area – has been lost to soil erosion.

Ongoing loss and degradation of agricultural soil and water assets due to increasing extremes in precipitation will continue to challenge both rainfed and irrigated agriculture unless innovative conservation methods are implemented (Larson et al., 1997; Gleick et al., 2011; Lal, 2013b). Bossio et al. (2010) reviewed and highlighted the connection between land degradation and WUE largely because soil degradation decreases soil organic matter (SOM) content, thus reducing water infiltration and water holding capacity (Noellemeyer et al., 2008). Use of Conservation Agriculture (CA) has increasingly been endorsed and promoted as a type of Climate Smart Agriculture, contributing to both climate change adaptation and mitigation (Jarvis et al., 2008; Harvey et al., 2013; Pretty and Bharucha, 2014; Kassam et al., 2017b; Lal 2013b). This will require a global orientation and perspective (Thierfelder and Wall, 2009; Hobbs and Govaerts, 2010; Kassam et al., 2009, 2019; Kirkegaard et al., 2014), and a concerted, strategic action to protect our soil, water, and air quality through new more efficient and ecological conservation practices.

8.1.3 OBJECTIVE – TO DESCRIBE AND REVIEW BENEFITS OF CONSERVATION AGRICULTURE INCREASING WATER USE EFFICIENCY IN DRYLAND AGRICULTURE

The objective of this chapter is to describe and review the benefits of CA increasing WUE in dryland agriculture and the impacts of SOM on soil hydrological properties and crop yields. The benefits help demonstrate the Conservation Agriculture Systems (CAS) umbrella of ecological conservation

Conservation Agriculture   183

practices that encompass aspects of carbon (C)-centric management and soil fertility on WUE for food production and security. The three primary principles of CA and associated conservation practices are integrated to enhance the "system synergies" required for optimizing WUE in agricultural systems within dryland agriculture. Further, this review will continue paving the way for implementing CA globally by bringing together an updated description and review for scientists, policymakers, and practitioners to share different knowledge, experiences and competencies and discuss opportunities, tools, and adaptations in agricultural drylands around the world at a time when climate extremes are increasing. Other reviews on broad aspects of CA and CAS are provided by Hobbs et al. (2008), Kassam (2019, 2020), Reicosky and Janzen (2019), Reicosky (2020), and Reicosky and Kassam (2021). Further emphasis on the details on WUE of CAS in dryland agriculture is provided by Stewart (2008a, b), Stewart et al. (2008), Kassam et al. (2014a), Basche et al. (2016), Basche and DeLonge M. (2017) and Basche and Edelson (2017).

8.2 CONSERVATION AGRICULTURE AS SUSTAINABLE MANAGEMENT OF WATER RESOURCES

8.2.1 EFFICIENT USE OF WATER

Benefits of CA are either directly or indirectly related to enhanced C management. CA systems provide clean air and water, biodiversity, genetic resources, pollinator and wildlife habitat, pest control, natural medicinal products, and recreational and other aesthetic benefits. CA uses minimum soil disturbance procedures and its prime aim is to use available water efficiently (Stewart and Steiner, 1990). CA is a major player in our survival through C cycling and soil C storage. Global food security requires efficient C cycling and global environmental preservation may require soil C storage and improved C flow management in sustainable cropping and farming systems (Bonan et al., 1994) through lands while, at the same time, maintaining sufficient reserves of energy to achieve long-term resilience. The goals of CA are to improve long-term productivity, profits, and food security, particularly under the threat of climate change in dryland areas. Because CA avoids tillage and places emphasis on agroecology (Valenzuela 2016; Altieri 2018), it is less time-consuming and can be more cost-effective than conventional farming methods. These procedures are also likely to conserve the water required by the microbial flora for efficient carbon (C) and nutrient cycling. WUE is an important concept for understanding soil–crop systems and designing practices for water conservation. Aspects of WUE were addressed earlier (Tanner and Sinclair, 1983; Sinclair et al., 1984; Hatfield et al., 2001; Hatfield, 2011), loosely defined as a given level of plant biomass or grain yield per unit of water used by the crop. Limited water resources in dryland agriculture require a better understanding of how WUE can be improved and how agricultural production can be changed to be more efficient in crop water use (Heitholt, 1989). Initial reviews on CA and WUE in the US are provided by Stewart (2008a, b) and Stewart and Steiner (1990) that demonstrate that

> Conservation Agriculture has become more widely practiced in dryland areas and particularly where fallow periods of 9 to 16 months are practiced between crops. Keeping vegetative mulch on the soil surface controls wind and water erosion, increases stored soil water by increasing infiltration and reducing runoff and evaporation, halts and often reverses SOM decline, and usually increases yields of subsequent crops. Small increases in seasonal water use by plants, particularly grain crops, can increase yields significantly.

Wahbi (2008) also concluded that CA had a positive effect on the plant root system, resulting in better water and nutrient uptake, enhanced hydraulic conductivity, and increased yield and biomass production (Brown et al., 1987). According to Hatfield et al. (2001), overall, WUE in semiarid environments can be enhanced through adoption of more intensive CA cropping systems and efficient nutrient management practices.

The terms "dryland agriculture" and "dryland farming" are universally used but the meaning of the terms can only be discovered by understanding the context in which they are used. These terms are also often used interchangeably with rainfed agriculture, but they are vastly different. Rainfed agriculture includes dryland farming, but dryland farming is generally defined as agriculture in regions where the lack of moisture limits crop production for part of the year. Two major constraints limit dryland agriculture: first, the shortage of precipitation and the high temperatures during the grain-filling period as well as large differences between day and night temperature; and, second, low soil fertility and poor soil structure resulting in soil loss by erosion and low crop productivity that is not profitable. Stewart and Burnett (1987) stated that dryland farming emphasizes water conservation in every practice throughout the year. It is within this context that dryland farming is used in this work.

There are three basic strategies for increasing crop yields in dryland cropping systems. The first is to increase the capture of precipitation by reducing runoff and to store it in the soil profile for later use by the crop for evapotranspiration (ET). The second is to increase the portion of ET that is used for transpiration (T) relative to that lost by evaporation (E) from the soil surface. The third is to ration plant water use so that there is water available during the reproductive and grain-filling periods, particularly for grain crops.

8.2.2 DRYLAND AGRICULTURE

The semiarid lands are where dryland farming is primarily practiced. Dryland agriculture is that part of rainfed agriculture where water is the most limiting factor. Mathews and Cole (1938) stated that dryland farming in its broadest aspects is concerned with all phases of land use under semiarid conditions. Not only how to farm, but also how much to farm, and whether to farm are questions that should be addressed for all semi-arid regions. Dryland farming systems must emphasize water conservation, sustainable crop yields, limited inputs, and wind and water erosion constraints. Dryland farming generally occurs in areas where the average crop water supply limits potential yield to <40% of full (water-unlimited) potential (Stewart and Koohafkan, 2004). CA systems enable the farmer to proactively manage their land using regenerative, no-till (NT) practices, continuous vegetative cover, and biodiversity that mimic a functional, natural ecosystem in order to maximize rainfall capture. Implementing the primary principles and the presence of living plant roots in soils allows for increased water infiltration, increasing C storage that contributes to storing approximately 90% of rainfall, which recharges the water table during rainfall events. This is a stark contrast to the tilled, bare fields that experience erosion and loss of topsoil from rainwater runoff (Choudhary et al., 1997). Noellemeyer et al. (2013) concluded from a 15-year study that the synergy between NT and water-efficient crops could be a promising step toward improving food production in semiarid regions. Their results indicated an improved water productivity of all crops under NT compared with that of CT; however, the response of cereals (corn and wheat) was higher than that of sunflower and soybean. Crop type had a higher impact on water productivity than did tillage system with cereal crops more efficient.

Stewart and Koohafkan (2004) reviewed crop growing in areas where water supply constitutes the major constraint and is widely practiced in semiarid and dry-subhumid regions. With associated tillage, soil degradation is a widespread problem in dry lands and largely results from wind, water and tillage erosion, organic matter depletion, chemical deterioration, and salinization. They described CA as the integration of practices that avoids mechanical soil disturbance, maintains a soil cover by a growing crop or biomass of previous crops, and rotates crops that reduce soil degradation and, in some cases, even restores many of the favorable chemical, physical, and biological properties present in the soil initially. The sparse rainfall and high temperatures in dry land regions are major constraints along with the demand of crop biomass for feed and fuel. Preliminary estimates are that the average yield of cereals in dryland regions can be increased by 30–60% annually by increasing

Conservation Agriculture 185

crop water use by 25–35 mm achievable using CA systems (Stewart and Koohafkan, 2004). In sparse rainfall and high temperature dryland regions, CA is gaining momentum in many countries of the world but has been most successful in favorable rainfall regions where it is relatively easy to maintain soil cover and rotate a wide variety of crops including legumes (Kassam et al., 2013).

Changes induced by CA principles and practices in soil properties related to soil water, fertility, and erosion, and the erosion processes as affected by CA practices have been researched in many dry areas. Most of the studies were conducted at research stations, on a limited number of soil types, and only few studies have referred to long-term experiments or to on-farm designs. A number of properties have been investigated (e.g., soil structure and porosity, aggregates stability, soil infiltration and hydraulic conductivity, soil compaction, earthworm population, SOM and soil organic carbon (SOC)), but the studies rarely addressed all the properties simultaneously (Logan et al., 1991; Kool et al., 2019; Riggers et al., 2021). This makes it difficult to understand the functioning of CA systems and to build a comprehensive knowledge base regarding the long-term impact of CA systems on soil and water in dryland agroecosystems (Stewart and Robinson 1997; Arrúe et al., 2007b).

8.3 IMPACT OF CARBON MANAGEMENT PRACTICES ON HYDROLOGICAL PROCESSES

Water is a critical resource for all life, yet drought is pervasive in terrestrial ecosystems. Microorganisms inhabiting soil ecosystems perform reactions and entail ecological interactions that influence the soil functions from the microscale up to the landscape scale (Tisdall 1994). Many soil functions in agricultural ecosystems relate to C management, C cycling, and C energy flow through the soil–plant–atmosphere system. CA has evolved as a resilient production system that naturally integrates many aspects of C management (Reicosky and Janzen, 2019; Reicosky and Kassam, 2021). Carbon dynamics in agriculture is a complex interaction between how growing plants capture C in photosynthesis and the long list of factors that release C, such as tillage, to the atmosphere. SOM is a small fraction of the soil, but a very critical component. SOM is a very heterogeneous mixture of substances, ranging from plant and root fragments, root exudates through the living bodies of the soil organisms, to brown amorphous humic substances produced by their activity. These materials have very different rates of decomposition in the soil and very different effects on the soil tilth, structure and nutrient status. SOM is treated as a continuum of organic material continuously processed by the decomposer community and with increasing oxidation and solubility protected from decomposition through mineral aggregation and adsorption (Wiesmeier et al., 2019). This approach also looked at how various food sources have different energy costs to access the energy in the food source, bringing ideas of energy consumption by the microbes into the soil. This energy is consumed in the "living soil" through C and nutrient cycling. Wiesmeier et al. (2019) summarized that soil microbial activity was the most important factor across time and space scales in controlling soil dynamics.

The awareness of CA impacts on WUE can pave the way for further implementing CA globally by bringing together an updated description and review for scientists, producers, policymakers, and consumers to share different knowledge, experiences, and competencies and discuss opportunities for sustainable agriculture, tools, and adaptations in agricultural drylands. As a result, CA has been highlighted as a key path toward food security (FAO 2008, 2011; Jarvis et al., 2008; Verhulst et al., 2010; Kassam et al., 2009, 2014a,b; 2017b; Kassam and Friedrich, 2012; Corsi et al., 2012; Pretty and Bharucha, 2014; Farooq and Siddique, 2015; Jat et al., 2014; Lal, 2015a,b; Reicosky and Janzen, 2019; Mitchell et al., 2019; Kassam, 2019; Reicosky, 2020; Reicosky and Kassam, 2021). Much of the literature on different types of agriculture is in what may be called the "grey literature" as reports that are not generally peer-reviewed and are typically published in nonscientific magazines. At the same time, the three core principles of CA are generally agreed to and included in other types of agriculture, particularly mulch cover and crop diversification. All three principles of CA need to be strategically integrated into all different types of agricultural approaches. Some promoters of different types of agriculture do not have a science background that occasionally makes it challenging to link the value systems to explicit conservation practices that align with the CA principles.

The coupled C and water cycling is a powerful planetary force that impacts the circle of all life and associated processes and activities, which remains poorly understood behind a mixture of disconnected facts, problems, and assumptions. The water cycle is also fundamental to the understanding of the C cycle as well as the nitrogen cycle (Lal and Stewart, 2012). This integration and interaction require enhanced management with cover crops to maintain a balance between the water required for plant growth and yield and C as the energy source for the microbiomes to cycle nutrients and produce optimum soil structure required for microbial activity, root function, and watershed hydrologic characteristics. The additional benefits of legumes capturing free nitrogen from the atmosphere cannot be ignored in any agricultural system.

The soil is the fundamental foundation of our life, economy, and environmental quality. Soil is alive and as vital to human survival as sun, air, water, intellect, and biodiversity; its protection and enrichment with energy and organic C are needed for the future sustainability of our planet. Carbon is a major player in the greenhouse effect and climate mitigation, in soil health and ecosystem services, and in our food security. The multiple synergistic benefits of cover crop mixes and C management are required for continued sustainable production. Soil C stands at the forefront of CA and soil health due to its critical importance in regulating physical, chemical, and biological processes and properties of soils (Reicosky et al., 2021).

Plant C is transient with continuous C energy flow through the soil–plant–atmosphere system and the soil food web, meaning that plant C is constantly changing as it is transformed into new organisms or converted into different compounds (Janzen, 2015; Kane, 2015). Soil scientists have typically classified C into general pools based on how long the C remains in the soil, a figure often referred to as "mean residence time." The commonly used simplistic pools include three different groupings: the fast (labile or active pool) pool, the slow pool, and the stable pool, based on the function of its physical and chemical stability (Jenkinson and Rayner, 1977; O'Rourke et al., 2015). The fast pool of fresh organic C decomposition results in a large proportion of the initial biomass being lost in one to two years (Jenkinson and Rayner, 1977; Voroney et al., 1989; Angers and Chenu, 1997). The intermediate pool is comprised of microbially processed organic C that is partially stabilized on mineral surfaces and/or protected within aggregates, with turnover times in the range of 10–100 years. A stable pool (slow or refractory pool) with highly stabilized SOC enters a period of very slow turnover of 100 to > 1,000 years and ultimately into humic acids and humus.

There has been a growing interest in studying the labile C pool in order to promote the sequestration and stabilization of SOC. Lehmann and Kleber (2015) argue that SOM should no longer be seen as large, persistent, and chemically unique substances, but as a continuum of progressively decomposing organic compounds. An emerging alternate view suggests that "SOM is a continuum of progressively decomposing organic compounds" and that recalcitrant "humic substances" do not exist in soils (Lehmann and Kleber, 2015). The complexity and interactions of these segments of the C cycle contribute to the contentious nature of SOM that is something to think about. Although labile SOC fractions have emerged as standardized indicators because of their potential to detect early SOC trends over time, the relationship between microbial attributes and labile SOC remains poorly understood.

Management of soil C is also central to maintaining soil health and ensuring global food security. The OC content of soil is a key indicator of soil health that indicates the efficient functioning of many ecosystem processes (e.g., C cycling and storage, nutrient and waste cycling, water storage, biodiversity, etc.). C comes primarily from plant materials that are created through the capture of atmospheric CO_2 through the process of photosynthesis. These organic materials are cycled through the soil, and used by organisms as a source of energy and nutrients. A significant amount of CO_2 is

TABLE 8.1
Summary of Primary Benefits of Soil Organic Carbon in Agricultural Production Systems

- Energy supply for microbes, macrofauna, and earthworms
- Direct nutrient supply to plants (particularly nitrogen, phosphorus, and sulfur)
- The capacity of the soil to retain and exchange nutrients
- Aggregation of soil particles and stability of soil structure
- Water storage and water availability to plants
- Beneficial thermal properties
- pH buffering (helping to maintain acidity at a constant level)

returned to the atmosphere as a result of respiration. Increasing SOC leads to an increase in benefits listed in Table 8.1.

The dynamic nature of C and anthropogenic impacts on soil can turn it into a net sink or a net source of greenhouse gases (GHGs). Soil has become a focus to mitigate climate change (Hatfield et al., 2011; Lal, 2015a; Sanderman et al., 2017; Hatfield et al., 2017). Carbon in a healthy soil makes it more resilient in a changing environment as it nourishes the microbiome and allows the plants to remove CO_2 from the atmosphere. Soil works as a massive repository for C; it stores the C compounds of decaying plants and animals as well as everything that lives within the soil microbial community, from microbes to worms to plant root exudates essential for soil microbes to be able to flourish and continue C and nutrient cycling. The composition and breakdown rate of SOM affect the diversity and biological activity of soil organisms, plant nutrient availability, soil structure and porosity, water infiltration rate, and water-holding capacity (WHC). While each of the individual benefits may not seem critical, the enhanced benefits with the associated synergies become very demanding with respect to providing ecosystem services and sustainable food production.

Building soil C content is highly desirable for CA, but also, by storing C, some of the effects of climate change can be mitigated. However, most work has considered C on its own, whereas, in due course the next most common element in plants, N will become limiting. In an extensive study in southern Brazil, C and N contents of soil were investigated under CT and NT agriculture (Sisti et al., 2004). Improvement of soil N varied with legumes used and whether more N was removed with the crop than was fixed biologically. Grasslands with a legume component are particularly good at storing C. A more difficult problem will arise when soil P becomes limiting, although there are many plants, including nodulated legumes that are native to low phosphorus soils (Sprent, 1999).

8.3.1 SOIL CARBON MANAGEMENT TERMINOLOGY

Two terms often used interchangeably are "C sequestration" and "C storage," which can many times lead to confusion and misinformation. Soil organic carbon "storage," that is, an increase of SOC "stocks," should be clearly differentiated from SOC "sequestration," as the latter assumes a "long-term" net removal of atmospheric CO_2 (Krna and Rapson, 2013; Chenu et al., 2019; Reicosky and Janzen, 2019). The vision emerging, showing the prominent role of soil microorganisms in the stabilization of SOM, draws the attention to more exploratory potential levers, through changes in microbial physiology or soil biodiversity induced by agricultural tillage practices, that require in-depth research.

Policymaker concerns about the permanence of sequestered soil C present a significant barrier to the wide-scale funding of soil C conservation projects as a climate mitigation tool (Smith, 2005; Smith et al., 2019; Riggers et al., 2021). Different terminology and definitions used by scientists and policymakers to describe soil C longevity/permanence have both slowed down the transfer of new scientific understanding of soil C sequestration to policymakers and insulated scientists from targeting the research needs of policymakers (Krna and Rapson, 2013). A clear understanding of definitions and research gaps related to soil C lifespans is both timely and necessary to catalyze more ambitious soil conservation practices as a part of broader climate change mitigation efforts. Integrating new scientific findings regarding soil C longevity into data-based C policies is critical for broader adoption of agricultural soil C sequestration projects that could expand the contribution of soils to climate change mitigation on a global scale. Currently, policy, market-based and science-based definitions of soil C permanence do not align (Krna and Rapson, 2013). Chenu et al. (2019) suggested a relatively short time interval of 20 years and because some C trading mechanisms require that offsets be maintained for a minimum of 100 years, this translates to a requirement that any management practices to sequester soil C should be maintained for 100 years, a cumbersome and unrealistic expectation for two or three generations of land managers. These existing policy precedents, combined with a better scientific understanding of soil C permanence or persistence, can help policymakers evaluate potential performance of climate projects on soil C sequestration and inform necessary funding structures (Bossio et al., 2020).

Janzen et al. (2022) reviewed the photosynthetic limits on C storage in agricultural landscapes. Their approach recognized that photosynthesis, the source of C input into the soil, represents the most fundamental constraint to soil C storage. They distinguished two forms of soil C: "ephemeral C" denoting recently applied plant-derived C that is quickly decayed to CO_2, and "lingering C," which remains in the soil long enough to serve as a lasting repository for C derived from atmospheric CO_2. Their unique and complex analysis led them to conclude that the lingering SOC pool was significantly less than previous estimates, even allowing for acknowledged uncertainties. They suggest change in emphasis from soil processes toward a wider ecosystem perspective that starts with photosynthesis that brings CO_2 and water together to form carbohydrates and oxygen (O_2) released.

8.3.2 Soil Structure – Hydrologic Characteristics

Soil structure refers to the size, shape, and arrangement of solids and voids, continuity of bio-pores and voids, their capacity to retain and transmit fluids and organic and inorganic substances, and the ability to support vigorous root growth and development (Lal, 1991). Soil structure is a key factor in the functioning of soil, its ability to support plant and animal life, and moderate environmental quality with particular emphasis on C storage and water quality. Aggregate stability is used as an indicator of soil structure (Six et al., 2000). Soil structure is the spatial arrangement of solids and voids across different scales without considering the physical and chemical heterogeneity of the solid phase. Thus, the solid phase and pore space are complementary aspects of soil structure which can be approached from both perspectives. Aggregation results from the rearrangement of particles, flocculation, and cementation (Duiker et al., 2003). Aggregation is mediated by SOC, biota, ionic bridging, clay, and carbonates. The SOC acts as a "binding agent" and as a nucleus in the formation of aggregates. Complex biota and their organic products contribute to the development of soil structure; which in turn exerts a significant control over SOC dynamics (Nannipieri, 2020; Nannipieri et al., 2020). The SOC residence time and decomposition rate are key factors influencing its effectiveness in increasing aggregation.

Soil structure is recognized to control many processes in soils. It regulates water retention and infiltration, gaseous exchanges, SOM and nutrient dynamics, root penetration, and susceptibility to erosion. Soil structure also constitutes the habitat for a myriad of soil organisms, consequently driving their diversity and regulating their activity (Elliott and Coleman, 1988; Wardak et al., 2022). Soil structure is correlated with root mass, root morphology, fungal mass, and use of organic matter (OM) amendments such as mulch, manure, and green manure. While these factors contribute to aggregation in soils, a comprehensive understanding of how this occurs is lacking. Soil structure impacts water availability, nutrient uptake, and leaching, thereby affecting ground and surface water supplies through sedimentation and chemical contamination. As an important feedback, soil structure is actively shaped by these organisms, thus modifying the distribution of water and air in their habitats (Bottinelli et al., 2015; Freeney et al., 2006; Young et al., 2008; Warkentin, 2001). Many biological processes in soil are linked to soil structure and its potential to be used in the assessment of hydrologic properties on soil functions (Or et al., 2021).

The physical disturbance of tillage breaks down the soil aggregates and any associated change in physical and biological properties. Only recently, however, are we getting a better understanding of the tillage impact and soil structure loss (Pires et al., 2017). Soil functions, which develop through formation of soil aggregates as fundamental ecological units, are manifest at the earliest stages of structure evolution (Banwart et al., 2019). Rabot et al. (2018) reviewed the importance of soil structure and its impact on soil functions, suggesting that the intact pore network is more appropriate than analyzing disturbed aggregates. They identified porosity, macroporosity, microporosity, pore distances, pore diameters, and pore connectivity and continuity that can be better visualized in a schematic (Banwart et al., 2019, Figure 2). The relative size of the pores and aggregates and the connectivity of the aggregates and the pores enable a clear understanding of the soil functions that are impacted by the soil structure. Pores create a transportation network within a soil matrix, which controls the flow of air, water, nutrients, and movement of microorganisms. Large connected interaggregate pores allow water drainage and O_2 downward flow that supports aerobic respiration. The interactive flow of air, water, and movement of microbes, in turn, control soil C dynamics, especially C oxidation. The larger diameter vertical earthworm holes and former root bio-pores become a direct conduit for high rates of infiltration and storing water to greater depths that then open up for bulk flow of CO_2 out of the soil and O_2 into the soil to greater depths to maintain aerobic soil respiration (Edwards et al., 1990). Maintaining the functionality of these large pores can only be accomplished by avoiding any soil disturbance like tillage. A clear understanding of the physical and ecological functioning of these bio-pores is required to take advantage of the synergies that occur with the relative increase in root biomass contributing more to stored C than the above ground biomass (Balesdent and Balabane, 1996; Puget and Drinkwater, 2001; Kätterer et al., 2011; Jackson et al., 2017).

Soil structure contributes to soil functions such as biomass production, storage and filtering of water, storage and recycling of nutrients, C storage, habitat for biological activity, and physical stability and support (Warkentin, 2001). Soil structural properties are extracted and analyzed from soil profile description, visual soil assessment, aggregate size and stability analysis, bulk density, mercury porosimetry, water retention curve, gas adsorption, and imaging techniques identifying porosity, macroporosity, pore distances, and pore connectivity derived from imaging techniques as relevant indicators for several soil functions for a large range of soil types, which could form the basis to relate more easily available measures of pore structural attributes taking into account texture, SOM content, etc. (Warkentin, 2001; Or et al., 2021).

Wardak et al.'s (2022) review illustrated that NT can influence porosity depending on soil texture, pore size class, depth and time, and also influence important transport mechanisms that are likely to impact the fate of agrochemicals in soils. They found decreased macroporosity in surface layers of soil under NT when compared with CT. In addition, soil pore connectivity tended to increase in soil under NT though the associated effects on hydraulic transport were less clear. Their investigation reveals the value of a prospective examination of an evolving NT pore network understanding both visually and functionally across temporal and spatial scales and highlights the necessity for standardized methodology to aid in future data compatibility and quantitative analysis in agreement with conclusions of Derpsch et al. (2013) on other aspects of CA systems.

Wardak et al. (2022) reviewed the importance of soil structure and porosity as they impact the hydrologic characteristics of the soil and the impacts on WUE. The soil pore network structure, and particularly the arrangement of macropores that are characteristic of biological activity (i.e., long, cylindrical, non-tortuous, and surface-connected macropores), induces changes in functionality, specifically with regard to the transport of water, solutes, and mobile particulate matter to deeper layers. Many studies have reported increased pesticide leaching under NT, likely resulting from a combination of the contrasting origins of pore establishment (i.e., biotic and abiotic in NT vs. anthropogenic in CT, and the physicochemical properties of pore wall linings).

Soil porosity, numbers of macropores, and pore connectivity are well established as relevant and important indicators of several soil functions, including those of hydraulic and biochemical origin (Jarvis, 2020; Rabot et al., 2018; Landl et al., 2019). Kay and VandenBygaart (2002) reviewed early literature investigating the effects of NT on porosity and pore characteristics and concluded that the loss in porosity from converting to NT is linked with changes to the pore size distribution, with macroporosity increasing as mesopores collapse. They also posited that the largest differences occur after at least 15 years, but that pore size distribution and continuity, especially of meso- and micropores, are very rarely addressed in the literature.

The connectivity of the macropores allows rapid infiltration of water to greater depths in the profile following a rainfall event, and when drained, allows for flow of CO_2 from respiration out of the pore and O_2 from the atmosphere down through the macro-pores to supply the heterotrophic microbial population important in organic matter decomposition (Dal Ferro et al., 2014; Piccoli et al., 2017; Li et al., 2021). Depending on the soil water content, the micropores may be filled with water held by capillary forces limiting gas exchange to only the connected macropore network. Reduced O_2 concentrations in the center of the macroaggregates result in little organic matter oxidation in the undisturbed state (Sexstone et al., 1985). When aggregates are near equilibrium in soil air, steep O_2 gradients usually occur over very small distances from the aggregate surface, suggesting a large part of the interior of the aggregate is deprived of O_2. However, following an intensive tillage that fractures many of the macro- and micro-aggregates enable free O_2 to flow over the freshly exposed surface enabling the microbes to increase soil respiration. With higher tillage intensity, one can visualize more aggregate fracturing and as a result more rapid oxidation of the organic matter reducing CO_2 to eventually be exhausted from the soil (Banwart et al., 2019).

Across the evidence considered within this systematic review, a suite of other management practices was investigated. Farmers rarely make decisions based on single management practices, but rather consider their field management in a holistic way. However, the majority of the evidence base examined the effect of tillage as a single practice. Key knowledge gaps, therefore, exist around the combined effects of tillage and amendments (such as farmyard manure application and stubble management) on SOC. Similarly, the combined effects of tillage and fertilizer were poorly studied. These represent partial knowledge gaps where further investigation may be warranted. Smith (2004) provided quantitative evidence in support of the previously held view that changes in SOC cannot be detected within a 10-year timeframe. This evidence should further strengthen guidance to ensure experiments are in place for longer than a decade before measurements aiming to detect SOC change are made, and researchers should ensure that investigations of SOC seek funding to cover periods of more than 10 years of study to have the necessary power to detect significant change. Chemical analysis of field-collected samples for soil C often uses the dry combustion method regarded as the standard method. However, conventional sampling of soil and their subsequent chemical analysis are expensive and time consuming. Furthermore, these methods are not sufficiently sensitive to identify small changes over time in response to alterations in management practices or changes in land use (Chatterjee et al., 2009).

8.3.3 Role of Biology in Soil Physical Structure

Despite the vital importance of C cycling, understanding the microbial processes occurring within the soil matrix is still incomplete (Reinsch et al., 2019; Nannipieri, 2020; Nannipieri et al., 2020). One of the reasons is that SOM is one of nature's most versatile materials comprising a

Conservation Agriculture 191

continuum of microbial and plant C compounds in various stages of decomposition, which need to be characterized by a combination of methods (Kögel-Knabner, 2002, 2017). The provision of these ecosystem services relies on the good functioning of soil, in particular its microbial processes responsible for biological C cycling (Oades, 1993; Lemanceau et al., 2015; Lehmann, et al., 2020; Reicosky 2020). In this context, SOC storage is an important process, which captures CO_2 via photosynthesis and transfers organic materials via plant activity belowground where they can be used by the soil micro-organisms (Janzen et al., 2022). Microbial processes involve the decomposition, transformation, and turnover of plant-derived organic materials, providing nutrients, C, and energy for microbial activity and growth and more generally SOC accumulation. This microbial functioning leads to nutrient available to plants and C for aggregate formation, a soil structure-giving process allowing for enhanced/deeper water infiltration and storage, thereby reducing soil loss through water erosion.

The interactions between the physical habitat and biological and chemical processes are key determinants of ecosystem health. Understanding of the origin and consequences of the physical habitat of soils to a wide range of functions is important for the sustainable use of our soil resources (Reinsch et al., 2019). Freeney et al. (2006) determined the impact of microbiota on spatial structures and found that it was visually evident within 7 days, showing in the rhizosphere soil that large-scale stable structures are rapidly constructed. Their results indicate that microbiota act to significantly alter their habitat toward a more porous, ordered, and aggregated structure that has important consequences for functional properties, including hydraulic transport processes. These observations support the hypothesis that the soil–plant–microbe complex is self-organized and critically important in the dynamics of soil functions and suggests that just a few weeks of cover crop growth has the potential for providing significant benefits in increased water infiltration and storage that should not be ignored in dryland areas (Unger and Vigil, 1998; Kell, 2011; Nielsen et al., 2015; Basche and DeLonge, 2017; Vogel et al., 2022).

Biological, physical, and chemical transformation processes convert dead plant material into organic products that are able to form intimate associations with soil minerals, making it difficult to study the nature of SOM. SOM is treated as a continuum of organic material continuously processed by the complex living decomposer community and with increasing oxidation and solubility protected from decomposition through mineral aggregation and adsorption on mineral particles as mineral-associated organic matter (Wiesmeier et al., 2019). The continuum concept that focuses on the ability of decomposer organisms to access SOM and on the protection of OM from decomposition provided by soil minerals. Viewing SOM as a continuum spanning the full range from intact fresh plant material to highly oxidized C in carboxylic and humic acids represents robust science and will facilitate the way we communicate between disciplines and with the public. SOM and the diverse decomposer organisms create a complex continuum of progressively decomposing organic compounds (Lehmann and Kleber, 2015; Lehmann et al., 2020).

Soil biology in agriculture has historically dealt with the effect of agricultural practices on free-living organisms in the soil that are nurtured by plant C inputs. Ecological biodiversity is easily maintained with cropping sequences in a diverse rotation system approach that enables the available natural resources to be preserved and more efficiently utilized (Valenzuela 2016; Altieri, 2018; Reicosky et al., 2021). Emerging views indicate that when supplied with abundant solar energy for photosynthesis, soil biology acts as a self-organizing system as soil microbes forge their habitats into a porous, well-aggregated structure with high functionality (Oades, 1993; Lemanceau et al., 2015). It has been generally accepted that C captured in photosynthesis eventually becomes the main food or energy source for the soil microbiome. As our understanding of the C cycle and energy advances, the flow of C and energy to power these self-organizing plant processes should be considered and accounted for using new systems concepts, which do not focus on static pools of SOM and C and their relationship to soil functionality. These C and energy flows within the plant must be understood and expanded through soil–plant–atmosphere system.

Plant C is transient with continuous movement through the soil food biological web, meaning that plant C is constantly changing as SOM is decomposed and is transformed into new organisms or converted into diverse compounds (Oades, 1993; Janzen, 2015; Kane, 2015: Lemanceau et al., 2015; Reicosky and Janzen, 2019; Wiesmeier et al., 2019). Evidence is accumulating on the benefits of minimum soil disturbance/NT enhancing C accumulation and all the associated synergistic benefits (Allison, 1973; Reicosky and Saxton, 2007; Lal, 1991; Lal, 1993; Chenu et al., 2019). There is consensus among researchers that this long-term retention of OM in soils and its effect on the resilience of soil architecture (e.g., Chenu et al., 2019; Wiesmeier et al., 2019; Vogel et al., 2022) are essential to guarantee that soils will be able, in spite of climate change, to fulfill the key functions on which humanity depends (Reicosky et al., 2011; Reicosky, 2020).

Carbon as energy flowing into and out of the soil via plant exudates and the deposition of plant biomass used for the creation and maintenance of biological activity (Kuzyakov and Cheng, 2004; Janzen, 2015; Lemanceau et al., 2015; Reicosky and Janzen, 2019). The microbiome uses that energy for the creation and maintenance of the soil structure creating enhanced hydrological properties and micropore development to transport soluble C within macroaggregates (Freeney et al., 2006; Smucker et al., 2007; Reinsch et al., 2019). Soil micropore development and contributions to soluble C transport occur within macroaggregates. Microbial community and ecosystem dynamics regulate the exchange of both nutrients and C between the soil and the atmosphere through the mineralization of SOM. Williams and Plante (2018) suggested that these ecosystem dynamics are driven by net energy flows, and analysis of SOM bioenergetics can provide complementary constraints to SOM understanding and insights into the fundamental challenge of why thermodynamically unstable OM persists in soil. Wacha et al. (2022) described a concept of soil energetics as a framework to better understand changes in soil function and C dynamics. They expanded the framework that quantifies the net energy flows within a soil control volume using energetic components including mechanical, biogeochemical, and hydrological processes. Their integrated analysis indicated that over half of the energetics in the soil comes from the in-season deposition of root exudates through growing plants, supporting the soil health principle requiring a living plant as long as biologically possible. Management practices, especially intensive tillage, impact energy fluxes through tillage type and intensity enabling raindrop-induced erosion events. Wacha et al. (2022) applied the system analysis to three different tillage management practices to assess energy balances. They found that seasonal net energy balances for a CT, NT, and grassland system were negative, neutral, and positive, respectively (Wacha et al., 2022), suggesting that tillage not only physically destroys the soil structure but also has a major impact on C energy flow and cycling, resulting in the net soil C loss to the atmosphere (Reicosky and Lindstrom, 1993; Ellert and Janzen, 1999; Rochette and Angers, 1999; Dold et al., 2017). These works reinforce the negative impacts of intensive tillage on the biological processes in soils and the need for further recognition and acceptance of nature's laws.

8.3.4 TILLAGE IMPACTS ON SOIL STRUCTURE AND CARBON EMISSIONS

The key problem of tillage agriculture is the steady decline in C, soil fertility, and productivity, which is closely correlated with continuous use. The traditional annual tillage ritual is often driven by the relatively rapid collapse of the loose soil structure by wetting–drying cycles (Ghezzehei and Or, 2000), compaction by farm machinery (Keller et al., 2019), weed and pest control and preparation of a seedbed for the next crop (Håkansson and Lipiec, 2000; Hadas, 1997). CT is done in dry regions to prepare an adequate seedbed, control weeds, improve aeration, increase water infiltration, make furrows for irrigation, and incorporate biomass or fertilizers into the soil (Hill, 1990; Cornish and Pratley, 1991; Lal, 1991; Lal, 1993; Haddaway et al., 2017; Banwart et al., 2019; Rabot et al., 2018; Hassan et al., 2022). Warkentin (2001) discussed how alteration of soils by tillage changes the sustainability of soil functions. Soil tillage presents an enigma in thinking about soil sustainability in ecosystems (Blevins and Frye, 1993). Reduced tillage is any combination of tillage operations

Conservation Agriculture 193

that do less soil disruption than CT. Reduced tillage requires less fuel (about 50%) and less time, and causes less soil erosion and compaction. However, it requires the use of more chemicals (herbicides and pesticides) to control weeds and insects and may require heavier equipment for planting. The reduced tillage farmer must be a better farmer than the CT farmer to produce the same yield. Many farmers, especially in dry regions, still question about which system is better: NT, reduced, or CT?

Reicosky et al. (1995) reported that the impact of tillage on SOM varied by soil type, cropping systems, residue management, and climate. Cropping sequence affects crop performance not only grown on lands prepared in conventional land preparation but also crops grown under NT systems. Different crop sequences under NT affect the quantity, quality, and permanence of crop biomass, amplitude of fallow periods, and distribution and the type of root systems (Kell, 2011; Amanuel et al., 2000; Benjamin et al., 2007; Pierret et al., 2016). The accumulation of crop biomass with frequent inclusion of pulse crops in a rotation improves the biochemical and physical properties of the soil by increasing the level of OM (Biederbeck et al., 1994; Stevenson and Van Kessel, 1996). Krna and Rapson (2013) discussed the implications of tillage disturbance in agricultural ecosystems for C dynamics and suggested that it requires greater exploration, including the anthropogenic movements of C, both deliberately and as a byproduct of human activities and management (Van Oost et al., 2007). Part of this should include better categorization of other unique C storage locations where CAS need to contribute to increase the WUE and climate extreme litigation.

Viewing the soil as a complex "living biological system," soil tillage is an unnatural, intrusive, and destructive management practice. Over the last 10,000 years, tillage has been a major part of our agricultural production systems, with many reasons given (Lal, 1991; Young and Ritz, 2000; Lal et al., 2007). Through culture, tradition, and lack of knowledge, the unintended consequences of intensive tillage are either ignored or accepted as the "costs" of food production. These degrading side effects of plowing or tilling the soil are not often discussed in research papers and have not earned the attention they deserve. Table 8.2 summarizes some of the unintended consequences of tillage (without claiming the list to be complete).

While many of the negative impacts of tillage may seem small and inconsequential, the cumulative effects of small changes over time can evolve into major soil degradation that limits productivity and production efficiency requiring additional nutritional inputs. Soil lost or degraded in any type of erosion is difficult to restore or regenerate within a generation or two. Ignoring the long-term negative environmental effects of tillage leads to economic losses at the farm and landscape levels that threaten the food production system. The true cost of any type of complex environmental degradation is very challenging to quantify from an economic perspective. Many studies have documented that over the long term frequently tilled soils undergo losses in SOM and microbial activity, increases in net nitrate production, and deterioration of soil structure (Doran, 1980; Dexter, 1988). Tillage events are known to produce a temporary burst of CO_2 flux from the soil surface (Reicosky and Lindstrom, 1993; Reicosky, 1997a,b; Ellert and Janzen, 1999; Rochette and Angers, 1999). Until recently, however, little was known about the concomitant biological, chemical, and physical responses of soil to tillage events that accompany this short-lived release of soil CO_2. Biogeochemical changes may be associated with changes in soil microbial community structure, especially with the fragile fungal network that is more susceptible to tillage disturbance (Six et al., 2006; Jiang et al., 2011; De Vries et al., 2006; Haddaway et al., 2017).

The unintended consequences of tillage-based conventional (CT) agriculture have contributed to the "climate problem" and now C-centric CAS can contribute to the "climate solution" with enhanced water management. Efforts to control human-induced land degradation and soil erosion over the last 10,000 years have been building on the ruins of past tillage and monoculture concepts (Lal et al., 2007; Reicosky et al., 2011; Friedrich, 2020). Soil erosion remains a major problem in tillage-based agricultural production systems. According to Pimentel et al. (1995), about 430 million ha, almost one-third of the global arable land area, had already been lost to soil erosion. Other studies have indicated that some 500 million ha of land under tillage agriculture have been abandoned due to soil

TABLE 8.2
A Summary of Unintended Consequences of Intensive Tillage in Agricultural Production Systems

- Destroys water stable aggregates and soil structure
- Destroys soil mycelia and fungi
- Destroys the habitat of earthworms
- Destroys animals living in the soil
- Results in surface sealing which impedes water infiltration
- Causes runoff and soil erosion
- Water that does not infiltrate the soil is lost to planted crops
- Water erosion is unavoidable without biomass cover
- Wind erosion occurs only with high wind velocities and bare, pulverized, soil
- Leads to tillage-induced C losses
- Instead of C being stored in the soil, C is oxidized and lost
- CO_2 is a GHG that contributes to global warming and climate change
- Each operation reduces soil water content
- Leads to problems in drinking water cleaning facilities
- Sediments cause the siltation of creeks, rivers, lakes, dams, and highway ditches
- It leads to important increases in fuel costs and agriculture's C footprint
- Requires more synthetic chemicals to offset tillage effects
- Soil erosion and degradation occur because of tillage
- Biodiversity is reduced with tillage
- Less insects and wildlife are found with tillage
- Destroys the vertical earthworm burrows and root bio-pores
- Most time and energy consuming farm operation in agriculture
- Most fossil fuel consuming operation (66% more)
- Most unnatural operation in farming that has no parallel in nature
- Most negatives of conventional agriculture are related to tillage
- Biological damage is catastrophic.

degradation and erosion since World War II (Montgomery, 2007). At the landscape level, CA enables soil-mediated environmental services to be harnessed at a larger scale, particularly C storage, water cycling, cleaner water resources, reduced erosion and runoff, and, with this, better management and conservation of terrestrial freshwater systems and biodiversity (Kassam et al., 2013).

There are many knowledge gaps about soil as a biological system (Nannipieri, 2020; Nannipieri et al., 2020). Soil structure constitutes the habitat for a myriad of soil organisms, consequently driving their diversity and regulating their activity (Elliott and Coleman, 1988; Tisdall 1994). Microorganisms inhabiting soil perform reactions and entail ecological interactions that influence the soil functions from the microscale and up to the landscape. Many processes are linked to soil structure and are used in the assessment of soil functions. Rabot et al. (2018) identified porosity, macroporosity, pore distances, and pore connectivity derived from imaging techniques as being the most relevant indicators for several soil functions. They emphasize the greater relevance of the "connectivity" of bio-pore network characterization as compared to the aggregate perspective that enables water and air transport to greater depths in the soil profile.

Soil structure is an important visual trait that evolves from biological activity including bio-pores formed by plant roots and earthworms and aggregation of soil particles by biopolymers and hyphae (Tisdall and Oades, 1982; Young and Ritz, 2000; Or et al., 2021). A different type of soil structure results from the mechanical breakup and loosening of soil by intensive tillage. Soil structure is fragile and easily disrupted: decades of natural structure formation can be undone in an instant by the passage of heavy farm equipment and tillage. Traditional agricultural definitions of soil structure are biased toward characterization, primarily focusing on the favorable arrangement of soil particles with respect to their agronomical functions (Dexter, 1988). The delicate sensitivity of soil structure to particle arrangement and the centrality of soil structure in the functioning of natural lands call for unifying concepts. Soil structure is "the spatial arrangement and binding of soil particles and the legacy of biological agents that support physical, chemical and biological functions in soils." Emphasis is on the importance of visible spatial organization (solids and pores) and invisible traits (mechanical bonds and biological legacy) of soil structure.

Significant global activity (i.e., tillage) rests on a limited quantitative framework for gauging direct tillage benefits (Lal, 1991; Huggins and Reganold, 2008). The rapid expansion of CA in the past few decades (presently comprising 12.5% of the arable land surface; Kassam et al., 2019) demonstrates that, for some regions and crops, farmers can do well without CT. Moreover, the projected intensification of agriculture for feeding a rapidly growing global population (Tilman et al., 2011) and associated risk of soil structure degradation by compaction (Keller et al., 2019) add urgency to define and quantify the benefits of tillage. Growing concerns over the role of tillage in GHG emissions and soil C storage (Reicosky et al., 1995; Houghton, 1999; Post and Kwon, 2000; Lee et al., 2006; Poeplau et al., 2011; Reicosky et al., 2011; Haddaway et al., 2017; Friedrich, 2020) add new dimensions to tillage decisions beyond short-term agronomic return.

Soil structure is the habitat for the microbiome, consequently affecting their diversity and regulating their activity (Elliott and Coleman, 1988; Young and Ritz, 2000). As an important feedback, soil structure is actively shaped by these organisms, thus modifying water and air distribution in the soil profile (Bottinelli et al., 2015). The cumulative effects of biological activity result in increased soil stabilization by organic C deposited by biological agents. Or et al. (2021) defined "natural soil structure" as the cumulative ecological legacy and soil constituent architecture by natural aggregation and bioturbation that support soil functioning under given climatic conditions. Anecdotal evidence suggests that natural soil structure not only promotes SOC accumulation (Reicosky et al., 1995; Poeplau et al., 2011), but also allows for preservation of spatial ecological traits including legacy rhizobiomes preserved in locations that provide a bio-pore network for future roots (Garbeva et al., 2008; Poudel et al., 2019).

Macro-aggregates are known to provide physical protection to SOM, shielding it from rapid decomposition, and thus are regarded among the key elements enabling soil C storage (Beare et al., 1994; Paustian et al., 1997; von Lützow et al., 2006). Intra-aggregate physical protection is the leading driver of C storage occurring when land under intensive agricultural management is converted to conservation land use practices (Jastrow, 1996; Grandy and Robertson, 2007). In soils under conservation practices, for example, grasslands or soils abandoned from agriculture, macro-aggregates tend to have higher C concentrations and are richer in newer C than other soil fractions (Jastrow, 1996; Six, et al., 2004). The newly added C often serves as a binding agent holding the macro-aggregates together. When intra-aggregate physical protection is eliminated by crushing macro-aggregates, the intra-aggregate C accumulated by conservation practices is easily mineralized (Beare et al., 1994; Hassink and Whitmore, 1997). However, when macro-aggregates stay intact for prolonged time periods in undisturbed soils, decomposition of organic binding agents is sufficiently slow to allow for formation of micro-aggregates where physical protection is enhanced by physicochemical and chemical protection processes (Six et al., 2000; Denef et al., 2001; Chenu and Plante, 2006).

Tillage destroys soil structure and breaks down the soil aggregates and any associated change in physical and biological properties. Only recently are we getting a better understanding of tillage impact and soil structure and function loss (Wardle, 1995; Young and Ritz, 2000; Pires et al., 2017), which develop through the formation of soil aggregates as fundamental ecological units that are manifest at the earliest stages of structure evolution (Banwart et al., 2011; Banwart et al., 2019). Rabot et al. (2018) reviewed the importance of soil structure and its impact on soil functions suggesting that the intact pore network is more appropriate than analyzing disturbed aggregates.

The connectivity of the macropores allow rapid infiltration of water to greater depths following a rainfall event, and when drained, allows for the flow of CO_2 from respiration out of the pore and O_2 from the atmosphere down through the macropores to supply the heterotrophic microbial population important in OM decomposition. Depending on the soil water content, the micropores may be filled with water held by capillary forces limiting gas exchange to only the connected macropore network. Reduced O_2 concentrations in the center of the macroaggregates result in little OM oxidation in the undisturbed state (Sexstone et al., 1985). When aggregates are near equilibrium in soil air, steep O_2 gradients usually occur over very small distances from the aggregate surface, suggesting a large part of the interior of the aggregate is deprived of O_2. However, following an intensive tillage that fractures many of the macro- and micro-aggregates enables free O_2 to flow over the freshly exposed surface enabling the microbes to increase soil respiration. With higher tillage intensity, one can visualize more aggregate fracturing and as a result more rapid OM oxidation reducing soil C to eventually be exhausted from the soil (Wardle, 1995; Banwart et al., 2019).

Hamblin (1985) reviewed the role of the soil structure as a complex soil–bio-pore network through which water, air, and roots move, all critical to the soil microbiome. This intrinsic physical property is the one most easily, frequently, and drastically altered by tillage or any type of mechanical disturbance (Pires et al., 2017). Kravchenko and Guber (2017) and Kravchenko et al. (2019) characterized these pores as both drivers and products of a variety of soil biological activities that ultimately determine physical protection of SOM by influencing its accessibility to microorganisms and concluded that the connected pore-oriented perspective will contribute to better structuring of the research efforts in understanding the mechanisms of soil C protection/storage by the "living soils."

Tillage releases CO_2 and mixes soil and crop residues to allow rapid decomposition of SOM (Reicosky and Lindstrom, 1993; Reicosky and Lindstrom, 1995; Reicosky, 1997; Reicosky, 2002; Ellert and Janzen, 1999; Rochette and Angers, 1999; Lee et al., 2006; Quincke et al., 2007). In this way, tillage is a "double negative," rapidly releasing C from the soil and contributing to the increase in the atmospheric CO_2 and an enhanced GHG effect. Tillage under windy conditions loses soil C faster than it is under low wind speeds (Reicosky et al., 2008). Tillage affects wind, water, and tillage erosion, leaching and runoff, GHG emissions, pesticide sorption and degradation, as well as other biophysical processes. Tillage intensity, by affecting the amount of crop biomass on the soil surface and how that biomass is distributed on and anchored to the soil, and by affecting the size of soil aggregates and their stability, has a large impact on wind and water erosion. Tillage, through the action of soil disturbance and the downward force of gravity, causes the slow progressive downslope movement of soil, that is, tillage erosion (Lobb et al., 1995). Soil erosion results in the redistribution of soil within fields and losses from fields. Typically, in cultivated topographically complex landscapes soil loss from tillage erosion is most severe on hilltops.

Tillage creates a priming effect for some microbes with the destruction of the fungal hyphae network structure of the mycorrhizal fungi and micro arthropods (Kuzyakov, 2010), in addition to letting more oxygen into the soil to stimulate the heterotrophs (Reicosky and Lindstrom, 1993). Tillage further modifies the bacteria:fungi ratio with direct impacts on C and N storage and soil (Wilson et al., 2009; Six et al., 2006; Jiang et al., 2011; De Vries et al., 2006; Haddaway et al., 2017). As a result, tillage favors the development of bacteria-dominated communities, which contribute less to C storage in soil relative to fungal communities due to the key role of mycorrhizal fungi in soil aggregation (Wilson et al., 2009). No soil disturbance enables the earthworms and fungal activity to continue without disruption (Stinner and House, 1990; Kladivko, 2001; Lehman et al., 2015; Briones and Schmidt, 2017). Tillage results in a homogenization of SOC within the tilled layer, while SOC accumulates near the surface under NT (Reicosky et al., 1995). Meta-analyses of data from CT and NT (Angers and Eriksen-Hamel, 2008) show no significant differences in soil SOC accumulation, contrary to expectations in terms of soil C storage under NT (Powlson et al., 2014; Haddaway et al., 2017).

SOM indirectly contributes to agroecosystem productivity and the ability to adapt to a changing climate with less frequent and regular precipitation (Chenu et al., 2000; Pan et al., 2009). Within and around the soil particles forming the micropores exists another form of C called mineral associated OC (MAOC) described by Schrumpf et al. (2013), Rabot et al. (2018), Banwart et al. (2019), Kravchenko et al. (2015) and Kravchenko and Guber (2017). Kramer et al. (2017) found that the C:N ratio and 15N natural abundance are controlled by association with minerals. The SOC originates from plants, animals, and microorganisms, and their exudates that enhance aggregation through the bonding of primary soil particles (Bronick and Lal, 2005). Roots and hyphae can enmesh particles together and release organic compounds that hold particles together, a process with a positive impact on soil C storage. This unique combination of soil particles, microbial material, and OC in and around the micropores serves to protect C from oxidation due to the low diffusion rate of oxygen through water films.

Improved understanding of the nature and role of soil structure in arable lands is critical for rational guidance of tillage operations with their agronomic, environmental, and climatic consequences (Reicosky et al., 2011; Friedrich, 2020). There is a need to demystify the benefits and role of NT relative to tillage practices within the spectrum of land management options. A "strategic" application of minimum soil disturbance in CA may result in numerous benefits (Conyers et al., 2019) despite conflicts with the philosophical ideal of no disturbance. The term "managed soil structure" does not preclude natural contributions within disturbed soil volumes and in the undisturbed subsurface, with NT practices further blurring the lines between the binary classification of natural and tillage-managed soil structure.

The most visible effect of applying continuous NT is the rapid development of a mulch cover on the soil surface. Crops that produce less biomass and more easily degraded biomass, such as soybean, when grown in rotation with corn will result in a lower amount of mulch cover. If disturbance is avoided, the amount of mulch cover on the soil surface can be considered at equilibrium. However, even a slight disturbance, and one that may be considered an appropriate conservation tillage practice, can greatly decrease the amount of mulch cover and the length of time the soil remains covered.

Soil disturbances that repeatedly disrupt soil aggregates, such as frequent tillage, are likely to suppress C flows by shifting the balance of interacting mechanisms sharply toward decomposition, resulting in net C loss (Six et al., 2000; Lal, 1993; Van Oost et al., 2007; Janzen, 2004; Reicosky and Janzen, 2019). Evidence suggests, however, that infrequent and/or less destructive tillage practices do not cause the same disruption to soil aggregates and do not result in significant C loss (Conant et al., 2007; Cooper et al., 2016). The soil pore network structure, and particularly the arrangement of macropores that are characteristic of biological activity (i.e., long, cylindrical, non-tortuous, surface-connected macropores), induces changes in functionality, specifically with transport of water, solutes, and mobile particulate matter to deeper layers. Wardak et al. (2022) concluded the value of a prospective examination of an evolving NT pore network both visually and functionally across temporal and spatial scales and highlighted the necessity for standardized methodology to aid in future data compatibility and quantitative analysis (Derpsch et al., 2013).

Because of the myriad of species and ecological interactions in soil, it will be very difficult to gain a quantitative understanding of soil fauna for soil functioning following a reductionist approach (Nannipieri, 2020; Nannipieri et al., 2020). Soil microbial biomass C and soil enzyme activities are affected by NT in a manner similar to that of organic C (Franzluebbers, 2002; Dick, 1984). These soil properties become highly stratified and when the crop production system is changed to NT, the microbial biomass and the biologically active C and N pools respond rapidly and the changes are more easily measured than changes in total C and N. However, not only is there a change in the profile distribution of biological activity, but the biological community itself is also changed with fungi becoming more dominant under NT (Six et al., 2006). Fungal hyphal length in the surface soil layer was 1.9–2.5 times greater for NT compared with conventional tilled fields when evaluated across six widely different geographical sites (Frey et al., 1999). The increased enzymatic and biological activity at the soil surface associated with minimum soil disturbance, NT affects many important functions in soil, such as fertilizer use efficiency, pesticide efficacy, and C storage (Hirsch, 2018; Zuber and Villamil, 2016; Henneron et al., 2015).

Tillage disturbs all soil biology and ecological functions. The "living soil" is full of bacteria, fungi, algae, protozoa, nematodes, and many other fragile creatures affected by intensive tillage and reflects a fundamental shift in care for our soils. Soil is a complex, dynamic, living resource with many micro-, meso-, and macro-biota that are essential for the sustainable production of food and fiber and for the maintenance of global biogeochemical C and nutrient cycling and ecosystem functioning (Hirsch, 2018). Soil tillage is an "apocalyptic event" for the soil organisms causing mortality and C and water loss. Zuber and Villamil's (2016) meta-analysis of 62 studies showed that plow tillage caused an overall reduction in soil microbial biomass and enzyme activities and an increase in CO_2 evolution from increased respiration. Henneron et al.'s (2015) results showed that both CA and organic systems increased the abundance and biomass of all soil organisms, except predaceous nematodes. For example, macrofauna increased from 100% to 2,500%, nematodes from 100% to 700%, and microorganisms from 30% to 70%. CA also increased the number of anecic earthworms, and phytophagous and rhizophagous arthropods. Henneron et al.'s (2015) results suggested that long-term NT associated with cover crops use has greater enhancement potential for maintaining soil biodiversity. In general, larger organisms of the megafauna (organisms > 2 mm, earthworms and large invertebrates) are damaged more by intensive tillage than smaller organisms of the meso- and microfauna and microflora (Ball and Robertson, 1994; Barnes and Ellis, 1979; Black and Okwakol, 1997; Chan, 2001; Folgarait, 1998). This may also be a reason for a shift in the ground beetle population due to a change in the tillage regime, as the prey of some species of ground beetles appears more or less frequently after tillage. Physical interference with the soil, by plowing, results in larger organisms of higher trophic levels being disadvantaged, while small organisms of lower trophic levels are less affected or even benefit to a small extent (Wardle, 1995).

We are slowly understanding the ecological implications of intensive tillage on both the water and C cycles. Isbell et al. (2019) found that 91 years after tillage, formerly plowed fields still had only three quarters of the plant diversity and half of the plant productivity was observed in a nearby remnant ecosystem that had never been plowed. These findings are supported by the review of Reeves (1997) who concluded that long-term tillage studies are in their infancy and these findings shed new light on the implications of "long-term" tillage research for about 30 yrs, since the advent of conservation tillage techniques, and only in developed countries in temperate regions. In most tillage research, paired tillage data sets > 20 years are considered to be long term, often required for NT to show significantly more C storage than conventional tillage-based agriculture (Cusser et al., 2020). Dick et al. (1986a, b) discussed difficulties associated with accurately quantifying changes in soil C stocks and emphasized the importance of following well-conceived sampling and analytical strategies, and the need for carefully evaluating earlier measurements to ensure that they are not inadvertently biased by sampling methods. Their data demonstrated both a tillage and a soil type effect on the changing soil C content. The initial soil C content was lower in the well-drained soil and showed reasonable trends likely reflecting better aeration relative to the poorly drained soil, even though both sites were tile drained. The well-drained soil showed a slightly higher increase in soil C than the poorly drained soil, which was more erratic. At the end of the measurement period, the well-drained NT soil had 18 g kg^{-1} more C than the plowed soil, whereas in the poorly drained soil, NT had about 16 g kg^{-1} more C than the plowed soil. Other soil type differences such as clay mineral type, pH, salinity, etc., may have contributed to the soil differences. Reeves (1997) concluded the long-term SOC changes in continuous cropping studies showing NT enables more C storage than tilled soils include (Ismail et al., 1994; Wiesmeier et al., 2015; Nunes et al., 2015; Daigh et al., 2018; Wiesmeier et al., 2019), while Dimassi et al. (2014) showed no significant difference in soil C after 41 years.

Conservation Agriculture   199

The plant root system acts as a bridge between the crop management and plant growth responses (Klepper, 1990; Kell 2011; Pierret et al., 2016). Root growth and development are affected by soil strength characterized in tillage studies by bulk density (BD) and penetrometer resistance (PR). They are interrelated, but the use of only one may lead to misleading results (Campbell and Henshall, 1991). BD is inversely related to total porosity (Carter and Ball, 1993), and this provides a measure of the pore space remaining in the soil for air and water movement. The optimum BD for plant growth is different for each soil. In general, low BD (high porosity) leads to poor soil–root contact, and high BD (low porosity) reduces aeration and increases PR which limits root growth (Cassel, 1982). BD is related to natural soil characteristics such as texture, organic matter, structure (Chen et al., 1998), and gravel content (Franzen et al., 1994) and varies continuously due to the action of several processes, such as freezing and thawing (Unger, 1991), settling by desiccation and kinetic energy of rain (Cassel, 1982), and soil loosening action by root action and animal activity.

Tillage alters BD to reduce it by increasing the soil porosity; however, this is temporary because following tillage the soil rapidly settles, eventually recovering its former BD (Grant and Lafond 1993; Franzluebbers et al., 1995). In the first few years of NT, soil BD may be slightly increased due to the repeated passes of the tractor and lack of loosening action of tillage that results in better soil structure and an increase of macropores (Martino and Shaykewich, 1994) which benefits root growth (Lampurlanés et al., 2001). NT may result in the stratification of soil nutrients, particularly immobile elements such as P (Crozier et al., 1999). Gregory (1994) showed that this induced a higher root length density (RLD). Rasmussen (1991) and Wulfsohn et al. (1996) showed that roots in an NT system accumulated largely in the 0–5 cm layer when compared to CT. Chan and Mead (1992) showed that the opposite occurred in the lower layers. The root diameters may be indicative of the effect of soil strength on root growth and affect the utilization of nutrients. Sidiras et al. (2001) reported larger diameter barley roots under CT compared with NT. Diameter of the roots is mostly controlled by soil BD, and often compact soil produces thick roots. Slightly different results were obtained in a 5-year field trial of winter wheat at two locations in Switzerland – RLD and mean root diameter (MD) were studied for NT and CT treatments. The results showed that NT slightly lowers RLD and slightly increased MD compared with CT (Qin et al., 2004). However, compared with CT, the RLD was higher in the upper soil layer (0–5 cm), similar for the 5–10 cm depth and lower than the 10–30 cm depth for NT. The tillage effect disappeared below 30 cm (Qin et al., 2004). However, a study of several years using barley with NT and different tillage practices showed that NT produced higher RLD profiles that revealed a better soil condition for root growth (Lampurlanés and Cantero-Martínez, 2003).

8.3.5 Soil Health Metaphor

Soil health (SH) is defined as the suite of soil biological, chemical, and physical properties which enable soils to function as vital living ecosystems that support all life above and in the soil to sustain plant and animal health and productivity and to maintain or enhance water and air quality (Kibblewhite et al., 2008; Karlen, 2012). Soil "health" is a metaphor that captures an important parallel between soils and levels of intervention to deliver the functions required. This metaphor serves as a teaching tool that captures the complexity and an essential parallel between soils and our human health. There is evidence throughout the world that demonstrates the capability of CA as a sustainable system to overcome these adverse effects on soil health, to avoid soil degradation, and to ensure food security. CA has multiple beneficial effects on the physical, chemical, and biological properties of soil. In addition, CA can reduce the negative impacts of CT practices on soil health while conserving the production and provision of soil ecosystem services (Carceles et al. 2022).

Soil C accumulation with cover crops has been linked to soil texture, with an increase in soil C more likely to occur in clay soils with cover crops to improve soil health. In Argentina, studies have shown that cover crops grown on fine- and coarse-textured soils accumulate more soil C

(Alvarez et al., 2017). While cover crops can help eroded soils with low C content accumulate more C (Hassink and Whitmore, 1997; Berhe et al., 2007), the benefits are more visible with NT due to a slower rate of residue decomposition than with CT (Olson et al., 2014). Jian et al. (2020) found using a recent meta-analysis of 131 global studies that incorporating cover crops into the rotation significantly improved SOC, with fine-textured soils showing the greatest increase; the increase was greater in shallow soil (30 cm) than in subsurface soil (>30 cm). This increase in SOC was associated with improvements in soil quality and mineralizable N and C, and was influenced by the annual temperature, the number of years after cover cropping began, initial SOC concentration, and latitude (Jian et al., 2020).

The integration of the hydrologic cycle is also critical to the understanding of the C cycle (Lal and Stewart, 2012). Water is the critical resource for all life, yet drought is pervasive in terrestrial ecosystems. Most ecosystems experience at least occasional drought, and droughts appear to be intensifying faster than the total rainfall with climate warming (Sherwood and Fu, 2014; Soong et al., 2021). Microbial community responses to dry conditions and to subsequent wetting events are complex, and understanding these may prove key in understanding the microbial ecology of soil (Tecon and Or, 2017). Schimel (2018) reviewed how dry conditions affect microbial life and biogeochemical processes in soil and concluded that microbial responses may actually not be very important in regulating how SOM is processed through dry and rewetting cycles. Rather, as soil microbes are generally C starved, whatever biodegradable C becomes accessible will be rapidly metabolized. Hence, the diversity and community composition may be more a response to the shifts in moisture and C availability than an independent driver of the biogeochemical processes, supporting the need for ecosystem biodiversity implemented through cover crops (Basche et al., 2016a; Laban et al., 2018; Araya et al., 2022; Blanco-Canqui et al., 2020; Reicosky et al., 2021; Carceles et al., 2022).

The principles then become improving nutrient cycling; improving water storage, availability, and infiltration; and creating the connected soil bio-pore network so that the roots will explore and provide additional water for that plant to increase WUE. The soil functions and CA principles go hand in hand with all interaction between how soil is absorbing water and how nutrients are being cycled. CA and soil health are interconnected and pointing us to minimizing soil disturbance, keeping soil covered and maintaining a living plant as long as possible with crop diversity as primary tenets because they are all oriented toward feeding the biology within the soil. The "living root" is part of the plant that photosynthesizes by capturing C and putting sugars into the soil that are the food source for the microbes to do useful work. The longer producers can keep feeding microbes through C management, the more would be the opportunities to change and improve the functionality of the soil biology by implementing the soil health concepts.

8.3.6 ROLE OF COVER CROPS IN BALANCING CARBON AND WATER CYCLES

Cover crops are widely considered for enhanced water management at the watershed scale (Unger and Vigil, 1998). Cover crops can even, over time, change the logistics of a farm. With established cover crops, fall labor and tillage costs can be reduced, since those passes are not needed. Cover crop functions, originally meant to "cover the soil" to decrease soil erosion and to fix nitrogen, are now being expanded to incorporate plant biodiversity, increased SOM, increased water infiltration, weed and pest control, nitrogen scavenging, decreased compaction (Keller et al., 2019), generating vertical bio-pores, and being a marketable commodity (Gesch et al., 2014; Finney and Kaye, 2017; Blesh, 2018). The terminology evolution requires definition and clarification with expanded functions, perennials, and "continuous living cover" related to agricultural watershed management. Continuous living cover is used in discussions of water management and quality with enhanced storage and other soil hydrological properties (Basche and DeLonge, 2017). In the northern hemisphere, it is necessary to include the provision "within biological limits" such as extreme temperature

Conservation Agriculture

and water availability during the winter. Depending on the severity of the winter, some cover crops that normally die off in winter may survive. Some winter annual cover crops are terminated by a "killing frost," while others go dormant during extreme cold and recover when temperatures become favorable. Basche and DeLonge (2017) suggested that continuous living cover may be a potential adaptation strategy to combat extreme rainfall variability by changing soil porosity and field capacity (Basche, 2017). King and Blesh (2018) evaluated functional diversity as a way of increasing SOC and found that cover-cropped and perennial cropped rotations, relative to grain-only rotations, increased C input by 42% and 23% and SOC concentrations by 6.3% and 12.5%, respectively. King and Blesh (2018) further suggested increasing the "perenniality" of crop rotations. Functionally diverse perennial- and cover-cropped rotations increase both C input and SOC concentrations by exploiting niches in time that would be otherwise unproductive. Another goal might include for N contribution, which would argue for a later termination date. Producers may need to consider the effect of cover crop termination date on delayed planting and the potential for grain yield loss. These benefits don't come without challenges; one is that farmers don't generally see an immediate economic benefit from incorporating cover crops. A larger challenge, though, may simply be a perceived short growing season in northern latitudes.

Cover crops have major impacts on soil-water relationships (Unger and Vigil, 1998). Nielsen et al. (2015) characterized soil water extraction patterns and determined water use of cover crops grown in single-species plantings and in a 10-species mixture and compared cover crop water use with evaporative water loss from NT fallow with a dryland treatment and an irrigated treatment. They found no consistent significant differences in soil water contents or growing season crop water use with the single-species plantings compared with the 10-species mixture. Cover crop water use (216 mm) averaged 1.78 times greater than evaporative water loss (122 mm) from the NT fallow treatment with crop residue. There appears to be no evidence from data collected in this semiarid environment, even when irrigated to simulate higher rainfall environments, to support the conclusion that cover crops grown in multi-species mixtures use water differently than single–species plantings of cover crops.

Soils hold the largest biogeochemically active terrestrial C pool on Earth and are critical for stabilizing atmospheric CO_2 concentrations and climate extremes. Perennials with deeper roots put much of their biomass belowground and exhibit many positive benefits (Kell, 2011; Pierret et al. 2016; Ledo et al., 2020). Better understanding is needed to predict the C storage capacity of soils and to quantify the percentage of C inputs retained in SOM derived from belowground inputs compared with those from aboveground inputs. While the roots are out of sight, they are not out of mind. The relative contributions to soil C pools of roots versus shoots is one aspect that has been mostly overlooked, although it appears a key factor that drives the fate of plant-tissue C either as mineralized CO_2 or as stabilized SOM. Early work by Balesdent and Balabane (1996) showed that maize roots could provide about 1.5 times more C than the aboveground biomass. Puget and Drinkwater (2001) showed that hairy vetch roots could provide about 3.8 times more C than the aboveground biomass. Kätterer et al. (2011) studied the effect of mineral N fertilizers and different organic amendments on crop yields, SOM changes, and soil physical properties. Their findings support the hypothesis that root-derived C from six different crops contributed 2.3 times more C to relatively stable soil C pools than the same amount of aboveground crop biomass-derived C. They concluded with strong evidence from this experiment that roots contribute more to relatively stable soil C pools than the same amount of shoot-derived plant material. Jackson et al. (2017) summarized the work of researchers using primarily isotopic approaches and found that several different species root systems can put an average of about five times more C into the soil through root exudates, root biomass, and increased microbial biomass in the rhizosphere than the aboveground biomass. With equal to larger quantitative C contributions and longer-lived C in soils than shoots, roots potentially contributed most of the organic C currently stored in soils of most ecosystems. The contribution of diverse cover crop root systems with respect to depth and root density to soil C pools is likely to increase with soil depth where root C contribution should be a consideration (Van Noordwijk 1983; Kell, 2011; Pierret et al., 2016). All the factors, limitations, knowledge, and skills around commodity and cover crops will be needed to minimize the effects of climate change on SOM in the long term.

8.3.7 SOIL FUNCTIONS AND CARBON

Soil function can be any service, role, or task that soil performs, many of which are vital to society and its environment. Soil functions include key life support processes in global ecosystems and are viewed from a "natural" perspective (Banwart et al., 2011; Banwart et al., 2019). Soil functions include biomass production for food, fuel, fiber, and timber; C storage derived from OM that is biologically fixed from atmospheric CO_2; and the mass transformation of nutrients and their plant-available storage. SOM also provides agroecosystems with the ability to adapt to a changing climate with less frequent and regular precipitation and more extreme rainfall, with related infiltration and erosion problems (Chenu et al., 2000; Pan et al., 2009; Vogel et al., 2022). Many soil functions in agricultural ecosystems relate to C management, C cycling, and C energy flow through the soil–plant–atmosphere system as reported by Kuzyakov and Cheng (2004), Wander (2004), and Janzen (2015) and as summarized by Reicosky (2020, see Table 8.1).

Soils perform a broad range of functions, many of which are vital to society and the environment. Biophysical soil functions include C and nutrient cycling, water cycling, chemical filtering and buffering, physical stability, support of plant systems and human structures, and promotion of biodiversity and habitat (Wander, 2004; Dudley and Alexander, 2017; Hird, 2017; Tully and Ryals, 2017). Soils and their management also play a large role in influencing hydrologic cycles that are important to humankind (Reicosky et al., 2011; Hatfield et al., 2017; Friedrich, 2020). In this regard, soil management, especially with respect to intensive tillage, is at the center of water management in agriculture and has direct bearing on food production and security. Building resilient food production systems in the face of increasing population and climate change requires improved water and soil management (Hatfield et al., 2001) to underpin productivity improvements across the entire range of production environments from exclusively rainfed to supplementally and fully irrigated. Recent reviews of agricultural management practices in response to climate extremes of droughts and floods have been provided by Bouma (2009), Lal (2015a, b), Hatfield et al. (2017), Basche et al. (2016a, b), Basche and DeLonge (2017), and Basche and Edelson (2017). Our understanding of the linkages between soil properties and soil functions and the resultant ecosystem services they provide is incomplete (Daily, 1997; Swinton et al., 2006; Adhikari and Hartemink, 2016).

Soil functions are viewed with minimum soil disturbance and dependence on soil C for optimum results in CA systems. The primary soil ecosystem functions include the soil as a medium for plant growth and nutrient cycling (Tully and Ryals, 2017). Decreasing tillage-induced GHG emissions requires minimum soil disturbance and enhanced C input with cover crops to increase C storage (Reicosky and Lindstrom, 1993; Reicosky et al., 2021). Carbon is a major player in contributing to soil functions in agricultural landscapes that indirectly contribute to the greenhouse effect and climate mitigation, in soil health and ecosystem services, and in our food security. Additional soil functions include storing, filtering, and transmitting water, heat, and gases to mitigate flooding and supply plant-available water; attenuate pollution loads; transmit water to aquifers and streams; exchange gases with the atmosphere; provide thermal mass for energy storage; and decompose and recycle waste (Reicosky, 2020). The critical importance of C management in soil functions is supported by numerous studies (Rabot et al., 2018; Wiesmeier et al., 2019; Baveye et al., 2020; Banwart et al., 2011; Banwart et al., 2019; Bardgett and Van Der Putten, 2014; Kuzyakov and Cheng, 2004; Janzen, 2015; Lehmann et al., 2020).

The integrated effects of physical, chemical, and biological components of soil function as a rooting environment are effectively "activated," forming a fourth ecohydrologic component

Conservation Agriculture   **203**

reflecting the biological contribution to water management and improvements in WUE (Rockström et al., 2007; Lin, 2012; Zalewski, 2013; Lemordant et al., 2018; Hauser et al., 2020). Ecohydrology is an integrative science that focuses on the interaction between hydrology and biota (Brussaard, 2012) to reinforce ecosystem services to reduce anthropogenic impacts. CA incorporates holistic approaches that manage hydrology and biota aim to achieve sustainability in both ecosystems and human populations. Ecohydrology principles provide the scientific background for regulating the processes and interactions for enhancing water resources, maintaining and restoring biodiversity (Brussaard, 2012), providing ecosystem services for societies and building resilience to climatic and anthropogenic impacts (Palm et al., 2014; Lemordant et al., 2018; Hauser et al., 2020). Ecohydrology promotes the integration of a catchment and its biota into a single entity and the use of ecosystem properties becomes a management tool in which ecohydrology can address fundamental aspects of water resources management (Lin 2012; Zalewski 2013).

The potential usefulness of SOC levels as metrics of ecosystem functioning (Schmidt et al., 2011) is supported by observations of SOC dynamics following land-use conversions (Houghton, 1999; Post and Kwon, 2000; Poeplau et al., 2011). Often, the response of SOC levels to land-use conversion is asymmetric with rapid SOC loss for conversion of natural to tilled soil, relative to slow accumulation and long equilibration times following the abandonment of tillage (Novara et al., 2011). Ledo et al. (2020) have shown that SOC increases under perennial crops in comparison to annual crops, attributing the trend to the absence of mechanical disturbance by tillage in perennial crops. These studies paint a consistent picture of the effects of introduction or abandonment of tillage on SOC trends.

Crop biomass returns or removal, biological oxidation rates, and soil erosion control the SOC inputs and losses as they impact soil functions in agricultural systems. Reicosky et al. (1995) reviewed tillage and biomass production impacts on SOC in agricultural systems and concluded that SOC was controlled by crop biomass input that was largely determined by crop choice, fertilization, and climate. Reicosky et al. (2002) evaluated the impact of 30 years of continuous corn with moldboard plow tillage with and without biomass (silage removal) and at low and high nitrogen rates. The experimental land area was farmed with conventional tillage and rotation methods prior to the initiation of the experiment. Measurements of SOC changes over the 30 years showed the total C, total N, and C:N ratio of the soil remained unchanged over the 30-year study in the fertilized treatments. All four treatments produced essentially the same SOC content of 2.2% C in the 0–20 cm depth. The 30-year cumulative input of 241 Mg ha^{-1} of the added aboveground biomass yielded no differences in SOC when compared to those treatments with biomass (silage) removed. The results demonstrated the impact of intensive tillage with the moldboard plow that resulted in soil C losses irrespective of biomass removal as corn silage or addition of aboveground crop biomass where only the grain was harvested. These results suggest that no or minimum soil disturbance is required before any addition or removal of aboveground biomass can impact SOC accumulation in CA systems.

8.4 CONSERVATION AGRICULTURE IS CARBON CENTRIC INCREASING WATER USE EFFICIENCY

The CA strategy is based on a set of guiding principles, and practitioners use a variety of tactics that integrate biological and ecological processes with the objective of increasing production and restoring landscape functionality. CAS restore ecosystem services by minimizing soil disturbance, which leaves organic crop biomass on the soil surface reducing erosion, increasing C and WUE through enhanced infiltration, deep percolation, plant available water storage, groundwater recharge, stream baseflow and decreases in peak stream flows and downstream flooding. Soil C management is a critical part of CAS with specific impacts of C management on WUE. CA is a holistic concept for sustainable management of agricultural drylands increasing WUE through enhanced soil and C management that will help ensure a more stable food supply for future generations.

8.4.1 Water Use Efficiency in Dryland Agriculture

Conservation has been the key concept in sustainable production systems throughout the world, enhancing WUE in dryland areas (Power, 1983; Ritchie 1983; Klocke et al., 1985; Heitholt, 1989; Stewart and Koohafkan, 2004; Stewart, 2008a, b; Hatfield et al., 2001; Nielsen et al., 2005; Stewart 2009; Stewart et al., 2010). Conservation has been effective at sustaining crop production in semi-arid rainfed regions. Wherever potential evaporation exceeds precipitation during most months of the year, proper application of water-conserving CA technology is critical. The unwillingness of farmers to adopt CA, or to adopt it continuously, implies that it is either perceived to be unprofitable or that other significant constraints to adoption exist. Because the world is a diverse place and farmers face different constraints imposed by soils, climates, financial wealth, customs, and national policy, it is imperative to continue the evolution of CT into the adoption of integrated CA systems and the opportunities to overcome barriers to greater adoption. It is imperative that CA retains a flexible approach to addressing challenges while maintaining the common goal to sustain the soil resource and produce food for an ever-growing world population. The following discussion focuses on soil C management as a critical aspect of CA systems and specific impacts of C management on WUE. Most benefits of CA, either directly or indirectly, relate to enhanced C management (Reicosky and Janzen, 2019; Reicosky, 2020; Reicosky and Kassam, 2021).

8.4.2 Tillage Impacts on Water Use Efficiency

Reducing the amount of tillage or avoiding tillage in cropping systems on the prairies has long been recognized as an important strategy to reduce soil erosion and degradation and improve soil health and productivity (Hill, 1990; Lal, 1991; Lal et al., 2007). Soils with minimal levels of crop biomass on the surface are more susceptible to soil erosion, and the surface soil structure is less stable (Burnett et al., 1985; Chang and Lindwall, 1989, 1992). Tillage is the "mechanical manipulation of the soil for any purpose" (Soil Science Glossary Terms Committee, 2008; Reicosky and Allmaras, 2003). It has been perceived as important to seedbed preparation, weed control, and incorporation of agricultural chemicals or amendments. However, tillage destroys SOM through two interrelated processes. First, OM at depth in the soil is slower to decompose as soil temperature and moisture levels vary more slowly at depth and oxygen partial pressure is often lower. Inversion tillage with moldboard plowing brings this OM to the surface and decomposition through oxidation is speeded up by an order of magnitude. Tillage disturbance has upset nature's "power balance" emitting soil C as CO_2 resulting in soil C loss and adding to GHG emissions from fossil fuel consumption – a double negative for agriculture's C footprint.

Conventional intensive tillage agriculture (CT) competes with nature, disturbs the soil and soil biology; uses reduced crop diversification or monoculture techniques; uses a reductionist approach in management requiring higher inputs of fossil fuels, equipment expense, and maintenance, as well as synthetic fertilizers, herbicides, pesticides, and insecticides (Lal, 1991; Reicosky and Allmaras, 2003). Modern CT has done little to integrate conservation into agriculture, as the main focus has been on increasing crop and livestock production, in terms of both yields and total output in rainfed and irrigated agriculture based on the narrow Green Revolution approach. Farmers in different parts of the world have also been reacting to address and overcome the inherent degradation and loss in productivity. The farmers, along with some extension agronomists and few machine companies, began to replace CT with NT agriculture and introduced soil and water conservation practices that eventually led to an alternate paradigm now generally referred to as CA.

Tillage practices receive attention for their role in changing SOC stocks over seasonal and decadal time scales (Moreira et al., 2016); moldboard and chisel plowing are implicated in large losses of SOC at regional scales, whereas CT or NT practices are suggested as means for preserving SOM (Lal et al., 2007; Reicosky and Lindstrom, 1993; Reicosky, 1997a, b; Reicosky and Archer, 2007; Reicosky et al., 2011; Friedrich, 2020). The act of tillage creates an instantaneous change in the soil

Conservation Agriculture 205

environment by opening physical spaces between soil aggregates, breaking aggregates, allowing for greater exchange of GHGs between the soil and atmosphere, and inducing immediate abiotic oxidation of freshly exposed organic C-containing surfaces (Homyak et al., 2016). Recent work by Alcántara et al. (2016) documented that deep plowing (55–90 cm) techniques can increase SOC on decadal time scales, and other reviews concluded that reports of increased SOC from lower intensity plowing are overstated due to historical bias toward shallow soil sampling (Baker et al., 2007; Powlson et al., 2014). Hatfield et al. (2001) concluded that it is possible to increase WUE by 25–40% through soil management practices that involve tillage. Overall, precipitation use efficiency can be enhanced through adoption of more intensive cropping systems in semiarid environments and increased plant populations in more temperate and humid environments.

The ecological implications of inversion tillage with the moldboard plow in some instances appeared to last for a long time after the plowing is discontinued. For example, Isbell et al. (2019) found that nearly 100 years after agricultural abandonment, local plant diversity recovers only incompletely, and plant productivity does not significantly recover. By 91 years after agricultural abandonment, despite gaining many local species, formerly plowed fields still had only three-quarters of the plant diversity and half of the plant productivity observed in a nearby remnant ecosystem that had never been plowed. Clearly, soil ecological disturbance takes much time to recover. The general goal is to integrate into the very conception of CA farming systems an awareness of the influence of agronomic techniques and practices (e.g., crop rotations, crop associations, intercropping, cover crops, and crop biomass management) on microbiome abundance, biodiversity, and activity by maintaining energy flow through the system. One way is to increase the use of leguminous crops to get maximum ecosystem benefit from the entry of nitrogen into soils via biological fixation.

8.4.3 Soil Evaporation

Early studies have shown that mulch-covered NT soils reduced surface evaporation, maintained moisture near the soil surface at a higher level, and created a favorable environment for root development at or near the mulch–soil surface interface (Van Doren and Allmaras, 1978; Triplett and Van Doren, 1969; Jalota and Prihar, 1990a, b; Todd et al., 1991). Tanner et al. (1960) estimated that in widely spaced row crops such as corn, if the soil is wetted frequently by rain or irrigation, soil evaporation may be as high as 50% of ET, even when the canopy is fully developed. Blevins et al. (1971) reported less water use under mulch provided by a killed sod early in the growing season, compared with a tilled treatment. After the crop canopy developed, water use was similar for NT and tilled systems and reflected transpiration losses. Burns et al. (1971) and Papendick et al. (1973) showed that tillage disturbance increased soil water evaporation compared with NT areas. Tillage moves moist soil to the surface where water losses to drying are offset by short-term increased infiltration rates.

There is potential for increasing capture and storage of precipitation in dryland areas, and for utilizing it more efficiently for growing crops. During the past few decades, major advances have been made in capturing and storing more water, but there have been fewer advances in using the water more efficiently. In the USA Midwest with typical rainfall patterns, soil evaporation accounted for as much as 50% of total water use where the soil surface was frequently wet from seasonal rains (Peters and Russel, 1959; Peters, 1960), with similar estimates in Africa (Rockström et al., 2009). In row crops, sufficient energy reaches the ground surface to account for the observed soil water evaporation. The more intensive systems allow more crop production during the time when precipitation occurs and shorten the long summer fallow periods that lose large amounts of water by evaporation. Mitchell et al. (2012) found that coupling NT with high-biomass preservation practices could reduce soil water evaporative losses during the summer season by about 102 mm, or 13%, assuming a seasonal ET demand of 762 mm. The reductions in soil water evaporation that have been measured add to the list of benefits of CA systems for California, USA, producers.

The use of CA systems has in a few cases doubled soil water storage during the fallow periods and increased yields significantly and greatly reduced risk (Noellemeyer et al., 2013). Even so, the amount of total precipitation used as ET is in the range of 50–55%, and only 55–60% of the ET is used for T. Therefore, the potential for improvement is enormous, but challenging (Rockström et al., 2009). Transpiration varied in only a minor way with soil water supply. These studies reinforce the importance of continuous biomass cover protecting the soil surface that can result in increased WUE in CA systems.

8.4.4 EVAPOTRANSPIRATION

The hydrologic cycle is also fundamental to the understanding of C cycle as well as the nitrogen cycle (Lal and Stewart, 2012). In view of this integration and interaction, the use of cover crops requires enhanced management to maintain a balance between the water required for plant growth and yield and C exported as grain and as the energy source for the microbiomes to recycle nutrients and produce a soil structure required for optimum microbial activity, root function, and watershed hydrologic characteristics. The management challenge here is coordinating and synchronizing the water and C cycle with the thought to optimize soil functions for optimum WUE.

Crop biomass can reduce the evaporation from soil by half, even beneath an irrigated crop canopy. The goal is to reduce the energy reaching the evaporating surface to protect the soil surface from direct raindrop impact (Todd et al., 1991). We may be talking about seemingly small increments of water savings in the case of crop biomass only. No matter how efficient sprinkler irrigation applications become, the soil is left wet and subject to evaporation. Frequent irrigations and shading by the crop leave the soil surface in the state of energy-limited evaporation for a large part of the growing season. Research has demonstrated that evaporation from the soil surface is a substantial portion of ET (Peters, 1960; Rockström et al., 2009; Mitchell et al., 2012). These measurements have been 30% of ET for E during the irrigation season for corn on sandy and silt loam soils (Klocke et al., 1985; Klocke et al., 2009; Van Donk et al., 2010).

Dry land cropping systems can take advantage of stored soil moisture by alternating shallow and deep-rooted crops. For instance, alternate winter wheat, a shallow-rooted crop, with safflower, a deep-rooted crop. WUE of maize improved by 18–56% by including broadleaf crop in a grass-based rotation (Fisseha and Tewodros, 2015). Cropping systems in the northern Great Plains tend to be more diverse and research results have suggested that seed yield of flax (*Linum usitatissimum* L.) can be tripled with a safflower (*Carthamus tinctorius* L.) flax crop sequence versus a flax–flax crop sequence (Tanaka et al., 2005). Maillard et al. (2018) concluded from a 30-year study in Saskatchewan that there was a more pronounced tillage effect in a fallow–wheat rotation than in continuous wheat, while there was no increase in C stocks in the semiarid prairie. The increase of the apparent biomass decomposition with increased rainfall exceeded that of the plant biomass C inputs.

Crop biomass retained as vegetative mulch (practice-plant biomass, stubble, and cover crops) on the soil has effects on many soil-related structural components and processes of the agroecosystem. The effects of permanent vegetative cover as mulch vary according to climate conditions, and factors such as biomass amount and the type of biomass can influence crop responses to mulching. Enhancing and maintaining vegetative cover on the soil surface, using crops, cover crops, or crop biomass and stubble (roots stock with straw or stalk), protects the soil surface, conserves water and nutrients, promotes soil biological activity and habitat, and contributes to integrated weed and pest management. Ranaivoson et al. (2017) reviewed agroecological functions of crop biomass under CA and found that ~8 Mg ha^{-1} of biomass was needed to decrease soil water evaporation by about 30% compared to NT bare soil, and a minimum of at least 2 Mg ha^{-1} of biomass was required to achieve the maximum effect on soil water infiltration, water runoff, and soil erosion control. They found a weak response with increased vegetative cover on soil meso- and macrofauna abundance and soil macro-nutrient supply (N, P, and K). Noteworthy was the average increase in annual SOC

Conservation Agriculture

gain with increased amounts of biomass mulch, with a mean of 0.38 Mg C ha^{-1} year^{-1} and 4–5 Mg ha^{-1} of biomass supplied. Ranaivoson et al. (2017) suggested that optimal amounts of vegetative cover in CA will largely depend on existing constraints to crop production which can be addressed with mulching.

The water saved plus other synergistic benefits of increased C in CA systems may eventually lead to increased WUE, ecosystem services, and food security (Hatfield et al., 2011). Active roots near the soil surface may permit the use of water from light rain showers that don't soak into the soil profile. Van Donk et al. (2010) measured soil water content change to a depth of 1.68 m using a neutron probe to characterize the difference between biomass covered and bare soils in corn canopies and found that the crop biomass decreased evaporation between 65 and 100 mm in the 2007 growing season and 90 and 125 mm in the 2008 growing season. CT that was converted to NT under overhead irrigation yielded annual water savings of 203 mm (Van Donk et al., 2010). Surface coverage and amount of crop biomass influenced E directly (Klocke et al., 2009). They found that E was reduced nearly 50% compared with bare soil E when corn stover and wheat stubble nearly covered the soil under a corn canopy during the growing season. Partial surface coverage, from 25% to 75%, with corn stover caused small reductions in E compared with bare soil when there was no crop canopy. Full surface coverage reduced energy-limited E by 50–65% compared with E from bare soil with no shading.

There are two primary considerations: the goal for the cover crop and objectives related to a grower's management perspective. First, cover crops are generally being grown to achieve soil health objectives with an interest in root growth (Kell, 2011), erosion control (Montgomery, 2007), or both, which might argue for an early or timely termination, or they are being grown for forage production (top growth), which might call for a later termination date.

8.4.5 Infiltration – Compaction Effects

Water infiltration into soil is a complex process that involves saturated and unsaturated flow. The initial stage involves unsaturated flow that is driven primarily by the attraction of water to dry soil particles and the surface tension of water held in the spaces between the particles (Kirkham, 2014). Gravity and soil solute content also affect unsaturated water flow (Hillel, 1998; Unger et al., 2010). Unsaturated flow dominates infiltration as long as the precipitation rate does not cause water ponding on the surface. When the application rate exceeds the unsaturated flow rate, saturated flow becomes dominant. Saturated flow is dominant unless precipitation application is of low intensity or short duration, or for coarse-textured soils in which water flow is rapid (Unger et al., 2010). Rainfall energy strongly influences aggregate dispersion and surface sealing, thereby also strongly influencing infiltration (Eigel and Moore, 1983; Giménez et al., 1992; Loch, 1989). Bare soil surfaces resulting from plowing were unprotected against the impact and energy of falling raindrops, which disrupted soil aggregates and thereby led to reduced infiltration. Besides passing through the surface soil, water must penetrate to adequate depths for storage in the zone from which plants use it. Vertical distribution, namely, water penetration to depths below the surface layer, is part of the infiltration process.

Surface conditions influencing infiltration include soil texture, aggregate size and stability, and surface water content (Burns et al., 1971; Papendick et al., 1973; Unger et al., 2010). Subsurface conditions influencing infiltration include soil texture, water content, structural stability, and horizon characteristics. These influence infiltration through their effect on unsaturated and saturated water flow in the soil profile. When rainfall or irrigation causes water ponding on the surface, entrapped air in soil pores can also reduce infiltration (Wangemann et al., 2000). In contrast, infiltration under ponded water conditions can be greatly enhanced when bio-pores formed by soil fauna (e.g., worms, insects, spiders, etc.) and decayed roots are open to the surface (Cochran et al., 1994; Kladivko, 1994; Briones and Schmidt, 2017). Araya et al. (2022) studied the long-term impact of contrasting tillage and cover cropping systems on soil structure and hydraulic properties and found that NT and cover crop plots had lower water content at field capacity and lower plant-available water compared with standard tillage and plots without cover crops. They concluded that the long-term practices of NT and cover crop systems were beneficial in terms of changes to the bio-pore network and size distribution. Nouri et al. (2019) found that long-term (34 years) incorporation of cover crops in NT significantly improved the infiltration rate, and field-saturated hydraulic conductivity and increased the mean weight diameter of aggregates by promoting the macro-aggregation, less compaction (Keller et al., 2019), and enhanced water storage.

The combination of permanent soil cover and minimum soil disturbance in CA reduces runoff, leading to higher infiltration rates and more water available to crops. Morell et al. (2011) experimented in the high to low water-holding Mediterranean soils of Spain and reported that NT strongly enhances soil water, crop WUE by improving soil quality infiltration and reducing soil evaporation. NT improves the pore size distribution and enhances water availability due to higher OM retention and aggregation (Frey et al., 1999). Surface mulches maintained under NT increase water infiltration and reduce water loss by evaporation (Jarecki and Lal, 2006). Yang et al. (2020) compared the effects of CT and NT in a typical dryland region of China, and reported that soils under the NT system had higher moisture content and WUE than CT. López-Fando and Pardo (2011), in a semi-arid region of Central Spain, observed ~ 30% higher soil water content and WUE under NT than under CT. Other studies (Pikul and Aase, 1995; Shukla et al., 2003) have found higher infiltration rates under NT than CT because of the protection of the soil surface and effect of SOC. Kemper et al. (1987) found that less intense tillage not only kept the crop residue at the soil surface but it also increased the activity of surface-feeding earthworms, leaving the root channels undisturbed, which in turn leads to the presence of numerous surface-connected macro-pores and inter-pedal voids resulting in higher infiltration.

8.4.6 Water Storage – Pore Space and Bio-pores, and "Sponge Effect"

Soil water retention is favored by SOM (Hudson, 1994; Rawls et al., 2003), an important benefit in some drought-stressed soils (Allison, 1973). SOM also contributes to the soil cation exchange capacity (CEC), to the WHC (Bouyoucos, 1939; Hudson, 1994; Rawls et al., 2003), and to soil structure development (Bronick and Lal, 2005; Banwart et al., 2019), which all lead to increased WUE (Basche et al., 2016b), which makes it difficult to identify a specific nutrient and amount contributing to increased productivity.

Bio-pores, which are important to the infiltration of water in water-saturated soils and to soil aeration, include root channels and earthworm holes (Kemper et al. (1987); Kemper et al. 2011). Earthworms also affect the hydraulic properties of soil, e.g. by causing bypass flow to greater depths (Edwards et al., 1990). Earthworms have an important effect on soil structure; a large population of earthworms produces a better structure by aerating the soil and increases the availability of organic nutrients (Kladivko, 2001, 1994; Briones and Schmidt, 2017). Crop yield (quality and quantity of the crop) is affected by many soil properties and processes and is the most visible sign of the health of agricultural soils.

Those gains in soil structure can translate into better water infiltration and retention for the crops. Before the water can be stored in the soil, it must be infiltrated into the soil that requires good soil structure and bio-pore network development to enhance infiltration and the percolation to increase the volume of water stored in the pores and the additional volume of water stored within the organic matter as the "sponge" effect (Basche, 2017). The water is being stored in the soil for use sometime later in the year.

Tillage also destroys much of the biological soil microbiome with concurrent reduction of natural soil fertility and increased compaction (Johnson-Maynard et al., 2007; Pires et al., 2017; Piron et al., 2017; Wardak et al., 2022). Particularly, the permanent network of multi-directional bio-pores

Conservation Agriculture 209

produced by earthworms, termites, and decayed plant roots are destroyed and with this the capacity of soils to infiltrate and drain highly intensive rainfall into the ground water bodies (Edwards et al. 1990; Kemper et al. 1987; Kemper et al., 2011). Modern mechanized agriculture also leads to subsoil compaction beyond the reach of mechanical subsoilers to break them (Turner 2004). In the absence of deep reaching earthworm bio-pores and root channels, tillage-based systems have no means to eventually remove this subsoil compaction (Keller et al., 2019; Friedrich, 2020).

Soil pores control the movement of gasses, water, and the microbiome, thus potentially influencing new photo-assimilated C gains and losses. Quigley and Kravchenko, 2022, explored the associations between soil pores and additions and losses of root-derived C in young cereal rye (*Secale cereale* L.) plants grown in soil with inherent/developed pore architecture that was destroyed by "dry" sieving and in soil with intact pore architecture, with each rye-planted container having a section inaccessible to plant roots. They found root-derived C, fungi accessible pores, and pore associations all changed with "disturbance". If the "disturbance" through soil sieving is considered analogous to "intensive tillage disturbance" in CT agriculture, the significance of the primary principle of CA, "minimum soil mechanical disturbance," provides scientific evidence of the importance of mechanical disturbance with microscopic pores ranging from 30 to 90 μm that are not readily visible to the naked eye are critically important in C accumulation and management. Kravchenko and Guber (2017) concluded that soil pores are both drivers and products of a variety of soil processes that ultimately determine physical protection of OM in soil by influencing oxygen and C accessibility to microorganisms and that any mechanical disturbance to this microcosm requires a recovery interval to get back to "natural microbial activity in soil,"

8.4.7 RUNOFF/EROSION

Soil erosion is still a major problem in agricultural production systems. More than one civilization has fallen after failing to protect and nurture soils and their functions (Hillel, 1991). Soil erosion likely hastened the decline of the Greek empire more than 2,000 years ago because people did not understand the irreversibility of soil loss. More recently, huge amounts of N were lost from Great Plains soils before their native stocks had even been measured (Burnett et al., 1985). Our task is to ensure that our own unfolding society, and that of our successors, does not falter for want of understanding its foundation – the soil. Historical lessons from impacts of intensive agriculture and plow-based tillage provide modern civilization learning opportunities that can and must lead to a brighter future.

According to Pimentel et al. (1995), about 430 million ha – almost one-third of the global arable land area – has been lost to soil erosion. Efforts to control human-induced land degradation and soil erosion over the last 10,000 years have been building on the ruins of past tillage and monoculture concepts (Lal et al., 2007; Montgomery, 2007; Van Oost et al., 2007). Soil erosion on agricultural land is induced by tillage; soil that is loosened by any type of tillage is more easily transported by wind or water, increasing the rate of erosion. The environmental damage takes a number of forms: erosion and salinization of soils, deforestation as more land is brought into cultivation, fertilizer runoff that ultimately creates enormous "dead zones" around the mouths of many rivers, loss of biodiversity, fresh water scarcity and agrochemical pollution of water and soil (Lal et al., 2007). Tillage also destroys and disturbs much of the biological activity in the soil, leaving it lifeless and robbed of its biological fertility and susceptible to erosion by tillage, wind, and rain that all require additional fertilizer (Power, 1983).

Soil erosion on agricultural land is induced by tillage and other forms of disturbance, combined with inadequate soil vegetative cover, resulting in soils with poor aggregate stability and structure that are fragile and more easily transported by water and wind. However, water erosion does not only lead to soil loss, but also to water loss. The surface runoff causing the erosion is water which does not infiltrate and renew the underground water resources but instead runs directly down to the rivers, along with soil sediments, agrochemicals, and microorganisms, creating flooding and pollution in lower parts of the watershed. These situations are more common in tropical and subtropical areas, susceptible to torrential rains during the rainy seasons alternating with droughts due to reduced soil and underground water resources between rainfall events (Turner, 2004). However, tilled and unprotected soils in temperate regions are equally susceptible and suffer from extensive water and wind erosion, particularly under a climate change scenario (Fryrear, 1985). Where water from watersheds drains into a reservoir or a lake, tillage-based land use leads to filling up the water body with sediments, eutrophication, and asphyxiation of aquatic life. Where the water body is part of the hydroelectric generating complex, tillage-based agriculture in the watershed drastically reduces the useful operating life of the turbines and the reservoir-dam infrastructure (Mello et al., 2021).

Seta et al. (1993) found on a Typic Paleudalf in Kentucky, simulated rainfall (66 mm hr^{-1} during three events totaling 2 hr during a 26-hr period) on 0.01-ha bordered plots resulted in 34% of water as runoff under moldboard plowing, 22% as runoff under chisel plowing, and 6% as runoff under NT (Seta et al., 1993). The corresponding soil loss was 15.5 Mg ha^{-1} under moldboard plowing, 3.3 Mg ha^{-1} under chisel plowing, and 0.3 Mg ha^{-1} under NT.

8.5 CONSERVATION AGRICULTURE BENEFITS

The successful application of CA requires that all principles and practices should be applied and managed concurrently for maximum benefits. CA principles are guided by key conservation principles (Hobbs et al., 2008; Delgado et al., 2011; Kassam, 2019; Kassam and Kassam 2020 a,b). Emphasis is on conservation practices that have important regenerative and protective potential for developing continual production and environmental benefits in food and agricultural production systems. Applied together in a system context, they provide synergistic benefits of the production system, but the application of one or two practices without the others provides less than anticipated results. Within each of the principles, there can be any number of conservation practices, applied individually or in combination, providing that they do not contravene the basic tenets of the foundation principles.

While CA is considered C centric, we are still learning more about soil C storage, energy flow, and its central role in direct environmental benefits related to air quality, we must understand the secondary environmental benefits of CA's foundation of NT and what they mean to sustainable agriculture (Reicosky and Saxton, 2007; Reicosky and Kassam, 2021). Understanding these environmental benefits and getting the conservation practices implemented on the land will hasten the development of harmony between man and nature while increasing the production of food and fiber. Increasing soil C storage can increase infiltration, increase fertility, decrease wind and water erosion, minimize compaction, enhance water quality, decrease C emissions, impede pesticide movement, and enhance environmental quality. Accepting the challenges of maintaining food security by incorporating C storage in conservation planning demonstrates concern for our global resources. This concern presents a positive role for CA that will have a major impact on global sustainability and our future quality of life.

A major conservation benefit from CA is greater natural capacity and resilience to change, due to the synergy and benefits from several conservation practices incorporated into a unified and synchronized land management strategy (Reicosky, 2020). Conservation water management captures and holds more rainfall, particularly under torrential tropical storms. This decreases surface erosion, makes more soil water available to crops, and increases WUE and water productivity (Turner, 2004; Rockström et al., 2010; Basche et al., 2016a, b; Basche and DeLonge, 2017). Under irrigation, water requirements are reduced by 30–50% with CA due to increased water infiltration and retention (Basche and DeLonge, 2017) and reduced soil evaporation (Friedrich, 2020). In rice-based CA systems, flood irrigation can be replaced by other forms of irrigation, such as subsurface

Conservation Agriculture 211

drip irrigation, or bed-and-furrow system, to maintain aerobic conditions in the soil and reducing water use as well as emissions of methane and nitrous oxides resulting from anaerobic soils (Turner et al., 2015). At the watershed level, water services are disrupted by tillage-based agriculture to the extent that watershed hydrology does not function effectively and there is a significant loss in water storage recharge capacity of groundwater and deeper aquifers as well as of water quality. Watersheds with CA-based land use behave exactly the opposite (Mello et al., 2021). When new CA technologies and systems were extended to farmers through a farmer-led network, collaborating with the ITAIPU electric generating powerplant and the community made it possible to develop both agricultural and electrical power production practices toward greater productivity and economic, social, and ecological sustainability.

CA utilizes the soil profile as a "giant buffer" for water storage and utilization for optimum WUE. Good C management in CAS enhances WUE at the watershed scale (Lemordant et al., 2018) and suggests the need to better understand the coupling between the hydrologic and C cycles in agroecosystems. It is equally important to understand the benefits and disadvantages in relation to annual versus perennial crops (De Oliveira et al., 2018). The concept of integration in both CA and SH principles links soil (physical, chemical, and biological) properties and now ecological properties in performing the required critical ecosystem functions (Kibblewhite et al., 2008; Karlen, 2012).

With minimum soil disturbance as single conservation practice and foundational principle of CAS, significant increases in global dryland WUE are being noted (Hassan et al., 2022). Basamba et al. (2006) found higher volumetric water content and retention in Colombian soils under NT than under CT. Veiga et al. (2008) examined the improvement in the volumetric water content and WHC in Brazilian soils under NT compared to CT. Long-term NT and CT (deep plowing) effects were observed in the Albic Luvisols of Poland, and it was suggested that NT significantly enhances the water content and its availability compared to CT (Małecka et al., 2012). Rasmussen (1999) found higher water content, high infiltration in the plow layer, and low evaporation due to plant biomass on Scandinavian soils under NT. Qamar et al. (2012) showed in Pakistan that WUE under CT was less than NT. Khorami et al. (2018) investigated Haploxerepts soils and reported higher water, WHC, and WUE under NT than under the CT system, leading to the conclusion that NT is effective for improving water conservation, WUE, and mitigation of soil salinity (Lee et al., 2006) by reducing ET. Basamba et al. (2006) indicated that straw cover due to NT increases soil water content and prevents soil salinity, owing to less water loss through ET, with a potential yield increase. Multiple studies on CT and NT effects on volumetric water content and WUE and significant consistent improvement under NT are summarized by Hassan et al. (2022). As the other primary principles are incorporated into more complex CA systems around the world, further increases in WUE can be expected.

CA systems restore ecosystem services by minimizing soil disturbance, which leaves organic crop biomass on the soil surface reducing erosion, increasing C and WUE through enhanced infiltration, deep percolation, plant-available water storage, groundwater recharge, stream baseflow and decreases in peak stream flows and downstream flooding (Rabot et al., 2018; Sithole et al., 2019). Soil C management is a critical part of CA systems with specific impacts of C management on WUE. Benefits of CA are either directly or indirectly related to enhanced C management. CA systems provide C and energy flow, clean air and water, biodiversity, genetic resources, pollinator and wildlife habitat, pest control, natural medicinal products, and recreational and other aesthetic benefits. The enhanced soil structure in CA utilizes the entire soil profile as a "giant buffer" for water storage and utilization for optimum WUE (Rabot et al., 2018). Good C management in CA enhances WUE at the watershed level as reported by Lemordant et al. (2018) who suggested the need for better understanding the coupling between the water and C cycles in agroecosystems (Lal and Stewart 2012).

All C forms can have a direct or indirect effect on soil structure for efficient infiltration and WUE, and contribute to enhanced WHC (Hudson, 1994; Rawls et al., 2003). The major benefit of CA systems on soil health is the relationship between SOC and the soil's capacity to infiltrate and hold plant-available water through enhanced soil structure and bio-pore networks (Edwards et al., 1990; Rabot et al., 2018; Sithole et al. 2019). At the landscape level, CA enables environmental services on a larger scale, particularly C storage, cleaner water, reduced erosion and runoff, and enhanced biodiversity. Cover crops can do a lot to increase WUE in our production systems, with plant C as our best water management tool (Rabot et al., 2018). Using a meta-analysis, Basche et al. (2016b) found that long-term cover crops improve soil water content. The cover crop increased the water retained in the soil at water potentials associated with field capacity by 10–11%, as well as increasing plant-available water by 21–22%. They concluded that a winter rye cover crop, if managed properly, can improve soil water dynamics without impacting cash crop growth and yield. Basche and DeLonge (2017) used meta-analysis to compare intensively tilled systems with NT annual and perennial systems including cover crops. They found that with reduced soil disturbance, surface cover systems increased porosity and water storage capacity. They further suggested that continuous living cover may be an adaptation strategy to combat rainfall variability and intensity by allowing more water to infiltrate to a greater depth (Kell, 2011; Kemper et al., 2011).

8.5.1 MITIGATE CLIMATE EXTREMES

Agricultural activity must face the challenge of climate change, which is causing an increase in the severity and frequency of extreme weather events and natural disasters (Easterling et al., 2019; Anderson and Bausch, 2006), because the greatest impact of climate change will occur in global arid and semiarid regions. More specifically, increasing temperatures and the constantly changing magnitude and frequency of meteorological events are causing growing concerns about new strategies to react to these threats, which will be even more impactful in the growing arid and semiarid areas (Turner, 2004; Jarvis et al., 2008; Cabrera and Lee, 2018). Di Santo et al. (2022) summarized useful information for policymakers and stakeholders for the development of efficient strategies for management of dryland farming. Their results highlight the need to include farmers to implement participatory policy and to summarize the main adaptive and technical innovations implemented by farmers. The importance of the concept of societal resilience and the need to analyze agricultural systems by considering their multifunctionality and ecosystem services along with purely technical, agronomical, and ecological features that all point to sustainable CA systems are the foundation of successful policy. Although not mentioned explicitly, Di Santo et al.'s (2022) summary points to the importance of establishing principles of CA systems similar to Stewart (1988), Stewart (2008a, b), Stewart and Koohafkan (2004), Stewart et al. (2008), Thierfelder and Wall (2009), Thierfelder et al. (2017), Kassam (2019), Kassam et al. (2014a), and Kassam and Kassam (2020 a,b).

Concerns have also been raised about the increased homogeneity of food supply at a regional and global scale, resulting in a general decline in global food security with 85% of countries showing 3 of 11 marginal or low food self-sufficiency indices (Khoury et al., 2014; Puma et al., 2015). Calls have also been made to revisit issues of agricultural sustainability concerning the impending environmental impacts of climate change and its effect on agriculture (Sachs et al., 2010; Fedoroff et al., 2010; Vermeulen et al., 2013). With potential global crop yield losses of over 50%, calls have thus been made to develop more "resilient" production systems to better withstand the impending impacts of climate change (Ekström and Ekbom, 2011; Lin, 2011; Gomiero et al., 2011; Heinemann et al., 2014).

Concerns have also been raised that industrial agricultural practices are exacerbating the anthropogenic causes of climate change by contributing about 25% of global GHG, about 60% of nitrous oxide emissions from the use of synthetic chemical nitrogen fertilizers and pesticides, and from its adverse impact on biodiversity (Barker, 2014; Turner et al., 2015). Furthermore, new data indicate that previous estimates of nitrous oxide emissions from industrial agricultural systems may have been grossly underestimated, and that when "riverine" watershed emissions are considered, the

Conservation Agriculture 213

levels of nitrous oxide emissions from areas such as the Midwestern USA may be up to 40% greater than what was earlier estimated (Turner et al., 2015). The overall global environmental impact of these increased emissions on climate change could be significant as other similar regions of the world where intensive industrial farming practices are followed represent, globally, an area of over 230 million ha.

More conservation with emphasis on minimum soil disturbance and maximum C management will be required for sustainable production so important to future generations. Adaptive and mitigating solutions will continue to be complex and evolving. Climate change and climate variability present us with moving targets. Ecosystems are interconnected and interdependent natural resources. It is important to be holistic and look at the whole system to recognize stresses on individual components. We must develop collaborative, interdisciplinary approaches to be more efficient and to recognize the interconnectedness and need for conservation of all the resources (Delgado et al., 2011).

8.5.2 ENHANCED ECOSYSTEM SERVICES

Ecosystem services are defined as the important benefits for human beings that arise from healthily functioning ecosystems, notably in photosynthesis and production of oxygen, soil genesis, and water detoxification. Soil ecosystem services are vital components to all aspects of life and support the production of ecosystem goods and services, such as food and fiber production, water storage, and climate and natural hazards regulation, among many others. Nature's warehouse of ecosystem services as a community of living organisms, including humans, interact as a system with nonliving components of their environment (things like air, water, and minerals). Since ecosystems are a network of interactions they can be of any size, but are usually referred to as specific types found in certain places (Palm et al., 2014). An agricultural landscape may contain a mosaic of interconnected ecosystems (Banwart et al., 2019).

Agriculture, which is practiced on 40% of the Earth's land surface, both provides and depends on ecosystem services (Daily, 1997; Swinton et al., 2006; Lal, 2013a, b; Adhikari and Hartemink, 2016). For example, crop production, a provisioning ecosystem service, depends on supporting services, such as nutrient and water cycling, pest regulation, and maintenance of soil quality and biodiversity (Wander and Bollero,1999; Jackson et al., 2003; Power 2010; Verhulst et al., 2010). In addition, production practices influence regulating services that provide benefits external to the farm, including regulating water and air quality, storing soil C, and supporting biodiversity (Power, 2010; Barrios, 2007; Brussaard et al., 2007; Brussaard, 2012; Janzen, 2004; Janzen, 2005).

Agriculture can have significant impacts on environmental quality. Substantial effort has been dedicated to identifying what influences farmers' decisions and incorporating that knowledge into projects, programs, and policies. Ecosystem services are functions provided by the environment that benefit humans and can be broadly classified into four groups: provisioning, regulating, supporting, and cultural services (MEA – Millennium Ecosystem Assessment, 2005). Crop production, a provisioning ecosystem service, depends on supporting services, such as C, nutrient and water cycling, pest regulation, and the maintenance of soil quality and biodiversity (Wander and Bollero, 1999; Jackson et al., 2003; Verhulst et al., 2010). In addition, production practices influence services that provide benefits external to the farm, including regulating water flow and air quality, storing SOC, and supporting biodiversity (Brussaard et al., 2007; Brussaard, 2012; Palm et al., 2014; Power, 2010). Appropriate agricultural management practices are critical to realizing the benefits of ecosystem services from agricultural activities. Farmers, consumers, and policymakers can now collaborate to achieve cover cropping improvements for soil, water, air quality, and climate-change adaptation and mitigation for the good of society.

Agricultural ecosystems provide humans with food, forage, bioenergy, and pharmaceuticals and are essential to human well-being. Ecosystem services are the many and varied benefits to humans

8.5.3 Water Quality

Soil and water resources are fundamental components of agriculture. Reicosky (2003) reviewed the key role of C in global environmental benefits in CA systems. While we learn more about soil C management and storage and its central role in direct environmental and water quality benefits, we must understand the secondary environmental benefits and what they mean to production agriculture. Enhancing water quality requires increasing soil C storage and infiltration, an increased fertility and nutrient cycling, decreased wind and water erosion, and minimized soil compaction (Keller et al., 2019). Decreased C emissions and managing pesticide application and movement generally enhance environmental quality. The sum of each individual benefit adds to a total package with major significance on a global scale. Incorporating C storage in conservation planning demonstrates concern for our global resources and presents a positive role for soil C that will have a major impact on water quality.

Achieving a balance between agricultural production and conservation of natural resources is a necessary goal for the development of sustainable agricultural systems and associated water quality (Sharpley et al., 1992; Nichols et al., 1994). A healthy watershed is where freshwater ecosystems, their biodiversity and their surrounding watersheds provide an equitable distribution of benefits through collaborative management. Ecosystems are central to healthy watersheds and provide a habitat for high concentrations of biodiversity, and a watershed's ecosystems exercise critical control over the pools and fluxes of water, both above and below the surface. Carbon-centric CA systems with minimum soil disturbance and high surface-soil organic C are highly effective in improving surface soil properties and associated hydrologic processes, thus reducing water runoff and soil erosion and improving water quality (Choudhary et al., 1997; Franzluebbers, 2008). SOM is a key property that drives many important soil water functions, for example, supplying and cycling of nutrients; infiltrating, filtering, and storing water; storing C from the atmosphere; and decomposing OM. Tillage perturbations of concern still remain with excessively high nutrient applications from fertilizer and manure inputs that can cause leaching of nitrate to groundwater and runoff of dissolved P to surface water bodies. As ecosystems become degraded and lose function, this negatively impacts freshwater biodiversity and can impair ecosystem service delivery as well. Franzluebbers (2008) demonstrated that CA and perennial pastures can mitigate sediment and nutrient loss to the environment. Lack of soil disturbance in grazed pastures and abundant application of animal manure to meet N requirements can lead to highly stratified depth distribution of soil P (Sharpley, 2003).

Research projects on direct comparisons of CT and NT in agricultural systems are limited. Tan et al. (2002) investigated the effect of CT and NT practices on water quality characterizing soil structure, hydraulic conductivity, and earthworm population. For both tillage systems, approximately 80% of tile drainage and NO_3-N loss in tile drainage water occurred during the November–April nongrowing season. Long-term NT improved wet aggregate stability, increased near-surface hydraulic conductivity and increased both the number and mass of earthworms relative to long-term CT. The greater tile drainage and NO_3-N loss under NT were attributed to an increase in continuous soil macropores, as implied by greater hydraulic conductivity and greater numbers of earthworms that reflect the negative impact of tillage on soil organisms (Edwards et al., 1990; Kladivko, 2001; Kemper et al. 2011; Briones and Schmidt 2017) reflecting a major benefit of CA systems.

Conservation Agriculture   215

Water quality concerns from CT agriculture are primarily from sediment, nutrient, and pesticide runoff from cropland and fecal-borne pathogen and nutrient runoff from pastureland and livestock operations (Chesters and Schierow, 1985; Myers et al., 1985). Both cropland and pastureland must be addressed with long-term solutions. Development and adoption of CA systems on cropland have revealed a key fundamental linkage between intensive soil tillage and water quality. Surface biomass cover and undisturbed soil are key factors in limiting soil and nutrient losses from cropland (Pelegrin et al., 1988; Holt, 1979; Lindstrom et al., 1979), as well as from pastureland (Jones et al., 1985; Harmel et al., 2004), which reflects the importance of enhanced plant C management in CA systems with minimum soil disturbance. To meet the human nutritional needs of the rapidly expanding global population while sustaining our invaluable natural resources, a multidisciplinary approach is needed to develop, characterize, and implement alternative, highly productive management systems that also conserve soil and water resources for the future (Franzlubbers, 2008).

8.6 SUMMARY AND CONCLUSIONS

CA has been proposed as a resilient type of agriculture to protect our natural resources that points to the need for better C management. CA systems minimize soil disturbance, which leave organic crop biomass on the soil surface reducing erosion and increasing soil C and WUE through enhanced infiltration, deep percolation, plant-available water storage, groundwater recharge, stream baseflow, and decreases in evaporation, peak stream flows and downstream flooding. Soil C management is a critical part of CA systems with specific impacts on WUE. CA makes better use of agricultural resources through the integrated management of available resources such as solar energy, sun, soil, water, air, and biodiversity resources, combined with limited external inputs. CA integrates system concepts based on three key interlinked principles: (1) continuous crop biomass cover on the soil surface, (2) continuous minimum soil disturbance (NT), and (3) diverse crop rotations and cover crop mixes with location-specific complementary practices, all important elements of CA. Enhanced C management results in interactive synergies between the biological, physical, and chemical properties and processes with multiple social, economic, ecosystem, and environmental benefits.

To meet the global challenges of food security and environmental conservation, CA has been identified and promoted as the technological option for a sustainable intensification of agriculture. CA systems can be implemented to minimize negative socioeconomic and environmental consequences associated with soil degradation by enhancing soil health and promoting the sustainability and multifunctionality of agroecosystems. The main challenge of conserving and improving SH is guaranteeing its long-term productivity and environmental sustainability. Rigorous long-term farming system trials are needed to compare CT and CA systems in order to build knowledge about the benefits and mechanisms associated with CA on regional scales. CA systems have clear advantages over CT agricultural systems in improving soil health and efficient use of natural resources, reducing the environmental impacts of agricultural activities, saving inputs, and reducing the cost of production. In water-limited environments, CA increases soil water storage and WUE, which allows for cropping system intensification and diversification. System intensification and diversification allow growers to better manage various aspects of the cropping system, for example, the leaching of water and nutrients below the root zone or the control of a specific pest, which have constrained the adoption of CT in systems with a low level of crop diversity. Researchers around the world are working to develop such technologies, and scientific evidence is mounting that CA systems can help prevent soil degradation, improve soil health, and produce nutritionally rich food in dryland agriculture.

CA is more than avoidance of tillage; it is an ecosystem approach that involves progressive, systemwide change in cultural practices, along with a change in mindset to bypass the use of tillage equipment and synthetic chemicals that have evolved over the last 12,000 years. Tillage leads to drastic physical changes in soil structure, with subsequent reduction in a soil's capacity to absorb and hold the water and air needed for season-long plant growth, particularly in dry and drought-prone areas. Reduced infiltration of rainfall, in turn, causes greater runoff over the land surface, raising the risks of erosion, catchment degradation and more variable stream-flows and pollution. Loss of soil C also lessens the chemical and biological processes, very important in providing organic compounds which contribute to biological activity and stability of soil aggregates and release nutrients for uptake by plants. The reduction in soil C due to frequent tillage is particularly deleterious in tropical and subtropical conditions under which soil C is oxidized quickly causing decreased WUE and food security.

The slow recognition of the ecological damage done by intensive tillage disrupting the "living biological system" has caused environmental and ecosystem degradation and lost productivity and biodiversity. Soil organisms play a key role in C dynamics flowing through the soil–plant–atmosphere system. Conventional plow tillage destroys the soil structure by breaking up the aggregates and earthworms, exposing available C, and releasing more C to the atmosphere through enhanced microbial activity. Furthermore, tillage fragments the fungal hyphae networks and upsets the balance between the fungi and bacteria by enhancing soil C loss. Along with soil C content, tillage decreases the biodiversity provided by the microbial and plant population in agricultural systems that must be replaced and enhanced to provide resilient soil ecosystem services.

The concerns about the lack of sustainability and the lack of resiliency observed in modern industrial agricultural production have caused a paradigm shift in the design of agricultural systems that has led to CA systems. Agroecological approaches have been put forward as viable solutions to increase agricultural productivity, to increase economic well-being as well as the social and gender equity in rural communities, and to increase agricultural productivity while minimizing reliance on external proprietary technology, capital, and synthetic chemical inputs. Key features of CA systems include an agroecological approach to follow a holistic and integrated participatory approach, an emphasis on minimizing erosion and enhancing soil quality, the conservation of natural resources, the promotion of agrobiodiversity and ecosystem services both at the farm and watershed levels, and the need to fully integrate socioeconomic, social, and gender equity considerations in all phases of the agricultural research, extension, and developmental process.

CA is a holistic concept for sustainable management of agricultural lands. It achieves to a high degree environmental and economic sustainability of farming and provides many benefits for the nonfarming rural population. However, by reducing labor requirements and drudgery of farming and by increasing the farm income, CA improves significantly the livelihoods of farmers and their families. An important element for the adoption of CA, regardless of whether the farms are operating at manual level, with animal traction, or at tractor level, is the accessibility of affordable and quality equipment suited for the local needs of farmers, producers, and entrepreneurs with intention to go toward CA practices. CA is a major player in our survival through C cycling and soil C storage. Global food security requires efficient C cycling and natural resources and global environmental preservation may require soil C storage as the main goal for improved C flow management in sustainable agriculture systems. These major functions will require a practical compromise that may depend on society's decision to support climate mitigation or to support food security, or both. Future research and innovative farmers using CA principles and practices for improved C management must find a workable solution. CAS that are developed and that adopt ecologically diversified cropping systems are key factors for agricultural policy setting and a top priority for on-farm decision-making to increase crop productivity and enhance soil health, while reducing negative environmental impacts. Increasing WUE through enhanced soil management in dryland areas with CA systems will help ensure a more stable food supply for future generations.

REFERENCES

Adhikari. K., and Hartemink, A.E. 2016. Linking soils to ecosystem services – A global review. *Geoderma* 262:101–111.

Al-Agely, Abid and Wahbi, Ammar. 2003. Extractable soil water and transpiration rate of mycorrihizal corn. Presented (orally) at the 4th International Conference on Mycorrhizae, 10–15 August 2003 at Montreal, Canada.

Alcántara, V., Don, A., Well, R. and Nieder, R. 2016. Deep ploughing increases agricultural soil organic matter stocks. *Glob. Chang. Biol.* 22(8):2939–2956.

Allen, R.G., Pereira, L.S., Raes, D., and Smith, M. 1998. Crop evapotranspiration guidelines for computing crop water requirements. FAO Irrigation and Drainage Paper 56. FAO, Rome.

Allison, F.E. 1973. *Soil Organic Matter and Its Role in Crop Production.* Elsevier Scientific Publishing Company.

Altieri, M.A. 2018. *Agroecology: The Science of Sustainable Agriculture, Second Edition.* CRC Press, Boca Raton, FL, 448 pp.

Alvarez, R., Steinbach, H.S., and De Paepe, J.L. 2017. Cover crop effects on soils and subsequent crops in the pampas: A meta-analysis. *Soil Tillage Res.* 170:53–65. https://doi.org/10.1016/j.still.2017.03.005

Amanuel, G., Kefyalew, G., Tanner, D.G., Asefa, T. and Shambel, M. 2000. Effect of crop rotation and fertilizer application on wheat yield performance across five years at two locations in South-eastern Ethiopia. In: *Proceedings of the 11th Regional Wheat Workshop for Eastern, Central and Southern Africa.* CIMMYT, Addis Ababa, Ethiopia, pp. 264–274.

Anderson, J. and Bausch, C. 2006. Climate change and natural disasters: Scientific evidence of a possible relation between recent natural disasters and climate change. *Policy Dep. Econ. Sci. Policy.* 2:1–30.

Angers, D.A. and Chenu, C. 1997. Dynamics of soil aggregation and C sequestration. In R. Lal (Ed.), *Soil Processes and the Carbon Cycle.* CRC Press, Boca Raton, FL, pp. 199–206.

Angers, D.A. and Eriksen-Hamel, N.S. 2008. Full-inversion tillage and organic carbon distribution in soil profiles: a meta-analysis. *Soil Sci Soc Am J.* 72:1370–1374.

Araya, S.N., Mitchell, J.P., Hopmans, J.W., and Ghezzehei, T.A. 2022. Long-term impact of cover crop and reduced disturbance tillage on soil pore size distribution and soil water storage. *SOIL,* 8:177–198. https://doi.org/10.5194/soil-8-177-2022

Arrúe, J.L. and Cantero-Martínez, C. (Eds.). 2006. *Third Mediterranean Meeting on No-Tillage.* Options méditerranéennes, Série A, no. 69, IAMZ, Zaragoza, 210 pp.

Arrúe, J.L., Cantero-Martínez, C., Cardarelli, A., de Benito, A., et al. 2007a. Prospects for sustainable agriculture in the Mediterranean platform of KASSA. In: Lahmar, R., Arrúe, J.L., Denardin, J.E., Gupta, R.K., Ribeiro, M.F.F, and de Tourdonnet, S. (Eds.), *Knowledge Assessment and Sharing on Sustainable Agriculture.* CD-Rom, CIRAD, Montpellier-France. ISBN 978-2-87614-646-4, 27 pp.

Arrúe, J.L., Cantero-Martínez, C., Cardarelli, A., Kavvadias, V., et al. 2007b. Comprehensive inventory and assessment of existing knowledge on sustainable agriculture in the Mediterranean platform of KASSA. In: Lahmar, R., Arrúe, J.L., Denardin, J.E., Gupta, R.K., Ribeiro, M.F.F, and de Tourdonnet, S. (Eds.), *Knowledge Assessment and Sharing on Sustainable Agriculture.* CD-Rom, CIRAD, Montpellier-France. ISBN 978-2-87614-646-4, 24 pp.

Atwell, B.J. 1993. Response of roots to mechanical impedance. *Environ. Exp. Bot.* 33:27–40.

Baker, C.J., Saxton, K.E., and Ritchie, W.R. 1996. *No-Tillage Seeding.* Wallingford, Oxon, UK: Science and Practice. CAB International.

Baker, J.M., Ochsner, T.E., Venterea, R.T., and Griffis, T.J. 2007. Tillage and soil carbon sequestration-What do we really know? *Agric. Ecosyst. Environ.* 118(1):1–5.

Balesdent, J. and Balabane, M. 1996. Major contribution of roots to soil carbon storage inferred from maize cultivated soils. *Soil Biology and Biochemistry* 28(9):1261–1263. https://doi.org/10.1016/0038-0717(96)00112-5

Ball, B.C. and Robertson, E.A. 1994. Effects of uniaxial compaction on aeration and structure of ploughed or direct drilled soils. *Soil Tilt. Res.* 31:135–148. https://doi.org/10.1016/0167-1987(94)90076-0

Banwart, S., Bernasconi, S.M., Bloem, J., Blum, W., Brandao, M., Brantley, S., et al. 2011. Soil processes and functions in critical zone observatories: Hypotheses and experimental design. *Vadose Zone J.* 10(3):974–987.

Banwart, S.A., Nikolaidis, N.P., Zhu, Y.G., Peacock, C.L., Sparks, D.L. 2019. Soil functions: Connecting earth's critical zone. *Annu. Rev. Earth Planet. Sci.* 47:333–359. https://doi.org/10.1146/annurev-earth-063016-020544

Bardgett, R.D. and Van Der Putten, W.H. 2014. Belowground biodiversity and ecosystem functioning. *Nature* 515(7528):505–511.

Barker, D. 2014. Genetically engineered (GE) crops: A misguided strategy for the twenty-first century? *Development* 57:192–200.

Barnes, B.T. and Ellis, F.B. 1979. Effects of different methods of cultivation and direct drilling, and disposal of straw residues, on populations of earthworms. *J Soil Sci* 30:669–679. https://doi.org/10.1111/j.1365-2389.1979.tb01016.x

Barrios, Edmundo. 2007. Soil biota, ecosystem services and land productivity. *Ecological Economics* 64(2):269–285. https://doi.org/10.1016/j.ecolecon.2007.03.004

Basamba, T.A., Amézquita, E., Singh, B.R., and Rao, I.M. 2006. Effects of tillage systems on soil physical properties, root distribution and maize yield on a Colombian acid-savanna Oxisol. *Acta Agriculture Scand. Section B-Soil and Plant Sci.*, 56:255–262.

Basche, A.D. 2017. *Turning Soils in the Sponges: How Farmers Can Fight Floods and Droughts.* Union of Concerned Scientists, Cambridge, Massachusetts, US; Available from: www.ucsusa.org/sites/default/files/attach/2017/08/turning-soils-into-sponges-fullreport-august-2017.pdf

Basche, A.D., Archontoulis, S.A., Kaspar, T.K., Jaynes, D.B., Parkin, T.B., and Miguez, F.E. 2016a. Simulating long-term impacts of cover crops and climate change on crop production and environmental outcomes in the Midwestern United States. *Agriculture, Ecosystems and the Environment* 218:95–106. https://doi.org/10.1016/j.agee.2015.11.011

Basche, A.D. and DeLonge, M. 2017. The impact of continuous living cover on soil hydrologic properties: A meta-analysis. *Soil Science Society of America Journal* 81(5):1179–1190.

Basche, A.D. and Edelson, O.F. 2017. Improving water resilience with more perennially based agriculture. *Agroecology and Sustainable Food Systems* 41 (7):799–824.

Basche, A.D., Kaspar, T.K., Archontoulis, S.A., Jaynes, D.B., Parkin, T.B., et al. 2016b. Soil water improvements with the long-term use of a cover crop. *Agricultural Water Management* 172:40–50. https://doi.org/10.1016/j.agwat.2016.04.006

Baumhardt, R.L., Keeling, J.W., and Wendt, C.W. 1993. Tillage and residue effects on infiltration into soils cropped to cotton. *Agron. J.* 85:379–383.

Baveye, P.C., Schnee, L.S., Boivin, P., Laba, M., and Radulovich, R. 2020. Soil organic matter research and climate change: merely re-storing carbon versus restoring soil functions. *Front. Environ. Sci.* 8:579904. https://doi.org/10.3389/fenvs.2020.579904

Beare, M.H., Hendrix, P.F., and Coleman, D.C. 1994. Water-stable aggregates and organic matter fractions in conventional and no-tillage soils. *Soil Sci. Soc. Am. J.* 58:777–786.

Benjamin, J.G., Mikha, M., Nielsen, D.C., Vigil, M.F., Calderon, F., et al. 2007. Cropping intensity effects on physical properties of a no-till silt loam. *J. Soil Sci.* 71:1160–1165.

Berhe, A.A., Harte, J., Harden, J.W., and Torn, M.S. 2007. The significance of the erosion-induced terrestrial carbon sink. *Bioscience* 57:337–346. https://doi.org/10.1641/b570408

Biederbeck, V.O., Janzen, H.H., Campbell, C.A., and Zentner, R.P. 1994. Labile soil organic matter as influenced by cropping practices in an arid environment. *Soil Biol. Biochem.* 26:1647–1656.

Bilbro, J.D., Harris, B.L., and Jones, O.R. 1994. Erosion control with sparse residue. In: Stewart, B.A. and Moldenhauer, W.C. (Eds.), *Crop Residue Management to Reduce Erosion and Improve Soil Quality.* Conservation Research Report No. 37, Agricultural Research Service, United States Department Agriculture, Washington, D.C., pp. 30–32.

Black, H. and Okwakol, M. 1997. Agricultural intensification, soil biodiversity and agroecosystem function in the tropics: The role of termites. *J. Appl. Soil Eco.* 6:37–53. https://doi.org/10.1016/S0929-1393(96)00153-9

Blanco-Canqui, H. and Ruis, S.J. 2020. Cover crop impacts on soil physical properties: A review. *Soil Sci. Soc. Am. J.* 84:1527–1576. https://doi.org/10.1002/saj2.20129

Blesh, J. 2018. Functional traits in cover crop mixtures: Biological nitrogen fixation and multifunctionality. *J. Appl. Ecol.* 55(1):38–48. https://doi.org/10.1111/1365-2664.13011

Blevins, R.L., Cook, D., Phillips, S.H., and Phillips, R.E. 1971. Influence of no-tillage on soil moisture. *Agron. J.* 63:593–596.

Blevins, R.L. and Frye, W.W. 1993. Conservation tillage: An ecological approach to soil management. *Adv. Agron.* 51:33–78.

Bonari, E., Mazzoncini, M., and Caliandro, A. 1994. Cropping and farming systems in Mediterranean areas. pp. 636–644. In: *Proc. Eur. Soc. Agron.* Congress 3rd, Abano-Padova, Italy. 18–22 Sept. 1994. M. Borin and M. Sattin, Colmar Cedx, France.

Bossio, D., Cook-Patton, S.C., Ellis, P.W., Fargione, J., Sanderman, J., Smith, P., Wood, S., Zomer, R.J., von Unger, M., Emmer, I.M., and Griscom, B.W. 2020. The role of soil carbon in natural climate solutions. *Nat. Sustain.* 3(5):391–398.

Bossio, D., Geheb, K., and Critchley, W. 2010. Managing water by managing land: Addressing land degradation to improve water productivity and rural livelihoods. *Agric. Water Manag.* 97:536–542.

Bottinelli, N., Jouquet, P., Capowiez, Y., Podwojewski, P., et al. 2015. Why is the influence of soil macrofauna on soil structure only considered by soil ecologists? *Soil Tillage Res.* 146, Part A:118–124.

Bouma, J. 2009. Soils are back on the global agenda: Now what? *Geoderma* 150(1–2):224–225. https://doi.org/10.1016/j.geoderma.2009.01.015

Bouyoucos, G.J. 1939. Effect of organic matter on the water holding capacity and the wilting points of mineral soils. *Soil Sci.* 47:377–383. https://doi.org/10.1097/00010694-193905000-00005

Bowden, L. 1979. Development of present dryland farming systems. In: Hall, A.E., Cannell, G.H., and Lawton, H.W. (Eds.), *Agriculture in Semi-Arid Environments.* Springer-Verlag, Berlin, pp. 45–72.

Briones, M.J.I. and Schmidt, O. 2017. Conventional tillage decreases the abundance and biomass of earthworms and alters their community structure in a global meta- analysis. *Glob. Change Biol.* 23 (10):4396–4419.

Bronick, C.J. and Lal, R. 2005. Soil structure and management: A review. *Geoderma.* 124 (1–2), 3–22.

Brown, S.C., Gregory, P.J., and Wahbi, A. 1987. Root characteristics and water use in Mediterranean environments. In: Srivastava, J.P., Poreddu, E., Acevedo, E., and Varma, S. (Eds.), *Drought Tolerance in Winter Cereals.* John Wiley & Sons, Chichester, pp. 275–283.

Brussaard, L. 2012. Ecosystem services provided by the soilbiota. In: Wall, D.H., Bardgett, R.D., Behan-Pelletier, V., Herrick, J.E., Jones, T.H., Ritz, K., Six, J., Strong, D.R. and van der Putten, W.H. (Eds.), *Soil Ecology and Ecosystem Services.* Oxford University Press, Oxford, UK, pp. 45–58.

Brussaard, L., Pulleman, M.M., Ouédraogo, É., Mando, A., and Six, J. 2007. Soil fauna and soil function in the fabric of the food web. *Pedobiologia* 50:447–462.

Burnett, E., Stewart, B.A., and Black, A.L. 1985. Regional effects of soil erosion on crop productivity – Great Plains. pp. 285–304. In: *Soil Erosion and Crop Productivity.* American Society of Agronomy, Inc., Crop Science Society of America, Inc., Soil Science Society of America, Inc., Publishers. Madison, WI.

Burns, R.L., Cook, D.J., and Phillips, R.E. 1971. Influence of no tillage on soil moisture. *Agron. J.* 73:593–596.

Cabrera, J.S. and Lee, H.S. 2018. Impacts of Climate Change on Flood-Prone Areas in Davao Oriental, Philippines. *Water* 10, 893. https://doi.org/10.3390/w10070893

Campbell, D.J. and Henshall, J.K. 1991. Bulk density. In: Smith, K.A. and Mullins, C.E. (Ed.) *Soil analysis. Physical methods.* Marcel Dekker, New York, pp. 329–366.

Carceles, B., Duran-Zuazo, V., Soriano, M., Garcia-Tejero, I.F., Galvez-Ruiz, B. & Tavira, S. 2022. Conservation agriculture as a sustainable system for soil health: a review. *Soil Syst.* 6(4), article 87. https://doi.org/10.3390/soilsystems6040087

Carter, M.R. and Ball, B.C. 1993. Soil porosity. In: Carter, M.R. (Ed.), *Soil Sampling Methods of Soil Analysis.* Lewis Publ., Boca Raton, FL, pp. 581–588.

Cassel, D.K. 1982. Tillage effects on soil bulk density and mechanical impedance. In: Unger, P.W. and Van Doren, D.M. (Ed.), *Predicting Tillage Effects on Soil Physical Properties and Processes. ASA Spec. Publ. 44.* ASA and SSSA, Madison, WI, pp. 45–67.

Chan, K.Y. 2001. An overview of some tillage impacts on earthworm population abundance and diversity—implications for functioning in soils. *Soil Tillage Res* 57(4):179–191. https://doi.org/10.1016/s0167-1987(00)00173-2

Chan, K.Y. and Mead, J.A. 1992. Tillage-induced differences in the growth and distribution of wheat-roots. *Aust. J. Agric. Res.* 43:19–28.

Chang, C., and Lindwall, C.W. 1989. Effect of long-term minimum tillage practices on some physical properties of a Chernozemic clay loam. *Can. J. Soil Sci.* 69:443–449. https://doi.org/10.4141/cjss89-046

Chang, C., and Lindwall, C.W. 1992. Effects of tillage and crop rotation on physical properties of loam soil. *Soil Tillage Res.* 22:383–389. https://doi.org/10.1016/0167-1987(92)90051-C

Chatterjee, A., Lal, R., Wielopolski, L., Martin, M.Z. and Ebinger, M.H. 2009. Evaluation of different soil carbon determination methods. *Crit. Rev. Plant Sci.* 28(3):164–178. https://doi.org/10.1080/0735268090 2776556

Chen, Y., Tessier, S. and Rouffignat, J. 1998. Soil bulk density estimation for tillage systems and soil textures. *Trans. ASAE.* 41:1601–1610.

Chenu, C., Angers, D.A., Barré, P., Derrien, D., Arrouays, D. and Balesdent, J. 2019. Increasing organic stocks in agricultural soils: Knowledge gaps and potential innovations. *Soil Tillage Res.* 188:41–52. https://doi. org/ 10.1016/j.still.2018.04.011

Chenu, C., Le Bissonnais, Y., and Arrouays, D. 2000. Organic matter influence on clay wettability and soil aggregate stability. *Soil Sci. Soc. Am. J.* 64(4):1479–1486.

Chenu, C. and Plante, A.F. 2006. Clay-sized organo-mineral complexes in a cultivation chronosequence: Revisiting the concept of the 'primary organo-mineral complex'. *Eur. J. Soil Sci.* 57(4):596–607. https://doi.org/ 10.1111/j.1365-2389.2006.00834.x

Chesters, G. and Schierow, L.J. 1985. A primer on nonpoint pollution. *J. Soil Water Conserv.* 40:9–13.

Choudhary, M.A., Lal, R., and Dick, W.A. 1997. Long-term tillage effects on runoff and soil erosion under simulated rainfall for a central Ohio soil. *Soil Tillage Res.* 42:175–184.

Cochran, V.L., Sparrow, S.D., and Sparrow, E.B. 1994. Residue effects on soil micro- and macroorganisms. In: Unger, P.W. (Ed.), *Managing Agricultural Residues*. Lewis Publ., Boca Raton, FL, pp. 163–184.

Conant, R.T., Easter, M., Paustian, K., Swan, A., and Williams, S. 2007. Impacts of periodic tillage on soil C stocks: A synthesis. *Soil Tillage Res.* 95:1–10. https://doi.org/10.1016/j.still.2006.12.006

Conyers, M., van der Rijt, V., Oates, A., Poile, G., Kirkegaard, J., and Kirkby, C. 2019. The strategic use of minimum tillage within conservation agriculture in southern New South Wales, Australia. *Soil Tillage Res.* 193:17–26.

Cooper, J., Baranski, M., Stewart, G., Nobel-de Lange, M., Bàrberi, P., Fließbach, A., et al. 2016. Shallow non-inversion tillage in organic farming maintains crop yields and increases soil C stocks: A meta-analysis. *Agron. Sustain. Dev.* 36:22. https://doi.org/10.1007/s13593-016-0354-1

Cornish, P.S. and Pratley, J.E. 1991. Tillage practices in sustainable farming systems. In: Squires, V. and Tow, P.G. (Eds.), *Dry Farming – Australia*. Sydney University Press, pp. 76–101.

Corsi, S., Friedrich, T., Kassam, A., Pisante, M., and de M. Sà, J. 2012. *Soil Organic Carbon Accumulation and Greenhouse Gas Emission Reductions from Conservation Agriculture: A Literature Review*. Rome, Italy: Food and Agriculture Organization of the United Nations (FAO).

Crozier, C.R., Naderman, G.C., Tucker, M.R., and Sugg, R.E. 1999. Nutrient and pH stratification with conventional and no-till management. *Commun. Soil Sci. Plant Anal.* 30:65–74.

Cusser, S., Bahlai, C., Swinton, S.M., Robertson, G.P., and Haddad, N.M. 2020. Long-term research avoids spurious and misleading trends in sustainability attributes of no-till. *Glob. Chang. Biol.* 26(6):3715–3725.

Daigh, A.L.M., Dick, W.A., Helmers, M.J., Lal, R., Lauer, J.G., Nafziger, E., CH, Strock, J., Villamil, M., Mukherjee, A., and Cruse, R. 2018. Yields and yield stability of no-till and chisel-plow fields in the Midwestern US Corn Belt. *Field Crops Res.* 218:243–253.

Daily, G. 1997. Introduction: What are ecosystem services? In: Daily G. (Ed.), *Nature's Services. Societal Dependence on Natural Ecosystems*, Island Press, Washington, DC.

Dal Ferro, N., Sartori, L., Simonetti, G., Berti, A., and Morari, F. 2014. Soil macro- and microstructure as affected by different tillage systems and their effects on maize root growth. *Soil Tillage Res.* 140:55–65.

Delgado, Jorge A., Groffman, Peter M., Nearing, Mark, et al. 2011. Conservation practices to mitigate and adapt to climate change. *J. Soil Water Conserv.* 66(4):118A–129A. DOI: 10.2489/jswc.66.4.118A.

Denef, K., Six, J., Bossuyt, H., Frey, S.D., et al. 2001. Influence of dry–wet cycles on the interrelationship between aggregate, particulate organic matter, and microbial community dynamics. *Soil Biol. Biochem.* 33(12):1599–1611.

De Oliveira, G., Brunsell, N.A., Sutherlin, C.E., Crews, T.E., and De Haan, L.R. 2018. Energy, water and carbon exchange over a perennial Kernza wheatgrass crop. *Agric. For. Meteorol.* 249:120–137. https:// doi.org/10.1016/j.agrformet.2017.11.022

Derpsch, R., Franzluebbers, A.J., Duiker, S.W., Reicosky, D.C., Koeller, K., Friedrich, T., Sturny W.G., Sá, J.C.M., and Weiss, K. 2013. Why do we need to standardize no-tillage research? Letter to the Editor in: *Soil Tillage Res.* 137:16–22.

De Vries, F.T., Hoffland, E., van Eekeren, N., Brussaard, L., and Bloem, J. 2006. Fungal/bacterial ratios in grasslands with contrasting nitrogen management. *Soil Biol. Biochem.* 38(8):2092–2103.

Dexter, A.R. 1988. Advances in characterization of soil structure. *Soil Tillage Res.* 11(3–4):199–238.

Dick, W.A. 1984. Influence of long-term tillage and crop rotation combinations on soil enzyme activities. *Soil Sci. Soc. Am. J.* 48:569–574.

Dick, W.A., Van Doren, D.M. Jr., Triplett, G.B. Jr., and Henry, J.E. 1986a. Influence of long-term tillage and rotation combinations on crop yields and selected soil parameters: I. Mollic Ochraqualf. *Research Bulletin*, 1180. Ohio Agricultural Research & Development Center Library, Ohio State University, Wooster, OH.

Dick, W.A., Van Doren, D.M. Jr., Triplett, G.B. Jr., and Henry, J.E. 1986b. Influence of long-term tillage and rotation combinations on crop yields and selected soil parameters: II. Typic Fragiudalf. *Research Bulletin*, 1181. Ohio Agricultural Research & Development Center Library, Ohio State University, Wooster, OH.

Dimassi, B., Mary, B., Wylleman, R., Labreuche, J., Couture, D., Piraux, F., and Cohan, J.-P. 2014. Long-term effect of contrasted tillage and crop management on soil carbon dynamics during 41 years. *Agric. Ecosyst. Environ.* 188:134–146.

Di Santo, N., Russo, I., and Sisto, R. 2022. Climate change and natural resource scarcity: A literature review on dry farming. *Land* 11:2102. https://doi.org/10.3390/land11122102

Dold, C., Büyükcangaz, H., Rondinelli, W., Prueger, J., Sauer, T., and Hatfield, J. 2017. Long-term carbon uptake of agro-ecosystems in the Midwest. *Agric. For. Meteorol.* 232:128–140. doi:10.1016/j.agrformet.2016.07.012

Doran, J.W. 1980. Soil microbial and biochemical changes associated with reduced tillage. *Soil Sci. Soc. Am. J.* 44(4):765–771.

Dudley, N. and Alexander, S. 2017. Agriculture and biodiversity: A review. *Food Agric. Biodiv.* 18(2–3):45–49. Available from: www.tandfonline.com/eprint/bUdkJiBbgR8fdBGzcjIG/full

Duiker, S.W., Rhoton, F.E., Torrent, J., et al. 2003. Iron (hydr)oxide crystallinity effects on soil aggregation. *Soil Sci. Soc. Am. J.* 67:606–611.

Easterling, D.R., Meehl, G.A., Parmesan, C., Changnon, S.A., Karl, T.R., and Mearns, L.O. 2019. Climate extremes: Observations, modeling, and impacts. *Science.* 289:2068–2074. https://doi.org/10.1126/science.289.5487.2068

Edwards, W.M., Shipitalo, M.J., Owens, L.B., and Norton, L.D., 1990. Effect of *Lumbricus terrestris* L. burrows on hydrology of continuous no-till corn fields. *Geoderma* 46, 73–84.

Eigel, J.D. and Moore, I.D. 1983. Effect of rainfall energy on infiltration into bare soil. In: *Advances in infiltration, Proc. Natl. Conf. on Advances in Infiltration*, Chicago, IL. 12–13 Dec. 1983. ASAE, St. Joseph, MI, pp. 188–200.

Ekström, G. and Ekbom, B. 2011. Pest control in agro-ecosystems: An ecological approach. *Crit. Rev. Plant Sci.* 30:74–94.

Ellert, B.H. and Janzen, H.H. 1999. Short-term influence of tillage on CO_2 fluxes from a semi-arid soil on the Canadian prairies. *Soil Tillage Res.* 50(1):21–32.

Elliott, E.T. and Coleman, D.C. 1988. Let the soil work for us. *Ecol. Bull.* 39:23–32.

FAO. 1996. Prospects to 2010: Agricultural resources and yields in developing countries. In: Volume 1: Technical Background Documents 1-5. World Food Summit, Food and Agriculture Organization of the United Nations, Rome, pp. 26–36.

FAO. 2000. Land resource potential and constraints at regional and country levels. Food and Agriculture Organization. World Soil Resources Rep. 90. FAO of the United Nations, Rome.

FAO. 2002. *Intensifying Crop Production with Conservation Agriculture*. Food and Agriculture Organization of the United Nations, Rome. www.fao.org/ag/ags/AGSE/main.htm

FAO. 2008. Investing in Sustainable Crop Intensification: The Case for Soil Health. Report of the International Technical Workshop, July. *Integrated Crop Management* (vol. 6). Food and Agriculture Organization of the United Nations, Rome.

FAO. 2017. A report produced for the G20 Presidency of Germany. Food and Agriculture Organization of the United Nations, Rome. ISBN 978-92-5-109977-3

Farooq, M. and Siddique, K.H.M. 2015. Conservation agriculture: Concepts, brief history, and impacts on agricultural systems. In: Farooq, M. and Siddique, K.H.M. (Eds), *Conservation Agriculture*. Springer International Publishing, Switzerland, pp. 3–17. Chapter 1. https://doi.org/10.1007/978-3-319-11620-4_1

Fedoroff, N.V., Battisti, D.S., Beachy, R.N., Cooper, P.J.M., Fischhoff, D.A., Hodges, C.N., and Zhu, J.K. 2010. Radically rethinking agriculture for the 21st century. *Science* 327:833–834.

Finney, D.M. and Kaye, J.P. 2017. Functional diversity in cover crop polycultures increases multifunctionality of an agricultural system. *J. Appl. Ecol.* 54(2):509–17. https://doi.org/10.1111/1365-2664.12765

Fisseha, N. and Tewodros, M. 2015. Effect of sowing time and moisture conservation methods on maize at Goffa, south region of Ethiopia. *Sky J. Agric. Res.* 4:240–248.

Folgarait, P.J. 1998. Ant biodiversity and its relationship to ecosystem functioning: A review. *Biodivers. Conserv.* 7:1221–1244. https://doi.org/10.1023/A:1008891901953

Franzen, H., Lal, R. and Ehlers, W. 1994. Tillage and mulching effects on physical properties of a tropical Alfisol. *Soil Tillage Res.* 28:329–346.

Franzluebbers, A.J. 2002. Water infiltration and soil structure related to organic matter and its stratification with depth. *Soil Tillage Res.* 66(2):197–205.

Franzluebbers, A.J. 2008. Linking soil and water quality in conservation agricultural systems. *J. Integr. Biosci.* 6(1):15–29.

Franzluebbers, A.J., Hons, F.M., and Zuberer, D.A. 1995. Tillage and crop effects on seasonal dynamics of soil CO_2 evolution, water content, temperature, and bulk density. *Appl. Soil Ecol.* 2:95–109.

Freeney, Debbie S., Crawford, John W., Daniell, Tim, Hallett, Paul D. et al. 2006. Three-dimensional microorganization of the soil–root–microbe system. *Microbial Ecology* 52:151–158. DOI: 10.1007/s00248-006-9062-8

Frey, S.D., Elliott, E.T., and Paustian, K. 1999. Bacterial and fungal abundance and biomass in conventional and no-tillage agroecosystems along two climatic gradients. *Soil Biol. Biochem.* 31:573–585.

Friedrich, T. 2020. The role of no or minimum mechanical soil disturbance in Conservation Agriculture systems. Chap 4. In: Kassam, A. (Ed.), *Advances in Conservation Agriculture Volume 1: Systems and Science*, Burleigh Dodds Science Publishing, Cambridge, UK, 2020, pp. 155–178.

Fryrear, D.W. 1985. Soil cover and wind erosion. *Transactions of American Society of Agricultural Engineers* 28:781–784.

Garbeva, P., van Elsas, J.D., and van Veen, J.A. 2008. Rhizosphere microbial community and its response to plant species and soil history. *Plant Soil*, 302:19–32.

Gesch, R.W., Archer, D.W. and Berti, M.T. 2014. Dual cropping winter Camelina with soybean in the Northern Corn Belt. *Agron. J.* 106(5):1735–1745. https://doi.org/10.2134/ agronj14.0215. Available at: http://handle.nal.usda.gov/10113/60081

Ghezzehei, T.A. and Or, D. 2000. Dynamics of soil aggregate coalescence governed by capillary and rheological processes. *Water Resour. Res.* 36:367–379.

Giménez, D., Dirksen, C., Miedema, R., Eppink, L.A.A.J. and Schoonderbeek, D. 1992. Surface sealing and hydraulic conductances under varying-intensity rains. *Soil Sci. Soc. Am. J.* 56:234–242.

Gleick, P.H., Christian-Smith, J., and Cooley, H. 2011. Water-use efficiency and productivity: Rethinking the basin approach. *Water Internat.* 36(7):784–798.

Gomiero, T., Pimentel, D., and Paoletti, M.G. 2011. Is there a need for a more sustainable agriculture? *Crit. Rev. Plant Sci.* 30:6–23.

Grandy, A.S. and Robertson, G.P. 2007. Land-use intensity effects on soil organic carbon accumulation rates and mechanisms. *Ecosystems* 10:59–74.

Grant, C.A., and Lafond, G.P. 1993. The effects of tillage systems and crop sequences on soil bulk density and penetration resistance on a clay soil in southern Saskatchewan. *Can. J. Soil Sci.* 73:223–232.

Gregory, P.J. 1994. Root growth and activity. In K.J. Boote, J.M. Bennett, T.R. Sinclair, and G.M. Paulsen (Ed.), *Physiology and Determination of Crop Yield*. ASA, CSSA, and SSSA, Madison, WI, pp. 65–93.

Hadas, A. 1997. Soil tilth-the desired soil structural state obtained through proper soil fragmentation and reorientation processes. *Soil Tillage Res.* 43, 7–40.

Haddaway, N.R., Hedlund, K., Jackson, L.E. et al. 2017. How does tillage intensity affect soil organic carbon? A systematic review. *Environ. Evid.* 6, 30 (2017). https://doi.org/10.1186/s13750-017-0108-9

Håkansson, I. and Lipiec, J. 2000. A review of the usefulness of relative bulk density values in studies of soil structure and compaction. *Soil Tillage Res.* 53(2):71–85.

Hamblin, A.P. 1985. The influence of soil structure on water movement, crop root growth, and water uptake. *Adv. Agron.* 38:95–158.

Harmel, R.D., Torbert, H.H., Haggard, B.E., Haney, R., and Dozier, M. 2004. Water quality impacts of converting to a poultry litter fertilization strategy. *J. Environ. Qual.* 33:2229–2242.

Harvey, C.A., Chacón, M., Donatti, C.I., et al. 2013. Climate-smart landscapes: Opportunities and challenges for integrating adaptation and mitigation in tropical agriculture. *Conserv. Lett.* 7(2): 77–90. https://doi.org/org/10.1111/conl.12066

Hassan, Waseem, Li, Yu'e, Sab, Tahseen, Jabbi, Fanta, Wang, Bin, Cai, Andong, and Wu, Jianshuang. 2022. Improved and sustainable agroecosystem, food security and environmental resilience through no tillage with emphasis on soils of temperate and subtropical climate regions: A review. *Int. Soil Water Conserv. Res.* 10(3):530–545. https://doi.org/10.1016/j.iswcr.2022.01.005

Hassink, J. and Whitmore, A.P. 1997. A model of the physical protection of organic matter in soils. *Soil Sci. Soc. Am. J.* 61:131–139.

Hatfield, J.L. 2011. Soil management for increasing water use efficiency in field crops under changing climates. In: Jerry L. Hatfield and Thomas J. Sauer (Eds.), *Soil Management: Building a Stable Base for Agriculture*. Chapter 10, pp. 161–173. Copyright © 2011 by American Society of Agronomy and Soil Science Society of America.

Hatfield, J.L., Boote, K.J., Kimball, B.A., Ziska, L.H., et al. 2011. Climate impacts on agriculture: Implications for crop production. *Agron. J.* 103:351–370. https://doi.org/10.2134/agronj2010.0303

Hatfield, J.L., Sauer, T.J., and Cruse, R.M. 2017. Soil: The forgotten piece of the water, food, energy nexus. *Adv. Agron.* 143:1–46. ISBN: 978-0-12-812421-5

Hatfield, J.L., Sauer, T.J., and Prueger, J.H. 2001. Managing soils to achieve greater water use efficiency: A review. *Agron. J.* 93:271–280.

Hauser, E., Richter, D.D., Markewitz, D., Brecheisen, Z., and Billings, S.A., 2020. Persistent anthropogenic legacies structure depth dependence of regenerating rooting systems and their functions. *Biogeochemistry* 147(3):259–275.

Heinemann, J.A., Massaro, M., Coray, D.S., Agapito-Tenfen, S.Z., and Wen, J.D. 2014. Sustainability and innovation in staple crop production in the U.S. Midwest. *Int. J. Agric. Sustain.* 12:71–88.

Heitholt, J.J. 1989. Water use efficiency and dry matter distribution in nitrogen- and water-stressed winter wheat. *Agron. J.* 81:464–469.

Helmecke, M., Fries, E., and Schulte, C. 2020. Regulating water reuse for agricultural irrigation: Risks related to organic micro-contaminants. *Environ. Sci. Eur.* 32, 4. https://doi.org/10.1186/s12302-019-0283-0

Henneron, L., Bernard, L., Hedde, M., Pelosi, C., et al. 2015. Fourteen years of evidence for positive effects of conservation agriculture and organic farming on soil life. *Agronomy for Sustainable Development*, Springer Verlag/EDP Sciences/INRA, 2015, 35(1):169–181. ff10.1007/s13593-014-0215-8ff. ffhal-01173289f

Hill, R.L. 1990. Long-term conventional and no-tillage effects on selected soil physical properties. *Soil Sci. Soc. Am. J.* 54:161–166.

Hillel, D. 1991. *Out of the Earth: Civilization and the Life of the Soil*. Univ. of California Press, Berkeley, 321 pp.

Hillel, D. 1998. *Environmental Soil Physics*. Academic Press, San Diego, CA.

Hird, V. 2017. Farming systems and techniques that promote biodiversity. *Biodiversity* 18(2–3):71–74. DOI: 10.1080/14888386.2017.1351395 https://doi.org/10.1080/14888386.2017.1351395

Hirsch, P.R. 2018. Soil microorganisms: role in soil health. Chap 8, In: Reicosky, D.C. (Ed.), *Managing Soil Health for Sustainable Agriculture*. Volume 1: Fundamentals. Cambridge, UK Burleigh Dodds, pp. 169–196.

Hobbs, P.R. and Govaerts, B. 2010. How conservation agriculture can contribute to buffering climate change. In: Reynolds, M.P. (Ed.), *Climate Change and Crop Production*. CABI Series in Climate Change. CABI International, New York, NY, pp. 151–176.

Hobbs, P.R., Sayre, Ken and Gupta, Raj. 2008. The role of conservation agriculture in sustainable agriculture. *Philos. Trans. R. Soc. Lond. B. Biol. Sci.* 363(1491):543–555. https://doi.org/10.1098/rstb.2007.2169

Holt, R.F. 1979. Crop residue, soil erosion, and plant nutrient relationships. *J. Soil Water Conserv.* 34:96–98.

Homyak, P.M., Blankinship, J.C., Marchus, K., Lucero, D.M., Sickman, J.O., and Schimel, J.P. 2016. Aridity and plant uptake interact to make dryland soils hotspots for nitric oxide (NO) emissions. *PNAS USA*, 113, E2608–E2616. https://doi.org/10.1073/pnas.1520496113

Houghton, R.A. 1999. The annual net flux of carbon to the atmosphere from changes in land use 1850–1990. *Tellus B*, 51:298–313.

Hudson, B.D. 1994. Soil organic matter and available water capacity. *J. Soil Water Conserv.* 49(2):189–194.

Huggins, D.R., Reganold, J.P. 2008. No-till: the quiet revolution. *Sci. Am.* 299(1):70–77. doi.,101038/scientific american 0708-70.

Isbell, F., Tilman, D., Reich, P.B., and Clark, A.T. 2019. Deficits of biodiversity and productivity linger a century after agricultural abandonment. *Nature Ecol. Evol.* 3:1533–1538. www.naure.com/natecolevol.

Ismail, I., Blevins, R.L, and Frye, W.W. 1994. Long-term no-tillage effects on soil properties and continuous corn yields. *SSSAJ*, 58(1):193–198. https://doi.org/10.2136/sssaj1994.03615995005800010028x

Jackson, L.E., Calderon, F.J., Steenwerth, K.L., Scow, K.M., and Rolston, D.E. 2003. Responses of soil microbial processes and community structure to tillage events and implications for soil quality. *Geoderma* 114:305–317.

Jackson, R.B., Lajtha, K., Crow, S.E., Hugelius, G., Kramer, M.G. and Piñeiro, G. 2017. The ecology of soil carbon: Pools, vulnerabilities, and biotic and abiotic controls. *Annu. Rev. Ecol. Evol. Syst.* 48:419–445.

Jalota, S.K. and Prihar, S.S. 1990a. Bare soil evaporation in relation to tillage. *Adv. Soil Sci.* 12:187–216.

Jalota, S.K. and Prihar, S.S. 1990b. Effect of straw mulch on evaporation reduction in relation to rates of mulching and evaporativity. *J. Indian Soc. Soil Sci.* 38:728–730.

Janzen, H.H. 2004. Carbon cycling in earth systems—A soil science perspective. *Agric. Ecosyst. Environ.* 104(3):399–417.

Janzen, H.H. 2005. Soil carbon: A measure of ecosystem response in a changing world? *Can. J. Soil Sci.* 85, Special Issue: 467–480. https://doi.org/10.4141/S04-081

Janzen, H.H. 2015. Beyond carbon sequestration: Soil as conduit of solar energy. *Eur. J. Soil Sci.* 66(1):9–32.

Janzen, H.H., Groenigen, K.J., Powlson, D.S., Schwinghamer, T., and Groenigen, J.W. 2022. Photosynthetic limits on carbon sequestration in croplands. *Geoderma.* 416, 115810 https://doi.org/10.1016/j.geoderma.2022.115810

Jarecki, M.K. and Lal, R. 2006. Compost and mulch effects on gaseous flux from an Alfisol in Ohio. *Soil Sci.* 171:249–260.

Jarvis, A., Upadhyaya, H., Gowda, C.L.L., Aggarwal, P.K., et al. 2008. Climate change and its effect on conservation and use of plant genetic resources for food and agriculture and associated biodiversity for food security. Thematic Background Study. pp. 26. Food and Agriculture Organization of the United Nations (FAO) Report, Rome. http://oar.icrisat.org/5810/1/Climate_FAO_Report__2008.pdf

Jarvis, N.J. 2020. A review of non-equilibrium water flow and solute transport in soil macropores: principles, controlling factors and consequences for water quality. *Eur. J. Soil Sci.* 71(3):279–302.

Jastrow, J.D. 1996. Soil aggregate formation and the accrual of particulate and mineral-associated organic matter. *Soil Biol. Biochem.* 28:665–676.

Jat, R.A., K.L. Sahrawat, A.H. Kassam et al. 2014. Conservation agriculture for sustainable and resilient agriculture: global status, prospects and challenges. In: Jat, R.A., Sahrawat, K.L., and Kassam, A.H. (Eds.), *Conservation Agriculture: Global Prospects and Challenges*, 1–25. Wallingford, UK: CAB International.

Jenkinson, D.S. and Rayner, J.H. 1977. The turnover of soil organic matter in some of the Rothamsted classical experiments. *Soil Sci.* 123(5):298–305.

Jian, J., Du, X., Reiter, M.S., and Stewart, R.D. 2020. A meta-analysis of global cropland soil carbon changes due to cover cropping. *Soil Biol. Biochem.*, 143, 107735. https://doi.org/10.1016/j.soilbio.2020.107735

Jiang, X., Wright, A., Wang, X. and Liang, F. 2011. Tillage-induced changes in fungal and bacterial biomass associated with soil aggregates: A long-term field study in a subtropical rice soil in China. *Appl. Soil Ecol.* 48(2):168–173.

Johnson-Maynard, J., Umiker, K., and Guy, S. 2007. Earthworm dynamics and soil physical properties in the first three years of no-till management. *Soil Tillage Res.* 94(2):338–345.

Jones, O.R., Eck, H.V., Smith, S.J., Coleman, G.A., and Hauser, V.L. 1985. Runoff, soil, and nutrient losses from rangeland and dryfarmed cropland in the Southern High Plains. *J. Soil Water Conserv.* 40:161–164.

Kane, D. 2015. Carbon sequestration potential on agricultural lands: A Review of Current Science and Available Practices. In association with National Sustainable Agriculture Coalition Breakthrough Strategies and Solutions. https://sustainableagriculture.net/wp-content

Karlen, D.L. 2012. Soil health: The concept, its role, and strategies for monitoring. In: Wall, D., Bardgett, R.D., Behan-Pelletier, V., Herrick, J.E., Jones, T.H., Ritz, K., Six, J., Strong, D.R. and van der Putten, W.H. (Eds), *Soil Ecology and Ecological Systems*. Oxford University Press, New York, NY, pp. 331–336. Chapter 5.3.

Kassam, A. 2016. Reversing agricultural land degradation worldwide. *News & Opinion.* June 2016. World Day to Combat Land Degradation. Research Program on Dryland Systems. ICARDA. http://drylandsystems. cgiar.org/content/reversing-agricultural-land-degradation-worldwide

Kassam, A. 2019. Integrating conservation into agriculture. Chap 8. In Farooq, M. and Pisante, M. (eds.), *Innovations in Sustainable Agriculture.* Springer Nature, Switzerland AG, pp. 27–42. https://doi.org/ 10.1007/978-3-030-23169-9_2

Kassam, A., Basch, G., Friedrich, T., Shaxson, F., Goddard, T., Amado, T., Crabtree, B., Hongwen, L., Mello, I., Pisante, M., and Mkomwa, S. 2013. Sustainable soil management is more than what and how crops are grown. In: Lal, R. and Stewart, B.A. (Eds.), *Principles of Soil Management in Agro-Ecosystems. Advances in Soil Science.* CRC Press, Taylor & Francis Group, Boca Raton, FL, pp. 337–399..

Kassam, A., Derpsch, R. and Friedrich, T. 2020a. Development of conservation agriculture systems globally. In Kassam, A. (Ed.), *Advances in Conservation Agriculture,* Volume 1, Systems and Science. Burleigh Dodds, Cambridge, UK, pp. 31–86.

Kassam, A., Gonzalez-Sanchez, E.J., Goddard, T., et al. 2020b. Harnessing ecosystem services with conservation agriculture. In: Kassam, A. (Ed.), *Advances in Conservation Agriculture,* vol. 2. Burleigh Dodds, Cambridge, UK, pp. 391–418.

Kassam, A., Gonzalez-Sanchez, E.J., Cheak, S.C., et al. 2020c. Managing conservation agriculture systems: Orchards, plantations and agroforestry. In: Kassam, A. (Ed.), *Advances in Conservation Agriculture,* vol. 1. Burleigh Dodds, Cambridge, UK, pp. 327–358.

Kassam, A., Derpsch, R., and Friedrich, T. 2014a. Global achievements in soil and water conservation: The case of conservation agriculture. *Int. Soil Water Conserv. Res.* 2(1):5–13.

Kassam, A. and Friedrich, T. 2012. An ecologically sustainable approach to agricultural production intensification: Global perspectives and developments. *Field Actions Sci. Rep.* 6. http://factsreports.revues. org/1382

Kassam, A., Friedrich, T., and Derpsch, R. 2019. Global spread of conservation agriculture. *Internat. J. Environ. Stud.* 76(1):29–51. https://doi.org/10.1080/00207233.2018.1494927

Kassam, A., Friedrich, T., Shaxson, F., Bartz, H., Mello, I., Kienzle, J., and Pretty, J. 2014b. The spread of conservation agriculture: Policy and institutional support for adoption and uptake. *Field Actions Sci. Rep.* 7:1–12.

Kassam, A., Friedrich, T., Shaxson, F., and Pretty, J. 2009. The spread of conservation agriculture: Justification, sustainability and uptake. *Int. J. Agric. Sustain.* 7(4):292–320.

Kassam, A. and Kassam, L. 2020a. The need for conservation agriculture. In: Kassam, A. (Ed.), *Advances in Conservation Agriculture, Volume 1, Systems and Science.* Burleigh Dodds, Cambridge, UK, pp. 1–30.

Kassam, A. and Kassam, L. 2020b. Practice and benefits of conservation agriculture systems. In: Kassam, A. (Ed.), *Advances in Conservation Agriculture, Volume 2, Practice and Benefits.* Burleigh Dodds, Cambridge, UK, pp. 1–36.

Kassam, A., Mkomwa, S., and Friedrich, T. 2017. *Conservation Agriculture in Africa: Building Resilient Farming Systems in a Changing Climate.* CABI, Wallingford, UK.

Kätterer, T., Bolinder, M.A., Andrén, O., Kirchmann, H., and Menichetti, L. 2011. Roots contribute more to refractory soil organic matter than aboveground crop residues, as revealed by a long-term field experiment. *Agric. Ecosyst. Environ.* 141(1–2):184–192. http://dx.doi.org/10.1016/j.agee.2011.02.029

Kay, B.D. and VandenBygaart, A.J. 2002. Conservation tillage and depth stratification of porosity and soil organic matter. *Soil Tillage Res.* 66(2):107–118.

Kell, D.B. 2011. Breeding crop plants with deep roots: Their role in sustainable carbon, nutrient and water sequestration. *Ann. Bot.* 108(3):407–418. https://doi.org/10.1093/aob/mcr175

Keller, T., Sandin, M., Colombi, T., Horn, R., and Or, D. 2019. Historical increase in agricultural machinery weights enhanced soil stress levels and adversely affected soil functioning. *Soil Tillage Res.* 194:104293. https://doi.org/10.1016/j.still.2019.104293

Kemper, W.D., Schneider, N.N., and Sinclair, T.R. 2011. No-till can increase earthworm populations and rooting depths. *J. Soil Water Conserv.* 66(1):13A–7A. https://doi.org/10.2489/jswc.66.1.13A

Kemper, W.D., Trout, T.J., Segeren, A., and Bullock, M. 1987. Worms and water. *J. Soil Water Conserv.* 42:401–404.

Khorami, S.S., Kazemeini, S.A., Afzalinia, S., and Gathala, M.K. 2018. Changes in soil properties and productivity under different tillage practices and wheat genotypes: A short-term study in Iran. *Sustainability,* 10, 3273.

Khoury, C.K., Bjorkman, A.D., Dempewolf, H., Ramirez-Villegas, J., Guarino, L., Jarvis, A., Struik, P.C. 2014. Increasing homogeneity in global food supplies and the implications for food security. *Proc. Natl. Acad. Sci. USA* 2014, 111, 4001–4006.

Kibblewhite, M.G., Ritz, K. and Swift, M. J. 2008. Soil health in agricultural systems. *Philos. Trans. R. Soc. Lond. B. Biol. Sci.* 363(1492):685–701. https://doi.org/10.1098/rstb.2007.2178

King, A.E. and Blesh, J. 2018. Crop rotations for increased soil carbon: Perenniality as a guiding principle. *Ecological Applications: A Publication of the Ecological Society of America* 28(1):249–61. https://doi.org/10.1002/eap.1648

Kirkegaard, J.A., Conyers, M.K., Hunt, J.R., Kirkby, C.A., Watt, M., and Rebetzke, G.J. 2014. Sense and nonsense in conservation agriculture: Principles, pragmatism and productivity in Australian mixed farming systems. *Agric. Ecosyst. Environ.* 187:133–145.

Kirkham, M.B. 2014. Infiltration. In: *Principles of Soil and Plant Water Relations*, 2nd edn. Elsevier, Amsterdam; Boston, pp. 201–227.

Kladivko, E.J. 1994. Residue effects on soil physical properties. In: P.W. Unger (Ed.), *Managing Agricultural Residues*. Lewis Publishers, Boca Raton, FL, pp. 123–141.

Kladivko, E.J. 2001. Tillage systems and soil ecology. *Soil Tillage Res.* 61(1–2): 61–76.

Klepper, B. 1990. Root growth and water uptake. pp. 281–322. In: Stewart, B.A. and Nielsen, D.R. (Ed.), *Irrigation of Agricultural Crops. Agron. Monogr.* 30. ASA, CSSA, and SSSA, Madison, WI.

Klocke, N.L., Currie, R.S., and Aiken, R.M. 2009. Soil water evaporation and crop residues. *Trans. ASABE* 52(1):103–110.

Klocke, N.L., Heermann, D.F., and Duke, H.R. 1985. Measurement of evaporation and transpiration with lysimeters. *Trans. ASAE* 28(1):183–189 & 192.

Kögel-Knabner, I. 2002. The macromolecular organic composition of plant and microbial residues as inputs to soil organic matter. *Soil Biol. Biochem.* 34(2):139–162.

Kögel-Knabner, I. 2017. The macromolecular organic composition of plant and microbial residues as inputs to soil organic matter: Fourteen years on. *Soil Biol. Biochem.* 105:A3–A8.

Kool, D., Tong, B., Tian, Z., Heitman, J.L., Sauer, T.J., and Horton, R. 2019. Soil water retention and hydraulic conductivity dynamics following tillage. *Soil Tillage Res.* 193:95–100.

Kramer, M.G., Lajtha, K., and Aufdenkampe, A.K. 2017. Depth trends of soil organic matter C:N and 15N natural abundance controlled by association with minerals. *Biogeochemistry* 136:237–248. https://doi.org/10.1007/s10533-017-0378-x

Kravchenko, A.N. and Guber, A.K. 2017. Soil pores and their contributions to soil carbon processes. *Geoderma* 287:31–39.

Kravchenko, A.N., Negassa, W.C., Guber, A.K., and Rivers, M.L. 2015. Protection of soil carbon within macroaggregates depends on intra-aggregate pore characteristics. *Sci. Rep.*, 5(1):1–10. 10.1038/srep16261

Kravchenko, A.N., Otten, W., Garnier, P., Pot, V., and Baveye, P.C. 2019. Soil aggregates as biogeochemical reactors: Not a way forward in the research on soil–atmosphere exchange of greenhouse gases. *Glob. Chang. Biol.* 25, 2205–2208. https://doi.org/10.1111/gcb.14640

Krna, Matthew A. and Rapson, Gillian L. 2013. Clarifying 'carbon sequestration', *Carbon Manag.* 4(3):309–322. https://doi.org/10.4155/cmt.13.25

Kuzyakov Y. 2010. Priming effects: interactions between living and dead organic matter. *Soil Biol. Biochem.* 42:1363–1371.

Kuzyakov, Y. and Cheng, W. 2004. Photosynthesis controls of CO_2 efflux from maize rhizosphere. *Plant Soil* 263:85–99.

Laban, P., Metternicht, G., and Davies, J. 2018. *Soil Biodiversity and Soil Organic Carbon: Keeping Drylands Alive*. IUCN, Gland, Switzerland. viii + 24p. www.iucn.org/resources/publications

Lal, R. 1991. Tillage and agricultural sustainability. *Soil Tillage Res.* 20(2–4):133–146.

Lal, R. 1993. Tillage effects on soil degradation, soil resilience, soil quality and sustainability. *Soil Tillage Res.* 27:1–8. https://doi.org/10.1016/0167-1987(93)90059-X

Lal, R. 2013a. Food security in a changing climate. *Ecohydrol. Hydrobiol.* 13(1):8–21.

Lal, R. 2013b. Intensive agriculture and the soil carbon pool. In: Kang, M.S. and Banga S.S. (Eds.), *Combating Climate Change: An Agricultural Perspective*. CRC Press, Taylor & Francis Group, Boca Raton, FL, pp. 59–72.

Lal, R. 2015a. A system approach to conservation agriculture. *J. Soil Water Conserv.* 70(4):82A–88A. https://doi.org/10.2489/jswc.70.4.82A

Conservation Agriculture

Lal, R. 2015b. Sequestering carbon and increasing productivity by conservation agriculture. *J. Soil Water Conserv.* 70(3):55A–62A. https://doi.org/10.2489/jswc.70.3.55A

Lal, R., Reicosky, D.C., and Hanson, J.D. 2007. Evolution of the plow over 10,000 years and the rationale for no-till farming. *Soil Tillage Res.* 93(1):1–12.

Lal, R. and Stewart, B.A. (Eds.), 2012. *Soil Water and Agronomic Productivity. Advances in Soil Science*, V. 19. CRC Press, Taylor & Francis Group, Boca Raton, FL, pp. 594. https://doi.org/10.1201/b12214

Lampurlanés, J., Angás, P., and Cantero-Martínez, C. 2001. Root growth, soil water content, and yield of barley under different tillage systems on two soils in semiarid conditions. *Field Crops Res.* 69:27–40.

Lampurlanés, J. and Cantero-Martínez, C. 2003. Soil bulk density and penetration resistance under different tillage and crop management systems and their relationship with barley root growth. *Agron. J.* 95:526–536.

Landl, M., Schnepf, A., Uteau, D., Peth, S., Athmann, M., Kautz, T., Perkons, U., Vereecken, H., and Vanderborght, J., 2019. Modeling the impact of biopores on root growth and root water uptake. *Vadose Zone J.* 18(1):1–20.

Larson, W.E., Lindstrom, M.J., and Schumacher, T.E. 1997. The role of severe storms in soil erosion: A problem needing consideration. *J. Soil Water Conserv.* 52(2):90–95.

Lazarova, Valentina and Asano, Takashi. 2013. Milestones in water reuse: Main challenges, keys to success and trends of development: An overview. Introductory Chapter. In: Lazarova and Asano (Eds.), *Milestones in Water Reuse*, IWA Publishing, London, pp. 1–20. https://library.oapen.org/handle/20.500.12657/43790

Ledo, Alicia, Smith, Pete, Zerihun, Ayalsew, Whitaker, Jeanette, et al. 2020. Changes in soil organic carbon under perennial crops. *Glob. Chang. Biol.* 26:4158–4168.

Lee, J., Six, J., King, A.P., Kessel, C.V., and Rolston, D.E. 2006. Tillage and field scale controls on greenhouse gas emissions. *J. Environ. Qual.* 35:714–725.

Lehmann, J., Hansel, C., Kaiser, C., et al. 2020. Persistence of soil organic carbon caused by functional complexity. *Nat. Geosci.* 3:529–534.

Lehmann, J. and Kleber, M. 2015. Contentious nature of soil organic matter. *Nat. Commun.* 528(7580):60–68. https://doi.org/10.1038/nature16069

Lemanceau, P., Maron, P.A., Mazurier, S., et al. 2015. Understanding and managing soil biodiversity: A major challenge in agroecology. *Agron. Sustain. Dev.* 35(1):67–81. https://doi.org/10.1007/s13593-014-0247-0

Lemordant, L., Gentine, P., Swann, A.S., Cook, B., and Scheff, J. 2018. Critical impact of vegetation physiology on the continental hydrologic cycle in response to increasing CO2. *Proc. Natl. Acad. Sci. USA.* 115(16):4093–4098. https://doi.org/10.1073/pnas.1720712115. 29610293

Li, S., Lu, J., Liang, G., Wu, X., Zhang, M., Plougonven, E., et al. 2021. Factors governing soil water repellency under tillage management: The role of pore structure and hydrophobic substances. *Land Degrad. Dev.* 32(2):1046–1059.

Lin, B.B. 2011. Resilience in agriculture through crop diversification: Adaptive management for environmental change. *BioSci.* 61:183–193.

Lin, H.S. (Ed.). 2012. *Hydropedology: Synergistic Integration of Soil Science and Hydrology*. Elsevier, Waltham, MA.

Lindstrom, M.J., Gupta, S.C., Onstad, C.A., Larson, W.E., and Holt, R.F. 1979. Tillage and crop residue effects on soil erosion in the Corn Belt. *J. Soil Water Conserv.* 34:80–82.

Lobb, D.A., Kachanoski, R.G., and Miller, M.H. 1995. Tillage translocation and tillage erosion on shoulder slope landscape positions measured ^{137}Cs as a tracer. *Can. J. Soil Sci.* 75:2ll–218.

Loch, R.J. 1989. Aggregate breakdown under rain: Its measurement and interpretation. PhD thesis. University of New England, QLD, Australia.

Logan, T.J., Lal, R., and Dick, W.A. 1991. Tillage systems and soil properties in North America. *Soil Tillage Res.* 20:241–270.

López-Fando, C. and Pardo, M.T. 2011. Soil carbon storage and stratification under different tillage systems in a semi-arid region. *Soil Till. Res.* 111:224–230.

Maillard, E., McConkey, B.G., St. Luce, M., Angers, D.A., and Fan, J. 2018. Crop rotation, tillage system, and precipitation regime effects on soil carbon stocks over 1 to 30 years in Saskatchewan. *Can. Soil Tillage Res.* 177:97–104.

Małecka, I., Blecharczyk, A., Sawinska, Z., and Dobrzeniecki, T. (2012). The effect of various long-term tillage systems on soil properties and spring barley yield. *Turk. J. Agricul. For.* 36:217–226.

Mangalassery, S., Sjogersten, S., Sparkes, D.L., Sturrock, C.J., and Mooney, S.J. 2013. The effect of soil aggregate size on pore structure and its consequence on emission of greenhouse gases. *Soil Tillage Res.* 132:39–46.

Martino, D.L. and Shaykewich, C.F. 1994. Root penetration profiles of wheat and barley as affected by soil penetration resistance in field conditions. *Can. J. Soil Sci.* 74:193–200.

Mathews, O.R. and Cole, J.S. 1938. Special dry-farming problems. In: *Soils and Men: Yearbook of Agriculture.* U.S. Department of Agriculture, Washington, D.C., pp. 679–692.

MEA – Millennium Ecosystem Assessment. 2005. *Ecosystems and Human Well-Being: Synthesis.* Island Press, Washington, DC.

Mello, I., Laurent, F., Kassam, A., Marques, G.F., Okawa, C.M.P., and Monte, K. 2021. Benefits of conservation agriculture in watershed management: Participatory governance to improve the quality of no-till systems in the Paraná 3 Watershed, Brazil. *Agronomy* 11(12):2455. https://doi.org/10.3390/agronomy1 1122455

Mitchell, J.P., Reicosky, D.C., Kueneman, E.A., et al. 2019. Conservation agriculture systems. *CAB Rev.* 14. www.cabi.org/cabreviews/review/20193184383

Mitchell, J.P., Singh, P.N., Wallender, W.W., et al. 2012. No-tillage and high-residue practices reduce soil water evaporation. *Calif. Agric.* 66(2):55–61. https://doi.org/10.3733/ca.v066n02p55

Montgomery, D.R. 2007. Soil erosion and agricultural sustainability. *Proc. Natl. Acad. Sci. USA.* 104(33): 13268–13272.

Moreira, W.H., Tormena, C.A., Karlen, D.L., da Silva, Á.P., Keller, T., and Betioli, E. 2016. Seasonal changes in soil physical properties under long-term no-tillage. *Soil Tillage Res.* 160:53–64. https://doi.org/10.1016/ j.still.2016.02.007

Morell, F.J., Cantero-Martínez, C., Lampurlanés, J., Plaza-Bonilla, D., and Alvaro-Fuentes, J. 2011. Soil carbon dioxide flux and organic carbon content: Effects of tillage and nitrogen fertilization. *Soil Sci. Soc. Am. J.* 75:1874–1884.

Myers, C.F., Meek, J., Tuller, S. and Weinberg, A. 1985. Nonpoint sources of water pollution. *J. Soil Water Conserv.* 40:14–18.

Nannipieri, P. 2020. Soil is still an unknown biological system. *Appl. Sci.* 10:3717. https://doi.org/10.3390/ app10113717

Nannipieri, P., Ascher-Jenull, J., Ceccherini, M.T., Pietramellara, G., Renella, G., and Schloter, M. 2020. Beyond microbial diversity for predicting soil functions: A mini review. *Pedosphere* 30(1):5–17.

Nichols, D.J., Daniel, T.C., and Edwards, D.R. 1994. Nutrient runoff from pasture after incorporation of poultry litter or inorganic fertilizer. *Soil Sci. Soc. Am. J.* 58:1224–1228.

Nielsen, D.C., Lyon, D.J., Hergert, G.W., Higgins, R.K., et al. 2015. Cover crop mixtures do not use water differently than single-species plantings. *Agron. J.* 107(3):1025–1038. https://doi.org/10.2134/agron j14.0504

Nielsen, D.C., Unger, P.W., and Miller, P.R. 2005. Efficient water use in dryland cropping systems in the Great Plains. *Agron. J.* 97:364–372.

Noellemeyer, E., Fernández, R. and Quiroga, A. 2013. Crop and tillage effects on water productivity of dryland agriculture in Argentina. *Agriculture*, 3:1–11. https://doi.org/10.3390/agriculture3010001

Noellemeyer, E., Frank, F., Alvarez, C., Morazzo, G., and Quiroga, A. 2008. Carbon contents and aggregation related to soil physical and biological properties under a land-use sequence in the semiarid region of central Argentina. *Soil Tillage Res.* 99:179–190.

Nouri, A., Lee, J., Yin, X., Tyler, D.D., and Saxton, A.M. 2019. Thirty-four year no-tillage and cover crops improve soil quality and increase cotton yield in Alfisols, Southeast, USA. *Geoderma* 345:51–62.

Novara, A., Gristina, L., La Mantia, T., and Rühl, J. 2011. Soil carbon dynamics during secondary succession in a semi-arid Mediterranean environment. *Biogeosci. Discuss.*, 8:11107–11138. https://doi.org/10.5194/ bgd-8-11107-2011

Nunes, M.R., Denardin, J.E., Pauletto, E.A., Faganello, A. and Pinto, L.F.S. 2015. Effect of soil chiseling on soil structure and root growth for a clayey soil under no-tillage. *Geoderma* 259:149–155.

Oades, J.M. 1993. The role of biology in the formation, stabilization and degradation of soil structure. In: L. Brussaard and M.J. Kooistra (Editors), Int. Workshop on Methods of Research on Soil Structure/Soil Biota Interrelationships. *Geoderma*, 56:377–400.

Olson, K., Ebelhar, S.A., Lang, J.M. 2014. Long-term effects of cover crops on crop yields, soil organic carbon stocks and sequestration. *Open J. Soil Sci.* 4:284–292. https://doi.org/10.4236/ojss.2014.48030

Or, D., Keller, T. and Schlesinger, W.H. 2021. Natural and managed soil structure: On the fragile scaffolding for soil functioning. *Soil Tillage Res*. 208, April 2021, 104912. https://doi.org/10.1016/j.still.2020.104912

O'Rourke, S.M., Angers, D.A., Holden, N.M., and McBratney, A.B. 2015. Soil organic carbon across scales. *Glob. Chang. Biol.*:1–14. https://doi.org/10.1111/gcb.12959

Palm, C., Blanco-Canqui, H., De Clerck, F., et al., 2014. Conservation agriculture and ecosystem services: An overview. *Agricul. Ecosyst. Environ*. 187:87–105.

Pan, G.X., Smith, P., and Pan, W.N. 2009. The role of soil organic matter in maintaining the productivity and yield stability of cereals in China. *Agricul. Ecosyst. Environ*. 129:344–348.

Papendick, R.I., Lindstrom, M.J., and Cochran, V.L. 1973. Soil mulch effect on seedbed temperature and water during fallow in eastern Washington. *Soil Sci. Soc. Am. Proc*. 37:307–314.

Paustian, K., O. Andrén, H. H. Janzen et al. 1997. Agricultural soil as a sink to mitigate CO2 emissions. *Soil Use Manag*. 13, 230–244. doi.org/10.1111/j.1475-2743.1997.tb00594.x.

Pelegrin, F., Moreno, F., Martin-Aranda, J., and Camps, M. 1988. The influence of tillage methods on soil-water conservation in SW Spain. pp. 803–808. In: *Tillage and Traffic in Crop Production. Proc. Int. Conf. Int. Soil Tillage Res. Organ., 11th*, Edinburgh, Scotland. 11–15 July 1988. ISTRO, Edinburgh, Scotland.

Peters, D.B. 1960. Relative magnitude of evaporation and transpiration. *Agron. J*. 52:536–538.

Peters, D.B. and Russel, M.B. 1959. Relative water losses by evaporation and transpiration in field corn. *Soil Sci. Soc. Amer. Proc*. 23:170–176.

Piccoli, I., Camarotto, C., Lazzaro, B., Furlan, L., and Morari, F. 2017. Conservation agriculture had a poor impact on the soil porosity of veneto low-lying plain silty soils after a 5-year transition period. *Land Degrad. Dev*. 28(7):2039–2050.

Pierret, A., Maeght, J.L., Cl'ement, C., Montoroi, J.P., Hartmann, C., and Gonkhamdee, S. 2016. Understanding deep roots and their functions in ecosystems: An advocacy for more unconventional research. *Ann. Bot*. 118(4):621–635.

Pikul Jr., J.L. and Aase, J.K. 1995. Infiltration and soil properties as affected by annual cropping in the northern Great Plains. *Agron. J*. 87:656–662.

Pimentel, D., Harvey, C., Resosudarmo, P., et al. 1995. Environmental and economic cost of soil erosion and conservation benefits. *Science* 267(5201):1117–1123.

Pires, L.F., Borges, J.A., Rosa, J.A., Cooper, M., Heck, R.J., Passoni, S., and Roque, W.L., 2017. Soil structure changes induced by tillage systems. *Soil Tillage Res*. 165:66–79.

Piron, D., Boizard, H., Heddadj, D., Pérès, G., Hallaire, V., and Cluzeau, D. 2017. Indicators of earthworm bioturbation to improve visual assessment of soil structure. *Soil Tillage Res*. 173:53–63.

Poeplau, C., Don, A., Vesterdal, L., Leifeld, J., et al. 2011. Temporal dynamics of soil organic carbon after land-use change in the temperate zone – carbon response functions as a model approach. *Glob. Chang. Biol*. 17:2415–2427.

Post, W.M. and Kwon, K.C. 2000. Soil carbon sequestration and land-use change: Processes and potential. *Glob. Chang. Biol*. 6:317–328.

Poudel, R., Jumpponen, A., Kennelly, M.M., et al. 2019. Rootstocks shape the rhizobiome: Rhizosphere and endosphere bacterial communities in the grafted tomato system. *Appl. Environ. Microbiol*. 85(2019):e01765–18.

Power, A.G. 2010. Ecosystem services and agriculture: Tradeoffs and synergies. *Phil. Trans. R. Soc. B* 365:2959–2971. https://doi.org/10.1098/rstb.2010.0143

Power, J.F. 1983. Soil management for efficient water use: Soil fertility.. In: Taylor, H.M., Jordan, W.R., and Sinclair, T.R. (Eds.), *Limitations to Efficient Water Use in Production*. American Society of Agronomy and Soil Science Society of America, Madison, WI, pp. 461–470.

Powlson, D.S., Stirling, C.M., Jat, M.L., Gerard, B., et al. 2014. Limited potential of no-till agriculture for climate change mitigation. *Nat. Clim. Chang*. 4:678–683.

Pretty, Jules and Bharucha, Zareen Pervez. 2014. Sustainable intensification in agricultural systems. *Ann. Bot*. 114(8):1571–1596. https://doi.org/10.1093/aob/mcu205

Puget, P. and Drinkwater, L.E. 2001. Short-term dynamics of root- and shoot-derived carbon from a leguminous green manure. *Soil Sci. Soc. Am. J*. 65:771–779.

Puma, M.J., Bose, S., Chon, S.Y., and Cook, B.I. 2015. Assessing the evolving fragility of the global food system. *Environ. Res. Lett*. 10:024007.

Qamar, R., Ehsanullah, A., Ahmad, R., and Iqbal, M. 2012. Response of wheat to tillage and nitrogen fertilization in rice-wheat system. *Pak. J. Agricul. Sci*. 49:243–254.

Qin, R., Stamp, P., and Richner, W. 2004. Impact of tillage on root system of winter wheat. *Agron. J.* 96:1523–1530.

Quigley, M.Y. and Kravchenko, A.N. 2022. Inputs of root-derived carbon into soil and its losses are associated with pore-size distributions. *Geoderma* 410:115667. https://doi.org/10.1016/j.geoderma.2021.115667

Quincke, J.A., Wortmann, C.S., Mamo, M., Franti, T., and Drijber, R.A. 2007. Occasional tillage of no-till systems: Carbon dioxide flux and changes in total and labile soil organic carbon. *Agron. J.* 99(4):1158–1168.

Rabot, E., Wiesmeier, M., Schlüter, S., and Vogel, H.J. 2018. Soil structure as an indicator of soil functions: A review. *Geoderma* 314:122–137. ISSN 0016-7061. https://doi.org/10.1016/j.geoderma.2017.11.009

Ranaivoson, L., Naudin, K., Ripoche, A., et al. 2017. Agro-ecological functions of crop residues under conservation agriculture. A review. *Agron. Sustain. Dev.* 37(26):1–17. doi:10.1007/s13593-017-0432-z

Rasmussen, K.J. 1991. Reduced soil tillage and Italian ryegrass as catch crop: II. Soil bulk density, root development and soil chemistry. (In Danish.) *Tidsskr. Planteavl.* 95:139–154.

Rasmussen, K.J. 1999. Impact of ploughless soil tillage on yield and soil quality: A Scandinavian review. *Soil Tillage Res.* 53:3–14.

Rawls, W.J., Pachepsky, Y.A., Ritchie J.C., Sobecki, T.M., and Bloodworth, H. 2003. Effect of soil organic carbon on soil water retention. *Geoderma* 116(1): 61–76.

Reeves, D. 1997. The role of soil organic matter in maintaining soil quality in continuous cropping systems. *Soil Tillage Res.* 43(1):131–167.

Reicosky, D.C. 1997a. Tillage methods and carbon dioxide loss: Fall versus spring tillage. In: Lal, R., Kimble, J.M., Follett, R.F., and Stewart, B.A. (Eds.), *Management of Carbon Sequestration in Soil*. CRC Press, Boca Raton, FL, pp. 99–111.

Reicosky, D.C. 1997b. Tillage-induced CO2 emission from soil. *Nutr. Cycl. Agroecosystems* 49:273–285.

Reicosky, D.C. 2003. Conservation agriculture: Global environmental benefits of soil carbon management. In: García-Torres, L., Benites, J., Martínez-Vilela, A., and Holgado-Cabrera, A. (Eds.), *Conservation Agriculture*. Springer, Dordrecht, pp. 3–12. https://doi.org/10.1007/978-94-017-1143-2_1

Reicosky, D.C. 2020. Conservation agriculture systems: Soil health and landscape management. Ch. 03. In: Kassam, A. (Ed.), *Advances in Conservation Agriculture, Volume 1, Systems and Science*. Burleigh Dodds Science Publishing Limited, Cambridge, UK, pp. 87–154.

Reicosky, D.C. and Allmaras, R.R. 2003. Advances in tillage research in North American cropping systems. In: Shrestha, A. (Ed.), *Cropping Systems: Trends and Advances*. Haworth Press, Inc., New York, NY, pp. 75–125.

Reicosky, D.C. and Archer, D.W. 2007. Moldboard plow tillage depth and short-term carbon dioxide release. *Soil Tillage Res.* 94:109–121.

Reicosky, D.C., Calegari, A., dos Santos, D.R., and Tiecher, T. 2021. Cover crop mixes for diversity, carbon and conservation agriculture. Chap 11. pp. 169–208. In: Islam, R. and B. Sherman (Eds.), *Cover Crops and Sustainable Agriculture*. CRC Press, Taylor & Francis Group, Boca Raton, FL, pp. 326.

Reicosky, D.C., Evans, S.D., Cambardella, C.A., et al. 2002. Continuous corn with moldboard tillage: Residue and fertility effects on soil carbon. *J. Soil Water Conserv.* 57(5):277–284.

Reicosky, D.C. and Forcella, F. 1998. Cover crop and soil quality interactions in agroecosystems. *J. Soil Water Conserv.* 53(3):224–229.

Reicosky, D.C., Gesch, R.W., Wagner, S.W., Gilbert, R.A. Wente, C.D. and Morris, D.R. 2008. Tillage and wind effects on soil CO2 concentrations in muck soils. *Soil Tillage Res.* 99:221–231.

Reicosky, D.C. and Janzen, H.H. 2019. Conservation agriculture: Maintaining land productivity and health by managing carbon flows, chap. 4. In: Lal, R. and Stewart, B.A. (Eds.). *Soil and Climate. Advances in Soil Science*. Taylor & Francis Group, LLC, pp. 31–161.

Reicosky, D.C. and Kassam, A. 2021. Conservation agriculture: Carbon and conservation centered foundation for sustainable production. Chap 02. pp. 19–64. In: Lal, R. (Ed.), *Advances in Soil Science, Soil Organic Matter and Feeding the Future: Environmental and Agronomic Impacts*. Taylor & Francis Group, LLC, Boca Raton, FL, pp. 428.

Reicosky, D.C. and Kemper, W.D. 1995. Effects of crop residue on infiltration, evaporation, and water-use efficiency. In: Moldenhauer, W.C. and Mielke, L.N. (Eds.), *Crop Residue Management to Reduce Erosion and Improve Soil Quality*. USDA-ARS, pp. 65–72, Conservation Research Report Number 42, North Central Region.

Reicosky, D.C., Kemper, W.D., Langdale, G.W., Douglas, C.L., and Rasmussen, P.E. 1995. Soil organic matter changes resulting from tillage and biomass production. *J. Soil Water Conserv.* 50(3):253–261.

Reicosky, D.C. and Lindstrom, M.J. 1993. Fall tillage method: Effect on short-term carbon dioxide flux from soil. *Agron. J.* 85:1237–1243.

Reicosky, D.C. and Lindstrom, M.J. 1995. Impact of fall tillage on short-term carbon dioxide flux. In: Lal, R., Kimbal, J., Levine, E., and Stewart, B.A. (Eds.) *Soils and Global Change.* Lewis Publishers, pp. 177–187.

Reicosky, D.C., Sauer, T.J., and Hatfield, J.L. 2011. Challenging balance between productivity and environmental quality: Tillage impacts. In: Jerry L. Hatfield and Thomas J. Sauer (Eds.), *Soil Management: Building a Stable Base for Agriculture.* pp. 13–38. Copyright © 2011 by American Society of Agronomy and Soil Science Society of America.

Reicosky, D.C. and Saxton, K., 2007. The benefits of no-tillage. In: Baker, C.J., Saxton, K.E., Ritchie, W.R., Chamen, W.C.T., Reicosky, D.C., Ribeiro, M.F.S., Justice, S.E., and Hobbs, P.R. (Eds.), *No-Tillage Seeding in Conservation Agriculture.* 2nd edn., UK, pp. 11–20.

Reinsch, S., Jarvis, N., and Tuller, M., 2019. Global environmental changes impact soil hydraulic functions through biophysical feedbacks. *Glob. Chang. Biol.* 25(6):1895–1904. https://doi.org/10.1111/gcb.14626

Reusser, P., Bierman, P., and Rood, D. 2015. Quantifying human impacts on rates of erosion and sediment transport at a landscape scale. *Geology* 43(2):171–174. https://doi.org/10.1130/G36272.1

Riggers, C., Poeplau, C., Don, A., Fruhauf, C., and Dechow, R. 2021. How much carbon input is required to preserve or increase projected soil organic carbon stocks in German croplands under climate change? *Plant Soil* 460(1–2):417–433. https://doi.org/10.1007/s11104-020-04806-8

Ritchie, J.T. 1983. Efficient water use in crop production: Discussion on the generality between biomass production and evapotranspiration. In: Taylor, H.M. Jordan, W., and Sinclair, T.R. (Eds.), *Limitations to Efficient Water Use in Crop Production.* American Society of Agronomy, Madison, WI, pp. 29–44.

Rochette, P. and Angers, D.A. 1999. Soil surface carbon dioxide fluxes induced by spring, summer, and fall moldboard plowing in a sandy loam. *Soil Sci. Soc. Am. J.* 63(3):621–628.

Rockström, J., Karlberg, L., Wani, S.P., Barron, J., Hatibu, N., Oweis, T., Bruggeman, A., Farahani, J., and Qiang, Z. 2010. Managing water in rainfed agriculture—The need for a paradigm shift. *Agric. Water Manag.* 97:543–550. https://doi.org/10.1016/j.agwat.2009.09.009

Rockström, J., Kaumbutho, P., Mwalley, J., Nzabi, A.W., Temesgen, M., Mawenya, L., Barron, J., Mutua, J., and Damgaard-Larsen, S. 2009. Conservation farming strategies in East and Southern Africa: Yields and rain water productivity from on-farm action research. *Soil Tillage Res.* 103:23–32.

Rockström, J., Wani, S.P., Oweis, T., and Hatibu, N. 2007. Managing water in rainfed agriculture. In: *Water for Food, Water for Life: A Comprehensive Assessment of Water Management in Agriculture.* Earth Scan/ IWMI; London, UK/Colombo, Sri Lanka, pp. 315–348.

Sachs, J., Remans, R., Smukler, S., Winowiecki, L., Andelman, S.J., Cassman, K.G., and Sanchez, P.A. 2010. Monitoring the world's agriculture. *Nature* 466:558–560.

Sanderman, J., Creamer, C., Baisden, W.T., Farrell, M., and Fallon, S., 2017. Greater soil carbon stocks and faster turnover rates with increasing agricultural productivity. *Soil* 3(1):1–16. https://doi.org/10.5194/soil-3-1-2017

Schimel, J.P. 2018. Life in dry soils: Effects of drought on soil microbial communities and processes. *Annu. Rev. Ecol. Evol. Syst.* 49:409–432. https://doi.org/10.1146/annurev-ecolsys-110617-062614

Schmidt, M.W.I., Torn, M.S., Abiven, S., Dittmar, T., Guggenberger, G., Janssens, I.A., Kleber, M., Kogel-Knabner, I., Lehmann, J., Manning, D.A.C., Nannipieri, P., Rasse, D.P., Weiner, S., and Trumbore, S.E. 2011. Persistence of soil organic matter as an ecosystem property. *Nature* 478:49–56. https://doi.org/10.1038/Nature10386

Schrumpf, M., Kaiser, K., Guggenberger, G., Persson, T., Kögel-Knabner, I., and Schulze, E.-D. 2013. Storage and stability of organic carbon in soils as related to depth, occlusion within aggregates, and attachment to minerals. *Biogeosciences* 10:1675–1691. https://doi.org/10.5194/bg-10-1675-2013

Seta, A.K., Blevins, R.L., Frye, W.W., and Barfield, B.J. 1993. Reducing soil erosion and agricultural chemical losses with conservation tillage. *J. Environ. Qual.* 22:661–665.

Sexstone, A.J., N.P. Revsbech, T.B. Parkin, and J.M. Tiedje. 1985. Direct measurement of oxygen profiles and denitrification rates in soil aggregates. *Soil Sci. Soc. Am. J.* 49:645–651.

Sharpley, A.N. 2003. Soil mixing to decrease surface stratification of phosphorus in manured soils. *J. Environ. Qual.* 32:1375–1384.

Sharpley, A.N., Smith, S.J., Jones, O.R., Berg, W.A., and Coleman, G.A. 1992. The transport of bioavailable phosphorus in agricultural runoff. *J. Environ. Qual.* 21:30–35.

Sherwood, S. and Fu, Q. 2014. A drier future? *Science* 343:737–739. DOI: 10.1126/science.1247620

Shukla, M.K., Lal, R., Owens, L., and Unkefer, P. 2003. Land use management impacts on structure and infiltration characteristics of soils in the north Appalachian region of Ohio. *Soil Sci.* 168:167–177.

Sidiras, N., Bilalis, D., and Vavoulidou, E. 2001. Effects of tillage and fertilization on some elected physical properties of soil (0–30 cm depth) and on the root growth dynamic of winter barley (Hordeum vulgare cv. Niki). *J. Agron. Crop Sci.* 187:167–176.

Sinclair, T.R., Tanner, C.B., and Bennet, J.M. 1984. Water use efficiency in crop production. *Bioscience* 34:36–40.

Sisti, D.P.J., dos Santos, H.P., Kohhann, R., Alves, B.J.R., Urquiaga, S., and Boddey, R.M. 2004. Change on carbon and nitrogen stocks in soil under 13 years of conventional or zero tillage in southern Brazil. *Soil Tillage Res.* 76: 39–58.

Sithole, N.J., Magwaza, L.S., and Thibaud, G.R. 2019. Long-term impact of no-till conservation agriculture and N-fertilizer on soil aggregate stability, infiltration and distribution of C in different size fractions. *Soil Tillage Res.* 190:147–156.

Six, J., Bossuyt, H., Degryze, S., and Denef, K., 2004. A history of research on the link between (micro) aggregates, soil biota, and soil organic matter dynamics. *Soil Tillage Res.* 79(1):7–31.

Six, J., Elliott, E., and Paustian, K. 2000. Soil macroaggregate turnover and microaggregate formation: A mechanism for C sequestration under no-tillage agriculture. *Soil Biol. Biochem.* 32(14):2099–2103.

Six, J., Frey, S.D., Thiet, R.K., and Batten, K.M. 2006. Bacterial and fungal contributions to carbon sequestration in agroecosystems. *Soil Sci. Soc. Am. J.* 70(2):555–569.

Smith, P. 2004. How long before a change in soil organic carbon can be detected? *Glob. Chang. Biol.* 10(11):1878–1883.

Smith, P. 2005. An overview of the permanence of soil organic carbon stocks: Influence of direct human-induced, indirect and natural effects. *Eur. J. Soil Sci.* 56:673–680. https://doi.org/10.1111/j.1365-2389.2005.00708.x

Smith, P., Soussana, J.-F., Angers, D., Schipper, L., Chenu, C., Rasse, D.P., et al. 2019. How to measure, report and verify soil carbon change to realize the potential of soil carbon sequestration for atmospheric greenhouse gas removal. *Glob. Chang. Biol.* 26:219–241. https://doi.org/10.1111/gcb.14815

Smucker, A.J.M., Park, E.-J., Dorner, J., and Horn, R. 2007. Soil micropore development and contributions to soluble carbon transport within macroaggregates. *Vadose Zone J.* 6(2):282–290.

Soil Science Glossary Terms Committee. 2008. *Glossary of Soil Science Terms.* Soil Science Society of America, Madison, WI.

Soong, J.L., Castanha, C., Pries, C.E.H., Ofiti, N., Porras, R.C., Riley, W.J., et al., 2021. Five years of whole-soil warming led to loss of subsoil carbon stocks and increased CO2 efflux. *Sci. Advan.* 7(21):eabd1343.

Sprent, J.I. 1999. Nitrogen fixation and growth of non-crop species in diverse environments. *Perspect. Plant Ecol. Evol. Syst.* 2:149–162.

Stevenson, F.C. and Van Kessel, C. 1996. A landscape scale assessment of the nitrogen and non-nitrogen rotation benefits of pea. *Can. J. Plant Sci.* 76:735–745.

Stewart, B.A. 1988. Dryland farming: The north American experience. In: Unger, P.W., Sneed, T.V., Jordan, W.R., and Jensen, R. (Eds.), *Challenges in Dryland Agriculture a Global Perspective*. pp. 54–59. *Proceedings of the International Conference on Dryland Farming*, August 15–19, 1988, Amarillo-Bushland, TX. Texas Agricultural Experiment Station, College Station, TX.

Stewart, B.A. 2008a. Experiences with conservation agriculture in semiarid regions of the USA. Workshop proceedings "Conservation Agriculture for Sustainable Land Management to Improve the Livelihood of People in Dry Areas". pp. 288. FAO, Rome, Italy. FAO Book 2008, pp. 141–155.

Stewart, B.A. 2008b. Water conservation and water use efficiency in drylands. In: B.A. Stewart, A. Fares Asfary, A. Belloum, K. Steiner, and T. Friedrich (Eds.), *Conservation Agriculture for Sustainable Land Management to Improve the Livelihood of People in Dry Areas.* © 2008 FAO Rome, Italy. ACSAD & GTZ. pp. 57–66.

Stewart, B.A. 2009. Manipulating tillage to increase stored soil water and manipulating plant geometry to increase water-use efficiency in dryland areas. *J. Crop Improve.* 23(1):71–82. https://doi.org/10.1080/15427520802418319

Stewart, B.A., Baumhardt, R.L., and Evett, S.R. 2010. Major advances in soil and water conservation in the US Southern Great Plains. Chap. 4. In: Zobeck, T.M. and Schillinger, W.F. (Eds.), *Soil and Water Conservation Advances in the United States*. SSSA Special Publication 60. pp 103–129.

Stewart, B.A. and Burnett, E. 1987. Water conservation technology in rainfed and dryland agriculture. In: W.R. Jordan (Ed.), *Water and Water Policy in World Food Supplies*. pp. 355–359. *Proceedings of the Conference*, May 26–30, 1985. Texas A&M University, College Station, Texas.

Stewart, B.A., Fares Asfary, A., Belloum, A., Steiner, K. and Friedrich, T. (Eds.), 2008. *Workshop proceedings "Conservation Agriculture for Sustainable Land Management to Improve the Livelihood of People in Dry Areas"*. pp. 288. FAO. Rome, Italy.

Stewart, B.A. and Koohafkan, P. 2004. Dryland agriculture: Long neglected but of worldwide importance. In: Rao, S.C. and Ryan, J. (Eds.), *Challenges and Strategies Ford Dryland Agriculture*. pp. 11–23. CSSA Special Publication 32, Crop Science Society of America, Madison, WI. https://doi.org/10.2135/cssaspecpub32.c2

Stewart, B.A. and Lal, R. 2012. Manipulating crop geometries to increase yields in dryland areas. In: Lal, R. and Stewart, B.A. (Eds.), *Soil Water and Agronomic Productivity*. CRC Press, Boca Raton, FL, pp. 409–425.

Stewart, B.A., Lal, R., and El-Swaify, S.A. 1991. Sustaining the resource base of an expanding world agriculture. In: Lal, R., and Pierce, F.J. (Eds.), *Soil Management for Sustainability*. Soil and Water Conservation Society, Ankeny, pp. 125–144.

Stewart, B.A., and Peterson, G.A. 2015. Managing green water in dryland agriculture. *Agron. J.* 107:1544–1553. https://doi.org/10.2134/agronj14.0038

Stewart, B.A. and Robinson, C.A. 1997. Are agroecosystems sustainable in semiarid regions? *Adv. Agron.* 60:191–228.

Stewart, B.A. and Steiner, J.L. 1990. Water-use efficiency. In: Singh, R.P., Parr, J.F., and Stewart, B.A. (Eds.) *Advances in Soil Science. Advances in Soil Science*, vol 13. Springer, New York, NY. https://doi.org/10.1007/978-1-4613-8982-8_7

Stewart, B.A. and Thapa, S. 2016. Dryland farming: Concept, origin and brief history. In: Farooq, M. and Siddique, K. (Eds.), *Innovations in Dryland Agriculture*. Springer, Cham. pp. 3–29. https://doi.org/10.1007/978-3-319-47928-6_1

Stinner, B.R. and House, G.J. 1990. Arthropods and other invertebrates in conservation-tillage agriculture. *Annu. Rev. Entomol.* 35:299–318.

Swinton, S.M., Lupi, F., Robertson, G.P., and Landis, D.A. 2006. Ecosystem services from agriculture: Looking beyond the usual suspects. *Am. J. Agric. Econom.* 88:1160–1166.

Tan, C.S., Drury, C.F., Reynolds, W.D., Gaynor, J.D., Zhang, T.Q., and Ng, H.Y. 2002. Effect of long-term conventional tillage and no-tillage systems on soil and water quality at the field scale. *Water Sci Technol.* 46(6–7):183–190.

Tanaka, D.L., Anderson, R.L., and Rao, S.C. 2005. Crop sequencing to improve use of precipitation and synergize crop growth. *Agron. J.* 97:385–390.

Tanner, C.B., Peterson, A.E., and Love, J.R. 1960. Radiant energy exchange in a cornfield. *Agron. J.* 52:373–379.

Tanner, C.B. and Sinclair, T.R. 1983. Efficient water use in crop production: Research or re-search? pp. 1–27. In: Taylor, H.M., et al. (Eds.), *Limitations to Efficient Water Use in Crop Production*. ASA, Madison, WI.

Tecon, R. and Or, D. 2017. Biophysical processes supporting the diversity of microbial life in soil. *FEMS Microbiol. Rev.* 41:599–623.

Thierfelder, C., Chivenge, P., Mupangwa, W., Rosenstock, T.S., Lamanna, C., Eyre, J.X. 2017. How climate-smart is conservation agriculture (CA)? – Its potential to deliver on adaptation, mitigation and productivity on smallholder farms in southern Africa. *Food Sec.* 9:537–560.

Thierfelder, C. and Wall, P.C. 2009. Effects of conservation agriculture techniques on infiltration and soil water content in Zambia and Zimbabwe. *Soil Tillage Res.* 105:217–227.

Tilman, D., Balzer, C., Hill, J., and Befort, B.L. 2011. Global food demand and the sustainable intensification of agriculture. *Proc. Natl. Acad. Sci.* 108(50):20260–20264. doi.org/10.1073/pnas.1116437108

Tisdall, J.M. 1994. Possible role of soil microorganisms in aggregation in soils. *Plant Soil.* 159:115–121.

Tisdall, J.M. and Oades, J.M. 1982. Organic matter and water stable aggregates in soils. *J. Soil Sci.* 33:141–163.

Todd, R.W., Klocke, N.L., Hergert, G.W., and Parkhurst, A.M. 1991. Evaporation from soil influenced by crop shading, crop residue and wetting regime. *Trans. ASAE* 34(2):461–466.

Triplett, G.B., Jr., and Dick, W.A. 2008. No-tillage crop production: A revolution in agriculture! *Agron. J.* 100:S-153–S-165. https://doi.org/10.2134/agronj2007.0005c

Triplett, G.B., Jr. and Van Doren, Jr. D.M. 1969. Nitrogen, phosphorus, and potassium fertilization of non-tilled maize. *Agron. J.* 61:637–639.

Tully, K. and Ryals, R. 2017. Nutrient cycling in agroecosystems: Balancing food and environmental objectives. *Agroecol. Sustain. Food Syst.* 41(7):761–798. DOI: 10.1080/21683565.2017.1336149

Turner, N.C. 2004. Agronomic options for improving rainfall-use efficiency of crops in dryland farming systems. *J. Exp. Bot.* 55:2413–2425. https://doi.org/10.1093/jxb/erh154

Turner, P.A., Griffis, T.J., Lee, X., Baker, J.M., Venterea, R.T., and Wood, J.D. 2015. Indirect nitrous oxide emissions from streams within the U.S. Corn Belt scale with stream order. *Proc. Natl. Acad. Sci. USA.* 112:9839–9843.

Unger, P.W. 1978. Straw-mulch rate effect on soil water storage and sorghum yield. *Soil Sci. Soc. Am. J.*, 42:486–491.

Unger, P.W. 1991. Overwinter changes in physical properties of no-tillage soil. *Soil Sci. Soc. Am. J.* 55:778–782.

Unger, P.W. and Baumhardt, R.L. 1999. Factors related to dryland grain sorghum yield increases. *Agron. J.* 91:870–875.

Unger, P.W., Kirkham, M.B., and Nielsen, D.C. 2010. Water conservation for agriculture. In: Zobeck T.M. and Schillinger, W.F. (Eds.), *Soil and Water Conservation Advances in the United States*. SSSA Special Publication 60. pp. 1–45.

Unger, P.W. and Parker Jr., J.J. 1976. Evaporation reduction from soil with wheat, sorghum, and cotton residues. *Soil Sci. Soc. Am. J.* 40:954–957.

Unger, P.W. and Vigil, M.F. 1998. Cover crop effects on soil water relationships. *J. Soil Water Conserv.* 53:200–207.

Valenzuela, H. 2016. Agroecology: A global paradigm to challenge mainstream industrial agriculture. *Horticulturae* 2:2. https://doi.org/10.3390/horticulturae2010002

Van Donk, S.J., Martin, D.L., Irmak, S., Melvin, S.R. Peterson, J., and Davison, D. 2010. Crop residue cover effects on evaporation, soil water content, and yield of deficit-irrigated corn in west-central Nebraska. *Trans. ASABE* 53(6):1787–1797.

Van Doren, D.M., Jr., and Allmaras, R.R. 1978. Effects of residue management practices on the soil physical environment, microclimate, and plant growth. In: Oschwald, W.R. (Ed.), *Crop Residue Management Systems*. American Society of Agronomy Special Publication 31. American Society of Agronomy, Madison, WI, pp. 49–83.

Van Noordwijk, M. 1983. Functional interpretation of root densities in the field for nutrient and water uptake. In: *Root Ecology and Its Practical Application*. pp. 207–226. *Int. Symp. Gumpenstein*, 1982, Bundesanstalt Gumpenstein, Iradning.

Van Oost, K., Quine, T.A., Govers, G., De Gryze, S., Six, J., et al. 2007. The impact of agricultural soil erosion on the global carbon cycle. *Science* 318:626–629.

Veiga, M., Reinert, D.J., Reichert, J.M., and Kaiser, D.R. 2008. Short and long-term effects of tillage systems and nutrient sources on soil physical properties of a southern Brazilian hapludox. *R. Bras. Ci. Solo* 32:1437–1446.

Verhulst, N., Govaerts, B., Verachtert, E., et al. 2010. Conservation agriculture, improving soil quality for sustainable cropping systems? In: Lal, R., and Stewart, B.A. (Eds.), *Advances in Soil Science: Food Security and Soil Quality*. CRC Press, Boca Raton, FL, pp. 137–408.

Vermeulen, S.J., Challinor, A.J., Thornton, P.K., Campbell, B.M., Eriyagama, N., Vervoort, J.M., and Smith, D.R. 2013. Addressing uncertainty in adaptation planning for agriculture. *Proc. Natl. Acad. Sci. USA* 110:8357–8362.

Vogel, H.J., Balseiro Romero, M., Kravchenko, A., Otten, W., Pot, V., Schlüter, S., Weller, U., and Baveye, P.C. 2022. A perspective on soil architecture is needed as a key to soil functions. *Eur. J. Soil Sci.* 73(1):e13152. https://doi.org/10.1111/ejss.13152

von Lützow, M., Kögel-Knabner, I., Ekschmitt, K., Matzner, E., et al. 2006. Stabilization of organic matter in temperate soils: mechanisms and their relevance under different soil conditions—a review. *Eur. J. Soil Sci.* 57:426–445.

Voroney, R.P., Paul, E.A., and Anderson, D.W. 1989. Decomposition of wheat straw and stabilization of microbial producers. *Can. J. Soil Sci.* 69(1):63–77.

Wacha, K., Philo, A., and Hatfield, J.L. 2022. Soil energetics: A unifying framework to quantify soil functionality. *Agrosyst. Geosci. Environ.* 5:e20314. https://doi.org/10.1002/agg2.20314

Wahbi, Ammar. 2008. Crop roots and water use efficiency in conservation agriculture and conventional tillage systems in drylands. In: Bobby A. Stewart, A. Fares Asfary, Abdelouahab Belloum, Kurt Steiner, and Theodor Friedrich (Eds.), *Conservation Agriculture for Sustainable Land Management to Improve the Livelihood of People in Dry Areas.* © 2008 ACSAD & GTZ. pp. 67–76.

Wander, M. 2004. Soil organic matter fractions and their relevance to soil function. In: Magdoff, F. and Weil, R.R. (Eds.), *Soil Organic Matter in Sustainable Agriculture.* CRC Press, Boca Raton, FL, pp. 67–102.

Wander, M. and Bollero, G.A. 1999. Soil quality assessment of tillage impacts in Illinois. *Soil Sci. Soc. Am. J.* 63:961–971.

Wangemann, S.G., Kohl, R.A., and Molumeli, P.A. 2000. Infiltration and percolation influenced by antecedent soil water content and air entrapment. *Trans. ASAE* 43:1515–1523.

Wardak, D., Luke R., F.N. Padia, M.I. de Heer, C.J. Sturrock, and S.J. Mooney. 2022. Zero tillage has important consequences for soil pore architecture and hydraulic transport: A review. *Geoderma* 422:115927. https://doi.org/10.1016/j.geoderma.2022.115927

Wardle, D.A. 1995. Impacts of disturbance on detritus food webs in agroecosystems of contrasting tillage and weed management practices. In: Begon, M. and Fitter, A.H. (Eds.), *Advances in Ecological Research,* vol. 26. Academic Press, pp. 105–185. https://doi.org/10.1016/s0065-2504(08)60065-3

Warkentin, B.P. 2001. The tillage effect in sustaining soil functions. *J. Plant Nutr. Soil Sci.,* 164(4):345–350.

Wiesmeier, M., Urbanski, L., Hobley, E., et al. 2019. Soil organic carbon storage as a key function of soils – A review of drivers and indicators at various scales. *Geoderma* 333:149–162.

Wiesmeier, M., von Lützow, Margit, Spörlein, Peter, Geuß, Uwe, et al. 2015. Land use effects on organic carbon storage in soils of Bavaria: The importance of soil types. *Soil Tillage Res.* 146, Part B:296–302.

Williams, E.K. and Plante, A.F. 2018. A bioenergetic framework for assessing soil organic matter persistence. *Frontiers in Earth Science* 6:143. https://doi.org/10.3389/feart.2018.00143

Wilson, G.W., Rice, C.W., Rillig, M.C., Springer, A. and Hartnett, D.C., 2009. Soil aggregation and carbon sequestration are tightly correlated with the abundance of arbuscular mycorrhizal fungi: results from long-term field experiments. *Ecol. Lett.* 12(5):452–461.

Wulfsohn, D., Gu, Y., Wulfsohn, A. and Mojlaj, E.G. 1996. Statistical analysis of wheat root growth patterns under conventional and no-till systems. *Soil Tillage Res.* 38:1–16.

Yang, H., Wu, G., Mo, P., Chen, S., Wang, S., Xiao, Y., Ma, H., Wen, T., Guo, X., and Fan, G. 2020. The combined effects of maize straw mulch and no-tillage on grain yield and water and nitrogen use efficiency of dry-land winter wheat (*Triticum aestivum* L.). *Soil Tillage Res.* 197:104485.

Young, I., Crawford, J.W., Nunan, N., Otten, W., and Spiers, A. 2008. Chap 4. pp 81–125. *Microbial Distribution in Soils: Physics and Scaling.* Advances in Agronomy, Volume 100(4): # 2008 Elsevier Inc.

Young, I. and Ritz, K. 2000. Tillage, habitat space and function of soil microbes. *Soil Tillage Res.* 53(3–4):201–213.

Zalewski, M. 2013. Ecohydrology: Process-oriented thinking towards sustainable river basins. *Ecohydrol. Hydrobiol.* 13(2):97–103.

Zuber, S.M. and Villamil, M.B. 2016. Meta-analysis approach to assess effect of tillage on microbial biomass and enzyme activities. *Soil Biol. Biochem.* 97:176–187. https://doi.org/10.1016/j.soilbio.2016.03.011

9 Physiological Mechanisms for Improving Crop Water Use Efficiency in the US Southern Great Plains

Qingwu Xue[1], Sushil Thapa[2], Shuyu Liu[1], Jourdan Bell[1], Thomas H. Marek[1], and Jackie Rudd[1]
[1]Texas A&M AgriLife Research and Extension Center, Amarillo, TX 79106, USA
[2]Department of Agriculture, University of Central Missouri, Warrensburg, MO 64093, USA

CONTENTS

9.1 Introduction 236
9.2 Definition of Water Use Efficiency...240
9.3 Yield, ET, and Yield–ET Relationship 241
9.4 Improving Crop Water Use Efficiency in the Southern Great Plains 243
    9.4.1 Biomass and Harvest Index...246
    9.4.2 Effective Use of Soil Water...247
    9.4.3 Transpiration Efficiency...248
    9.4.4 Irrigation Management...249
9.5 Future Perspectives...250
References...251

9.1 INTRODUCTION

The US Southern Great Plains (SGP) is one of the most productive agricultural regions in the world. The region includes the Texas High Plains (THP) and Rolling Plains, western Oklahoma, eastern New Mexico, southwestern Kansas, and southeastern Colorado (Musick et al., 1994; Howell et al., 1995; Baumhardt and Salinas-Garcia, 2006). From 2013 to 2017, the THP (top 26 counites) produced 44% of corn, 29% of wheat, 14% of sorghum, 13% of cotton, 69% of ensilage, and 10% of forage crops in the state. Together with other crops and forages, the economic value of annual total crop production in the area is nearly $1.6 billion. The area also serves as a major livestock center for fed cattle and, increasingly, other feeding industries such as dairy and swine, which largely contribute to total agricultural cash receipts of $5.7 billion (Benavidez et al., 2019). Combined with the neighboring states of Kansas and Oklahoma, the region produces a significant amount of winter wheat (40%) and grain sorghum (80%) in the US (Figure 9.1, USDA-NASS, 2022).

Although corn production in the SGP is less than the US Corn Belt, the county level corn yields are greater than the national average (Xue et al., 2017). Corn provides a significant feeding source to area feeding industries (Benavidez et al., 2019).

Physiological Mechanisms for Crop Water Use Efficiency 237

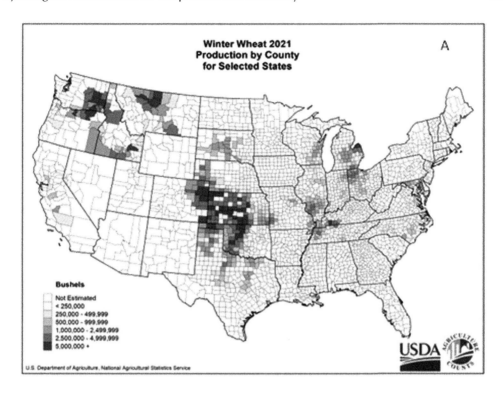

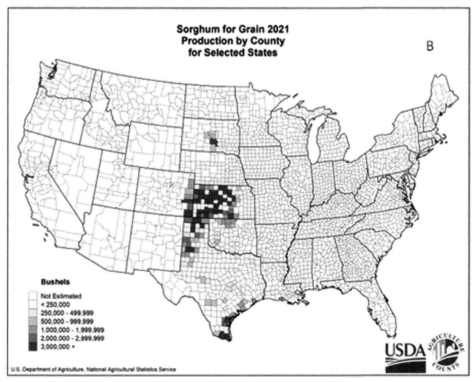

FIGURE 9.1 The 2021 winter wheat (a) and grains sorghum (b) productions by county in the US selected states (USDA-NASS, 2022).

The Köppen climate classification (Kottek et al., 2006) for the southern Great Plains is cold semiarid (BSk) in the west and humid subtropical (DFa) in the east. The annual precipitation varies little from north to south but decreases significantly from east (>600 mm) to west (~350 mm). In the most eastern part of the region with humid subtropical climate in Oklahoma and Texas Rolling Plains, annual precipitation can be 900 mm (Baumhardt and Salinas-Garcia, 2006). Precipitation can vary over 100% from year to year but, on average, growing season precipitation is about 250 mm in the semiarid areas (500–600 mm annual precipitation) (Musick et al., 1994). There is high evaporation demand across the SGP because of the combination of high temperatures, strong winds, and significant solar radiation. Annual pan evaporation ranges from 1,600 mm in the north to 2,400 mm in the south (Baumhardt and Salinas-Garcia, 2006). Periodic wet cycles in the region are often countered by persistent dry periods, such as the decade-long Dust Bowl in the 1930s. Severe and prolonged drought recurred in the early 1950s, early 1960s, mid-1970s, and again in the 2010s. Highly variable, short-term, recurring dry periods within the growing season are also common across the region (Hansen et al., 2012).

In the SGP, the high crop productivity has relied on irrigation. Among the major crops, corn is grown nearly 100% under irrigation and uses 53% of the entire water budget annually for irrigation in the region (1.76 km^3). Even though winter wheat and grain sorghum are relatively more drought tolerant than corn, the acreage of irrigated wheat and sorghum is still nearly 40% (Colaizzi et al., 2009; Xue et al., 2017; Evett et al., 2020). Almost all irrigation in the High Plains is from the Ogallala Aquifer. The Ogallala Aquifer, which spans ~450,000 km^2 in parts of South Dakota, Wyoming, Colorado, Nebraska, Kansas, Oklahoma, Texas, and New Mexico, is the primary source of groundwater in the US Great Plains (Figure 9.2).

In the Texas portion of the Ogallala Aquifer, irrigation development has significantly increased since the 1950s. However, the dramatic increase in water extraction for crop irrigation resulted in a significant decline in the water table. Some areas have experienced over 50% reduction in predevelopment saturated thickness. Irrigated land area has decreased from a peak of 2.4 million ha in 1974 to 1.9 million ha in 2000 (Colaizzi et al., 2009; McGuire, 2017) and to approximately 1.0 million ha in 2022 (USDA-FSA, 2022). The available water storage of the aquifer in the THP region decreased by 31% from 1950 to 2010 (560 km^3 in 1950 vs. 387 km^3 in 2010) (Steward and Allen, 2016). With declining water supplies from the aquifer and increasing pumping costs, sustainable crop production in the area faces challenges (Colaizzi et al., 2009; Xue et al., 2017; Evett et al., 2020).

Currently, the SGP region faces considerable challenges related to the domestic and global economy, climate change, and societal concerns over the impacts of agriculture on environments. First, the growing world population continuously requires more production of food, forage, fiber, and fuel, particularly the major food crops such as wheat and corn. Second, the declining water table in the Ogallala Aquifer and increasing pumping costs will inevitably reduce irrigation (Musick et al. 1994; Colaizzi et al., 2009; Evett et al., 2020). Third, the possibility for increasing the frequency and severity of drought stress as well as other abiotic and biotic stresses under climate change will likely reduce crop yields more frequently. Projections of greater annual and inter-annual climate variability, such as increasing temperature, longer and more intense drought period, and more extreme rainfall events, present growing challenges for crop production in the SGP (Steiner et al., 2018; Evett et al., 2020).

Since water is the most important factor affecting crop production in the SGP, development of crop management practices to conserve water, optimize water use, and improve water use efficiency (WUE) becomes critical, particularly under a changing climate condition. The objective of this chapter is to review the progress and challenges in the research for improving yield and WUE in three major crops (winter wheat, grain sorghum, and corn) in the SGP region. The major data source is from the research conducted in THP over the last 50–60 years along with some research conducted in other areas of the SGP region.

Physiological Mechanisms for Crop Water Use Efficiency 239

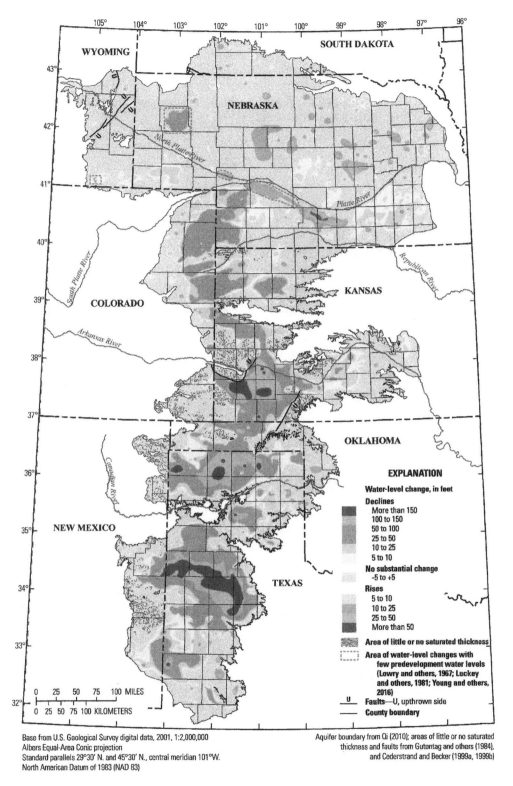

FIGURE 9.2 The map of the Ogallala or High Plains Aquifer, and the water-level changes in the aquifer from predevelopment (about 1950) to 2015 (McGuire, 2017).

9.2 DEFINITION OF WATER USE EFFICIENCY

WUE is the ratio of plant carbon gain to water use and can be defined at a range of scales from cell to organ to canopy levels. It is a complex trait and related to physiological, agronomic, and engineering processes, and management practices (Tanner and Sinclair, 1983; Howell, 2001; Richards et al., 2002; Leakey et al., 2019; Kang et al., 2021). In this chapter, we focus on field management practices for improving WUE and define it as follows:

$$WUE = \frac{Y}{ET} \qquad \text{(Eq. 9.1)}$$

where Y is crop grain or economic yield and ET is evapotranspiration. ET is the sum of plant transpiration (T) and water lost through soil evaporation (Es). The unit of WUE used in this chapter is kg m^{-3} (Musick et al., 1994; Howell, 2001; Xue et al., 2012).

Since Y is determined by biomass production and harvest index, and soil evaporation cannot be avoided during growing season, WUE can be expressed as:

$$WUE = \frac{BM \times HI}{T\left(1 + \dfrac{Es}{T}\right)} \qquad \text{(Eq. 9.2)}$$

where BM is biomass production, HI is harvest index, T is transpiration, and Es is soil evaporation (Howell, 2001).

Because the ratio of BM and T is defined as plant transpiration efficiency (TE) (Tanner and Sinclair, 1983; Richards et al., 2002), Eq. (9.2) can be rewritten as:

$$WUE = \frac{TE \times HI}{\left(1 + \dfrac{Es}{T}\right)} \qquad \text{(Eq. 9.3)}$$

If a crop is harvested for biomass production such as forage, HI can be removed from Eq. (9.2) and Eq. (9.3). As such, WUE becomes water use efficiency for biomass (WUEbm) as:

$$WUEbm = \frac{BM}{T\left(1 + \dfrac{Es}{T}\right)} \qquad \text{(Eq. 9.4)}$$

$$WUEbm = \frac{TE}{\left(1 + \dfrac{Es}{T}\right)} \qquad \text{(Eq. 9.5)}$$

For the studies with different soil water levels and multiple seasons, linear regression between Y or BM and ET has been used as:

$$Y \text{ or } BM = a \times ET + b \qquad \text{(Eq. 9.6)}$$

Physiological Mechanisms for Crop Water Use Efficiency

where the slope of Eq. (9.6) becomes WUE or WUEbm within a range of environmental conditions. The linear relationship between Y or BM and ET has been found in most of the field studies in wheat, sorghum, and corn in the SGP (Musick et al., 1994; Xue et al., 2012, 2017; Bell et al., 2018; O'Shaughnessy et al., 2019).

9.3 YIELD, ET, AND YIELD–ET RELATIONSHIP

Winter wheat and grain sorghum are two crops grown under both dryland and irrigated conditions in the SGP. Due to the highly variable annual precipitation, dryland crop production must rely on soil water storage as well as growing season precipitation. Therefore, fallow is widely practiced in dryland cropping systems. Winter wheat–fallow and winter wheat–fallow–sorghum–fallow are two common dryland cropping systems in the region (Jones and Popham, 1997; Unger, 2001). Soil water content at crop planting and precipitation amount and distribution during the growing season strongly affect crop yields under dryland conditions. The major challenge for dryland crop production is the development of management practices for efficient use of soil water storage as well as growing season precipitation (Unger, 2001; Stone and Schlegel, 2006; Stewart et al., 2010; Hansen et al., 2012).

In the SGP, the long-term dryland winter wheat grain yield at Bushland, Texas (1984–1993) ranged 1–2 Mg ha^{-1} under wheat–fallow and wheat–sorghum–fallow systems (Jones and Popham, 1997). In western Kansas, the long-term dryland wheat yields from 1973 to 2004 averaged 2.5 Mg ha^{-1} (Stone and Schlegel, 2006). Grain yields in dryland treatment from irrigation studies were frequently over 3 Mg ha^{-1} (Musick et al., 1994; Xue et al., 2003, 2006). In another long-term study, Shrestha et al. (2020) showed that dryland wheat yield has increased from about 1.0 Mg ha^{-1} in 1950s to 5.0 Mg ha^{-1} in 2016 at Bushland, Texas. In the recent wheat trials (2010–2019) under wheat–fallow conditions at the same location, dryland wheat grain yield ranged from 0.7 Mg ha^{-1} in an extremely dry year (2011) to 4.7 Mg ha^{-1} in a relative wet year (2019) (Xue, 2020). Genetic improvement has played an important role in dryland wheat yield increase. For example, yield among wheat genotypes varied from 1.3 Mg ha^{-1} to 4.7 Mg ha^{-1}, and newer cultivars yielded 40% more than an older cultivar (Xue et al., 2020). The seasonal ET for dryland wheat ranged from about 200 mm to nearly 600 mm and WUE from 0.08 to 0.83 kg m^{-3} in the long-term field studies (Musick et al., 1994; Jones and Popham, 1997; Xue et al., 2014). Irrigated wheat yields in the THP ranged from 3.0 to 7.7 Mg ha^{-1} and ET from 400 to over 900 mm, depending on irrigation timing and frequency (Musick et al., 1994; Howell et al., 1995a; Xue et al., 2003, 2006; Schneider and Howell, 2001; AgriPartners, 2007). Wheat yields under full irrigation were in the range of 5.3–7.7 Mg ha^{-1} and required about 700–950 mm seasonal ET (Howell et al., 1995; Schneider and Howell, 2001; AgriPartners, 2007). The WUE for irrigated wheat was higher than for dryland wheat and ranged from 0.5 to 1.2 kg m^{-3} (Musick et al., 1994; Schneider and Howell, 2001; Xue et al., 2006, 2014).

In the THP, the county-level dryland grain sorghum yield has increased from about 1.9 Mg ha^{-1} in the 1970s to over 3.1 Mg ha^{-1} in recent years, with an increase of 20 kg ha^{-1} year^{-1} (USDA-NASS, 2021). Sorghum yield increase over the years was greater at research plots. At Bushland, Texas, dryland grain sorghum yield ranged 1–5 Mg ha^{-1} (Steiner, 1986; Jones and Popham, 1997; Unger and Baumhardt, 1999). Dryland grain sorghum yields increased 139% between 1939 and 1997, with an increase of 50 kg ha^{-1} year^{-1}. Among the percentage of yield increases for that period, 46% resulted from genetic improvement and 93% was attributed to other factors, primarily to improved soil water at planting (Unger and Baumhardt, 1999). The seasonal ET and WUE for dryland sorghum at Bushland, Texas, ranged from 200 to 500 mm and 0.10 to 1.72 kg m^{-3}, respectively (Steiner, 1986; Jones and Popham, 1997). The irrigated sorghum yield at county level ranged 4.4–6.4 Mg ha^{-1} in the THP (USDA-NASS, 2021). Similarly, irrigated sorghum yields were greater in research and demonstration plots than those at county level. Under full irrigation level (300–700 mm, depending on seasonal precipitation), grain sorghum yields ranged 8.2–11.0 Mg ha^{-1} under center pivot system at Bushland, Texas, as well as in the top producers' field demonstrations (AgriPartners, 2007; Tolk and Howell, 2008; Bell et al., 2018). Depending on the irrigation levels, seasonal ET for irrigated sorghum ranged 400–660 mm and WUE 0.56–1.79 kg m^{-3} (Tolk et al., 1997; Tolk and Howell, 2003; Howell et al., 2007; Bell et al., 2018).

Corn is an irrigated crop and dryland corn production has high risks in the SGP (Bean, 2007; Xue et al., 2017). As such, we focus on irrigated corn production in this chapter. In the last 50 years from 1970s to the current time, corn grain yield in the THP has increased linearly, with an annual increase of about 120 kg ha^{-1} (Xue et al., 2017). At the county level, average yield increased from about 6–8 Mg ha^{-1} in the 1970s to about 13 Mg ha^{-1} in 2013. Corn yields at research and demonstration plots were higher than those of the county level, ranging from about 8 Mg ha^{-1} in the 1970s to 16 Mg ha^{-1} in recent years (e.g., 2011–2013). The higher yield of the research and demonstration plots is generally due to the smaller, more intensively managed production areas, as compared to field-based production practice (Xue et al., 2017). Corn has a high evapotranspiration (ET) requirement (both daily and seasonally) in the SGP (Howell et al., 1995b; Howell et al., 1996). Daily ET measured in large, monolithic weighing lysimeters at Bushland, Texas often exceeds 10 mm d^{-1} for significant periods of time (Howell et al., 1996). Early studies under furrow irrigated conditions showed that seasonal ET ranged 667–984 mm under full irrigation (Musick and Dusek, 1980; Eck, 1984). Field studies using sprinkler irrigation showed that seasonal ET ranged from 750–973 mm (mostly between 800 and 900 mm) in the 1990s at irrigation of 100% ET requirement (Howell et al., 1995b; Howell et al., 1996; Schneider and Howell, 1998; Yazar et al., 1999; Evett et al., 2000). Corn ET levels in producer's fields had a larger variation (750–1200 mm) (AgriPartners, 2007). In recent studies at both Bushland and Etter, Texas, a lower seasonal ET was recorded under full irrigation at Etter than at Bushland. Colaizzi et al. (2011) conducted a 3-year experiment using a subsurface drip irrigation system and seasonal corn ET ranged from 711 to 818 mm at Bushland at irrigation level of 100% ET requirement. Hao et al. (2015a) showed that corn under irrigation of 100% ET requirement ranged from 634 to 796 mm in drought years. Corn seasonal ET has a decreasing trend while WUE varied linearly over the years under full irrigation (100% ET requirement or greater) in the THP (Xue et al., 2017). In the THP, WUE values were affected by irrigation amount, frequency, and management methods (Howell et al. 1995b; Schneider and Howell, 1998; Hao et al., 2015a). Under full irrigation conditions (100% ET requirement or greater) in the THP, corn WUE has increased from about 1.00 kg m^{-3} in the 1975–1977 study to about 2.00 kg m^{-3} in the 2011–2013 study (Xue et al., 2017). The increased WUE in recent years (2011–2013) reflected the yield improvement over the last four decades. Studies have also shown that the maximum WUE is generally achieved at less than the 100% ET requirement level (Howell et al., 1995, 2001; Payero et al., 2008; Colaizzi et al., 2011; Hao et al., 2015a). Corn WUE at the 75% and 100% ET levels ranged from 1.80 to 2.17 kg m^{-3} at Bushland (Colaizzi et al., 2011). The WUE values at the same two irrigation levels ranged from 1.51 to 2.57 kg m^{-3} with a relatively low WUE (1.51–2.15 kg m^{-3}) in 2011 and a relatively high WUE (1.70–2.57 kg m^{-3}) in 2012 and 2013 in a study at Etter, Texas (Hao et al., 2015a). In a two-year (1992 and 1993) field study at Bushland, Texas, a lower corn WUE of 1.58–1.75 kg m^{-3} was reported by Howell et al. (1995b). However, WUE decreased significantly as irrigation levels were reduced to 50% ET requirement or less (Schneider and Howell, 1998; Hao et al., 2015a). For example, irrigation at the 25% ET level can result in crop failure (Schneider and Howell, 1998).

The yield–ET relationship in wheat, sorghum, and corn has been reported in different studies in the SGP under both dryland and irrigated conditions (Musick and Dusek, 1980; Steiner, 1986; Musick et al., 1994; Howell et al., 1995a, b; Howell et al., 1996; Jones and Popham, 1997; Yazar et al., 1999; Schneider and Howell, 2001; Stone and Schlegel, 2006; Colaizzi et al., 2011; Xue et al., 2012, 2014; Hao et al., 2015a; Bell et al., 2018; Zhao et al., 2018a). Xue et al. (2012) analyzed the yield–ET relationship in wheat based on the field studies at Bushland, Texas, from the 1960s to 1999 and found that there was a significant linear relationship between yield and ET, and the regression

Physiological Mechanisms for Crop Water Use Efficiency 243

resulted in a slope of 1.06 kg m^{-3} and a threshold ET of 164 mm (Y = 0.0106X–1.7393, R^2 = 0.75, P < 0.001) (Figure 9.3a). Threshold ET is the ET level at which the first grains are initiated (Musick et al., 1994; Schneider and Howell, 1998). Stone and Schlegel (2006) summarized dryland wheat data from 1974 to 2004 and also found a linear relationship between wheat yield and ET (Y = 0.01X–1.838, R^2 = 0.64, P < 0.0001). The slope was 1.0 kg m^{-3} and the ET threshold was 183 mm, which was close to the results from Bushland, Texas. The yield–ET relationship in grain sorghum is shown in Figure 9.3b, based on the dryland and irrigated studies from the 1980s to 2012 at Bushland, Texas (dryland: Steiner, 1986; Jones and Popham, 1997; irrigated: Tolk et al., 1997; Tolk and Howell, 2003; Howell et al., 2007; Bell et al., 2018). The linear regression between sorghum yield and ET has a slope of 1.59 kg m^{-3} and a threshold ET of 121 mm (Y = 0.0159X–1.9257, R^2 = 0.56, P < 0.01). Similarly, Stone and Schlegel (2006) showed a slope of 1.66 kg m^{-3} and a threshold ET of 136 mm in the linear relationship between grain sorghum yield and ET (Y = 0.0166X– 2.252, R^2 = 0.60, P < 0.001), based on dryland sorghum data from 1974 to 2004. The more generalized WUE based on the above linear regression analysis was about 1.0 kg m^{-3} for winter wheat and 1.6 kg m^{-3} for grain sorghum in the SGP region.

For the yield–ET relationship in corn, Xue et al. (2017) selected three data sets, representing different time periods in the THP. The first data set was from Musick and Dusek (1980) who conducted field studies at different furrow irrigation levels and frequencies from 1975 to 1977. The second data set was compiled from a field study using center pivot sprinkler irrigation in 1994 and 1995 (Schneider and Howell, 1998). The third data set was from recent field studies conducted at Etter, Texas, from 2011 to 2013 using different hybrids and planting densities (Hao et al., 2015a). During all three time periods, corn yield increased linearly as seasonal ET increased (Figure 9.4). Linear regression between corn yield and ET resulted in a slope of 2.43, 2.77, and 2.85 kg m^{-3} in the studies of 1975–1977, 1994–1995, and 2011–2013, respectively. The threshold ET value was 346 mm for the 1975–1977 study, 310 mm for the 1994–1995 study, and 214 mm for the 2011–2013 study. The smaller threshold ET value in the 2011–2013 study (214 mm) may indicate management improvement over the decades as compared to the 1975–1977 study (346 mm). Comparing yield and ET in the three studies, corn yield increased at any ET level from 1975 to 2013, indicating overall corn yield improvement in the last four decades. Greater yields in the studies of 1994–1995 and 2011–2013 at lower ET levels (<600 mm) indicate that drought tolerance technology in new corn hybrids has improved in the last four decades. The improvement of drought and other stress tolerance in corn hybrids has also been reported in the US Corn Belt region and Canada (Tollenaar and Lee, 2002; Hammer et al., 2009).

Comparing the yield–ET relationship among the three crops in SGP, all crops require high seasonal ET (up to 900 mm) for the highest grain yield. Corn and grain sorghum had greater yield than wheat at any ET level, i.e., greater WUE than winter wheat due to their C$_4$ photosynthesis characteristics. However, corn had a greater threshold ET (214–346 mm) than grain sorghum (121 mm) and winter wheat (164 mm) (Figures 9.3 and 9.4). Although corn had the highest WUE, it requires more water than sorghum and wheat to produce initial grain. As such, sorghum and wheat are more suitable for producing grain under water- limited conditions. Akbar et al. (1997) compared the dryland corn and sorghum yield and seasonal ET in the SGP using a simulation study and their results clearly showed that sorghum yielded more than corn when the seasonal ET is less than 430 mm, and south part of the Southern Great Plains had more risks for growing dryland corn than sorghum (Akbar et al., 1997).

9.4 IMPROVING CROP WATER USE EFFICIENCY IN THE SOUTHERN GREAT PLAINS

Based on Eq. (9.1)–(9.5), WUE or WUEbm can be improved by (i) increasing the biomass production, (ii) enhancing the harvest index (for WUE only), (iii) reducing the soil evaporation and

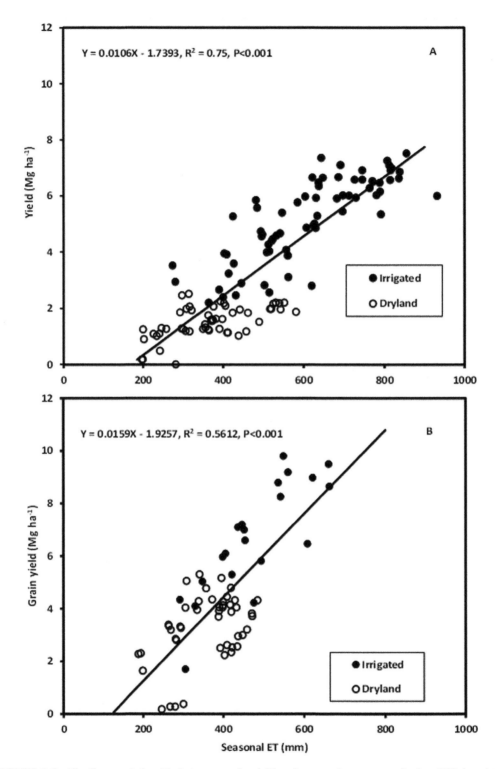

FIGURE 9.3 The linear relationship between grain yield and seasonal evapotranspiration (ET) in winter wheat (a) and grain sorghum (b) at Bushland, Texas. (Wheat: Xue et al., 2012; Sorghum: Steiner, 1986; Jones and Popham, 1997; Tolk et al., 1997; Tolk and Howell, 2003; Howell et al., 2007; Bell et al., 2018).

Physiological Mechanisms for Crop Water Use Efficiency

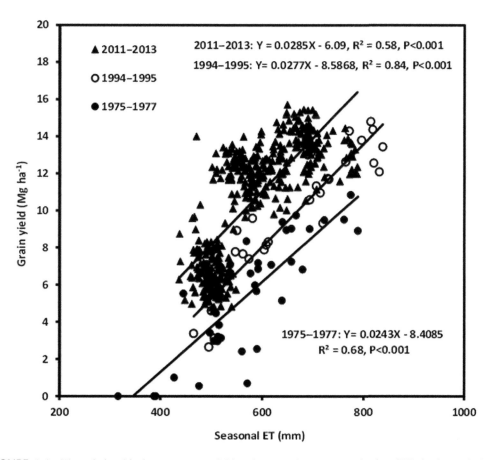

FIGURE 9.4 The relationship between corn yield and seasonal evapotranspiration (ET) in the periods of 1975–1977, 1994–1995, and 2011–2013 in the Texas High Plains (Musick and Dusek, 1980; Schneider and Howell, 1998; Hao et al., 2015a).

increasing proportion of T in ET, and (iv) increasing the TE (Howell, 2001; Richards et al., 2002). Generally, the improved WUE and WUEbm can be achieved by two ways: breeding and management practices. Over the decades, genetic improvement through breeding has played an important role for increasing yield and WUE in major crops in the SGP. For example, wheat yield has been increased by conventional breeding under both irrigated and dryland conditions at Bushland, Texas (Xue et al., 2014; Shrestha et al., 2020). Multi-location field trials showed that newer wheat cultivars consistently had 10–14% higher yield than the cultivars released 10–15 years ago under both dryland and irrigated conditions in the THP (Trostle and Bean, 2010; Xue et al., 2014). Irrigated corn yields have increased from about 8 Mg ha^{-1} in 1970s to about 16 Mg ha^{-1} in recent years (Xue et al., 2017). Currently, advances in genomics, phenomics, and data analytics have been used to assist breeding programs for breeding crops better adapted to water-limited environment (Hammer et al., 2021). Management practices are equally as important as breeding to improve crop yield and WUE under abiotic stress (Passioura and Angus, 2010; Richards et al., 2010; Hammer et al., 2021). Optimizing major management practices is important to maximize crop yields under water-limited conditions. Because the ultimate goal for field crop production is to maximize yield, increased WUE is of greatest interest to producers when yields are maximized for the available water supply during growing season. High WUE may not be important if it is not associated with high yield (Sinclair and Muchow, 2001).

9.4.1 Biomass and Harvest Index

Crop grain yield is determined by biomass production as well as harvest index (HI) under water-limited conditions in SGP (Howell, 1990; Xue et al., 2012). For dryland crops at Bushland, Texas, the average biomass at maturity (1984–1993) ranged 3.94–6.25 Mg ha^{-1} for winter wheat and 8.10–9.97 Mg ha^{-1} for grain sorghum. The average HI was relatively stable (0.30 for wheat and 0.40 for grain sorghum). Grain sorghum generally had greater biomass and HI than winter wheat (Jones and Popham, 1997). A more recent study showed that dryland wheat biomass and HI ranged 6.40–7.50 Mg ha^{-1} and 0.26–0.31 in different genotypes, respectively. Genotypes with higher yield had greater biomass at anthesis and maturity. The high WUE was more related to biomass because genotypic differences in ET were not significant (Xue et al., 2014). The biomass in irrigated wheat and grain sorghum was much higher than that in dryland and was up to 18 Mg ha^{-1} at full irrigation. However, irrigated sorghum also had greater HI (0.47) than wheat (0.39) (Xue et al., 2006; Bell et al., 2018). Howell (1990) analyzed grain yield, biomass, and HI based on field studies from the 1950s to the early 1980s at Bushland, Texas. In general, grain yield is linearly related to aboveground biomass for both crops. The irrigated wheat yield increased from 3 Mg ha^{-1} in the 1950s to 6 Mg ha^{-1} in the 1980s, mainly due to the increased biomass from about 9 Mg ha^{-1} to 16 Mg ha^{-1}. The HI was 0.35 for wheat and 0.47 for sorghum, and was largely unaffected by management practices (Howell, 1990). Although WUE was positively correlated to both biomass and HI, the correlation coefficient between WUE and HI was greater than that between WUE and biomass for both wheat and sorghum (Xue et al., 2006; Bell et al., 2018). The closer relationship between WUE and HI in wheat was also reported in North China Plains and Mediterranean environment (Zhang et al., 1998; Oweis et al., 2000). The dryland corn biomass (5.9–8.3 Mg ha^{-1}) was within the range of dryland grain sorghum. However, the HI of dryland corn (0.27–0.40) was generally smaller than grain sorghum (0.40) (Thapa et al., 2018). Irrigated corn had much greater biomass and HI than irrigated sorghum and wheat. Depending on hybrids, corn biomass at full irrigation ranged 16–25 Mg ha^{-1} and HI ranged 0.58–0.63 in the THP (Hao et al., 2015a; Zhao et al., 2018a). Similar to wheat and sorghum, WUE is also more correlated to HI than biomass in corn (Zhao et al., 2018a).

Under water-limited conditions during crop growing season, soil water must be conserved before flowering so that it can be used for grain filling and maintaining HI (Richards et al., 2002). As such, low planting populations are frequently used for grain sorghum and corn under dryland or limited irrigation conditions. However, low planting density can result in more tillers in sorghum and corn as compared to high planting density, which resulted in more vegetative biomass but low HI (Thapa et al., 2018). Manipulation of planting geometry may affect biomass and HI. Thapa et al. (2017a, 2018) Asseng and van Herwaarden, 2003 investigated the effects of different planting geometries (evenly spacing planting, ESP, clump, cluster, and skip-row) on grain sorghum and corn yield and HI in the THP. In comparison to ESP, alternative planting geometries such as clump and cluster planting significantly increased HI grain sorghum and corn (Thapa et al., 2017, 2018). In wheat, the remobilization of pre-anthesis carbon can be important for yield and HI under drought conditions (Foulkes et al., 2002; Richards et al., 2002; Xue et al., 2006, 2014). Xue et al. (2006) showed that the contribution of remobilization of pre-anthesis carbon reserves to yield was up to 80%. Lopez-Castaneda and Richards (1994) showed that grain yield increased linearly as stem loss between anthesis and maturity increased in wheat, barley, triticale, and oats under rainfed conditions in Australia. Foulkes et al. (2002) showed that winter wheat cultivars with more drought resistance had higher HI and more remobilization of pre-anthesis carbon reserves under drought in the UK. The contribution of remobilized pre-anthesis carbon reserves to grain yield in wheat varied from 5% to 90%, depending on the amount and distribution of precipitation, nitrogen supply, crop growth, and seasonal ET (Foulkes et al., 2002; Yang et al., 2000). For remobilization of pre-anthesis carbon, irrigation at early stage (e.g., jointing and booting) is more beneficial than at late stage (e.g., anthesis and grain filling) because early irrigation increased the number of spikes/stems per square meter and the increased sink demand in turn increased the amount of remobilized carbon per square meter (Xue et al., 2006).

9.4.2 Effective Use of Soil Water

Under water-limited conditions, crop yield is largely dependent on soil water storage at planting as well as growing season precipitation (Musick et al., 1994; Unger, 2001; Xue et al., 2012). Greater yield is associated with a greater ability to effectively use soil water, particularly the soil water at deeper profile. At Bushland, Texas, Winter and Musick (1993) compared the soil water extraction and yield in winter wheat in three planting dates (August, October, and November) and found that wheat planted in October had the highest yield and the soil water extraction was as deep as 2.4 m. In contrast, wheat planted in November had the lowest yield and only extracted soil water at 0–1.2 m profile. In a recent dryland wheat study at the same location, Thapa et al. (2017b) showed that grain yield is largely related to soil water extraction among different growing seasons and cultivars. In a season with high yield (3.55 Mg ha^{-1} in 2016), soil water extraction occurred in the whole 0–2.4 m profile. In a season with very low yield (0.65 Mg ha^{-1} in 2011), soil water extraction was limited to only 0–1.2 m profile. The newer drought-tolerant (DT) cultivars such as TAM 111 and TAM 112 were able to extract more water from deeper soil profile (particularly between jointing and maturity) than a relatively older cultivar (TAM 105) in dry seasons of 2011 and 2012 (Figure 9.5, Thapa et al.,

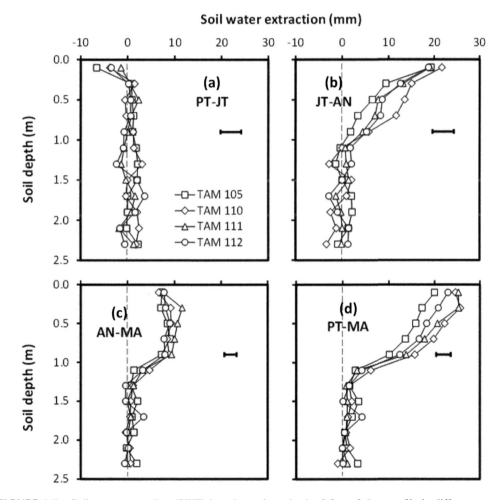

FIGURE 9.5 Soil water extraction (SWE) by wheat plants in the 0.0- to 2.4-m profile in different growth periods in a dry season (2010–2011). (PT, planting; JT, jointing; AN, anthesis; MA, maturity; The horizontal bar is the LSD at P = 0.05; Thapa et al., 2017b).

2017b). Root growth is critical for crops to use soil water and obtain high yield under water deficit conditions. Xue et al. (2003) showed a yield of 3.2 Mg ha^{-1} and WUE of 0.78 kg m^{-3} in the dryland treatment in an irrigation study. However, dryland wheat yield ranged 1.2–1.6 Mg ha^{-1} and WUE 0.37–0.43 kg m^{-3} in the same growing season and same soil under a 3-year wheat–sorghum–fallow rotation system (Jones and Popham, 1997). Comparing the seasonal ET in two studies, dryland crops in Xue et al.' (2003) used 52 mm more water than those in Jones and Popham (1997), indicating that the root system in dryland crops in Xue et al. (2003) was larger or deeper than that in Jones and Popham (1997). The relatively deep root system was due to higher available soil water at early developmental stage (Xue et al., 2003). The depth of soil water extraction was about 1.6–2.0 m in dryland grain sorghum (Jones and Popham, 1997; Moroke et al., 2005) but about 1.3 m in irrigated sorghum (Bell et al., 2020).

Hao et al. (2015b) investigated the soil water extraction, ET, WUE, and grain yield in two corn hybrids (one conventional and one DT hybrid) in the THP. The DT hybrid (P1151HR) and the conventional hybrid (33D49) were grown in three water regimes (I_{100}, I_{75}, and I_{50}, referring to 100%, 75%, and 50% of the ET requirement, respectively). The depth of soil water extraction was not affected by hybrid or water regime with the maximum extraction depth of 1.2–1.4 m. The maximum soil water extraction at I_{50}, I_{75}, and I_{100} occurred in 0.6–0.8, 0.6–1.0, and 0.8–1.0 m soil profile, respectively. Under water stress conditions at I_{50}, DT hybrid had 24% higher yield and 30% higher WUE than the conventional hybrid. Meanwhile, DT hybrid had less soil water extraction in the upper soil layers but more water extraction in the deeper layers than a conventional hybrid. Clearly, DT hybrid was able to capture water from deeper soil layers under water stress. In the US Corn Belt, Sinclair and Muchow (2001) showed that corn yields in a 20-year simulation were benefited from the increased depth of soil water extraction based on weather conditions of Columbia, Missouri. In another simulation study in the US Corn Belt, Hammer et al. (2009) showed that root system architecture and water capture directly affected biomass accumulation and yield in corn under water-limited conditions.

9.4.3 Transpiration Efficiency

TE is the ratio of biomass and transpiration. Increasing TE can improve WUE under water-limited conditions. In a Mediterranean environment like Australia, TE is particularly important for dryland wheat which largely relies on soil water storage. Genetic improvement of TE in wheat using carbon isotope discrimination ($\Delta^{13}C$) as a selection tool has been successful in Australia, and a low Δ resulted in higher yield under low seasonal rainfall (250 mm) environment (Richards et al., 2002). However, the carbon isotope discrimination has not been used widely in the US Great Plains. In a field study in central Great Plains, there was a linear negative relationship between Δ and TE at leaf level in wheat. However, low Δ was not correlated to high grain yield (Xue et al., 2002).

The TE and WUEbm in grain sorghum have been reported in different studies over the years in the SGP (Balota et al., 2008; Tolk and Howell, 2009; Xin et al., 2008, 2009; Narayanan et al., 2012; Thapa et al., 2017a). Tolk and Howell (2009) reported a relatively stable TE (3.5 kg m^{-3}) in sorghum in three soil types at different soil water levels. Thapa et al. (2017a) found that crop growth stage did not affect TE in sorghum and TE ranged from 4.10 to 5.85 kg m^{-3} in two greenhouse studies and two growth chamber studies. However, genetic variation in TE and WUEbm among sorghum genotypes was found in both greenhouse and field studies (Balota et al., 2008; Xin et al., 2008, 2009). Balota et al. (2008) showed that TE among 12 sorghum hybrids ranged from 5.04 to 7.33 kg m^{-3}. The TE range was 4.7–7.7 kg m^{-3} among 39 sorghum genotypes in three greenhouse studies conducted by Xin et al (2008, 2009). For all the above studies, greater TE was correlated to greater biomass production because genotypes did not differ in transpiration. In the western Kansas, Narayanan et al. (2012) investigated the variation of biomass production, WUEbm, and radiation use efficiency (RUE) in eight grain sorghum genotypes in 2009 and 2010 growing seasons under relatively well-watered conditions. The WUEbm ranged 3.39–5.42 kg m^{-3} in 2009 and 4.04–7.63 kg m^{-3} in 2010.

Physiological Mechanisms for Crop Water Use Efficiency   249

Similar to the greenhouse studies (Xin et al., 2009), WUEbm was more correlated to biomass production than to ET. In addition, WUEbm was linearly related to RUE for both years (Narayanan et al., 2013). The results from the above studies showed that it is possible to increase TE or WUEbm without compromising biomass production in sorghum. These studies also provided useful plant materials for identifying the physiological mechanisms for further improving yield and WUE in grain sorghum (Narayanan et al., 2013).

Zhao et al. (2018b) investigated the shoot and root traits, water use, and WUEbm of four corn hybrids (one conventional and three DT hybrids) under two water regimes. There were significant differences in most of the shoot and root traits, water use, and WUEbm among the four hybrids in both water regimes. The WUEbm varied from 5.82 to 7.89 kg m^{-3}. Under water stress, the three DT hybrids showed greater drought tolerance than conventional hybrid. There are two different mechanisms to respond to water stress among the DT hybrids. Compared with the conventional hybrid (33D53AM), the two commercial DT hybrids had smaller leaf area, shoot dry weight, and root system. As a result, the two hybrids used less water but had higher WUEbm. In contrast, the experimental DT hybrid had greater leaf area, shoot biomass, and root dry weights compared with the commercial DT hybrids. The hybrid used more water under water stress due to the larger root system (Zhao et al., 2018b).

9.4.4 IRRIGATION MANAGEMENT

Although many management factors affect crop production in the SGP, irrigation remains the most effective way to increase crop yield and maximize the WUE. Yield of irrigated crop can be 2–4 times higher than that of dryland crops (Figure 9.3). In the SGP, irrigation systems and accommodating agricultural practices have changed significantly in the last four decades, from furrow irrigation (1950s–1970s) to center pivot sprinkler and to (a lesser degree) subsurface drip irrigation systems are currently being used. The history and trends of irrigation research and development in the region have been reviewed in different eras from the 1990s to recent time (Musick et al., 1990; Colaizzi et al., 2009; Evett et al., 2014; Evett et al., 2020). The details of irrigation history, economic impact, research, and development trends can be found in the aforementioned four review papers. Based on these reviews, there are three generally inferred statements regarding irrigation in the SGP. First, the irrigation supply from the Ogallala Aquifer is continuously declining. Second, irrigation efficiency through system conversion has dramatically increased over the last few decades. Third, the future challenge is how to efficiently use reduced amounts of irrigation while sustaining irrigated crop yields. The third factor is further challenged by the fact that future increases in other regional needs (such as municipal demand, power generation, etc.) will be added to the expense of the irrigation supply (Panhandle Water Planning Group, 2011). Nevertheless, preserving irrigation water is crucial to sustainable crop production in SGP and other water-limited regions (Musick et al., 1994; Howell, 2001; Xue et al., 2012; Evett et al., 2020).

To increase the efficiency of irrigation applications, irrigation events should be scheduled using measurements of ET, soil moisture depletion, and/or plant-based measurements. Irrigation scheduling to enhance the WUE includes the management of both water and soils. Crop ET is typically calculated using a crop-specific coefficient (Kc) and reference evapotranspiration (ET$_o$). The ET$_o$ is calculated evapotranspiration from a reference crop (turf grass or alfalfa) using the meteorological parameters of temperature, relative humidity, precipitation, wind speed, and solar radiation. Irrigation can also be scheduled according to changes in soil moisture. Soil moisture sensors can provide information on soil moisture fluxes in the root zone, which provide information on when to initiate and terminate irrigation events. Irrigation scheduling using ET, a soil water balance method, or a combination of the two can be successfully employed in irrigated corn production (Colaizzi et al., 2009; Evett et al., 2014; Evett et al., 2020).

The seasonal average irrigation requirement is approximately 500 mm for full irrigated corn in the THP (Top 26 counties of Texas, Marek et al., 2011). However, irrigation demand can vary and be much higher during a severe drought year such as in 2011, where a seasonal ET of 900 mm with an irrigation requirement of 754 mm (Hao et al., 2015a) was observed. Based on multiple-year studies at Bushland and Etter, Texas, using irrigation to meet a 75–80% of ET demand level can result in similar yields as compared to years with 100% of the ET demand when seasonal rainfall is average or above average. Also, WUE is generally maximized at the 75–80% irrigation level (Xue et al., 2017). Unless adequate seasonal rainfall (normally > 250 mm) and excellent soil water storage exist at the time of planting, lowering irrigation levels to a 50% ET requirement or less significantly reduced corn yield (Schneider and Howell, 1998; Colaizzi et al., 2011; Hao et al., 2015a, b; Zhao et al., 2018a, b). Irrigation levels at the 25% ET level can result in yield failures (Schneider and Howell, 1998; Colaizzi et al., 2011).

In general, irrigation requirement is less for winter wheat and grain sorghum as compared to corn. At full irrigation level (i.e., the water application to meet the 100% ET requirement), irrigation ranged from 300 to 521 mm for winter wheat (Howell et al., 1995; Schneider and Howell, 2001), and from 300 to 628 mm for grain sorghum (Bell et al., 2018). However, irrigated wheat and sorghum are mostly managed under deficit irrigation in the THP (Musick et al., 1994; Schneider and Howell, 2001; Xue et al., 2006; Howell et al., 2007; Evett et al., 2020). In a wheat study, Xue et al. (2006) showed that 220 mm of irrigation at jointing and anthesis (45% irrigation reduction from full irrigation) resulted in 13% yield reduction as compared to the full irrigation of 400 mm, and the deficit irrigation did not reduce WUE. Similarly, Schneider and Howell (2001) investigated wheat responses to irrigation levels under center pivot in two growing seasons and found that wheat grain yield at 50% irrigation level reduced 5% in 1998 and 14% in 1999 as compared to 100% irrigation level. Also, the deficit irrigation did not affect wheat WUE in both years (Schneider and Howell, 2001). Howell et al. (2007) reported that deficit irrigation at 50% ET did not affect grain sorghum yield. Sorghum yield was 9 Mg ha^{-1} at both full (100% ET, 369 mm) and deficit irrigation (50%, 171 mm) levels. However, the deficit irrigation treatment had greater WUE (1.45 kg m^{-3} at full irrigation vs. 1.64 kg m^{-3} at deficit irrigation) (Howell et al., 2007). The above results demonstrated that deficit irrigation can be an effective irrigation management strategy for wheat and sorghum in the SGP. The efficiency of deficit irrigation is related to effective use of soil water. In the wheat study of Schneider and Howell (2001), the depth of soil water extraction was 1.1 m for 100% irrigation and 1.5 m for 50% irrigation. For the sorghum study, the depth of soil water extraction for deficit irrigation was 1.7 m as compared to 1.2 m for full irrigation treatment (Howell et al., 2007). In addition, improved HI under deficit irrigation significantly contributed to increased wheat yield and WUE under deficit irrigation (Xue et al., 2006).

9.5 FUTURE PERSPECTIVES

Although progress has been made to improve crop yield and WUE in the last few decades in the US SGP, crop production still faces more challenges due to changes in environmental (e.g., climate and water resources) as well as socioeconomic (e.g., growing population and changes in rural communities) conditions. In the region, declining irrigation water and climate change are the two major threats to sustainable crop production. As mentioned in several reports, irrigation supply from the Ogallala Aquifer is continuously decreasing (Steward and Allen, 2016; McGuire, 2017; Evett et al., 2020). This will lead to more dryland production in the future. A recent study by Deines et al. (2020) showed that 22,000 km^2 (24%) of currently irrigated lands in the Ogallala Aquifer region may not be able to support irrigated agriculture by 2100, and 13% of these areas are not suitable for dryland crop production primarily due to low-quality soils. The transitions from irrigated to dryland crop production will significantly affect the crop production. Climate change is another important threat to the future crop production. Under climate change scenario,

Physiological Mechanisms for Crop Water Use Efficiency

air temperatures are expected to increase in the entire Great Plains, and precipitation is expected to increase in the northern Great Plains but decrease in the central and southern Great Plains (Steiner et al., 2018; Evett et al., 2020). As such, the possibility for increasing frequency and severity of drought stress as well as other abiotic and biotic stresses will likely reduce crop yields more frequently. Dryland crop production may face more challenges than irrigated crops due to increasing temperature and extreme weather events. For example, Shrestha et al. (2020) examined the climate and wheat variety trial data at Bushland, Texas, from 1940 to 2016 and concluded that the changing climate is making the SGP less suitable for wheat grain production because the reproduction and grain filling stages are becoming less synchronized with the precipitation patterns. For irrigated crop production, limited or deficit irrigation will be the common practice in the future. Although irrigation technologies have significantly improved, the technologies only can increase irrigation efficiency and may still be limited under reduced irrigation conditions. As such, improving crop yield and WUE will need to interact with other management strategies and genetic improvement under limited or deficit irrigation. In recent years, cotton acreage has been increasing while corn acreage has been decreasing in the THP and western Kansas. One reason is that cotton production risks are smaller than corn under limited irrigation conditions (Evett et al., 2020). Regardless of dryland or irrigation conditions, how to enhance the effective use of available water remains a critical research question for crop production in water-limited environments. A key challenge for researchers and policymakers is to develop and deliver technologies that could lend greater resilience to agricultural production under environmental stress. A better understanding of genotype by management by environment (G×M×E) interaction will be important for crop adaptation to future climate (Hammer et al., 2020). Currently, advances in genomics, phenomics, and data analytics have been used to assist breeding programs for breeding crops better adapted to water-limited environment (Hammer et al., 2021). It is also envisioned by many that additional supporting data using unmanned aerial systems (UAS) should aid in the assessment of plant traits and improve management practices and breeding efforts. Kothari et al. (2019) assessed the impacts of climate change on winter wheat production in the THP and evaluated potential adaptation strategies using the CERES-Wheat model. Among the virtual genotypes tested for climate change adaptation, increasing potential number of grains and vigorous root system were found to be the most desirable traits. Enhancing yield potential traits and root architecture may be considered for screening genotypes for climate change adaptation (Kothari et al., 2019). The past and current research on crops in the SGP have been focused on aboveground traits and soil water dynamics. More research is needed to characterize crop rooting traits and their relationship with yield and WUE.

REFERENCES

AgriPartners. 2007. Irrigation and Cropping Demonstrations. Available at: http://amarillo.tamu.edu/amarillo-center-programs/agripartners/

Akbar, M.A., B.A. Stewart, and C.D. Salisbury. 1997. Evaluation of dryland corn and grain sorghum production in the southern plains using computer simulation. WTAMU Research Report 97-4.

Asseng, S. and A. F. van Herwaarden. 2003. Analysis of the benefits to yield from assimilates stored prior to grain filling in a range of environments. *Plant Soil* 256: 217–229.

Balota, M., W. A. Payne, W. Rooney, and D. Rosenow. 2008. Gas exchange and transpiration ratio in sorghum. *Crop Sci.* 48: 2361–2371.

Baumhardt, R. L., and J. Salinas-Garcia. 2006. Mexico and the U.S. southern Great Plains. In: G. A. Peterson et al. (Ed.), *Dryland Agriculture* (2nd ed., pp. 341–364). Madison, WI: ASA, CSSA, and SSSA.

Bean, B. 2007. Dryland corn in the Texas Panhandle. Texas A&M AgriLife Extension Service. http://agrilife.org/amarillo/files/2010/11/DrylandCorn.pdf

Bell, J. M., R. Schwartz, K. J. McInnes, T. A. Howell, and C. L. S. Morgan. 2018. Deficit irrigation effects on yield and yield components of grain sorghum. *Agricultural Water Management* 203: 289–296. doi.org/10.1016/j.agwat.2018.03.002

Bell, J. M., R. Schwartz, K. J. McInnes, T. A. Howell, and C. L. S. Morgan. 2020. Effects of irrigation level and timing on profile soil water use by grain sorghum. *Agricultural Water Management* 232: 106030. https://doi.org/10.1016/j.agwat.2020.106030

Benavidez, J., B. Guerrero, R. Dudensing, D. Jones, and S. Reynolds. 2019. The impact of agribusiness in the Texas High Plains trade area. Texas AgriLife Extension Service. https://amarillo.tamu.edu/files/2019/12/Impact-of-AgriBusiness.pdf

Colaizzi, P. D., S. R. Evett, and T. A. Howell. 2011. Corn production with spray, LEPA, and SDI. In: *Proceedings of the 23rd Annual Central Plains Irrigation Conference,* Burlington, CO. 22–23 Feb. 2011. Central Plains Irrigation Association, Colby, KS. pp. 52–67.

Colaizzi, P., P. Gowda, T. Marek, and D. Porter. 2009. Irrigation in the Texas High Plains: A brief history and potential reductions in demand. *Irrigation and Drainage* 58: 257–274. doi:10.1002/ird.418

Deines, J. M., M. E. Schipanski, B. Golden, S. C. Zipper, S. Nozari, C. Rottler, and V. Shardah. 2020. Transitions from irrigated to dryland agriculture in the Ogallala Aquifer: Land use suitability and regional economic impacts. *Agricultural Water Management*. Vol. 233: 106061. https://doi.org/10.1016/j.agwat.2020.106061

Eck, H. 1984. Irrigated corn yield response to nitrogen and water. *Agronomy J.* 76: 421–428.

Evett, S. R., P. D. Colaizzi, F. R. Lamm, S. A. O'Shaughnessy, D. M. Heeren, T. J. Trout, W. L. Kranz, and X. Lin. 2020. Past, present and future of irrigation on the U.S. Great Plains. *Transactions of the ASABE.* 63: 703–729.

Evett, S. R., P. D. Colaizzi, S. A. O'Shaughnessy, F. R. Lamm, T. J. Trout, and W. L. Kranz. 2014. The future of irrigation on the U.S. Great Plains. In: *Proceedings of the 26th Annual Central Plains Irrigation Conference*, Burlington, CO, February 25–26, 2014.

Foulkes, M. J., R. K. Scott, and R. Sylvester-Bradley. 2002. The ability of wheat cultivars to withstand drought in UK conditions: Formation of grain yield. *J. Agric. Sci.* 138: 153–169.

Hammer, G. L., M. Cooper, and M. P. Reynolds. 2021. Plant production in water-limited environments. *J. Exp. Bot.* 72: 5097–5101. doi:10.1093/jxb/erab273

Hammer, G. L., Z. Dong, G. McLean, A. Doherty, C. Messina, J. Schussler, C. Zinselmeier, S. Paszkiewicz, and M. Cooper. 2009. Can changes in canopy and/or root system architecture explain historical maize yield trends in the U.S. Corn Belt? *Crop Sci.* 49: 299–312.

Hammer, G. L., G. McLean, E. van Oosterom, S. Chapman, B. Zhang, A. Wu, A. Doherty, and D. Jordan. 2020. Designing crops for adaptation to the drought and high-temperature risks anticipated in future climates. *Crop Sci.* 60: 605–621.

Hansen, N. C., B. L. Allen, R. L. Baumhardt, and D. J. Lyon. 2012. Research achievements and adoption of no-till, dryland cropping in the semi-arid U.S. Great Plains. *Field Crops Res.* 132: 196–203. https://doi.org/10.1016/j.fcr.2012.02.021

Hao, B., Q. Xue, T. H. Marek, K. E. Jessup, J. Becker, X. Hou, W. Xu, E. D. Bynum, B. W. Bean, P. D. Colaizzi, and T. A. Howell. 2015a. Water use and grain yield in drought-tolerant corn in the Texas High Plains. *Agronomy J.* 107: 1922–1930.

Hao, B., Q. Xue, T. H. Marek, K. E. Jessup, X. Hou, W. Xu, E. D. Bynum, and B. W. Bean. 2015b. Soil water extraction, water use and grain yield in drought-tolerant maize in the Texas High Plains. *Agricultural Water Management* 155: 11–21.

Howell, T. A. 1990. Grain, dry matter yield relationships for winter wheat and grain sorghum – Southern High Plains. *Agronomy Journal* 82: 914–918.

Howell, T. A. 2001. Enhancing water use efficiency in irrigated agriculture. *Agron. J.* 93: 281–289.

Howell, T. A., S. R. Evett, J. A. Tolk, A. D. Schneider, and J. L. Steiner. 1996. Evapotranspiration of corn - - Southern High Plains. In: *Evapotranspiration and Irrigation Scheduling*, C. R. Camp, E. J. Sadler, and R. E. Yoder (Eds.). *Proceedings of the International Conference*, Nov. 3–6, 1996, San Antonio, TX, ASAE, St. Joseph, MI, pp. 158–166. www.nass.usda.gov/Charts_and_Maps/Field_Crops/sorg acm.php

Howell, T. A., J. L. Steiner, A. D. Schneider, and S. R. Evett. 1995a. Evapotranspiration of irrigated winter wheat. Southern High Plains. *Transactions of the ASAE* 38: 745–759.

Howell, T. A., J. A. Tolk, S. R. Evett, K. S. Copeland, and D. A. Dusek. 2007. Evapotranspiration of deficit irrigated sorghum and winter wheat. In: A. J. Clemmens, editor, *USCID Fourth International Conference on Irrigation and Drainage. The Role of Irrigation and Drainage in a Sustainable Future*, Sacramento, CA. 3–6 Oct. 2007. U.S. Committee on Irrigation and Drainage (USDID), Denver, CO. pp. 223–239.

Howell, T. A. Yazar, A. Schneider, D. Dusek, and K. Copeland. 1995b. Yield and water use efficiency of corn in response to LEPA irrigation. *Transactions of the ASAE* 38: 1737–1747.

Jones, O. R. and T. W. Popham. 1997. Cropping and tillage systems for dryland grain production in the Southern High Plains. *Agronomy J.* 89: 222–232.

Kang, J., X. Hao, H. Zhou, and R. Ding. 2021. An integrated strategy for improving water use efficiency by understanding physiological mechanisms of crops responding to water deficit: Present and prospect. *Agricultural Water Management* 255: 107008.

Kothari, K., S. Ale, A. Attia, N. Rajan, Q. Xue, and C. L. Munster. 2019. Potential climate change adaptation strategies for winter wheat production in the Texas High Plains. *Agricultural Water Management* 225: 105764.

Kottek, M., J. Grieser, C. Beck, B. Rudolf, and F. Rubel. 2006. World map of the Köppen-Geiger climate classification updated. *Meteorologische Zeitschrift* 15: 259–263.

Leakey, A. D. B., J. N. Ferguson, C. P. Pignon, A. Wu, Z. Jin, G. L. Hammer, and D. B. Lobell. 2019. Water use efficiency as a constraint and target for improving the resilience and productivity of C3 and C4 crops. *Annu. Rev. Plant Biol.* 70: 781–808.

Lopez-Castaneda C. and R. A. Richards. 1994. Variation in temperate cereals under rainfed conditions. II. Phasic development and growth. *Field Crops Res.* 37: 63–75.

Marek, T. H., D. P. Porter, N. P. Kenny, P. H. Gowda, T. A. Howell, and J. E. Moorhead. 2011. Educational enhancements to the Texas High Plains Evapotranspiration (ET) Network. Technical Report for the Texas Water Development Board, Austin, Texas. Texas A&M AgriLife Research, Amarillo, Texas. AREC Publication 2011-8.

McGuire, V. L. 2017. Water-level and recoverable water in storage changes, high plains aquifer, predevelopment to 2015 and 2013-15. U.S. Scientific Investigations Report 2017–5040.USGS.

Moroke, T. S., R. C. Schwartz, K. W. Brown, and A. S. R. Juo. 2005. Soil water depletion and root distribution of three dryland crops. *Soil Sci. Soc. Am. J.* 69:197–205.

Musick, J. T. and D. A. Dusek. 1980. Irrigated corn yield response to water. *Transactions of the ASAE* 23(1): 92–98.

Musick, J. T., O. R. Jones, B. A. Stewart, and D. A. Dusek. 1994. Water–yield relationships for irrigated and dryland wheat in the U.S. Southern Plains. *Agronomy Journal* 86: 980–986.

Musick, J. T., F. B. Pringle, W. L. Harman, and B. A. Stewart. 1990. Long-term irrigation trends: Texas High Plains. *Applied Engineering in Agriculture* 6: 717–724.

Narayanan, S., R. M. Aiken, P. V. Vara Prasad, Z. Xin, and J. Yu. 2013. Water and radiation use efficiencies in sorghum. *Agron. J.* 105: 649–656.

O'Shaughnessy, S. A., M. Kim, M. A. Andrade, P. D. Colaizzi, and S. R. Evett. 2019. Response of drought-tolerant corn to varying irrigation levels in the Texas High Plains. *Transactions of the ASABE* 62: 1365–1375.

Oweis, T., H. Zhang, and M. Pala. 2000. Water use efficiency of rainfed and irrigation bread wheat in a Mediterranean environment. *Agronomy J.* 92: 231–238.

Panhandle Water Planning Group. 2011. Regional Water Plan for the Panhandle Water Planning Area. Available at: www.twdb.texas.gov/waterplanning/rwp/plans/2011/A/Region_A_2011_RWP.pdf?d=9005.76.

Passioura, J. B. and J. F. Angus. 2010. Improving productivity of crops in water-limited environments. *Advances in Agronomy* 106: 37–75.

Payero, J. O., D. D. Tarkalson, S. Irmak, D. Davison, and J. L. Petersen. 2008. Effect of irrigation amounts applied with subsurface drip irrigation on corn evapotranspiration, yield, water use efficiency, and dry matter production in a semiarid climate. *Agricultural Water Management* 95: 895–908. doi:10.1016/j.agwat.2008.02.015.

Richards, R. A., G. J. Rebetzke, A. G. Condon, and A. F. van Herwaarden. 2002. Breeding opportunities for increasing the efficiency of water use and crop yield in temperate cereals. *Crop Sci.* 42: 111–121.

Richards, R. A., G. J. Rebetzke, M. Watt, A. G. Condon, W. Spielmeyer, and R. Dolferus. 2010. Breeding for improved water productivity in temperate cereals: phenotyping, quantitative traits loci, markers and the selection environment. *Functional Plant Bio.* 37: 85–97.

Schneider, A. D. and T. A. Howell. 1998. LEPA and spray irrigation of corn—Southern High Plains. *Transactions of ASAE* 41: 1391–1396.

Schneider, A. D. and T. A. Howell. 2001. Scheduling deficit irrigation with data from an evapotranspiration network. *Trans. ASAE* 44: 1617–1623.

Shrestha, R., S. Thapa, Q. Xue, B. A. Stewart, B. C. Blaser, E. K. Ashiadey, J. C. Rudd, and R. N. Devkota. 2020. Winter wheat response to climate change under irrigated and dryland conditions in the US southern High Plains. *J. Soil and Water Conservation* 75: 112–122.

Sinclair, T. R. and R. C. Muchow. 2001. System analysis of plant traits to increase grain yield on limited water supplies. *Agronomy J.* 93: 263–270. doi: 10.2134/ agronj2001.932263x.

Steiner, J. L. 1986. Dryland grain sorghum water use, light interception, and growth responses to planting geometry. *Agronomy J.* 78: 720–726.

Steiner, J. L., D. D. Briske, D. P. Brown, and C. M. Rottler. 2018. Vulnerability of Southern Plains agriculture to climate. *Climate Change* 146: 201–218.

Steward, D. R. and A. J. Allen. 2016. Peak groundwater depletion in the High Plains Aquifer, projections from 1930 to 2110. *Agric. Water Manage.* 170: 36–48.

Stewart, B. A., R. L. Baumhardt, and S. R. Evett. 2010. Major advances of soil and water conservation in the U.S. Southern Great Plains. In *Soil and Water Conservation Advances in the United States,* Special Publication 60, eds. T. M. Zoebeck and W. F. Schillinger. Soil Science Society of America, Madison, WI, pp. 103–129.

Stone, L. R. and A. J. Schlegel. 2006. Yield–water supply relationships of grain sorghum and winter wheat. *Agronomy J.* 98:1359–1366.

Tanner, C. B. and T. R. Sinclair. 1983. Efficient water use in crop production: research or re-search? In *Limitation to Efficient Water Use in Crop Production*, eds. H. M. Taylor, W. R. Jordan, and T. R. Sinclair. American Society of Agronomy, Madison, WI, pp. 1–27.

Thapa, S., B. A. Stewart, and Q. Xue. 2017a. Grain sorghum transpiration efficiency at different growth stages. *Plant, Soil and Environment* 63: 70–75.

Thapa, S., B. A. Stewart, Q. Xue, M. B. Rhoades, B. Angira, and J. Reznik. 2018. Canopy temperature, yield, and harvest index of corn as affected by planting geometry in a semi-arid environment. *Field Crops Res.* 227: 110–118.

Thapa, S., Q. Xue, K. E. Jessup, J. C. Rudd, S. Liu, G. P. Pradhan, R. N. Devkota, and J. Baker. 2017b. More recent wheat cultivars extract more water from deeper soil profile in the Texas High Plains. *Agronomy J.* 109: 2771–2780.

Tolk, J. A. and T. A. Howell. 2003. Water use efficiencies of grain sorghum grown in three southern Great Plains soils. *Agricultural Water Management* 59: 97–111.

Tolk, J. A. and T. A. Howell. 2008. Field water supply relationships of grain sorghum grown in three USA Southern Great Plains soils. *Agricultural Water Management* 95: 1303–1313.

Tolk, J. A. and T. A. Howell. 2009. Transpiration and yield relationships of grain sorghum grown in a field environment. *Agronomy J.* 101: 657–662.

Tolk, J. A., T. A. Howell, J. L. Steiner, and S. R. Evett. 1997. Grain sorghum growth, water use, and yield in contrasting soils. *Agricultural Water Management* 35: 29–42.

Tollenaar, M. and E. A. Lee. 2002. Yield potential, yield stability and stress tolerance in maize. *Field Crops Research* 75: 161–169.

Trostle, C. and B. Bean. 2010. Wheat Grain Variety 2006-2009 Yield Summary— Texas High Plains. Texas A&M AgriLife Extension. http://agrilife.org/lubbock/files/2011/10/wheattrialreport0609_5.pdf

Unger, P. W. 2001. Alternative and opportunity dryland crops and related soil conditions in the southern Great Plains. *Agronomy Journal* 93: 216–226.

Unger, P. W. and R. L. Baumhardt. 1999. Factors related to dryland grain sorghum yield increases: 1939 through 1997. *Agronomy J.* 91: 870–875.

USDA-FAS. 2022. Crop Acreage Data. https://www.fsa.usda.gov/news-room/efoia/electronic-reading-room/frequently-requested-information/crop-acreage-data/index#

USDA-NASS. 2021. Texas Sorghum. https://quickstats.nass.usda.gov/

USDA-NASS. 2022. Charts and County Maps. www.nass.usda.gov/Charts_and_Maps/Crops_County/index.php

Winter, S. R. and J. T. Musick, 1993. Wheat planting date effects on soil water extraction and grain yield. *Agron. J.* 85: 912–916.

Xin, Z., R. Aiken, and J. J. Burke. 2009. Genetic diversity of transpiration efficiency in sorghum. *Field Crops Res.* 111: 74–80.

Xin, Z., C. Franks, P. Payton, and J. J. Burke. 2008. A simple method to determine transpiration efficiency in sorghum. *Field Crops Res*. 107: 180–183.

Xue, Q., W. Z. Liu, and B. A. Stewart. 2012. Improving wheat yield and water-use efficiency under semiarid environment: The US Southern Great Plains and China's Loess Plateau. In: *Advances in Soil Science: Soil Water and Agronomic Productivity*, eds. R. Lal and B. A. Stewart. Taylor & Francis, New York.

Xue, Q., S. Madhavan, A. Weiss, T. J. Arkebauer, and P. S. Baenziger. 2002. Genotypic variation of gas exchange parameters and carbon isotope discrimination in winter wheat. *J. Plant Physiol*. 159: 891–898.

Xue, Q., T. H. Marek, W. Xu, and J. Bell. 2017. Irrigated corn production and management in the Texas High Plains. *J. Contemporary Water Research & Education* 162: 31–41.

Xue, Q., J. C. Rudd, S. Liu, K. E. Jessup, R. N. Devkota, and J. R. Mahan. 2014. Yield determination and water use efficiency of wheat under water-limited conditions in the US Southern High Plains. *Crop Sci*. 54: 34–47.

Xue, Q., S. Thapa, K. Jessup, S. Liu, J. C. Rudd, J. M. Bell, S. Baker, J. Baker, and R. N. Devkota. 2020. Genetic improvement contributed to increased yield and water use efficiency in wheat under water-limited conditions – A long-term study. ASA-CSSA-SSSA, International Annual Meetings, Nov. 9–13, 2020 (Virtual).

Xue, Q., Z. Zhu, J. T. Musick, B. A. Stewart, and D. A. Dusek. 2003. Root growth and water uptake in winter wheat under deficit irrigation. *Plant and Soil* 257: 151–161.

Xue, Q., Z. Zhu, J. T. Musick, B. A. Stewart, and D. A. Dusek. 2006. Physiological mechanisms contributing to the increased water-use efficiency in winter wheat under deficit irrigation. *J. Plant Physiol*. 163: 154–164.

Yang J., J. Zhang, Z. Huang, Q. Zhu, and L. Wang. 2000. Remobilization of carbon reserves is improved by controlled soildrying during grain filling of wheat. *Crop Sci*. 40:1645–1655.

Yazar, A., T. A. Howell, D. A. Dusek, and K. S. Copeland. 1999. Evaluation of crop water stress index for LEPA irrigated corn. *Irrigation Science* 18: 171–180.

Zhang, J. H., X. Z. Sui, B. Li, B. Su, J. M. Li, and D. X. Zhou. 1998. An improved water-use efficiency for winter wheat grown under reduced irrigation. *Field Crops Res*. 59: 91–98.

Zhao, J., Q. Xue, X. Hou, B. Hao, K. E. Jessup, T. H. Marek, W. Xu, S. R. Evett, S. O'Shaughnessy, and D.K. Brauer. 2018b. Shoot and root traits in drought tolerant maize (*Zea mays* L.) hybrids. *J. Integrative Agriculture* 17(5): 1093–1105.

Zhao, J., Q. Xue, K. E. Jessup, B. Hao, X. Hou, T. H. Marek, W. Xu, S. R. Evett, S. O'Shaughnessy, and D. K. Brauer. 2018a. Yield and water use of drought-tolerant maize hybrids in a semiarid environment. *Field Crops Res*. 216: 1–9.

10 Improving Water Storage through Effective Soil Organic Matter Management Strategies under Dryland Farming in India

Ch. Srinivasarao[1], S. Rakesh[1], G. Ranjith Kumar[1],*
M. Jagadesh[2], K.C. Nataraj[3], R. Manasa[1], S. Kundu[4],
S. Malleswari[3], K.V. Rao[4], J.V.N.S. Prasad[4], R.S. Meena[5],
G. Venkatesh[4], P.C. Abhilash[5], J. Somasundaram[6], and R. Lal[7]

[1]ICAR-National Academy of Agricultural Research Management,
Rajendranagar – 500030, Telangana, India
[2]Tamil Nadu Agricultural University, Coimbatore – 641003,
Tamil Nadu, India
[3]Agriculture Research Station, Acharya N.G. Ranaga Agricultural
University, Ananthapuramu – 515 001, Andhra Pradesh, India
[4]ICAR-Central Research Institute for Dryland Agriculture,
Hyderabad – 200 059, Telangana, India
[5]Banaras Hindu University, Varanasi – 221005, Uttar Pradesh, India
[6]Indian Institute of Soil Science, Nabibagh, Bhopal – 462038,
Madhya Pradesh, India
[7]CFAES Rattan Lal Center for Carbon Management and Sequestration,
the Ohio State University, Columbus, OH 43210, USA
*Corresponding author: cherukumalli2011@gmail.com

CONTENTS

10.1 Introduction 257
10.2 Challenges in Achieving Water Productivity in Rainfed Soils of India 258
10.3 Water Resources Availability in Indian Rainfed Drylands 260
    10.3.1 Surface Water Resources...260
    10.3.2 Groundwater Resources 261
    10.3.3 Rainfall Trends in Rainfed Drylands 262
    10.3.4 Rainwater Management 262
        10.3.4.1 In-Situ Moisture Conservation 262
        10.3.4.2 Ex-Situ Rainwater Harvesting through Farm Ponds 262
10.4 Improving Soil Moisture Storage through SOC Buildup...262
10.5 Soil Organic Matter and Key Contributing Factors..265
10.6 Soil Water Storage through Building SOM 266
    10.6.1 Mulch Cum Manuring of Soils 266
    10.6.2 Drip Irrigation with Organic Mulch..267

DOI: 10.1201/b22954-10

	10.6.3 Green Manuring for SOC and Moisture Storage	267
	10.6.4 Intercropping as an Efficient Production System under Droughts	268
	10.6.5 Contour Farming for Erosion Control	270
	10.6.6 Conservation Agriculture for Carbon Storage and Water Use Efficiency	270
	10.6.7 Integrated Nutrient Management (INM): Win-Win for SOC and Water Storage in Soil Profile	271
	10.6.8 Organic Amendments	272
	10.6.9 Biochar	273
	10.6.10 Tank Silt	273
	10.6.11 Cover Crops and Market Wastes	274
10.7	National and State Government Initiations for Soil Carbon and Water Storage	275
10.8	Conclusion	275
10.9	Way Forward	276
References		277

10.1 INTRODUCTION

Population of India increased from 451 million in 1960 to 1.39 billion in 2021, which reflects a growth of 209.3% over the past 61 years. The highest annual increase in population of India was recorded in 1974 at 2.36%, and the lowest increase of 0.97% was observed in 2021. India accounts for 2.4% of the global land surface but hosts more than 17.73% of the world population (ICAR - Data Book 2022). Due to population pressure, India has faced several problems like food insecurity, low income, and vulnerability to environmental and biotic factors. The total geographical area of the country is 328.73 million ha (M ha), in which net area sown is 139.35 M ha and net irrigated area is 71.55 M ha. Cropping intensity of the country increased from 131% in 2000–01 to 144.9% in 2018–19. Fertilizer consumption increased from 26.75 million tons in 2015–16 to 32.54 million tons in 2020–21. As per the 2020–21 statistics, food grain, horticultural, nutri-cereals, pulses, and oilseeds production were 308.6, 331, 51.15, 25.72, and 36.1 million tons (Mt), respectively (ICAR - Data Book 2022). Total food grain production increased from 251.6 (2015–16) to 308.6 Mt (2020–21) in conjunction with the nutrient consumption of 136 kg ha^{-1} (2015–16) to 161 kg ha^{-1} (2020–21), showing a direct cause–effect relation (Figure 10.1).

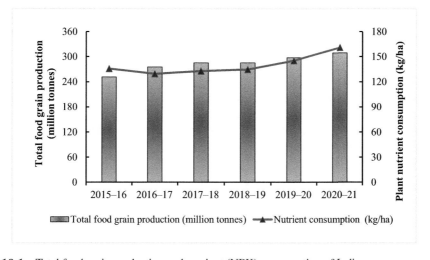

FIGURE 10.1 Total food grain production and nutrient (NPK) consumption of India.

In the 2021–22 crop year, total food grain production of 315.7 Mt has been the highest till now, and the Government of India (GOI) has set a food grain production target of 328 Mt for the 2022–23 crop year, which is 4% higher than production for 2020–21. Due to population pressure, by 2050 the total food grain production should be increased to 377 Mt (Srinivasarao et al., 2021a). As per the 2019–20 estimates, per capita availability of milk and eggs was 406 gram/day and 86/annum, respectively. Globally, the country's milk and egg production has achieved 1st and 3rd places. However, despite phenomenal growth in food grain production, the country still needs more food grains to feed the growing population, especially women and children.

In India, small and marginal farmers constituted 86.1% of the total landholdings in 2015–16 against 85.0% in 2010–11 and contribute about 60% of the total food grain production and over 50% of the country's fruits and vegetables production (All report on Agricultural Census, 2015–16). The average size of the land holdings of small and marginal farmers is about 0.38 ha when compared to large farmers (17.37 ha) and the country's total average size of the land holding declined from 2.28 ha (1970–71) to 1.08 ha (ICAR-Data Book 2022) which cannot efficiently contribute to employment and income of the nation. Nonetheless, small and marginal farmers are highly efficient in food production as compared to large farmers (Chand et al. 2011); their contribution to total food grain production and poverty reduction is rather high compared with that of large farmers (Gururaj et al., 2017).

Rainfed farming is complex, highly diverse, and risk prone. In India, the rainfed area covers almost 55% of the net sown area (139.42 M ha) and about 61% of the farmers are cultivating crops under rainfed conditions. Thus, rainfed farming is crucial for food security and economy. At present, rainfed farming contributes around 40% of the total food grain production (85, 83, 70, and 65% of nutri-cereals, pulses, oilseeds, and cotton, respectively). Further, rainfed farming impacts the livelihoods of 80% of small and marginal farmers in the country (NRAA, 2022). Alfisols and Vertisols are predominant soil orders in the peninsular plateau of India. Aridisols exist in hot dry climates along with Entisols and Inceptisols. Alluvial (Inceptisols) soils are most dominant (93.1 M ha) when it comes to agricultural production and land use and management, followed by red (Alfisols, 79.7 M ha), black (Vertisols, 55.1 M ha), desert (Entisols, Aridisols, 26.2 M ha), and lateritic (Plinthic horizon, 17.9 M ha) soils. However, in the rainfed regions, Inceptisols are dominant followed by Entisols, Alfisols, Vertisols, Mixed soils, Aridisols, Mollisols, Ultisols, and Oxisols (Srinivasarao et al., 2013a). Generally, most of the soils belonging to the rainfed areas are coarse textured, except Vertisols. Accordingly, their ability to retain water and nutrients is low, and crops cultivated on these soils are susceptible to drought and nutrient deficiencies compared to irrigated soils. Rainfed soils have multiple nutrient deficiencies. The SOC levels are as low as 5 g/kg, whereas the desired level should be more than 11 g/kg (Srinivasarao et al., 2009; NRAA, 2022). Inherent fertility is low in most rainfed soils except in some Vertic group which are rich in bases. Low aggregate stability, insufficient SOM content, and high erosion rates are the serious problems in rainfed soils which decline the crop yields and threaten the nation's food security (Srinivasarao et al., 2009, 2015, 2019b) (Figure 10.2). Nonetheless, large yield gaps remain in several crops grown under rainfed farming. There was a modest increase in productivity from 0.6 tons in the 1980s to 1.1 tons at present. In general, the productivity of rainfed areas is around 1.1 tons ha^{-1}, compared with 3 tons ha^{-1} in irrigated areas (NRAA, 2022). Thus, there is a strong need for the uplifting of the rainfed farming community to meet the food demand of the country.

10.2 CHALLENGES IN ACHIEVING WATER PRODUCTIVITY IN RAINFED SOILS OF INDIA

India is the world's 12th most water-stressed country. As per "Prioritization of Districts for Development Planning in India – A Composite Index Approach (2020)" report by NRAA, about 15 M ha of rainfed (arid, semiarid, dry, sub-humid, and humid regions) cropped area in the arid region

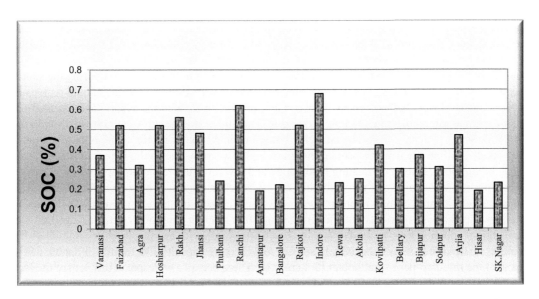

FIGURE 10.2 Organic carbon status of various soil types under diverse rainfed production systems of India.

receives less than 500 mm rainfall per annum. Another 15 M ha receives 500–750 mm, and about 42 M ha receives 750–1100 mm. About 73 out of 127 agroclimatic zones in India are predominantly rainfed. This variability requires identification of location-specific cultivation approaches. Climate change can influence the rainfed farming. Long-term data indicate that rainfed areas experience 3–4 drought years per decade which are moderate to severe in intensity. Rainfed crops are often affected by irregular patterns of precipitation and high temperatures, which result in frequent modifications in sowing time and harvesting dates.

The Indian economy is severely affected by variability in monsoons. Rainfall reduction causes drought, which is a climatic disaster affecting the farming activities. Prolonged dry spells in rainfed regions initially hamper the nutrient uptake, diminish crop yield, and can eventually collapse the country's agrarian economy (Singh et al., 2021). Rainfed areas are more vulnerable to extreme climatic events. Thus, water resource management and mitigation of hydrological disasters are highly dependent on climate variability in rainfed regions (Neog et al., 2019). Rainfall deficiency causes depletion of surface and ground water levels which leads to deleterious effects on cultivation practices because of inadequate availability of water for the crops, particularly during the critical stages of plant growth. Water scarcity impacts in rainfed regions are determined by the climate change, soil and hydrological profiles, soil moisture availability, crop selection and agricultural operations, etc. Poor rainfall in sequential years adversely affects the ground water table by reducing the scope of surface and ground water recharge, and replenishment of soil moisture. Increased depletion rates of groundwater and declined recharge may deplete aquifers, adversely affect water quality (e.g., salt concentration, acidity, etc.), and jeopardize the soil's biological productivity. Too much exploitation of groundwater resources by tube wells has depleted this finite resource. In rainfed areas, crops are affected by delays in the onset of monsoons, which narrow the sowing window. A wider sowing window reduces the risks of any major yield loss. Delayed monsoons reduce the crop growth duration and hampers agronomic yield. Delays in the onset of monsoon may result in poor crop growth and development due to lack of soil moisture during critical growth stages. Similarly, an early withdrawal of monsoon may adversely affect the reproductive stage of crop, which is crucial to attaining the desired yield. Mid-season drought occurring during the vegetative phase may result in stunted growth. It adversely affects crop yield when it occurs at flowering or early reproductive stage. Nonetheless, nutrient top-dressing, plant protection, and intercultural operations are most useful during mid-season drought. Among all types of droughts, terminal drought is critical to farmers due to contingency plans not working out and grain yield is strongly related to water availability during the reproductive stage. Sometimes terminal drought leads to forced maturity which declines the quality of the yield (Srinivasarao et al., 2019a, b). Most of the rainfall is received during the southwest monsoon, and India is already witnessing extreme weather events. In 2021, rainfed regions of central and northwest India received 6% above normal rain throughout the country and received 1% below normal rainfall despite the longer duration. Then there was month-to-month variation in the 2021 monsoon (Nayar, 2021). The changing pattern of monsoon in rainfed regions of the country has taken a toll on the lives of people, livestock, and also crops due to the high intensity of rains in short durations. Due to undulating topography and low moisture retention capacity of the soil, a major portion of the rainwater is lost through runoff, accelerating losses by soil erosion. Deficiency/uncertainty in rainfall of high intensity causes excessive loss of soil through erosion, which leads to the loss of carbon and nutrients. Insufficient soil moisture is left in the profile to support plant growth and grain production from rainfed drylands. Due to erratic behavior and inappropriate rainfall distribution, agriculture is risky especially in soils of low carrying capacity like rainfed drylands, where most of the resource-poor smallholder farmers exist. There occurs a widespread problem of the lack of resource tools which are inefficient and are the cause of a low productivity.

10.3 WATER RESOURCES AVAILABILITY IN INDIAN RAINFED DRYLANDS

The principal source of water in rainfed regions of India is precipitation (rainfall and snow stocks from the Himalayan mountains). Out of total rainfall received only a part of it is stored as groundwater and the remaining is lost as runoff and evaporation. On an average, India receives a total annual precipitation of around 4,000 km^3. Out of this, the majority share of rainfall is from the southwest monsoon, and it contributes 3,000 km^3. The rainfed regions in India comprise predominantly arid, semiarid, and dry sub-humid areas (Figure 10.3) with an annual rainfall ranging from 100 to 1000 mm, whereas moist sub-humid, humid, and per humid regions receive an annual rainfall ranging from 1000 to >2500 mm (Sharma and Kumar, 2014).

The distribution of water resources in India is highly uneven and skewed and it varies from dry and semiarid Rajasthan in the western India to wet and water-excess states of West Bengal and Orissa (Kumar et al., 2005). Yet, there has been a paradigm shift in water resources used in India since the 1950s from communities (tanks and small water structures) to government (major and medium irrigation projects), and the private domain (groundwater). Groundwater is the source for 70% of the irrigation and 80% of the drinking water. Average annual per capita availability of water for major part of India is around 2251 m^3/year. About 40% of the population in India has 70% of India's water resources, while the remaining 60% has only 30% indicating extreme disparities in distribution (Majumdhar, 2008). Out of the total annual precipitation of around 3880 km^3 (10^9 m^3) in India, the groundwater and surface water are about 184.56 km^3, comprising 690 km^3 (37%) of surface water, and 436 km^3 of groundwater (CWC, 2022). By 2050, the total per annum consumption of national water may exceed the utilizable water resources, unless significant changes occur through potential water management. Thus, available water resources need judicious management through the artificial recharge of groundwater, desalination of brackish water, rainwater harvesting, etc..

10.3.1 SURFACE WATER RESOURCES

India's average annual surface runoff through rainfall and snowmelt is estimated to be 1869 km^3. Yet, only about 213 km^3 (11.4%) of surface water can be harnessed as 90% of the annual flow of the Himalayan rivers occurs for only 4 months' period and there is no potential to capture due to

Effective Soil Organic Matter Management Strategies

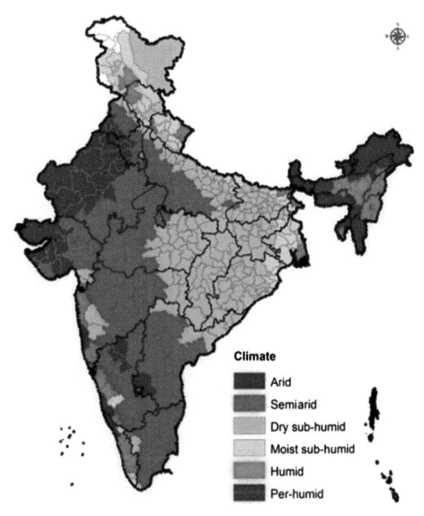

FIGURE 10.3 Predominant rainfed dryland regions in India (arid, semiarid, and dry subhumid). (Source: Adopted from Srinivasarao et al., 2020a).

limited suitable storage reservoir sites. Soon (by 2050), this capacity may double to about 21%. The utilization in the peninsular basins (e.g., Godavari, Krishna, Cauvery, Mahanadi, Tapti, and Narmada) is more than 70% of the present capacity. The Krishna River basin has the highest storage capacity (49 km^3) and can store 64% of the mean annual river flow, or about 220 days of average flow (CWC 2022).

10.3.2 Groundwater Resources

The annual replenishable groundwater resource is contributed by rainfall (280 km^3) and other sources (156.1 km^3) from canal seepage return flow from irrigation, seepage from water bodies, and artificial recharge. Rainfall contributes about 67% to the country's annual replenishable groundwater. The central groundwater board has drafted a model bill for ensuring sustainable and equitable development and use of groundwater resources that can be adopted by different states of India (Srinivasarao et al., 2015).

10.3.3 RAINFALL TRENDS IN RAINFED DRYLANDS

Rainfall is the ultimate source of water in drylands (Srinivasarao et al., 2014a, b). Coping with extreme variability in rainfall, high-intensity storms, and high frequency of dry spells are the key challenges for rainfed agriculture. The water retention and length of crop growing period in diverse production systems under varied soil types and climatic conditions of India is presented in Table 10.1.

10.3.4 RAINWATER MANAGEMENT

10.3.4.1 In-Situ Moisture Conservation

This technique is more practical and feasible as a large part of rainfed agriculture consists of marginal and smallholdings. Hence, location-specific in-situ moisture conservation practices were developed based on rainfall, soil types, and overall agro-ecological conditions (Table 10.2) (Srinivasarao et al., 2014b). Measures such as ridge/furrow, sowing across the slope, and paired row sowing are observed to be important water conservation measures for effective water conservation besides draining out the excess rainwater.

10.3.4.2 Ex-Situ Rainwater Harvesting through Farm Ponds

India receives about 4,000 km^3 rainfall per annum, of which about 1869 km^3 flows as runoff every year in the country. Due to geographical limitations, we are unable to utilize a good amount of surface water. Farm ponds constructed in coarse-textured soils (Alfisols) require lining to minimize seepage, but in clayey soils (Vertisols) having negligible seepage may not require lining. The unlined farm ponds which have higher seepage can be utilized to recharge the aquifers (Srinivasarao et al., 2017). In this context, the importance of rainwater harvesting has increased in recent years because of depletion of groundwater levels and increased rainfall variability. Watershed management is the flagship program in India to enhance the water resource availability and improve agricultural production in a sustainable manner (Ravindra Chary and Gopinath, 2022)). As per the projections, about 27.5 M ha of Indian rainfed area can contribute an amount of 114 km^3 of water for water harvesting that will be adequate to supply water for one supplementary irrigation of 100 mm depth to 20 M ha during drought years and 25 M ha during normal years (Sharma et al., 2010; Srinivasarao and Gopinath, 2016). However, harvesting the runoff water and storing it in farm ponds in rainfed areas depend on the amount of rainfall received per annum, topography, and the soil type. The harvested water during the rainy season can be utilized for supplemental irrigation during dry spells coinciding with critical growth stages in rainy season or for establishment of winter crops. Half of the Indian agriculture is rain-dependent and, therefore, in-situ and ex-situ rainwater conservation in terms of farm or community ponds is highly prioritized (Srinivasarao et al., 2019a, b).

10.4 IMPROVING SOIL MOISTURE STORAGE THROUGH SOC BUILDUP

SOM has a great potential for water retention, thus becoming a key factor in contributing to available water-holding capacity (AWHC) (Figure 10.4). SOM has a practical implication for water management in agriculture. Increasing 1% SOM increases the AWHC range to varying degrees (Hudson, 1994). SOM could potentially contribute to available water by 2.2–12.5% (Emerson, 1995). Soils from the USDA Natural Resources Conservation Service (NRCS) soil survey reported that a 1% increase in SOM resulted in 2 to >5% increase in AWHC (soil water retention at –1,500 and –33 kPa), however, the magnitude of increase depended on soil textural properties (Olness and Archer, 2005). Several old studies reported that enhancement of AWHC under SOM increments was observed to be more prominent in lighter than in heavy soils (Rawls et al., 2003). In the case of fine textured soils, SOM improves the soil structure and aggregation and decreases bulk density (BD) that improves water storage closer to field capacity than at the wilting point (Rawls et al., 2003).

Effective Soil Organic Matter Management Strategies 263

TABLE 10.1

Mean Annual Rainfall, Climate, Soil Type, Production System Length of Growing Period and Water Retention (% wt. Basis) (Mean of 7 Layers of Soil Profile) of Rainfed Drylands in India

location / state	Mean Annual Rainfall (mm)	Climate	Soil type	Production system	Water retention at (bar)			Growing period (days)
					1/3 (%)	15 (%)	Available (%)	
Hissar, Haryana	412	Arid	Alluvial Deep -Aridisols	Pearl millet	15.9	5.7	10.3	60–90
Bellary Karnataka	500	Arid	Black deep Vertisols	Rabi sorghum	35.8	23.0	12.8	90–120
Sardar Krushinagar Gujarat	550	Arid	Desert Deep -Aridisols	Pearl millet	3.1	1.7	1.4	60–0
Anantapur, Andhra Pradesh	560	Arid	Red Shallow- Alfisols	Groundnut	18.3	8.3	10.0	90–120
Rewa, Madhya Pradesh	590	Semiarid	Black medium deep Vertisols	Soybean	29.6	16.0	13.6	150
Rajkot, Gujarat	615	Arid	Black Deep -Vertisols	Groundnut	35.1	23.5	11.6	60–90
Arija, Rajasthan	656	Semiarid	Black shallow deep Vertisols	Maize	6.6	2.5	4.0	90–120
Agra, Uttar Pradesh	665	Semi-arid	Alluvial deep Inceptisol	Pearl millet	21.1	8.4	13.0	90–120
Bijapur Karnataka	680	Semiarid	Black medium deep Vertisols	Rabi sorghum	45.7	24.7	21.0	90–120
Solapur, Maharashtra	723	Semiarid	Black medium deep vertic/ Vertisols	Rabi sorghum	42.6	30.5	12.1	90–120
Kovilpatti Tamilnadu	743	Semiarid	Black deep Vertisols	Cotton	40.4	26.7	13.8	120
Akola, Maharashtra	825	Semiarid	Black medium deep vertic/ Vertisols	Cotton	36.8	25.3	11.5	120–150
Bengaluru, Karnataka	926	Semiarid	Red deep Alfisols	Finger millet	14.1	8.6	5.5	120–150
Indore Madhya Pradesh	944	Semiarid	Black deep Vertisols	Soybean	33.0	19.9	13.1	120
Ballowal Sunkari, Punjab	1000	Semiarid	Alluvial deep Inceptisol	Maize	9.3	4.1	5.2	120–150
Jhansi, Uttar Pradesh	1017	Semiarid	Alluvial deep Inceptisol	Rabi Sorghum	20.3	3.8	16.4	120
Faizabad, Uttar Pradesh	1057	Sub humid	Alluvial deep Inceptisol	Rice	24.6	10.4	14.2	90–20
Varanasi, Uttar Pradesh	1080	Sub humid	Alluvial deep Inceptisol	Rice	21.1	7.5	13.6	150–180
Rakh Dhinsar, J&K	1180	Sub humid	Alluvial deep Inceptisol	Maize	5.1	2.0	3.0	150–210
Ranchi, Jharkhand	1299	Sub humid	Red shallow Alfisols	Rice	15.8	10.9	4.9	150–180
Phulbani, Odisha	1378	Sub humid	Red yellow deep Alfisols	Rice	18.2	9.8	8.4	180–210

Source: Complied and modified from Srinivasarao et al., 2009a; Srinivasarao et al., 2013a.

TABLE 10.2
Recommended Soil Moisture Conservation Measures / Practices Based on Rainfall Received in Rainfed Regions of India

Mean annual rainfall (mm) received in rainfed region

< 500 mm	500–750	750–1000	>1000
• Conservation furrows	• Conservation furrows	• Broad bed furrow (for Vertisols)	• Broad bed furrow (for Vertisols)
• Contour bunds	• Contour cultivation	• Conservation furrows	• Field bunds
• Contour cultivation	• Ridging	• Sowing across slopes	• Vegetative bunds
• Ridging	• Sowing across slopes	• Tillage	• Graded bunds
• Sowing across slopes	• Scoops	• Lock and spill drains	• Level terrace
• Mulching	• Tied ridges	• Small basins	
• Off season tillage	• Mulching	• Field bunds	
• Scoops	• Zing terrace	• Vegetative bunds	
• Tied ridges	• Off season tillage	• Graded bunds	
• Inter row water harvesting systems	• Broad bed furrow	• Zing terrace	
• Small basins	• Inter row water harvesting systems		
• Field bunds	• Small basins		
	• Field bunds		

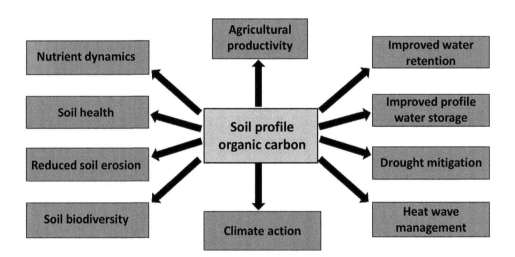

FIGURE 10.4 Critical role of SOC in improved profile water storage and drought mitigation.

Libohova et al. (2018) evaluated the effect of SOM on AWHC using the National Cooperative Soil Survey (NCSS) database. In their study, it is observed that AWHC correlated positively with silt content ($r = 0.56$). Decreased BD values increased the AWHC ($r = -0.34$), but this relation was observed to be extremely variable with the SOM and soil textural properties. Porosity is another important factor that governs the relationship between SOC and soil water storage. A unit increase of SOC would enable a large increase in porosity (0.2–5 and 480–720 μm diameter classes) that ultimately enhances the AWHC and unsaturated hydraulic conductivity (Fukumasu et al., 2022). A positive correlation recorded between SOC and mean weight diameter of water-stable aggregates may be ascribed to the fact that water wettability of aggregates decreased with larger SOC content (Chenu et al., 2000). Also, the variations between SOC and soil water storage (SWS) are more highly influenced by the land use system than the local environmental conditions. Chen et al. (2022)

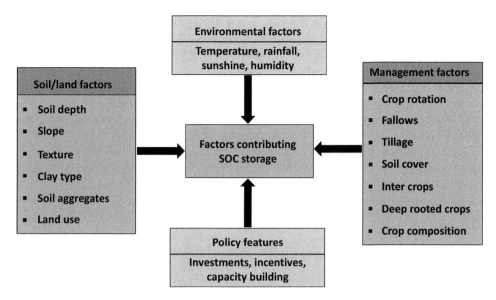

FIGURE 10.5 Factors contributing to SOC storage in particular agro-ecosystem.

observed that the interaction of SOC and SWS was maximum at surface depth (0–20 cm) and the topography was a predominant factor that influenced the SOC and SWS in the deep soil layers (100–200 cm). Soil texture was a stable driving factor which influenced these in the whole soil profile (0–200 cm). Some of the recent studies (Zhou et al., 2020) also reported that factors like soil texture, BD, SOM, and plant roots and litters determine the soil water retention.

Additionally, while positive correlations between SOC and AWHC were reported by Olness and Archer (2005), recent global and continental-scale studies found limited effects of SOC on AWHC(Minasny and McBratney, 2018). The relationships between SOC and porosity have implications for soil water dynamics. A 10 g kg^{-1} (1%) increase in SOC increased the AWHC of about 3 mm to 100 mm^{-1} soil depth (Fukumasu et al., 2022) and this was larger than the results of a meta-analysis (1.1–1.9 mm 100 mm^{-1} soil depth) (Minasny and McBratney, 2018). Nevertheless, the farmland practices like crop residue managements and tillage practices also influence the SOC, SWS, and soil structure (Zhao et al., 2020). Amalgamation of improved SOC and SWS and soil structure should be promoted to ensure sustainable food production and national food security. Because a good soil structure generally has better water-holding capacities, hydraulic conductivity, and adequate aeration that are extremely helpful for better plant growth and development (Karami et al., 2012). The critical role of SOC in improved profile water storage and drought mitigation is illustrated in Figure 10.4.

Hence, it is paramount to increase the SOC storage in order to significantly increase the soil water storage in the soil profile. However, there are several important factors that are highly influencing the SOC storage (Figure 10.5) and, therefore, it is important to critically consider all these factors to efficiently manage and enhance the SOC in the agriculture system.

10.5 SOIL ORGANIC MATTER AND KEY CONTRIBUTING FACTORS

SOM has an impact on several factors that are essential for maintaining soil productivity. As the "lifeline" of the soil, it significantly affects qualities like aggregate stability, water-holding capacity, buffering capacity, cation exchange capacity (CEC), acidification, solidification, sodification, salinization, etc. However, soils in rainfed regions are highly diverse and include Vertisols and Vertic sub-groups, Inceptisols, Entisols, Alfisols, Oxisols, Aridisols, etc., in different agroecological zones of India. The main constraints on soil health in rainfed regions are moisture stress, unfavorable permeability, nutrient P-fixation, poor nutrient retention, erosion, and slope. Additionally, there is a significant variation in annual rainfall (between 400 and 1500 mm). In rainfed agroecological regions, the SOM status of soil in various locations was found to be low which is an emerging issue (Srinivasarao et al., 2009b). In most cases, the SOM concentration measured under various crop production techniques in rainfed (non-irrigated) agro-ecological zones was estimated to be < 5 g kg^{-1} soil. Additionally, surface soils (0–30 cm) in many Indian regions under rainfed crop production systems are significantly low in SOC, which may be as low as 0.15% in some of the rainfed agroecological locations (Srinivasarao et al., 2011a; Srinivasarao and Vittal, 2007). Therefore, the reduction in SOM content can be slowed and crop yield can be greatly increased by adopting appropriate soil–crop management strategies. Overall, SOC building up in dryland agriculture depends on soil factors (clay, depth of soil, and slope), environment (temperature, rainfall, humidity, and sunshine) as well as the management practices in agriculture (Ramesh et al., 2019).

10.6 SOIL WATER STORAGE THROUGH BUILDING SOM

Soils in dry regions are degraded and have low quantities of SOM, which is vital for key soil processes: nutrient dynamics, water interactions, and biological and physical soil health. Due to the quick oxidation process that occurs in the dry regions, most dryland soils have low quantities of SOM (Srinivasarao et al., 2011a; Srinivasarao et al., 2009b; Srinivasarao et al., 2008; Srinivasarao and Vittal, 2007). The limited biomass production and loss of the top layer of soil rich in SOM during heavy rains are two important factors that govern the SOM status in tropical regions. A low water infiltration and porosity affect local and regional water cycles, agroecosystem resilience, plant productivity, and global carbon cycles, which have significant negative implications on water use efficiency (WUE) (Wani et al., 2009; Vineela et al., 2008). In drylands, where there are climatic extremes and unpredictability, SOM performs a critical role in maintaining optimum WHC and infiltration rates, hence restricting yield losses. To prevent further harm and improve soil health, carbon sequestration, greenhouse gas (GHG) mitigation, and the country's food security, management strategies that enhance SOM and its maintenance at a threshold level are urgently needed at this hour (Srinivasarao et al., 2011a). Improved management suggests sustainable productivity along with improved soil quality (physical, chemical, and biological parameters), increased carbon sequestration, and other benefits as shown by long-term research at the International Crops Research Institute for the Semi-Arid Tropics (ICRISAT), India (Wani et al., 2003). To maintain or boost SOM, amendments such as farmyard manure, plant residues, and farm compost must be applied continuously. However, it is challenging to apply these to soil at the appropriate rates due to a lack of organic manures and enormous uses of farm waste as feed and fuel. There are several technologies that significantly improve the SOC and water storage in the soil profile.

10.6.1 MULCH CUM MANURING OF SOILS

Evaporation in mulched soil is minimal when compared to bare soil. Total porosity, texture, and soil structure are the critical factors affecting the overall and accessible soil moisture storage capacity (Prem et al., 2020). It can be achieved by using organic mulching material. Soil wetting depth increases with increase in mulching materials. Mulching with straw has the capacity to store more soil water from light precipitation (Ji and Unger, 2001). Using straw mulch can increase soil moisture storage by 55% compared to that under control (Sood and Sharma, 1996). The application of wheat residue mulch at a rate of 6,730 kg ha^{-1} has been shown to improve AWHC to a depth of 1.5 m of soil in comparison to a bare soil (Black, 1973). In *Rabi* (winter) sorghum, total soil moisture loss from planting to harvest was 9.14%, 11.33%, and 11.92%, respectively, at 15, 30, and 45 cm

Effective Soil Organic Matter Management Strategies

of soil depth. The percentage increases in soil moisture for the sugarcane bagasse mulch, soybean straw mulch, wheat straw mulch, and intercultural operation were 7.54, 12.26, 17.81, and 28.19, respectively, over the control (without mulch). Average soil temperatures were 19.58 °C, 20.04 °C, 20.37 °C, 20.73 °C, and 21.33 °C, respectively, in the intercultural operation, sugarcane waste mulch, wheat straw mulch, soybean straw mulch, and control (without mulch) (Ranjan et al., 2017). In contrast, fields with mulched grass have the lowest observed soil temperature.

Agricultural wastes used as mulching material (e.g., grass clippings, bark chips, wheat or paddy straw, rice hulls, plant dry leaves, compost, and sawdust) are examples of materials with an organic origin that can decay naturally. Mulches decomposes over time, increasing the soil's ability to store water (Yang et al., 2015). As it decomposes, it also adds nutrients to the soil. Additionally, it indirectly increases WUE (Ossom et al., 2001). There are about 38 trillion Mt of organic waste produced annually due to various human activities across the world. In India alone, there are about 600–700 million Mt of agricultural waste each year (along with 272 million Mt of crop residues), but most of these materials are not utilized (Suthar, 2009).

10.6.2 DRIP IRRIGATION WITH ORGANIC MULCH

Drip or trickle irrigation is the frequent, slow distribution of water to soils by mechanical emitters or applicators located at certain points along delivery lines. A small amount of water can discharge from the emitters because they dissipate the pressure from the distribution system via orifices, vortices, and long or tortuous flow routes. Sub-surface drip irrigation is a unique and efficient method of irrigation which provides flexible and light irrigation, especially in dryland or arid conditions. It eliminates surface water evaporation and avoids potential surface runoff, thereby improving soil moisture availability. With the use of this technique, a little amount of soil may be kept moist by applying small amounts of water frequently (Du et al., 2015), thereby decreasing the drainage of water from the root zone soil and restricting the rooting zone to moist soil (Du et al., 2010). Subsurface drip irrigation significantly reduces evaporation from topsoil in comparison to surface drip irrigation and enhances irrigation WUE (Lamm and Trooien, 2003). Micro irrigation like drip and sprinkler along with organic mulches improves SOC levels and WUE further. Frequency of critical irrigation required in rainfed dryland crops is reduced by one-third with micro-irrigation with organic mulches.

10.6.3 GREEN MANURING FOR SOC AND MOISTURE STORAGE

Regardless of the cropping pattern and the climate, it is generally known that adding biomass from green manuring crops to soil improves its SOM and other nutrients, particularly nitrogen (N) (Walia and Kler, 2007). Cluster beans, cowpea, green gram, sesbania, sunhemp, dhaincha, etc., are common green manuring crops in various rainfed ecosystems since they are leguminous in nature and have a high ability to fix atmospheric N in the root nodules. A considerable research on this topic, done under the umbrella of irrigated ecology (Rao et al., 2017), showed that various green manuring crops, especially dhaincha and sun hemp, had significant ability to augment SOM content and enhance other related soil functions. Green manuring with cowpea, green gram, and sun hemp crops significantly increased the activities of various enzymes (dehydrogenase and phosphatase activity) when compared to the control (Yogesh and Hiremath, 2014; Mrunalini et al., 2022).

Dryland soils with light sandy loam are ideal for growing green manuring crops, viz., lobia, guar, green gram, and black gram (Meena, 2019). Cowpea and sun hemp green manure (GM) plots recorded significantly higher dehydrogenase activity (5.58 and 4.79 µg TPF/g of soil/day, respectively) followed by the green gram GM (3.55 µg TPF/g of soil/day) as compared to fallow land (2.78 µg TPF/g of soil/day). The higher dehydrogenase activity in cowpea and sun hemp was due to the addition of a higher quantity of easily mineralizable phytomass and biomass when compared to green gram GM. Additionally, the former GM crops had a higher leaf-to-stem ratio, which may have aided in the soil microbes' quicker multiplication. The cowpea GM plots had higher levels of phosphatase activity (40.02 mg of PNP/g of soil/h) than the sun hemp and green gram GM plots (34.70 mg of PNP/g of soil/h, 31.45 mg of PNP/g of soil/h, respectively). In fallow, phosphatase activity was noticeably reduced (Yogesh and Hiremath, 2014). The plants that produce green leaf manure acclimatize well to diverse soil types and are crucial for increasing the fertility and productivity of the soil. The leaves of some tree species, including *Azadirachta indica*, *Pongamia glabra*, *Delonix regia*, and *Peltophorum ferrugenum*, can be utilized as mulching material in the rainfed (non-irrigated) agroecology as a source of SOM. These species are easily cultivated in different parts of India (Srinivasarao et al., 2011b; Vineela et al., 2008). Tree green leaf manuring with the loppings of *Gliricidia sepium* has been proven to be a cost-effective and environment-friendly method. Another key characteristic of *Gliricidia* is its ease of growth in a variety of soil types, including acidic, dry, and even marginally damaged fields. Adopting green leaf manuring is one of the key strategies for improving SOM content. Interestingly, when biomass is used as mulching material, it aids in the conservation of moisture and lowers soil erosion losses (Photo 10.1).

10.6.4 INTERCROPPING AS AN EFFICIENT PRODUCTION SYSTEM UNDER DROUGHTS

Numerous studies across the world have shown how intercropping can increase the effectiveness of resource utilization (Martin-Guay et al., 2018). Intercropping is considered a vital strategy for ensuring food security, diversifying cropping systems, advancing agricultural progress in a sustainable manner, and making the most of smallholder farms' scarce labor resources (Ouma and Jeruto, 2010). Intercropping has the potential to significantly boost the primary yield of land per unit area by raising the use efficiency of resources like water, light, heat, and fertilizer. This would significantly improve global food security (Foley et al., 2011; Jensen et al., 2015) and raise WUE and economic profit for farmers (Yin et al., 2018). The goal of intercropping

PHOTO 10.1 Gliricidia nursery for green manuring to improve SOC and moisture storage of soil.

PHOTO 10.1 (Continued)

research is to increase resource use efficiency as agricultural production gradually shifts from a resource-intensive to a technology-efficient mode. The best ways to optimize the benefits of intercropping systems and generate stable yields while conserving water should continue to be the focus of research.

Intercropping can use less water if the intercrop strips are laid out optimally. As an illustration, strip intercropping of maize (*Zea mays* L.) and pea (*Pisum sativum* L.) (4:4 model, four rows of maize and four rows of pea) decreased water usage by 10.2–13.7% in comparison to solo cropping (Mao et al., 2012). When using an alternative irrigation technique to flood irrigation, intercropping's water consumption was decreased by 16.1% and 15.3% at high and low water supply levels, respectively (Yang et al., 2011). Water consumption is reduced by 4.4–8.5% (save

25.1–70.9 mm of soil water) and irrigation water is saved by 15% with limited water supply after the booting stage of wheat (*Triticum aestivum* L.) in wheat–maize intercropping without significantly reducing crop productivity (Wang et al., 2015). Because wheat and maize are intercropped, the amount of evaporation from the soil in the maize strips is decreased because wheat is more competitive than maize throughout the co-growth period and competes for soil water from the maize strips (Yin et al., 2018). Contrarily, intercropped maize receives compensatory soil water from the wheat strips after the wheat harvest, reducing soil evaporation in wheat strips (Yin et al., 2018). The overlapping of root spatial distribution is a natural result of diverse root morphologies of the numerous intercrops (Gao et al., 2010). Crop species with rapid water absorption and rapid growth have an advantage over those with efficient soil water usage but slow growth in the zone where crop roots overlap (Bramley et al., 2007). Legumes can obtain the water below the root zone of maize and increase the water supply of maize by water lifting during the intercropping between maize and legumes (Sekiya and Yano, 2004). The ability of intercropping systems to use water depends on soil water availability. Cowpea (*Vigna unguiculata*) and maize intercropping have higher WUE than the comparable solo cropping when soil water availability is high; however, when soil water availability is low, intercropping has higher WUE than sole cowpea but lower WUE than sole maize (Droppelmann et al., 2000).

10.6.5 Contour Farming for Erosion Control

One of the biggest challenges to efficient land management is limiting surface runoff and sediment loss. Soil loss along with carbon and nutrients is the key determinant for productivity loss in rainfed drylands. To enhance crop production while also efficiently protecting the soil, practical strategies are required. Terracing and contouring are two such techniques that aid with this. Today's improvements in precision conservation allow for better terrace location and field-based contour grade maintenance. Contour farming is one of the simplest and most efficient sustainable farming techniques used to minimize erosion. It is also known as strip farming, contour cropping, contour cultivation, terrace farming, or terracing (Dorren and Rey, 2004). The effects of erosion are dramatically reduced when agricultural techniques are carried out along the contours of a sloping land area. According to some estimates, contour farming can minimize erosion by up to 50% (Babubhai, 2018). Terracing is used as one of the ways of soil conservation because of its ability to control erosion. Reducing the chance of environmental damage through soil loss and land deformation is also advisable. Organic matter in soil increases because of contour farming (Abiye, 2022). This happens because of the conservative method, which enables organic matter to build up on (and in) soil without being lost to erosion or leaching. Contour cropping improves soil health by enhancing the carbon concentration and lowering soil disturbance. Moreover, the regional water quality is also improved because contour farming reduces runoff of pollutants into water bodies (Gathagu et al., 2018; Srinivasarao et al., 2019a).

10.6.6 Conservation Agriculture for Carbon Storage and Water Use Efficiency

Conservation agriculture (CA) with zero-tillage (ZT) basically involves three principles that include no-tillage, residue retention, and diversified crop rotations. Crop residue biomass is rich in carbon. Returning this C-rich biomass into soil or placing it on the surface would provide multiple benefits like soil surface protection, temperature moderation, restricting the evaporation loss at the initial stages of the farming, etc. Later it will be converted into SOC, and then would play a greater role in maintaining soil health like activating the soil microbial activity, reducing BD, soil water storage, enhancing the plant available nutrients, etc. Indeed, there are several benefits from the adoption of CA in the existing farming practice (Figure 10.6). CA practice plays effectively in sustainable agriculture, helps in mitigating the GHG emissions, protects the soil from erosion, and improves SOC

Effective Soil Organic Matter Management Strategies 271

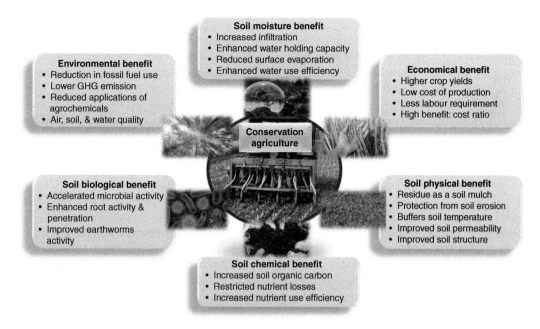

FIGURE 10.6 Multi-benefits obtained from the adoption of conservation agriculture.

and soil water storage (Singh et al., 2020). A large data meta-analysis on the impact of adoption of CA practices on soil C stock revealed the increase in soil C stocks at the surface soil depth (0–15 cm) with a potential of 15 Mg C ha^{-1} (Jat et al., 2022). The maximum total organic carbon (TOC) and its fractions at the surface layers in the soils under ZT practice are due to the crop residue retention with higher biomass residue and a slow decomposition rate due to minimum soil disturbance (Roy et al., 2022). Sinha et al. (2019) evaluated the effect of adoption of CA on the key soil parameters in the Indo-Gangetic plains of West Bengal and found that the soils under ZT were having greater amounts of SOC among all sites studied as compared to conventional tillage. Enhancement of SOC in soil certainly has a positive impact on water storage. Residue retention on the soil surface or return to the soil significantly increased the soil water storage (SWS) and decreased soil water consumption (Zhao et al., 2022). About 11.7% increments in SWS are noticed in case of residue return in the rainy years when compared with that under no residue in dry year (Zhao et al., 2022). Thus, leaving the crop residues on the soil surface reduces the C footprints and helps in achieving sustainability from an environmental perspective (Rakesh et al., 2021). Adoption of ZT in the wheat crop helped in conserving more soil residual moisture that supported better wheat growth than that without residue retention (Sahoo et al., 2022). Reduced tillage influenced the growth and performance of crops due to competition from weeds during the cropping season and also by influencing the water entry into the soil profile (Prasad et al. 2016).

10.6.7 Integrated Nutrient Management (INM): Win-Win for SOC and Water Storage in Soil Profile

Though organic fertilizers are slow in releasing plant available nutrients, they have a greater residual effect on subsequent crops; combining these with the inorganic fertilizers would provide multiple benefits to farmers. Recently, farmers and agriculture specialists changing their mindset to substitute a part of inorganic fertilizers with the eco-friendly, sustainable, and economic natural nutrients like farmyard manures (FYM), crop residues, green manures, vermicompost, soil amendments, and agroforestry (Selim, 2018). The key goals of integrated nutrient management (INM) are holistic and balanced management of existing natural resources with fertilizers, optimizing nutrient-use efficiency, minimizing nutrient losses, enhancing WUE (Wu and Ma, 2015), sustainability, grain superiority, and high economic returns (Selim 2020). Using biofertilizers along with organic and inorganic fertilizers helps in enhancing SOC content, aggregate stability, and moisture-retention capacity (Kumari et al., 2017).

Srinivasarao et al. (2011a, b; 2012a, b, c, d, e, f; 2020b; 2021b) have investigated the impact of various INM treatments on the SOC stocks under different cropping systems and showed that INM practice in pearl-millet cropping improved the profile SOC stock by 110–112% over the 100% inorganic treatment. The increase in SOC stock was 32% in pearl millet–cluster bean–castor, 43–47% in sole groundnut, 16–22% in finger millet, 16–22% in post-monsoon sorghum, 9.6–9.7% in groundnut–finger millet, 7.2–6%, in soybean-safflower, 3.5% in rice-lentil, and 3.5% in sole groundnut in comparison with the 100% inorganic treatment. The improved productivity of dry-land crops with various location-specific INM practices is due to improved water storage in the soil profile with enhanced SOC content along with soil fertility improvement. The enhanced storage of water due to SOC enrichment was attributed to higher water retention capacity parameters of soil estimated during the mid-season droughts. The improved water storage estimated with enhanced SOC content was equivalent to the water required for one or two critical irrigation through micro-irrigation systems (Srinivasarao et al., 2014a). The relation of water conserving or water use is largely dependent on the soil structure, quality, and quantity of SOM as it is considered the primary control among different soil properties. Thus, SOC becomes a basis of soil physical (structure), chemical (plant nutrients), and biological (microbial population) attributes. Brady and Weil (2005) observed that 1% of SOM (i.e., 1 g SOM per 100 kg of dry soil) has the potential to hold 30 kg of water. Many investigators have also reported similar values (Bastida et al., 2008) and highlighted the importance of the relationship between soil structure and soil productivity (Yadav and Meena 2009). Thus, improvement of soil structure through improvement in SOC facilitates the water absorption capacity of the soil. The INM can significantly improve the SOC which is an indicator of soil fertility and soil structure along with its greater role in nutrient availability, acceleration of microbial population and its activity, reducing BD, cation-exchange capacity, soil pH, and soil aeration (Mohammad et al., 2012). Such an improvement in overall soil quality leads to improvement in water infiltration, increase in soil field capacity, and WHC that ultimately results in achieving economic water use and WUE despite attaining the desirable crop yields (Nazli et al., 2015).

10.6.8 Organic Amendments

Most of the organic amendments are bulky in nature, eco-friendly, and cost-effective. Such amendments have a greater potential in holding soil moisture that helps in saving time, energy, and money in crop production. Improved water storage as contributed by the application of OM or FYM can solve the problems of water limitations during the crop growth and development. Several researchers have documented that the percent increase in SOM can hold the percent amount of water in soil. A field experiment revealed that every 1% increment in SOM can hold 16,500 gallons of plant-available water per acre of soil – that is equal to 1.5 quarts of water per cubic foot of soil (Gould, 2015). Application of vermicompost in agricultural fields results in improving crop productivity and curtailing the water stress problems as vermicomposts are porous with excellent water storage capacity (Abaranji et al., 2021). Use of vermicompost in agriculture helps in improving the soil physical structure including some macro- and micronutrients (Azarmi et al., 2008), while combining this with other organic or inorganic fertilizers effectively increases the growth and yield of crops (Javaad et al., 2013). The use of beneficial microbes in farming was started during the 1960s. In recent decades, bio-fertilizers played a commendable role in enhancing agricultural productivity through biotic and abiotic stress management, such as water and nutrient deficiency and heavy metal contamination (Wu et al., 2005). However, combining bio-fertilizers with organic amendments would further enhance its performance in overall soil quality buildup. Application of poultry manure and filter mud cake resulted in improving the status of N, P, and K, but the inclusion of bio-fertilizers significantly increased the decomposition of the organic waste (Bakr, 2016) which is the key mechanism behind the release of plant-available nutrients. Application of *Azolla caroliniana* compost significantly improved soil carbon sequestration by 1.1–1.4 folds and reduced the global warming potential by 1.2–1.4 folds when compared to cow dung and green manure treatments. Addition of *Azolla caroliniana* and rice husk dust was also observed to suppress methane (CH_4^+) by 30–36% through the enhancement of porosity, C-storage, and recalcitrant C fractions in soil (Bharali et al., 2021). Input of compost has some advantage because of its rich C biomass which is the key element used as an energy for their growth promotion activities (Ullah et al., 2021); consequently, it helps in improving soil quality and WHC (Alzamel et al., 2022). Compost coupled with bio-fertilizer and filter mud cake represents best amendments for improving soil porosity and soil fertility and harvesting healthy crops. The addition of filter mud cake into the soil also significantly enhanced nutrient and water movement (Alzamel et al., 2022).

10.6.9 BIOCHAR

The use of biochar is getting more attention in India for its role in improving SOC content and soil health (Samra and Srinivasarao, 2021). Biochar is a C-rich product produced from crop residues through controlled pyrolysis at temperatures of 400–600 °C under anaerobic conditions (Sinha et al., 2021). The concentration of total C and total N in different biochar materials shows that agroforestry-based biochar has higher levels (Wood and Eucalyptus) compared to paper mill, green waste, and poultry litter (Srinivasarao et al., 2013b). Thus, biochar can be effectively used for agricultural purposes (Lehman and Joseph, 2009). It helps in soil carbon sequestration, reduces farm waste, and improves the soil quality (Srinivasarao et al., 2012g, 2013b). It has highly concentrated carbon chains, hydrophilic characteristics, and high surface area (Qian et al., 2020; Wang et al., 2019) as it is produced by thermal decomposition (Beesley and Marmiroli, 2011). The crop residue coverted into biochar has shown in Figure 10.7. Biochar has numerous advantages in agriculture; most importantly, it helps in increasing carbon sequestration, improving soil WHC, increasing drought mitigation, accelerating the microbial population and their activity, leading to an overall improvement of soil physical condition, and increasing crop yields (Lal, 2008).

Increased porosity of biochar helps in maximizing the soil's WHC (Singh et al., 2010) and ultimately results in dissolving the soil nutrients in water that provides large amounts of nutrients for plant uptake (Lehman and Joseph, 2009). Application of biochar at a 2% mixture rate increased the WHC of a loamy sand soil (Novak et al., 2009). Yet, temperature also plays a crucial role in biochar performance. Varying pyrolysis temperature from 250°C to 750°C showed that WHC of the soil ranged from 7% to 16% (Yu et al., 2013a, b). Biochar can effectively minimize the external fertilizer additions, enhance the WHC, improve crop productivity, and reduce the nutrient leaching losses (Li et al., 2021). In fact, biochar fills the empty space between soil particles that increases total porosity and improves water retention parameters such as field capacity (FC) and permanent wilting point (PWP) (Alghamdi et al., 2020). In rainfed maize production systems on Alfisols of southern India, addition of biochar significantly improved biomass and economic yield during season with mid-season droughts (Srinivasarao et al., 2013a, b).

10.6.10 TANK SILT

"Tank silt" (TS) is rich in SOC and other plant nutrients, with greater potentiality in enhancing the WHC, and is highly suitable to cope with the mid-season droughts under rainfed agriculture. SOC helps in soil aggregation that restores the degraded lands (Rakesh et al., 2022). Addition of TS, which is rich in microbial biomass, would enhance crop production and restore degraded soils

FIGURE 10.7 Crop residue coverted into biochar for SOC improvement and water storage in the soil profile.

(Tiwari et al., 2014). Application of TS in the farmlands improves soil physicochemical and biological properties that ultimately increase crop productivity (Indoria et al., 2018). Addition of TS in the Alfisol significantly enhances the profitability of rainfed agriculture (Osman et al., 2007). TS-amended plots showed a greater residual impact on Horse gram biomass production besides mitigating N_2O emissions in the degraded semiarid Alfisol (Sharan et al., 2022). However, the direct and residual influence of TS application in *Kharif* (summer) and *Rabi* (winter) crops is highly influenced by soil texture and moisture availability and management practices (Sharma et al., 2015).

10.6.11 Cover Crops and Market Wastes

Cover cropping is highly effective in preventing nutrient losses through leaching and percolation, restricting the weed growth and enhancing soil C sequestration (Srinivasarao et al., 2021c). A field study conducted by Mohanty et al. (2015) to evaluate the impact of combining tillage with different cropping systems revealed that involving CA practice that included tillage (conventional and minimum) with cropping systems (sole maize and maize + cowpea intercrop) and followed by cover crops [fallow, horse gram, and toria (rapeseed)] significantly improved SOC storage under the inclusion of cover crops when compared with no cover crop. Venkateswarlu et al. (2007) reported that incorporation of crop biomass into soil improved SOC, MBC, and available nutrients over 10 years. Market-waste materials are rich in plant nutrients. Market wastes contain the mixture of vegetables, fruits, flowers, animal wastes, etc., that are easily decomposable and carbon-rich compost. Using such wastes as manure enables the effective utilization of wastes in agriculture and helps in enhancing SOC and other available nutrients that ultimately maximize the crop yields and soil health.

10.7 NATIONAL AND STATE GOVERNMENT INITIATIONS FOR SOIL CARBON AND WATER STORAGE

The National Mission for Sustainable Agriculture (NMSA) was launched under the eight missions outlined under National Action Plan on Climate Change (NAPCC) in 2010 to sustain agricultural productivity through the conservation of natural resources like soil and water in conjunction with development of rainfed agriculture in India. The *Mahatma Gandhi National Rural Employment Guarantee Act* (MGNREGA) was launched in 2005 aimed to improve soil fertility, carbon sequestration, rainwater harvesting, and restoration of degraded lands, etc., targeting the rural communities of India. The Soil Health Card (SHC) Mission was initiated in 2015 to improve the soil fertility status through soil test-based fertilizer recommendations. This mission made a revolution in India, distributed about 10.48 crores of soil health cards. The National Policy for Management of Crop Residues (NPMCR) program came into existence in 2014 that widely restricted the crop residue burning and promoted its utilization through innovative farm mechanization and this scheme also helped in monitoring the residue burning through satellite-based technologies. The Paris Climate Change Agreement committed by India also contributes to the mitigation of GHG emissions and enhances carbon sink. The Neem Coated Urea (NCU) launched in 2015 promoted widely to use this fertilizer that helps in maintaining soil health besides protecting the ground water from excess nitrate nutrient pollution. The Rashtriya Krishi Vikas Yojana (RKVY) scheme was launched in 2007 for the holistic development of agriculture and allied sectors through soil and water conservation and to strengthen the farmer's effort, risk mitigation, and promoting agribusiness entrepreneurship.

The Integrated Watershed Management Programme (IWMP) was implemented by the Department of Land Resources of the Ministry of Rural Development. This program aimed to restore ecological balance by harnessing, conserving, and developing degraded soil and water while increasing the overall biodiversity. The Desert Development Programme (DDP) was started during 1977–78 aimed with the objectives to restore the natural resources like land, water, and vegetative cover and implement watershed approach, water resource development, etc. The scheme Watershed Development Project in Shifting Cultivation Areas (WDPSCA) was implemented during 1995–96 to protect the hill slopes of jhum (shifting cultivation) areas through soil and water conservation measures on a watershed basis. The Accelerated Irrigation Benefits Programme (AIBP) was launched during 2009–10 and sponsored by the Ministry of Water Resources to increase the area under irrigation for increasing the crop productivity and socioeconomic condition of the people. The NABARD Loan-Soil & Water Conservation Scheme under the Rural Infrastructure Development Fund (RIDF) was launched during 2000–01 to enhance the agricultural productivity in small river valleys of rural areas. The synergy requirements of technologies, policy, and programs along with community participation for enhancing SOC storage in Indian agriculture are illustrated in Figure 10.8.

10.8 CONCLUSION

The barriers to achieving the potential productivity of rainfed farming in India can be broadly categorized as being connected to knowledge and institutional, technology/resource, and socioeconomic factors. Of these, the technological and resource-related constraints can be efficiently handled using the technologies already in use, which can greatly boost productivity in rainfed conditions. The secret to increasing productivity and narrowing or even eliminating the production gaps is from effective soil and water management. Building the SOM for soil health restoration should be the focus. Water is a precious natural resource, and rainfed farming can only be sustained by managing rainfall in situ or by collecting runoff and recycling it. Effective use of water, soil, and farm management practices in an integrated strategy is both necessary and a requirement for making rainfed farming more economic and sustainable. To achieve gains in productivity and widespread effects, it is necessary to scale these demand-driven approaches through farm science

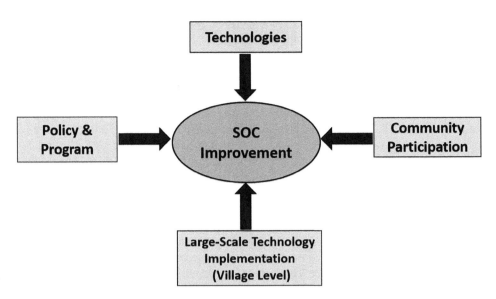

FIGURE 10.8 Critical requirements for enhancing SOC storage in Indian agriculture.

centers, agricultural technology management agencies, and various national and state government initiatives, which are present in every region of the developing nations. Yet, there is no "one-size-fits-all" solution to the complex problems of water scarcity in diverse agricultural systems. Considering the current scenario, agriculture waste management technologies like CA, mulching, organic amendments, novel approaches like biochar, TS, green manuring, integrated nutrient management options, etc., are some of the "win–win" strategies to unlock the potential of rainfed dryland soils without jeopardizing the quality of SOM content. As these areas are crucial for India's food security, publishing of this chapter will further motivate and enhance efforts to eliminate yield gaps and unleash the potential of rainfed agriculture.

10.9 WAY FORWARD

- Coordinating policies of integrating soil carbon with national climate commitments to achieve more carbon sequestration in India.
- Stopping crop residue burning to restrict further carbon emissions.
- Training programs to educate farmers about the importance of carbon and its greater role in agriculture and food security.
- Developing innovative methodologies for a quick, cheaper, and authentic measurement of soil carbon changes.
- Remapping is critical to find out the actual SOC depleted areas at regional and national levels for diverse land use and management for the year 2021–22.
- Integrating of water, agriculture, and forests policies is vital to bring coordination and collective achievements.
- Involving community is critical in collecting and sharing the information.
- Promoting carbon neutrality policy at national level to strengthen carbon finance.
- Encouraging farmers to shift to organic agriculture/carbon smart agriculture by adopting diversified cropping systems and minimizing fertilizer usage.
- Providing a strong support is needed to R&D activities to bring innovations and dissemination of feasible technologies related to carbon sequestration.
- Focusing is also needed on the groundwater recharge and water management programs.

Effective Soil Organic Matter Management Strategies

- Provisioning of knowledge and skills on soil carbon improvement and soil condition monitoring for land managers and planners.
- Converging of inter-ministries with a group of secretaries for the effective monitoring of national programs.
- Developing national mission on SOC sequestration and cross-learning platform for SOC sequestration at micro level.
- Financing the farmers who adopt climate-resilient agricultural technologies.

REFERENCES

Abaranji, S., Panchabikesan, K., & Ramalingam, V. 2021. Experimental study on the direct evaporative air-cooling system with vermicompost material as the water storage medium. *Sustainable Cities and Society* 71, 102991.

Abiye, W. 2022. Soil and water conservation nexus agricultural productivity in Ethiopia. *Advances in Agriculture*. Article ID 8611733, 1–10. https://doi.org/10.1155/2022/8611733

Alghamdi, A.G., Alkhasha, A. & Ibrahim, H.M. 2020. Effect of biochar particle size on water retention and availability in a sandy loam soil. *Journal of Saudi Chemical Society* 24(12), 1042–1050. https://doi.org/10.1016/j.jscs.2020.11.003

All report on agricultural census 2015–16, Agriculture Census Division, Department of Agriculture, Cooperation and Farmers Welfare, Ministry of Agriculture and Farmers Welfare, GOI.

Alzamel, N., M., Taha, E.M., Bakr, A.A., & Loutfy, N. 2022. Effect of organic and inorganic fertilizers on soil properties, growth yield, and physiochemical properties of sunflower seeds and oils. *Sustainability* 14(19), 12928. https://doi.org/10.3390/su141912928

Azarmi, R.M., Giglou, T., & Taleshmikail, R.D. 2008. Influence of vermicompost on soil chemical and physical properties in tomato (*Lycopersicum esculentum*) field. *African Journal of Biotechnology* 7, 2397–2401. www.academicjournals.org/AJB

Babubhai, S.H. 2018. Cover Cropping and Contour Cultivation. *Biotech Articles.*

Bakr, A.A. 2016. Dynamic of some plant nutrients in soil under organic farming conditions. PhD Thesis, Faculty of Agriculture, Assiut University, Assiut, Egypt.

Bastida, F., Kandeler, E., Hernández, T., & García, C. 2008. Long term effect of municipal solid waste amendment on microbial abundance and humus-associated enzyme activities under semiarid conditions. *Microbial Ecology* 55(4), 651–661. https://doi.org/10.1080/03650340.2013.766721

Beesley, L., & Marmiroli, M. 2011. The immobilisation and retention of soluble arsenic, cadmium and zinc by biochar. *Environmental Pollution* 159(2), 474–480. https://doi.org/10.1016/j.envpol.2010.10.016

Bharali, A., Baruah, K.K., Bhattacharya, S.S., & Kim, K.H. 2021. The use of Azolla caroliniana compost as organic input to irrigated and rainfed rice ecosystems: Comparison of its effects in relation to CH4 emission pattern, soil carbon storage, and grain C interactions. *Journal of Cleaner Production* 313, 127931.

Black, A.L. 1973. Crop residue, soil water, and soil fertility related to spring wheat production and quality after fallow. *Soil Science Society of America Journal* 37, 754–758.

Brady N.C. & R.R. Weil. 2005. *Nature and Properties of Soil*. MacMillan Publishing Co. Ltd., New York, NY, USA, 13th edition. https://doi.org/10.1155/2020/2821678

Bramley, H., Turner, D.W., Tyerman, S.D., & Turner, N.C. 2007. Water flow in the roots of crop species: The influence of root structure, aquaporin activity, and waterlogging. pp. 133–196. https://doi.org/10.1016/S0065-2113(07)96002-2

Chand, R., Prasanna, L.P., & Singh, A. 2011. Farm size and productivity: Understanding the strengths of smallholders and improving their livelihoods. *Economic and Political Weekly* 5, 11.

Chen, Y., Wei, T., Ren, K., Sha, G., Guo, X., Fu, Y., & Yu, H. 2022. The coupling interaction of soil organic carbon stock and water storage after vegetation restoration on the Loess Plateau, China. *Journal of Environmental Management* 306, 114481. https://doi.org/10.1016/j.jenvman.2022.114481

Chenu, C., Le Bissonnais, Y., & Arrouays, D. 2000. Organic matter influence on clay wettability and soil aggregate stability. *Soil Science Society of America Journal* 64, 1479–1486.

CWC. 2022. Water Resources at a glance-2022. Central Water Commission (CWC), Ministry of Jal Shakti, Government of India. Cwc.gov.in

Dorren, L., & Rey, F. 2004. A review of the effect of terracing on erosion. In: Briefing Papers of the 2nd SCAPE Workshop. *Citeseer,* pp. 97–108.

Droppelmann, K.J., Lehmann, J., Ephrath, J.E., & Berliner, P.R. 2000. Water use efficiency and uptake patterns in a runoff agroforestry system in an arid environment. *Agroforestry Systems* 49, 223–243.

Du, T., Kang, S., Sun, J., Zhang, X., & Zhang, J. 2010. An improved water use efficiency of cereals under temporal and spatial deficit irrigation in north China. *Agricultural Water Management* 97, 66–74.

Du, Z., Ren, T., Hu, C., & Zhang, Q. 2015. Transition from intensive tillage to no-till enhances carbon sequestration in microaggregates of surface soil in the North China Plain. *Soil Tillage & Research* 146, 26–31.

Emerson, W.W. 1995. Water retention, organic C and soil texture. *Australian Journal of Soil Research* 33, 241–251. https://doi.org/10.1071/SR9950241

Foley, J.A., Ramankutty, N., Brauman, K.A., Cassidy, E.S., Gerber, J.S., Johnston, M., Mueller, N.D., O'Connell, C., Ray, D.K., & West, P.C. 2011. Solutions for a cultivated planet. *Nature* 478, 337–342.

Fukumasu, J., Jarvis, N., Koestel, J., Kätterer, T., & Larsbo, M. 2022. Relations between soil organic carbon content and the pore size distribution for an arable topsoil with large variations in soil properties. *European Journal of Soil Science* 73(1), e13212. https://doi.org/10.1111/ejss.13212

Gao, Y., Duan, A., Qiu, X., Liu, Z., Sun, J., Zhang, J., & Wang, H. 2010. Distribution of roots and root length density in a maize/soybean strip intercropping system. *Agricultural Water Management* 98, 199–212.

Gathagu, J.N., Mourad, K.A., & Sang, J. 2018. Effectiveness of contour farming and filter strips on ecosystem services. *Water* (Basel) 10, 1312.

Gould, C.M. 2015. Compost increases water holding capacity of droughty soils. Michigan State University, East Lansing, MI, USA.

Gururaj, B., Hamsa, K.R., & Mahadevaiah, G.S. 2017. Doubling of small and marginal farmer's income through rural non-farm and farm sector in Karnataka. *Economic Affairs* 62(4), 581–587.

Hudson, B.H. 1994. Soil organic matter and available water capacity. *Journal of Soil and Water Conservation* 49(2), 189–194.

ICAR Data Book. Indian Council of Agricultural Research. Agricultural Research Data Book 2022. http://apps.iasri.res.in/agridata/22data/HOME.HTML

Indoria, A.K., Sharma, K.L., Reddy, K.S., Srinivasarao, Ch., Srinivas, K., Balloli, S.S., Osman, M., Pratibha, G., & Raju, N.S. 2018. Alternative sources of soil organic amendments for sustaining soil health and crop productivity in India–impacts, potential availability, constraints and future strategies. *Current Science* 115, 2052–2062. http://krishi.icar.gov.in/jspui/handle/123456789/32401

Jat, M.L., Chakraborty, D., Ladha, J.K., Parihar, C.M., Datta, A., Mandal, B., Nayak, H., Maity, P., Rana, D.S., Chaudhary, S.K., & Gerard, B. 2022. Carbon sequestration potential, challenges, and strategies towards climate action in smallholder agricultural systems of South Asia. *Crop and Environment* https://doi.org/10.1016/j.crope.2022.03.005

Javaad, S., & Panwar, A. 2013. Effect of biofertilizer, vermicompost and chemical fertilizer on different biochemical parameters of Glycine max and Vigna mungo. *Recent Research in Science and Technology* 5, 40–44.

Jensen, E.S., Bedoussac, L., Carlsson, G., Journet, E.-P., Justes, E., & Hauggaard-Nielsen, H. 2015. Enhancing yields in organic crop production by eco-functional intensification. *Sustainable Agricultural Research* 4, 42–50.

Ji, S., & Unger, P.W. 2001. Soil water accumulation under different precipitation, potential evaporation, and straw mulch conditions. *Soil Science Society of America Journal* 65, 442–448.

Karami, A., Homaee, M., Afzalinia, S., Ruhipour, H., & Basirat, S. 2012. Organic resource management: Impacts on soil aggregate stability and other soil physico-chemical properties. *Agriculture Ecosystem. Environment.* 148, 22–28. https://doi.org/10.1016/j.agee.2011.10.021

Kumar, R., Singh, R.D., & Sharma, K.D. 2005. Water resources of India. *Current Science.* 89(5), 794–811.

Kumari, R., Kumar, S., Kumar, R. et al. 2017. Effect of long-term integrated nutrient management on crop yield, nutrition and soil fertility under rice-wheat system. *Journal of Applied and Natural Science* 9(3), 1801–1807. https://doi.org/10.31018/jans.v9i3.1442

Lal, R. 2008. Carbon sequestration. *Philosophical Transactions of the Royal Society B: Biological Sciences* 363(1492), 815–830. https://doi.org/10.1098/rstb.2007.2185

Lamm, F.R., & Trooien, T.P. 2003. Subsurface drip irrigation for corn production: a review of 10 years of research in Kansas. *Irrigation Science* 22, 195–200.

Lehmann, J., & Joseph, S. 2009. *Biochar for Environmental Management: Science and Technology*. Earthscan/ James & James, London.

Li, L., Zhang, Y.-J., Novak, A., Yang, Y., & Wang, J. 2021. Role of biochar in improving sandy soil water retention and resilience to drought. *Water* 13(4), 407. https://doi.org/10.3390/w1304 0407

Libohova, Z., Seybold, C., Wysocki, D., Wills, S., Schoeneberger, P., Williams, C., & Owens, P.R. 2018. Reevaluating the effects of soil organic matter and other properties on available water-holding capacity using the National Cooperative Soil Survey Characterization Database. *Journal of Soil and Water Conservation* 73(4), 411–421. https://doi.org/10.2489/jswc.73.4.411

Mao, L., Zhang, L., Li, W., van der Werf, W., Sun, J., Spiertz, H., & Li, L. 2012. Yield advantage and water saving in maize/pea intercrop. *Field Crops Research* 138, 11–20.

Martin-Guay, M.-O., Paquette, A., Dupras, J., & Rivest, D. 2018. The new green revolution: Sustainable intensification of agriculture by intercropping. *Science of the Total Environment* 615, 767–772.

Meena, R. 2019. Green manuring. An approach to improve soil fertility and crop production. German National Library. ISBN: 9783668948594. www.grin.com/document/468117

Minasny, B., & McBratney, A.B. 2018. Limited effect of organic matter on soil available water capacity. *European Journal of Soil Science* 69, 39–47. https://doi.org/10.1111/ejss.12475

Mohammad, W., Shah, S.M., Shehzadi, S., & Shah, S.A. 2012. Effect of tillage, rotation and crop residues on wheat crop productivity, fertilizer nitrogen and water use efficiency and soil organic carbon status in dry area (rainfed) of north-west Pakistan. *Journal of Soil Science and Plant Nutrition* 12(4), 715–727. http://dx.doi.org/10.4067/S0718-95162012005000027

Mohanty, A., Mishra, K.N., Roul, P.K., Dash, S.N., & Panigrahi, K.K. 2015. Effects of conservation agriculture production system (CAPS) on soil organic carbon, base exchange characteristics and nutrient distribution in a tropical rainfed agro-ecosystem. *International Journal of Plant, Animal and Environmental Sciences* 5, 310–314.

Mrunalini, K., Behera, B., Jaraman, S., Abhilash, P.C., Dubey, P.K., Narayanaswamy, G., Prasad, J.V.N.S., Rao, K.V., Krishnan, P., Pratibha, G., & Srinivasarao, Ch. 2022. Nature based solutions in soil restoration for improving agricultural productivity. *Land Degradation and Development*. https://doi.org.10.1002/ldr.4207

Mujumdar, P.P. 2008. Implications of climate change for sustainable water resources management in India. *Physics and Chemistry of the Earth, Parts A/B/C* 33(5), 354–358.

National Rainfed Area Authority (NRAA) 2022. Accelerating the Growth of Rainfed Agriculture – Integrated Farmers Livelihood Approach. Department of Agriculture and Farmers' Welfare, Ministry of Agriculture & Farmers' Welfare.

Nayar, L. 2021. *Outlook India Magazine*. www.outlookindia.com/author/lola-nayar-40

Nazli, R.I., Inal, I., Kusvuran, A., Demirbas, A., & Tansi, V. 2015. Effects of different organic materials on forage yield and nutrient uptake of silage maize (*Zea mays* L.). *Journal of Plant Nutrition* 39(7), 912–921. 10.3906/tar-1302-62

Neog, P., Sarma, P.K., Saikia, D., Borah, P., Hazarika, G.N., Sarma, M.K., Sarma, D., Ravindra Chary, G., & Srinivasarao, Ch. 2019. Management of drought in *Sali* rice under increasing rainfall variability in the North Bank Plains Zone of Assam, North East India. *Climatic Change*. https://doi.org/10.1007/s10584-019-02605-4, 1-12

Novak, J.M., Lima, I., Xing, B., Gaskin, J.W., Steiner, C., Das, K., Ahmedna, M., Rehrah, D., Watts, D.W., & Busscher, W.J. 2009. Characterization of designer biochar produced at different temperatures and their effects on a loamy sand. *Annals of Environmental Science* 3(1), 195–206. www.aes.northeastern.edu/

Olness, A., & Archer, D. 2005. Effect of organic carbon on available water in soil. *Soil Science* 170(2), 90–101. https://doi.org/10.1097/01.ss.0000155496.63323.35

Osman, M., Ramakrishna, Y.S., & Haffis, S. 2007. Rejuvenating tanks for self-sustainable rainfed agriculture in India. *Agriculture Situation India* 64, 67–70.

Ossom, E.M., Pace, P.F., Rhykerd, R.L., & Rhykerd, C.L. 2001. Effect of mulch on weed infestation, soil temperature, nutrient concentration, and tuber yield in Ipomoea batatas (L.) Lam. in Papua New Guinea. *Tropical Agriculture* 78, 144.

Ouma, G., & Jeruto, P. 2010. Sustainable horticultural crop production through intercropping: The case of fruits and vegetable crops: A review. *Agriculture and Biology Journal of North America* 1, 1098–1105.

Prasad, J.V.N.S., Srinivasarao, Ch., Srinivas, K., Naga Jyothi, Ch., Venkateswarlu, B., Ramachandrappa, B.K., Dhanapal, G.N., Ravichandra, K., & Mishra, P.K. 2016. Effect of ten years of reduced tillage and recycling of organic matter on crop yields, soil organic carbon and its fractions in Alfisols of semi-arid tropics of southern India. *Soil & Tillage Research* 156, 131–139.

Prem, M., Ranjan, P., Seth, N., & Patle, G.T. 2020. Mulching techniques to conserve the soil water and advance the crop production—A review. *Current World Environment* 15, 10–30.

Qian, Z., Tang, L., Zhuang, S., Zou, Y., Fu D., & Chen, X. 2020. Effects of biochar amendments on soil water retention characteristics of red soil at south China. *Biochar* 2(4), 479–488. https://doi.org/10.1007/s42773-020-00068-w

Rakesh, S., Sarkar, D., Sinha, A.K., Shikha, Mukhopadhyay, P., Danish, S., Fahad, S., & Datta, R. 2021. Carbon mineralization rates and kinetics of surface-applied and incorporated rice and maize residues in Entisol and Inceptisol soil types. *Sustainability* 13(13), 7212. https://doi.org/10.3390/su13137212

Rakesh, S., Sinha, A.K., Juttu, R., Sarkar, D., Jogula, K., Reddy, S.B., Raju, B., Danish, S., & Datta, R. 2022. Does the accretion of carbon fractions and their stratification vary widely with soil orders? A case study in Alfisol and Entisol of sub-tropical eastern India. *Land Degradation & Development* 33(12), 2039–2049. https://onlinelibrary.wiley.com/doi/10.1002/ldr.4291

Ramesh, T., Bolan, N.S., Kirkham, M.B., Wijesekara, H., Manjaiah, K.M., Srinivasarao, Ch., Sandeep, S., Rinklebe, J., Ok, Y.S., Choudhury, B.U., Want, H., Tang, C., Song, Z., & Freeman II, O.W. 2019. Soil organic carbon dynamics: Impact of land use changes and management practices: A review. *Advances in Agronomy* 156, 1–125.

Ranjan, P., Patle, G.T., Prem, M., & Solanke, K.R. 2017. Organic mulching-a water saving technique to increase the production of fruits and vegetables. *Current Agriculture Research Journal* 5(3), 371–380.

Rao, S., Indoria, A.K., & Sharma, K.L. 2017. Effective management practices for improving soil organic matter for increasing crop productivity in rainfed agroecology of India. *Current Science* 112, 1497. https://doi.org/10.18520/cs/v112/i07/1497-1504

Ravindra Chary, G., & Gopinath, K.A. 2022. Agro-ecology specific rainwater management interventions for higher productivity and income in rainfed areas: In B. Krishna Rao, S. Annapurna, B. Renuka Rani, Z. Srinivasa Rao, K. Sunitha, M. SchinDutt, S.K. Jamanal & V. Ramesh (Eds.), *Soil and Water Conservation Techniques in Rainfed Areas* (e-book). National Institute of Agricultural Extension Management (MANAGE) & Water and Land Management Training and Research Institute (WALAMTARI), Hyderabad.

Rawls, W.J., Pachepsky, Y.A., Ritchie, J.C., Sobecki, T.M., & Bloodworth, H. 2003. Effect of soil organic carbon on soil water retention. *Geoderma* 116, 61–76. https://doi.org/10.1016/S0016-7061(03)00094-6

Roy, D., Datta, A., Jat, H.S., Choudhary, M., Sharma, P.C., Singh, P.K., & Jat, M.L. 2022. Impact of long-term conservation agriculture on soil quality under cereal based systems of North West India. *Geoderma*. 405: 115391. https://doi.org/10.1016/j.geoderma.2021.115391

Sahoo, S., Mukhopadhyay, P., Sinha, A.K., Bhattacharya, P.M., Rakesh, S., Kumar, R., ... & Kumar, U. 2022. Yield, nitrogen-use efficiency, and distribution of nitrate-nitrogen in the soil profile as influenced by irrigation and fertilizer nitrogen levels under zero-till wheat in the eastern Indo-Gangetic plains of India. *Frontier Environmental Science* 10, 970017. https://doi.org/10.3389/fenvs.2022.970017

Samra, J.S., & Srinivasarao, Ch. 2021. Circular carbon economy in India: Efficient crop residue management for harnessing carbon, energy and manure with co-benefits of greenhouse gases (GHGs) emissions mitigation. Policy Paper, ICAR-National Academy of Agricultural Research Management, Hyderabad, India, p. 20.

Sekiya, N., & Yano, K. 2004. Do pigeon pea and sesbania supply groundwater to intercropped maize through hydraulic lift?—Hydrogen stable isotope investigation of xylem waters. *Field Crops Research* 86, 167–173.

Selim, M. 2018. Potential role of cropping system and integrated nutrient management on nutrients uptake and utilization by maize grown in calcareous soil. *Egyptian Journal of Agronomy* 40(3), 297–312. 10.21608/AGRO.2018.6277.1134

Selim, M. 2020. Introduction to the integrated nutrient management strategies and their contribution to yield and soil properties. *International Journal of Agronomy*. https://doi.org/10.1155/2020/2821678

Sharan Bhoopal Reddy, Srinivasarao, Ch., Rao, P.C., Rattan L., Rakesh, S., Kundu, Singh, R.N., Dubey, P.K., Abhilash, P.C., Rao, K.V., Abrol, V., & Somasundaram, J. 2022. Greenhouse gas emission and agronomic productivity as influenced by varying levels of N fertilizer and tanksilt in degraded semi-arid Alfisol of Southern India. https://doi.org/10.1002/ldr.4507

Sharma, B.R., Rao, K.V., Vittal, K.P.R., Ramakrishna, Y.S., & Amarasinghe, U. 2010. Estimating the potential of rainfed agriculture in India: Prospects for water productivity improvements. *Agricultural Water Management*. 97(1): 23–30.

Sharma, P.K. and Kumar, M., 2014. Status and Management of Water in Rainfed Agriculture. *Efficient Water Management for Sustainable Agriculture*. 41.

Sharma, S.K., Sharma, R.K., Kothari, A.K., Osman, M., & Chary, G.R. 2015. Effect of tank silt application on productivity and economics of maize-based production system in southern Rajasthan. *Indian Journal of Dryland Agriculture Research and Development* 30(2), 24–29. https://doi.org/10.5958/2231-6701.2015.00021.4

Singh, B., Singh, B.P., & Cowie, A.L. 2010. Characterisation and evaluation of biochars for their application as a soil amendment. *Soil Research* 48(7), 516–525. https://doi.org/10.1071/SR10058

Singh, D., Lenka, S., Lenka, N.K., Trivedi, S.K., Bhattacharjya, S., Sahoo, S. et al. 2020. Effect of reversal of conservation tillage on soil nutrient availability and crop nutrient uptake in soybean in the vertisols of central India. *Sustainability*, *12*(16), 6608.

Singh, N.P., Anand, B., Singh, S., Srivastava, S.K., Srinivasarao, Ch., Rao, K.V. and Bal, S.K. (2021). Synergies and trade-offs for climate-resilient agriculture in India: an agro-climatic zone assessment. *Climatic Change*. 164, 11 https://doi.org/ 10.1007/ s10584-021-02969-6

Sinha, A.K., Ghosh, A., Dhar, T., Bhattacharya, P.M., Mitra, B., Rakesh, S., Paneru, P., Srestha, S.R., Manandhar, S., Beura, K., Dutta, S., Pradha, A.K., Rao, K.K., Hossain, A., Siddquie, N., Molla, M.S.H., Chaki, A.K., Gathala, M.K., Islam, M.S., Dalal, R.C., Gaydon, D.S., Laing, A.M. and Menzies, N.W. 2019. Trends in key soil parameters under conservation agriculture- ased sustainable intensification farming practices in the Eastern Ganga Alluvial Plains. *Soil Research*. 57(8), 883–893. https://doi.org/10.1071/SR19162 (IF: 1.878).

Sinha, A.K., Rakesh, S., Mitra, B., Roy, N., Sahoo, S., Saha, B.N., Dutta, S., & Bhattacharya, P.M. 2021. Agricultural waste management policies and programme for environment and nutritional security. In Rajan Bhatt et al. (Eds.), *Input Use Efficiency for Food and Environmental Security*. Springer Nature. Chapter 21.

Sood, B.R., Sharma, V.K. 1996. Effect of intercropping and planting geometry on the yield and quality of forage maize. *Forage Research* 24, 190–192.

Srinivasarao, Ch., Chary, G.R., Raju, B.M.K., Jakkula, V.S., Rani, Y.S., & Rani, N., 2014a. Land use planning for low rainfall (450-750 mm) regions of India. *Agropedology* 24(2), 197–221.

Srinivasarao, Ch., Deshpande, A.N., Venkateswarlu, B., Lal, R., Singh, A.K., Kundu, S., Vittal, K.P.R., Mishra, P.K., Prasad, J.V.N.S., Mandal, U.K., & Sharma, K.L. 2012a. Grain yield and carbon sequestration potential of post monsoon sorghum cultivation in Vertisols in the semi-arid tropics of central India. *Geoderma* 175–176, 90–97.

Srinivasarao, Ch., & Gopinath, K.A. 2016. Resilient rainfed technologies for drought mitigation and sustainable food security. *Mausam* 67(1), 169–182.

Srinivasarao, Ch., Lal, R., Prasad, J.V., Gopinath, K.A., Singh, R., Jakkula, V.S., Sahrawat, K.L., Venkateswarlu, B., Sikka, A.K., & Virmani, S.M. 2015. Potential and challenges of rainfed farming in India. In Donald L. Sparks (Ed.), *Advances in Agronomy* 133, 113–181. https://doi.org/10.1016/bs.agron.2015.05.004

Srinivasarao, Ch., Rao, K.V., Chary, G.R., Vittal, K.P.R., Sahrawat, K.L., & Kundu, S. 2009a. Water retention characteristics of various soil types under diverse rainfed production systems of India. *Indian Journal of Dryland Agricultural Research and Development* 24(1), 1–7.

Srinivasarao, Ch., Venkateswarlu, B., Lal, R., Singh, A.K., Kundu, S., Vittal, K.P.R., Patel, J.J., & Patel, M.M. 2011a. Long-term manuring and fertilizer effects on depletion of soil organic carbon stocks under pearl millet cluster bean–castor rotation in western India. *Land Degradation & Development*. http://dx.doi.org/10.1002/ldr.1158

Srinivasarao, Ch., Venkateswarlu, B., Dinesh Babu, M., Wani, S.P, Dixit, S, Sahrawat, K.L. and Sumanta Kundu (2011b). Soil Health Improvement with Gliricidia Green Leaf Manuring in Rainfed Agriculture, On Farm Experiences. Central Research Institute for Dryland Agriculture, Santoshnagar, PO. Saidabad, Hyderabad 500 059, Andhra Pradesh, p:16.

Srinivasarao, Ch., Venkateswarlu, B., Lal, R., Singh, A.K., & Kundu, S. 2013a. Sustainable management of soils of dryland ecosystems of India for enhancing agronomic productivity and sequestering carbon. *Advances in Agronomy* 121, 253–329.

Srinivasarao, Ch., Gopinath, K.A., Venkatesh, G., Dubey, A.K., Harsha Wakudkar, Purakayastha, T.J., Pathak, H., Pramod Jha, Lakaria, B.L., Rajkhowa, D.J., Sandip Mandal, Jeyaraman, S., Venkateswarlu, B., & Sikka, A.K. 2013b. Use of biochar for soil health management and greenhouse gas mitigation in India: Potential and constraints. Central Research Institute for Dryland Agriculture, Hyderabad, Andhra Pradesh. 51 p.

Srinivasarao, Ch., Prasad, R.S., & Mohapatra, T. 2019a. Climate change and Indian agriculture: Impacts, coping strategies, programmes and policy. Technical Bulletin/Policy Document. Indian Council of Agricultural Research, Ministry of Agriculture and Farmers' Welfare and Ministry of Environment, Forestry and Climate Change, Government of India, New Delhi. p. 25.

Srinivasarao, Ch., Kareemulla, K., Krishnan, P., Murthy, G.R.K., Ramesh, P., Ananthan, P.S., & Joshi, P.K. 2019b. Agro-ecosystem based sustainability indicators for climate resilient agriculture in India: A conceptual framework. *Ecological Indicators* 105 (2019), 621–633.

Srinivasarao, Ch., Rakesh, S., Ranjith Kumar, G., Manasa, R., Somashekar, G., Subha Lakshmi, C., and Kundu, S. 2021a. Soil degradation challenges for sustainable agriculture in tropical India. *Current Science* 120(3), 492–500.

Srinivasarao, Ch., Rattan, L., Kundu, S., Prasad Babu, M.B.B., Venkateswarlu, B., & Singh, A.K. 2014b. Soil carbon sequestration in rainfed production systems in the semiarid tropics of India. *Science of the Total Environment* 487, 587–603.

Srinivasarao, Ch, Rejani, R., Rama Rao, C.A., Rao, K.V., Osman, M., Srinivasa Reddy, K., Kumar, M. & Kumar, P. 2017. Farm pond for climate-resilient rainfed agriculture. *Current Science* 112(3), 471–477.

Srinivasarao, Ch., Singh, S.P., Sumanta K., Abrol, V., Lal, R., Abhilash, P.C., Chary G.R., Pravin, B.T., Prasad, J.V.N.S., Venkateswarlu, B. 2021b. Integrated nutrient management improves soil organic matter and agronomic sustainability of semiarid rainfed Inceptisols of the Indo-Gangetic Plains. *Journal of Plant Nutrition and Soil Science* 184(5), 562–572.

Srinivasarao, Ch., Subha Lakshmi, C., Sumanta Kundu, S., Ranjith Kumar, G., Manasa, R., & Rakesh, S. 2020a. Integrated nutrient management strategies for rainfed agro-ecosystems of India. *Indian Journal of Fertilizers* 16(4), 344–361.

Srinivasarao, Ch., Kundu, S., Rakesh, S., Lakshmi, C.S., Kumar, G.R., Manasa, R., ... & Prasad, J.V.N.S. 2021c. Managing Soil Organic Matter under Dryland Farming Systems for Climate Change Adaptation and Sustaining Agriculture Productivity. Chapter 10. Ratan Lal (Ed.): Soil Organic Carbon and Feeding the Future: Basic Soil Processes. First Edition. ISBN 9781032150673. *Advances in Soil Science*. Published by CRC press. https://doi.org/10.1201/9781003243090-10. 219-251

Srinivasarao, Ch., Kundu, S., Yashavanth, B.S., Rakesh, S., Akbari, K.N., Sutaria, G.S., Vora, V.D., Hirpara, D.S., Gopinath, K.A., Chary, G.R., Prasad, J.V.N.S., Bolan, N.S., & Venkateswarlu, B. 2020b. Influence of 16 years of fertilization and manuring on carbon sequestration and agronomic productivity of groundnut in vertisol of semi-arid tropics of Western India. *Carbon Management* 12(1), 13–24. https://doi.org/10.1080/17583004.2020.1858681

Srinivasarao, Ch., Venkateswarlu, B., Singh, A.K., Vittal, K.P.R., Kundu, S., Gajanan, G.N., Ramachandrappa, B., & Chary, G.R. 2012b. Critical carbon inputs to maintain soil organic carbon stocks under long term finger millet (*Eleusine coracana* (L.) Gaertn) cropping on Alfisols in semi-arid tropical India. *Journal of Plant Nutrition and Soil Science* 175(5), 681. https://doi.org/10.1002/jpln.201000429

Srinivasarao, Ch., Venkateswarlu, B., Lal, R., Singh, A.K., Vittal, K.P.R., Kundu, S., Singh, S.R., & Singh, S.P. 2012c. Long-term effects of soil fertility management on carbon sequestration in a rice-lentil cropping system of the Indo-Gangetic plains. *Soil Science Society of America Journal* 76(1), 168–178.

Srinivasarao, Ch., Venkateswarlu, B ., Lal, R., Singh, A.K., Kundu, S., Vittal, K.P.R., Basavapura, K.R., & Narayanaiyer, Gajanan G. 2012d. Yield sustainability and carbon sequestration potential of groundnut–finger millet rotation in Alfisols under semi-arid tropical India. *International Journal of Agricultural Sustainability* 10(3), 1–15.

Srinivasarao, Ch., Venkateswarlu, B., Lal, R., Singh, A.K., Kundu, S., Vittal, K.P.R., Balaguravaiah, G., Vijaya Shankar Babu, M., Ravindra Chary, G., Prasadbabu, M.B.B., & Yellamanda Reddy, T. 2012e. Soil carbon sequestration and agronomic productivity of an Alfisol for a groundnut-based system in a semi-arid environment in South India. *European Journal of Agronomy* 43, 40–48. http://dx.doi.org/10.1016/j.eja.2012.05.001

Srinivasarao, Ch., Venkateswarlu, B., Lal, R., Singh, A.K., Kundu, S., Vittal, K.P.R., Sharma, S.K., Sharma, R.A., Jain, M.P., & Chary, G.R. 2012f. Sustaining agronomic productivity and quality of a Vertisolic Soil

(Vertisol) under soybean-safflower cropping system in semi-arid central India. *Canadian Journal of Soil Science* 92, 771–785.

Srinivasarao, Ch., Venkateswarlu, B., Lal, R., Singh, A.K., Kundu, S., Vittal, K.P.R., Balaguruvaiah, G., Vijaya Shankar Babu, M., Ravindra Chary, G., Prasadbabu, M.B.B., & Yellamanda Reddy, T. 2012g. Soil carbon sequestration and agronomic productivity of an Alfisol for a groundnut-based system in a semiarid environment in southern India. *European Journal of Agronomy* 43, 40–48.

Srinivasarao, Ch., & Vittal, K.P.R. 2007. Emerging nutrient deficiencies in different soil types under rainfed production systems of India. *Indian Journal of Fertilisers* 3, 37.

Srinivasarao, Ch., Vittal, K.P.R., Gajbhiye, P.N., Kundu, S., & Sharma, K.L. 2008. Distribution of micronutrients in soils in rainfed production systems of India. *Indian Journal of Dryland Agricultural Research and Development* 23, 29–35.

Srinivasarao, Ch., Vittal, K.P.R., Venkateswarlu, B., Wani, S.P., Sahrawat, K.L., Marimuthu, S., & Kundu, S. 2009b. Carbon stocks in different soil types under diverse rainfed production systems in tropical India. *Communication in Soil Science and Plant Analysis* 40, 2338–2356. https://doi.org/10.1080/0010362090 3111277

Suthar, S. 2009. Impact of vermicompost and composted farmyard manure on growth and yield of garlic (*Allium stivum* L.) field crop. *International Journal of Plant Production* 3(1), 1735–6814. https://doi.org/10.22069/IJPP.2012.629

Tiwari, R., Ramakrishna Parama, V.R., Murthy, I.K. and Ravindranath, N.H. 2014. Irrigation tank silt application to croplands: Quantifying effect on soil quality and evaluation of nutrient substitution service. *International Journal of Agricultural Science Research,* 3, pp.1–10. http://academeresearchjournals.org/journal/ijasr

Ullah, N., Ditta, A., Imtiaz, M.; Li, X. Jan, A.A.; Mehmood, S., Rizwan, M.S., Rizwan, M. 2021. Appraisal for organic amendments and plant growth-promoting rhizobacteria to enhance crop productivity under drought stress: A review. *Journal of Agronomy and Crop Sciences.* 2021, 207, 1–20. https://doi.org/10.1111/jac.12502

Venkateswarlu, B., Ch. Srinivasarao., G. Ramesh., S. Venkateswarlu., and J.C. Katyal. 2007. Effects of long-term legume cover crop incorporation on soil organic carbon, microbial biomass, nutrient build-up and grain yields of sorghum/sunflower under rain-fed conditions. *Soil Use and Management* 23l:100–107.

Vineela, C., Wani, S. P., Srinivasarao, C. H., Padmaja, B., & Vittal, K. P. R. (2008). Microbial properties of soils as affected by cropping and nutrient management practices in several long-term manurial experiments in the semi-arid tropics of India. *Applied Soil Ecology* 40(1), 165–173.

Walia, S.S., Kler, D.S. 2007. Ecological Studies on Organic vs Inorganic Nutrient Sources Under Diversified Cropping Systems. *Indian Journal of Fertilisers* 3, 55.

Wang, D., Li C, Parikh, SJ., Scow, KM. 2019. Impact of biochar on water retention of two agricultural soils – a multi-scale analysis. *Geoderma* 340:185–191. https://doi.org/10.1016/j.geoderma.2019.01.012

Wang, X., Wang, J., Xu, M., Zhang, W., Fan, T., Zhang, J. 2015. Carbon accumulation in arid croplands of northwest China: Pedogenic carbonate exceeding organic carbon. *Scientific Reports* 5, 1–12.

Wani, S.P., Pathak, P., Jangawad, L.S., Eswaran, H., & Singh, P. 2003. Improved management of Vertisols in the semiarid tropics for increased productivity and soil carbon sequestration. *Soil Use and Management* 19, 217–222.

Wani, S.P., Sreedevi, T.K., Rockström, J., & Ramakrishna, Y.S. 2009. Rainfed agriculture–past trends and future prospects. *Rainfed Agriculture: Unlocking the Potential* 7, 1–33.

Wu, S.C., Cao, Z.H., Li, Z.G., Cheung, K.C., & Wong, M.H. 2005. Effects of biofertilizer containing N-fixer, P and K solubilizers and AM fungi on maize growth: a greenhouse trial. *Geoderma* 125(1–2), 155–166. https://doi.org/10.1016/j.geoderma.2004.07.003

Wu, W., & Ma, B. 2015. Integrated nutrient management (INM) for sustaining crop productivity and reducing environmental impact: a review. *Science of the Total Environment*, vol. 512–513, pp. 415–427. https://doi.org/10.1016/j.scitotenv.2014.12.101

Yadav R.L., & M.C. Meena. 2009. Available micronutrient status and their relation with soil properties of Degana soil series of Rajasthan. *Journal of the Indian Society of Soil Science* 57(1), 90–92. www.researchgate.net/publication/287778539

Yang, C., Huang, G., Chai, Q., Luo, Z. 2011. Water use and yield of wheat/maize intercropping under alternate irrigation in the oasis field of northwest China. *Field Crops Research* 124, 426–432.

Yang, N., Sun, Z. X., Feng, L. S., Zheng, M. Z., Chi, D. C., Meng, W. Z., ... & Li, K. Y. (2015). Plastic film mulching for water-efficient agricultural applications and degradable films materials development research. *Materials and Manufacturing Processes* 30(2), 143–154.

Yin, W., Guo, Y., Hu, F., Fan, Z., Feng, F., Zhao, C., Yu, A., & Chai, Q. 2018. Wheat-maize intercropping with reduced tillage and straw retention: A step towards enhancing economic and environmental benefits in arid areas. *Frontiers in Plant Science* 9, 1328.

Yogesh, T.C., & Hiremath, S.M. 2014. Incorporation of green manure crops on soil enzymatic activities under rainfed condition. *Karnataka Journal of Agricultural Sciences* 27, 300–302.

Yu, O.-Y., Raichle, B., & Sink, S. 2013a. Impact of biochar on the water holding capacity of loamy sand soil. *International Journal of Energy Environmental Engineering* 4(1), 44. https://doi.org/10.1186/2251-6832-4-44

Yu, O.Y., Raichle, B., & Sink, S. 2013b. Impact of biochar on the water holding capacity of loamy sand soil. *International Journal of Energy and Environmental Engineering* 4(1), 1–9. www.journal-ijeee.com/content/4/1/44

Zhao, H., Qin, J., Gao, T., Zhang, M., Sun, H., Zhu, S., & Ning, T. 2022. Immediate and long-term effects of tillage practices with crop residue on soil water and organic carbon storage changes under a wheat-maize cropping system. *Soil and Tillage Research* 218, 105309. https://doi.org/10.1016/j.still.2021.105309

Zhao, X., Virk, A.L., Ma, S.T., Kan, Z.R., Qi, J.Y., Pu, C & Zhang, H.L. 2020. Dynamics in soil organic carbon of wheat-maize dominant cropping system in the North China Plain under tillage and residue management. *Journal of Environmental Management* 265, 110549. https://doi.org/10.1016/j.jenvman.2020.110549

Zhou, H., Chen, C., Wang, D., Arthur, E., Zhang, Z., Guo, Z., & Mooney, S.J. 2020. Effect of long-term organic amendments on the full-range soil water retention characteristics of a Vertisol. *Soil and Tillage Research* 202, 104663. https://doi.org/10.1016/j.still.2020.104663

11 Ancient Infrastructure Offers Sustainable Agricultural Solutions to Dryland Farming

Matthew C. Pailes, Laura M. Norman,
Christopher H. Baisan, David M. Meko, Nicolas Gauthier,
Jose Villanueva-Diaz, Jeff Dean, Jupiter Martínez,
Nicholas V. Kessler, and Ron Towner

This chapter appears online at www.routledge.com/9781032286747.

This chapter has been made available under a CC-BY-NC-ND license. The funder is U.S. Geological Survey.

DOI: 10.1201/b22954-11

Index

Note: Page numbers in **bold** refer to tables, and those in *italics* refer to figures.

A

adaptation strategies 64–7; animal breed 65–6; animal numbers 64–5; animal size/species 65; stocking rates 64–5; tools and external inputs 66–7
Aeolian EROsion (AERO) model 56
aggregate stability 188
agricultural drought 159
agricultural ecosystems 213–4
agricultural production systems: SOC benefits **187**; tillage effects **194**
agroecology, definition of 164
agroecosystems management 163–73; Argentina 170; Brazil 165–9; intercropping systems 167; least limiting water range 165, *167*; soil organic carbon (SOC) stocks *166*, 168, *168*; water retention curves 169, *169*; Mexico 172–3; crop systems, Central Mexico 172–3; livestock systems, South-Southeast Mexico 173
agronomic drought 3
Argentine drylands 170
aridity 160
arid rangelands: carbon cycle 51–6; animal processes 53–4; biocrust 55; plant processes 52–3; soil components and processes 54–5; temporal and spatial scales 55; soil C dynamics 68–9; soil inorganic carbon **44**, 44–5; soil organic carbon **44**, 44–5, *45*
Artic Oscillation (AO) 26
atmospheric CO_2 concentration 47
Australian Grass Production (GRASP) model 53
available water-holding capacity (AWHC) 262, 264–5

B

base temperature 114
basin tillage 87, *87*
biochar 273, *274*
biocrust: carbon cycle 56; community changes in rangelands 63–4; in stocks/flows/ecosystem processes 48–50
biodiversity 62–3; ecological 191
biomass: crop 206; global drylands **6**; production 240, 246
biomass water use efficiency (WUEb) 118
bio-pores 208–9
black water 4
blue water 4
Brazilian agroforestry systems 165–9; intercropping systems 167; least limiting water range 165, *167*; soil organic carbon (SOC) stocks *166*, 168, *168*; water retention curves 169, *169*
brown water 4
bulk density 199

C

canopy temperature (CT) 114, *115*
canopy temperature depression (CTD) 114–5, *115*

carbon cycle: arid rangelands 51–6; animal processes 53; biocrust 55; plant processes 52–4; soil components and processes 54–5; temporal and spatial scales 55; soil redistribution 56–62; omitted SOC erosion 59–62, *61*; soil C erosion paradox 56; vertical emission 58–9; wind erosion 56–9, *58*, *60*
carbon management practices 185–203; cover crops: balancing carbon 200–2; continuous living cover 200, 201; water cycles 200–2; grey literature 185; hydrologic characteristics 188–90; soil biology 190–2; soil functions 189, 202–3; soil health metaphor 199–200; soil organic carbon sequestration 187–8; soil organic carbon storage 187–8; soil physical structure 190–2; soil porosity 189–90; soil structure 188–90; soil terminologies 187–8; tillage effects 192–9; agricultural production systems **194**; apocalyptic event 198; bulk density 199; climate solution 193; cropping sequence 193; ecological implications 198; living biological system 193; macro-aggregates 195; penetrometer resistance 199; reduced tillage 192–3; soil disturbances 197; soil structure 194–7
carbon sequestration 5, 7
cattle models 53
clean tillage 12
climate change: conservation agriculture 182, 212–3; global drylands 2–3; Latin America and the Caribbean (LAC) drylands 173; strategies 30; wheat yield 28–9
climate extremes 212–3
compaction effects 207–8
conservation agriculture 129–47, 181–216; agriculture use 182; benefits 210–5; climate extremes 212–3; ecosystem services 213–4; water quality 214–5; climate change 182; conservation tillage 131–2; cover crops 133–4, *134*; cropland area **132**; crop rotation 132–3; ecosystem services 182; environment 182; FAO principles 130–1; freshwater use 182; future advancement 147; Indian rainfed farming 270–1, *271*; soil and water resources 134–45; integrated crop–livestock systems 145; soil erosion 138–40; soil properties 134–7; soil water 140–5; stacked approach 147; water use efficiency 140–5; sustainable management: dryland agriculture 184–5; efficient water use 183–4; water use efficiency 203–10; bio-pores 208–9; compaction effects 207–8; dryland agriculture 182–3; in dryland agriculture 204; evapotranspiration 206–7; infiltration 207–8; pore space 208–9; soil erosion 209–10; soil evaporation 205–6; sponge effects 208–9; sustainable management 183–4; tillage impacts 204–5
conservation tillage 131–2
continuous living cover 200, 201
contour farming 270
corn yield *vs.* evapotranspiration *245*
cover crops: adoption 133–4, *134*; carbon management practices: balancing carbon 200–2; continuous living

287

288   Index

cover 200, 201; water cycles 200–2; conservation agriculture 133–4; definition of 133; in dryland agriculture 120–1; Indian rainfed farming 274; soil water conservation 21–2
crop biomass 206
crop grain 240
crop nutrition management in SSA 83–101; background studies 84–5; fertilizer use optimization 92–5; economically optimal rates 93, **95**; grain yield *96*, 97; mean yield response **95**; net returns to investment *92*, 93–4, *97*; profit to cost ratios 92–3, **95**, 97; synergisms and targeting 95–6; rainfall distribution 98–9; soil conditions/sustainable production fields 85–92; parkland farming *89*, 90; perennial grass 91–2, *92*; reduced tillage 85–6; runoff barriers *86*, 86–7, *87*; trees and shrubs 88–9; translating science 99–101
crop responses: planting geometry 115–8, *116–8*; soil moisture 111–3, *112*; temperature 114–5, *115*
crop rotation 132–3
cumulative evaporation 17, *18*
cumulative infiltration 13–4, *14*

D

Darcy-flux method 19–20
desertification 2
diversity 164
drought: agricultural 159; agronomic 3; elements: diversity 164; efficiency 164; resilience 164; hydrological 3, 159; management 4–5, *5*; meteorological 3, 159; socioeconomic 159; sociological 3; stress 8; types of 3, 159
dry grain yield 8
dryland(s): definition of 160; drought management 4–5, *5*; ecoregions 8; global 1–2; in South America 160–1; sustainable management 8–9
dryland agriculture: conservation *see* conservation agriculture; cover crops 120–1; soil–plant–water–environment interactions 108–22; US Great Plains *109*, 109–11; water use efficiency *see* water use efficiency (WUE)
dryland cropping: climate-informed management 26–8; clumping 25–6; intensification 22–4; plant density 25–6; spacing 25–6
dry land cropping systems 206
dryland farming: ancient infrastructure *see* North American Southwest (NAS); definition of 11; in India *see* Indian rainfed farming; soil water conservation *see* soil water conservation; in US Great Plains 109–11
dust storms *146*

E

ecological biodiversity 191
ecological drought 3
ecological restoration strategies 67–8
economic yield 240
ecosystem services 213–4
efficiency 164
El Nino-Southern Oscillation (ENSO) 26–8
ephemeral carbon 188

evapotranspiration (ET) 206–7; corn yield *vs. 245*; grain yield *vs. 244*; relationship with yield 241–3
ex-situ rainwater harvesting 262

F

fallow crop intensification 22–4
fertilizer use optimization 92–8; economically optimal rates 93, **95**; grain yield *96*, 97; mean yield response **95**; net returns to investment *92*, 93–4, *97*; profit to cost ratios 93, 94, **95**, 98; synergisms and targeting 95, 97–8
fire 47
Food and Agriculture Organization (FAO) 130
forage-livestock production models 52

G

geo-referenced yield 93–4
global drylands 1–2; biomass **6**; carbon sequestration 5, 7; climate change 2–3; distribution of *40*; soil carbon stocks 5, **6**, 7
glomalin 51
grain water use efficiency (WUEg) 118
grain yield *vs.* evapotranspiration *244*
green manuring 267–8, *268*, *269*
green water 4; supply in root zone 8
grey water 4
groundwater resources 261

H

harvest index (HI) 240, 246
heterotrophic microbial soil respiration 50
hydrological drought 3, 159
hydrologic characteristics 188–90

I

Indian rainfed farming 256–77; mean annual rainfall/climate/soil type/production system length/water retention **263**; National/State Government initiatives 275; rainwater management: ex-situ rainwater harvesting 262; in-situ moisture conservation 262; soil moisture conservation measures **264**; soil moisture storage 262–5; soil organic matter 265–6; total food grain production and nutrient consumption *257*, 257–8; water productivity challenges 258–60; water resources availability 260–2, *261*; groundwater resources 261; rainfall trends 262; surface water resources 260–1; water storage through SOM 266–75; biochar 273, *274*; conservation agriculture 270–1, *271*; contour farming 270; cover crops 274; erosion control 270; green manuring 267–8, *268*, *269*; integrated nutrient management 271–2; intercropping 268–70; market-waste materials 274; mulch cum manuring 266–7; organic amendments 272–3; sub-surface drip irrigation 267; tank silt 273–4; water use efficiency 270–1
in-situ moisture conservation 262
integrated crop–livestock systems 145
integrated nutrient management (INM) 271–2
intercropping 167, 268–70
in vitro dry matter digestibility (IVDMD) 63

Index

irrigation management 249–50
irrigation scheduling 249

L

Latin America and the Caribbean (LAC) drylands 157–75;
Argentina 170; Brazil 165–9; intercropping systems 167;
least limiting water range 165, *167*; soil organic carbon
(SOC) stocks *166*, 168, *168*; water retention curves 169,
169; climate changes 160–2; drought stress 163–73;
general background 158–9; Mexico 170–3; crop systems,
Central Mexico 172–3; livestock systems, South-
Southeast Mexico 173; water security 162–3; adaptation
emergency 162; adaptation measures 162–3; financing
163; governance 163
least limiting water range (LLWR) 165
lingering carbon 188

M

macro-aggregates 195
market-waste materials 274
mean residence time 186
meteorological drought 3, 159
Mexican drylands 170–3; crop systems, Central Mexico
172–3; livestock systems, South-Southeast Mexico 173
Millennium Ecosystem Assessment 40, 68
mitigation strategies 66–7
monocropping 132
mulch cum manuring 266–7

N

National Oceanic and Atmospheric Administration
(NOAA) 28
National/State Government rainfed farming initiatives 275
natural soil structure 195
net primary productivity (NPP) 41, *41*
Noble, Charles 12

O

Oceanic Nino Index (ONI) 27, **27**
Ogallala Aquifer 109, 141, 238, *239*, 249, 250
organic amendments 272–3

P

parkland farming *89*, 90
pedological drought 3
penetrometer resistance 199
perennial grass 91–92, *92*
permanent wilting point (PWP) 111
plant available soil water (PASW) 111, *112*
plant available water (PAW) 111
plant community composition 62–3
pore space 208–9
precipitation variability 13

R

radiation use efficiency (RUE) 248–9
rainfed farming in India *see* Indian rainfed farming

rainwater management: ex-situ rainwater harvesting 262;
in-situ moisture conservation 262
rangelands 44–51; arid/semiarid: soil inorganic carbon
43, 44–5; soil organic carbon **43**, 44–5, *45*; biocrust
community changes 63–4; fire 47; forage quality
interaction 63
reduced tillage 85–6, 192–3
resilience 164
resist-accept-direct (RAD) framework 62–3
reverse phenology 90
robust agriculture 142, *143*

S

semiarid rangelands: soil inorganic carbon **43**, 44–5; soil
organic carbon **43**, 44–5, *45*
simulation models 28
simulation of production and utilization of rangelands
(SPUR) model 52
socioeconomic drought 159
sociological drought 3
soil biology 190–2
soil carbon dynamics: arid/semiarid rangelands **43**, 44–5,
45; atmospheric CO_2 concentration 47; biocrust 48–50;
fire 47; net primary productivity 41, *41*; soil aggregation
50–1; soil inorganic carbon sequestration 50; soil
respiration 41, *43*; woody-shrub encroachment 45–6
soil carbon stocks **6**
soil degradation 129–30
soil ecosystem services 213
soil energetics 192
soil erosion 138–40, 209–10
soil evaporation 205–6
soil fauna 55
soil functions 189, 202–3
soil health 129–30, *130*
soil health metaphor 199–200
soil inorganic carbon (SIC): arid/semiarid rangelands **43**,
44–5; stocks in global drylands 5, **6**, 7; in world biomes **7**
soil-landscape inorganic carbon (SLIC) model 55
soil organic carbon (SOC): agricultural production system
benefits **187**; arid/semiarid rangelands **44**, 44–5, *45*;
binding agent 188; carbon cycling: omitted erosion 59–60,
61; vertical emission 58–9; wind erosion 56–9, *58*, *60*;
conservation agriculture adoption 131; contributing factors
in agro-ecosystem *265*; green manuring 267–8, *268*, *269*;
integrated nutrient management 271–2; profile water
storage and drought mitigation *264*; sequestration 187–8;
stocks in global drylands 5, **6**, 7; storage 187–8; storage
requirements, Indian agriculture *276*; in world biomes **7**
soil organic matter (SOM) 110, 185, 186; contributing
factors 265–6; water-holding capacity of soil 131; water
storage through 266–75; biochar 273, *274*; conservation
agriculture 270–1, *271*; contour farming 270; cover
crops 274; erosion control 270; green manuring 267–8;
integrated nutrient management 271–2; intercropping
268–70; market-waste materials 274; mulch cum
manuring 266–7; organic amendments 272–3; subsurface drip irrigation 267; tank silt 273–4; water use
efficiency 270–1

soil physical structure 190–2
soil–plant–water–environment interactions 108–22; cover crops 120–1; crop responses: planting geometry 115–8, *116–8*; soil moisture 111–3, *112*; temperature 114–5, *115*; in US Great Plains 109–11; water use efficiency 118–20, *120*
soil pores 208–9
soil porosity 189–90
soil properties 134–7
soil quality 129–30
soil respiration (SR) 41, *42*; heterotrophic microbial 50
soil stability kit 50
soil structure: carbon management practices 188–90; tillage effects 194–7
soil temperature 137
soil water 140–5
soil water conservation: climate change: strategies 30; wheat yield 28–9; cover crops 21–2; deep drainage 19–20; dryland cropping: climate-informed management 26–8; clumping 25–6; intensification 22–4; plant density 25–6; spacing 25–6; evaporation 17–9; residue effects 18–9; tillage effects 18–9; fallow crop intensification 22–4; precipitation variability 13; redistribution 19–20; water capture and infiltration 13–7; cumulative infiltration 13–4, *14*; deep soil loosening 15–6, *16*; residues 13; runoff observations 15; soil surface alterations 16–7; weed control 20–1
soil-water-crop nexus 5, *5*
soil water deep drainage 19–20
soil water evaporation 17–9; residue effects 18–9; tillage effects 18–9
soil water extraction (SWE) 113, *247*
soil water redistribution 19–20
soil water use efficiency *247*, 247–8
sponge effects 208–9
Stewart, Bobby A. 2
stubble mulch tillage 12, 14
sub-Saharan Africa (SSA) crop nutrition management 83–101
sub-surface drip irrigation 267
surface water resources 260–1
sustainable management: conservation agriculture: dryland agriculture 184–5; efficient water use 183–4; research and development 8–9; soil erosion 138; soil quality 130

T

tank silt 273–4
Texas High Plains (THP) 236, 238, 241–3, 245, 246, 248, 250, 251
tie-ridging 86, *86*
tillage effects: carbon management practices 192–9; agricultural production systems **194**; apocalyptic

event 198; bulk density 199; climate solution 193; cropping sequence 193; ecological implications 198; living biological system 193; macro-aggregates 195; penetrometer resistance 199; reduced tillage 192–3; soil disturbances 197; soil structure 194–7; conservation agriculture: water use efficiency 204–5
tillage management 131–2
transpiration 240
transpiration efficiency (TE) 112, 248–9

U

United Nations Environment Programme (UNEP) 160
United States Department of Agriculture (USDA) 12, 133
US Great Plains *109*, 109–11
US Southern Great Plains (SGP) 236, *237*, 238, *239*, 245; future perspectives 250–1; water use efficiency 243–50; biomass production 246; harvest index 246; irrigation management 249–50; soil water use efficiency *247*, 247–8; transpiration efficiency 248–9; yield–ET relationship 241–3

V

vapor pressure deficit (VPD) 114, 117–8

W

water capture and infiltration 13–7; cumulative infiltration 13–4, *14*; deep soil loosening 15–6, *16*; residues 13; runoff observations 15; soil surface alterations 16–7
water infiltration 207–8
water quality 214–5
water security, LAC drylands 162–3; adaptation emergency 162; adaptation measures 162–3; financing 163; governance 163
water use efficiency (WUE): conservation agriculture 140–5, 203–10; bio-pores 208–9; compaction effects 207–8; in dryland agriculture 182–3, 204; evapotranspiration 206–7; infiltration 207–8; pore space 208–9; soil erosion 209–10; soil evaporation 205–6; sponge effects 208–9; sustainable management 183–4; tillage impacts 204–5; definition of 240–1; Indian rainfed farming 270–1; soil–plant–water–environment interactions 118–20, *120*; in Southern Great Plains 243–50; biomass production 246; harvest index 246; irrigation management 249–50; soil water use efficiency *247*, 247–8; transpiration efficiency 248–9
weed control 20–1
wind erosion 56–7, *58*, *60*, 138–40
woody-shrub encroachment 45–6

Y

yield–ET relationship 241–3

9781032286747